CONTINUUM THEORY OF PLASTICITY

CONTINUUM THEORY OF PLASTICITY

AKHTAR S. KHAN
The University of Maryland Baltimore County
Baltimore, Maryland

SUJIAN HUANG
Materials Research
Smith International Inc.
Houston, Texas

A Wiley-Interscience Publication
JOHN WILEY & SONS, INC.

New York • Chichester • Brisbane • Toronto • Singapore

To Farhan, Meena, and Tanveer
for their patience, understanding, and encouragement
—Akhtar S. Khan

To my wife, children, and my other family members
—Sujian Huang

A NOTE TO THE READER
This book has been electronically reproduced from digital information stored at John Wiley & Sons, Inc. We are pleased that the use of this new technology will enable us to keep works of enduring scholarly value in print as long as there is a reasonable demand for them. The content of this book is identical to previous printings.

This text is printed on acid-free paper.

Library of Congress Cataloging in Publication Data:

Khan, Akhtar S.
Continuum theory of plasticity / Akhtar S. Khan, Sujian Huang.
p. cm.
Includes bibliographical references.
ISBN 0-471-31043-3
1. Plasticity. 2. Continuum mechanics. I. Huang, Sujian.
II. Title.
QA931.K478 1995
620.1'1233—dc20 94-13851

10 9 8

CONTENTS

PREFACE

Since Rodney Hill's classic text *The Mathematical Theory of Plasticity* published about a half-century ago, there have been several elementary books on this subject. However, the contents in these later books did not change much from Hill's superb treatise on this subject. During the half-century though, there has been phenomenal growth in this area, especially in the finite plastic deformation regime. Thus a modern textbook, using the continuum mechanics approach and incorporating the developments in the subject over the past 50 years, is generally lacking. Recently an excellent contemporary monograph, *Finite Plastic Deformation of Crystalline Solids*, by Kerry Havner has been published, but this has been largely limited to single crystals. We have attempted to provide a continuum mechanics based recent development of the subject, while retaining essential classical contributions. We have presented the material in such a way that the book can be used in a first graduate course on this subject.

First, we introduce some of the elementary concepts in Chapter 1. Then we provide the necessary continuum mechanics background in Chapters 2 and 3. We follow this by a discussion of the classical theory of plasticity in Chapters 4 and 5. Then we introduce recent developments on this topic in Chapters 6, 7, and 8. Our discussion includes some elementary insight to the micromechanics of plastic deformation in Chapters 9 and 10. Finally, in the last chapter contains analytical coupling between deformation of individual crystals and macroscopic material response of the polycrystal aggregate.

Since the book will be used primarily for the first graduate course on plasticity, we have assumed only minimal background; namely some knowledge in tensors, Miller indices, and crystal structure is assumed. Of course it would be helpful for the student to have an elementary course in continuum mechanics before starting to use this book.

The authors wish to gratefully acknowledge the influence and contribution of several individuals. The senior author wishes to thank Professors Clifford Truesdell and James Bell for introducing continuum mechanics and plasticity, respectively, in the late 1960s and early 1970s. He is grateful to the constant encouragement and support by the colleagues in the plasticity area, especially Professors Lallit Anand, Satya Atluri, Romesh Batra, Kerry Havner, David McDowell, and George Weng. The typing of the manuscript by Rose Benda is sincerely appreciated. The helpful comments of the graduate students at the University of Oklahoma are gratefully acknowledged.

AKHTAR KHAN
SUJIAN HUANG

CONTINUUM THEORY OF PLASTICITY

1

INTRODUCTION

Any material body deforms when it is subjected to external forces. The deformation is called *elastic* if it is reversible and time independent, that is, if the deformation vanishes instantaneously as soon as forces are removed. A reversible but time-dependent deformation is known as *viscoelastic*; in this case the deformation increases with time after application of load, and it decreases slowly after the load is removed. The deformation is called *plastic* if it is irreversible or permanent. A brittle material such as glass, concrete, or rock under low hydrostatic pressure can only have elastic deformation before it fails under ultimate load. On the other hand, metals and rocks under high confining pressure can undergo substantial plastic deformation before failure and therefore are known as *ductile* materials.

The theory of plasticity deals with the stress-strain and load-deflection relationships for a plastically deforming ductile material or structure. The establishment of these relationships should follow two steps: (1) the experimental observation and (2) the mathematical representation. The stress states that are normally achieved in any experiment are simple and uniform, but the ultimate goal of any plasticity theory is a general mathematical formulation that can predict the plastic deformation of materials under complex loading and boundary condition.

The history of plasticity theory dates back to 1864 when Tresca published his yield criterion based on his experimental results on punching and extrusion. Since then, over 130 years, tremendous progress has been made by many researchers, such as Saint-Venant, Levy, von Mises, Hecky, Prandtl, and Taylor, who have established the cornerstones for the theory. Now developments in plasticity theory is an active field of mechanics.

The theories of plasticity can be established into two categories: One group is known as mathematical theories of plasticity, and the other is physical theories of plasticity. Mathematical theories are formulated to represent the experimental observations as general mathematical formulations. This group of theories does not require a deep knowledge of the physics of plastic deformation and is based on hypotheses and assumptions from experimental results. Therefore the mathematical theories are phenomenological in nature and are sometimes referred to as *phenomenological theories*. The physical theories, on the other hand, attempt to quantify plastic deformation at the microscopic level and explain why and how the plastic deformation occurs. The movements of atoms and the deformation of the crystals and grains are important considerations. The responses of metals, viewed as aggregates of single crystals or polycrystals, to applied loads are derived from those of their building blocks, namely single crystals and single-crystal grains.

Most applications, such as structural design or metal forming, are on the macroscopic scale. Any plasticity theory must therefore contain mathematical variables that can be measured during experiments at macroscopic levels. To make reasonable hypotheses and assumptions on the basis of experiments and to fully understand the meaning and limitations of a proposed theory, a knowledge of the material structure and physics is very helpful. Keeping this rationale in mind, we will discuss both the mathematical theories and the physical theories of plasticity in this book, with equal emphasis on each.

1.1 PHENOMENOLOGICAL MACROSCOPIC NATURE OF METALS

As a starting point, this section will provide some fundamental experimental observations on metals that are plastically deformed. From these observations one can gain insight into the basic behavior of plastically deformed materials and establish several concepts useful for further study.

Uniaxial Tension Test

To describe this test, one can use the nominal stress defined by

$$\hat{\sigma} = \frac{P}{A_0} \tag{1.1}$$

and the nominal strain

$$\hat{\epsilon} = \frac{l - l_0}{l_0} \tag{1.2}$$

where P is the applied axial load, and A_0 and l_0 are, respectively, the

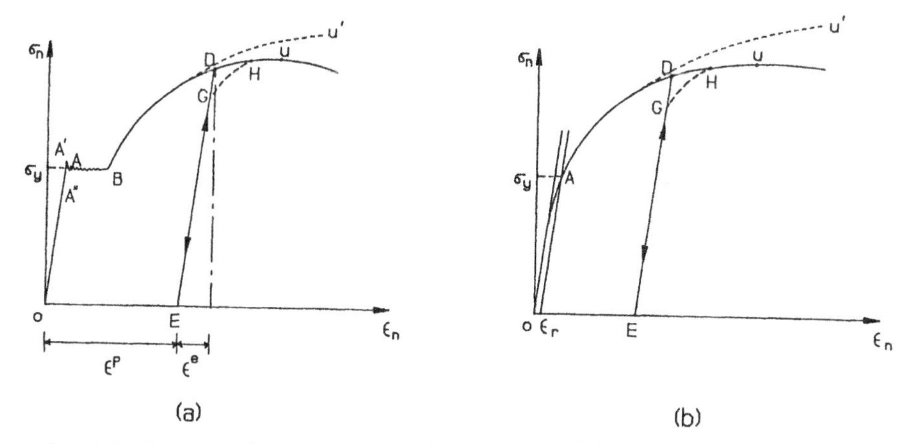

Fig. 1.1 Stress-strain curve under uniaxial loading: (*a*) mild steel; (*b*) aluminum

underformed area of the cross section and the initial length of the specimen. The typical $\hat{\sigma} - \hat{\epsilon}$ curve for mild steel is shown in Fig. 1.1(*a*). It is seen from the figure that initially the relation between stress and strain is linear until the stress level A'' is reached. Point A'' is called the *proportional limit*. The segment oA'' can be represented using Hooke's law, which forms the foundation of the theory of elasticity. Beyond this point increase in strain is not linear with increase in stress, but the deformation is still in the elastic region, in the sense that the removal of stress will restore the material to its original shape. This remains true until the point A', which is called the *upper yield point*, is reached. Further deformation is accompanied by a small local drop following by several oscillations of the stress level and an increased rate of strain. This forms a plateau until point B. The region AB is known as *perfectly plastic behavior*. The lower bound of the stress oscillation, or so-called lower yield point shown by point A, is usually chosen as the elastic limit, or yield stress σ_Y. If the stress keeps increasing at point B, the material hardens, and the stress increases until the ultimate stress is reached at point U. After that the stress decreases with increase in strain representing the behavior of instability. The region BU is known as *work hardening* or *material hardening*, while in the entire region *ABU the material is plastically deforming*. If the material is unloaded in this region, say, at point D, it will follow the path DE which is parallel to the initial elastic loading path oA'. As a result only part of the strain, called *elastic strain* ϵ^e, is recovered, while the other part of the strain will remain as the permanent strain or plastic strain ϵ^p. From the figure, it can be considered that the total strain in the plastic region consists of two parts:

$$\epsilon = \epsilon^e + \epsilon^p \tag{1.3}$$

When reloading from point E, the response will be along the elastic unloading line EG and can be described by Hooke's law. At point G the material will yield again and the plastic deformation will resume. Further deformation beyond H will generally follow the path HU as if no unloading and reloading happened. Point G is termed the *subsequent yield* point or stress, while point A, which represents the onset of initial yielding, is called *initial yield* point or stress. It should be noted that for real materials, the subsequent yield point G is lower than point D and that there is a transition region from G to H; H is on the stress-strain curve obtained during an uniaxial test without unloading.

Some metals, such as aluminum, copper, and stainless steel, do not exhibit sharp yield points as shown in Fig. 1.1(a). Instead, the yield of these materials is a gradual transition from a linear elastic to a nonlinear plastic behavior. (See Fig. 1.1(b) for aluminum subjected to simple tension.) This transition makes it very difficult to determine precisely where yield has taken place. In such cases one usually defines an offset yield stress σ_Y for which a definite plastic strain (e.g., 0.2%) will be left after unloading (see Fig. 1.1(b)). Except for the yield stress, unloading and reloading behavior in the plastic region and subsequent yield stress is similar to the behavior of mild steel, as discussed above.

1.1.1 True Stress-Strain Curve

Nominal stress $\hat{\sigma}$ and nominal strain $\hat{\epsilon}$ (introduced previously) are defined based on the original dimensions. For example, $\hat{\sigma}$ is defined as the load divided by the original cross-sectional area. As an alternative, one can use true stress σ and true strain or logarithmic strain ϵ defined as

$$\sigma = \frac{P}{A} \tag{1.4}$$

$$d\epsilon = \frac{dl}{l} \quad \text{or} \quad \epsilon = \int_{l_0}^{l} \frac{dl}{l} = \ln \frac{l}{l_0} \tag{1.5}$$

where A is the current or deformed cross-sectional area and l is the current or deformed length of the specimen. From Eqs. (1.2) and (1.5) one can easily obtain the following relationship between $\hat{\epsilon}$ and ϵ:

$$\epsilon = \ln(1 + \hat{\epsilon}) \tag{1.6}$$

In the plasticity theory the material is usually assumed to be incompressible based on experimental evidence. Therefore one can write

$$A_0 l_0 = Al \tag{1.7}$$

Hence the relationship between σ and $\hat{\sigma}$ can be derived:

$$\sigma = \frac{P}{A} = \frac{P}{A_0}\frac{l}{l_0} = \hat{\sigma}(1 + \hat{\epsilon}) \tag{1.8}$$

Using Eqs. (1.6) and (1.8), one can convert the nominal stress-nominal strain curve (Fig. 1.1) into one in terms of the true stress and true strain. This is shown by the dashed line in Fig. 1.1(a) and (b). Observe that in contrast to the nominal stress, the true stress keeps increasing until the material fails. This can also be verified analytically as follows: From Eqs. (1.6) and (1.8) we have

$$\hat{\epsilon} = e^{\epsilon} - 1 \tag{1.9}$$

$$\hat{\sigma} = \sigma e^{-\epsilon} \tag{1.10}$$

and

$$\frac{d\hat{\sigma}}{d\hat{\epsilon}} = \frac{d\hat{\sigma}}{d\epsilon}\frac{d\epsilon}{d\hat{\epsilon}} = \left[e^{-\epsilon}\frac{d\sigma}{d\epsilon} - e^{-\epsilon}\sigma\right]\frac{1}{1 + \hat{\epsilon}} = \left(\frac{d\sigma}{d\epsilon} - \sigma\right)e^{-2\epsilon} \tag{1.11}$$

It follows that

$$\frac{d\hat{\sigma}}{d\hat{\epsilon}} = 0 \Leftrightarrow \frac{d\sigma}{d\epsilon} = \sigma \tag{1.12}$$

Since σ is positive, $d\sigma/d\epsilon$ is also positive. This means the true stress–true strain curve keeps monotonically increasing beyond even the point at which $\hat{\sigma}$ reaches its maximum. Since $d\hat{\sigma}/d\hat{\epsilon} = 0$ corresponds to the onset of instability, Eq. (1.12) gives the true stress at this point.

One of the advantages of true strain over nominal strain is that true strain is an additive measure while nominal strain is not. Suppose that we have a specimen with initial length l_0. It is strained under uniaxial tension to l_1. Then $\epsilon_1 = \ln(l_1/l_0)$ and $\hat{\epsilon}_1 = (l_1 - l_0)/l_0$. If the specimen is now pulled to the length l_2, the additional true and nominal strains are $\epsilon_2 = \ln(l_2/l_1)$ and $\hat{\epsilon}_2 = (l_2 - l_0)/l_0$, respectively. The total strain, in terms of ϵ, can be obtained by

$$\epsilon = \epsilon_1 + \epsilon_2 = \ln\left(\frac{l_1}{l_0}\right) + \ln\left(\frac{l_2}{l_1}\right) = \ln\left(\frac{l_2}{l_0}\right) \tag{1.13}$$

on the other hand, the nonadditive nominal strain is given by

$$\hat{\epsilon} = \frac{l_2 - l_0}{l_0} \neq \hat{\epsilon}_1 + \hat{\epsilon}_2 \tag{1.14}$$

1.1.2 Incompressibility Effect of Hydrostatic Pressure

In a series of experiments Bridgman (1947, 1952) examined the material response to very high hydrostatic pressure. He performed tensile tests in the presence of hydrostatic pressure up to 24,000 atm (367,500 psi). The results can be summarized below.

1. The volume of the material does not change permanently even for very large pressure; thus the material can be assumed to be plastically incompressible.
2. The stress-strain curve is not affected by hydrostatic pressure in the small strain range.
3. The ductility of the material increases under hydrostatic pressure.
4. The effect of hydrostatic pressure on yield of any material is negligible. The first and last of the above conclusions—incompressibility and the independence of yield from the hydrostatic pressure—are fundamental to the plasticity theory.

Mathematically incompressibility can be expressed in terms of the principal values of the true strain

$$\epsilon_1 + \epsilon_2 + \epsilon_3 = 0 \tag{1.15}$$

or in terms of nominal strain

$$(\hat{\epsilon}_1 + 1)(\hat{\epsilon}_2 + 1)(\hat{\epsilon}_3 + 1) = 1 \tag{1.16}$$

Obviously it is more convenient to use true strain instead of nominal strain to express the incompressibility condition. For infinitesimal strain, by neglecting the higher order terms, one can represent the incompressible condition as

$$\hat{\epsilon}_1 + \hat{\epsilon}_2 + \hat{\epsilon}_3 = 0 \tag{1.17}$$

1.1.3 Bauschinger Effect

Consider a specimen of a material loaded in tension or compression into the plastic region. The load is removed, and the specimen is reloaded in the reverse direction until yield. It has been observed that the yield stress obtained in the reloading or reversed direction is substantially less than the yield stress in the original direction. This phenomenon is called the *Bauschinger effect* and is illustrated in Fig. 1.2.

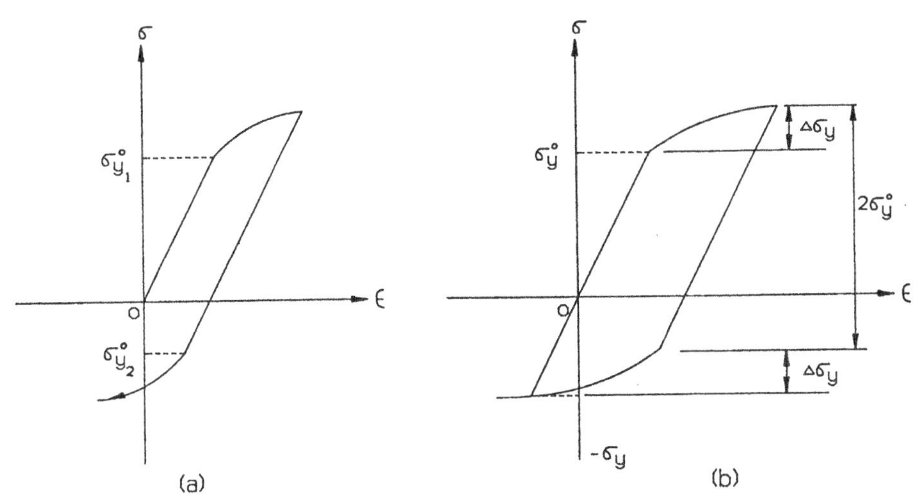

Fig. 1.2 The Bauschinger effect: (*a*) experimental observation; (*b*) kinematic hardening model

The Bauschinger effect has been observed in polycrystalline metals as well as in single crystals. It can be explained by the anisotropy of the dislocation fields caused by the previous loading. The presence of this effect greatly complicates the modeling of plastic deformation; therefore a simplified model, called the *kinematic hardening model*, is usually used to describe it. This model, shown in Fig. 1.2(*a*), assumes that the reduction of the yield stress in the reversed direction is equal to the stress increment by which the specimen is loaded beyond yield stress in the original direction. We will discuss this model in more detail later.

1.1.4 Rate Effect

So far we have only discussed the material response to one-dimensional loading that is applied very slowly, the so-called quasi-static loading. Usually loading with a strain rate of the order 10^{-2} to 10^{-3} per second can be considered quasi-static loading, even though there is no general agreement on the specific value. It has been found that the strain rate or loading rate has a pronounced effect on the material behavior in the plastic region. The effects of increasing the loading rate are generally the following:

1. The initial and subsequent yield stresses increase with loading rate (Fig. 1.3).
2. The ductility of the material decreases with an increase in loading rate.

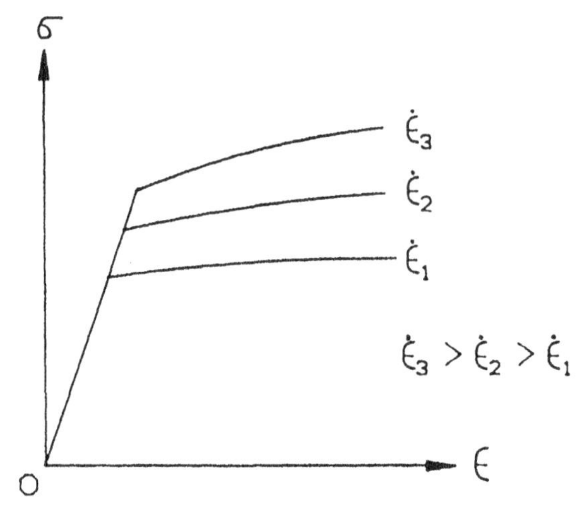

Fig. 1.3 Effect of strain rate

1.1.5. Temperature Effect

Temperature, as strain rate, also has a very important influence on the material properties. As temperature increases, the ductility of metals changes from brittle to ductile behavior. A metal that is ductile at room temperature can become brittle at temperature greatly below room temperature, and brittle metals at room temperature become ductile at very high temperature. This is shown in Fig. 1.4 (from Magnusson et al. 1957). From the figure, one can see that as the temperature increases, the ductility of the material

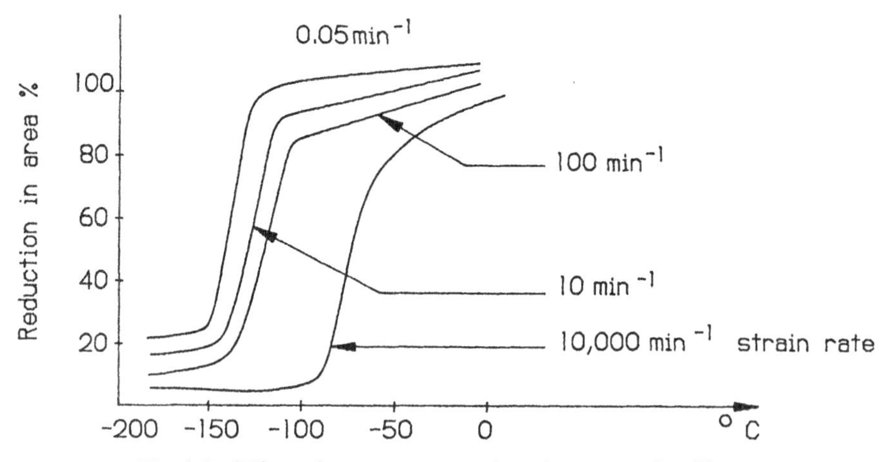

Fig. 1.4 Effect of temperature and strain rate on ductility

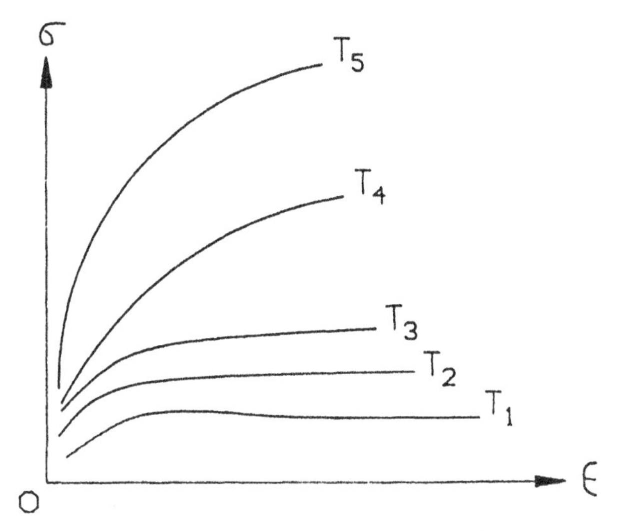

Fig. 1.5 Stress-strain curves at different temperature ($T_1 > T_2 \cdots > T_5$)

increases. The effect of the strain rate on the ductility of metals can also be seen from this figure.

Temperature influences the shape of the stress-strain curve, as shown in Fig. 1.5.

As Fig. 1.5 shows, the stress-strain curve is higher for lower temperatures. This means the material becomes stronger as the temperature decreases; however, lower temperatures reduce ductility.

1.2 IDEALIZATIONS OF MACROSCOPIC PLASTIC BEHAVIOR

Real material behavior in the plastic region is very complex. Sometimes in order to simplify the solution of a boundary value problem, one has to idealize or approximate the plastic behaviors described previously. The most commonly used idealizations are given in Fig. 1.6. The adequacy of a particular idealization depends on the actual application. In cases where the total strain is large, neglecting the elastic strain in a rigid-plastic idealization may give negligible errors in the solution of a boundary value problem. If the structure is made of mild steel and the maximum strains are expected to be less than point B (Fig. 1.1), then the use of the perfectly plastic model (i.e., no strain or work hardening) can provide acceptable solutions.

Many empirical equations have been suggested in order to describe the stress-strain curve shown in Fig. 1.1., especially the hardening portion AU'. These equations can be used to represent the nonlinear work hardening

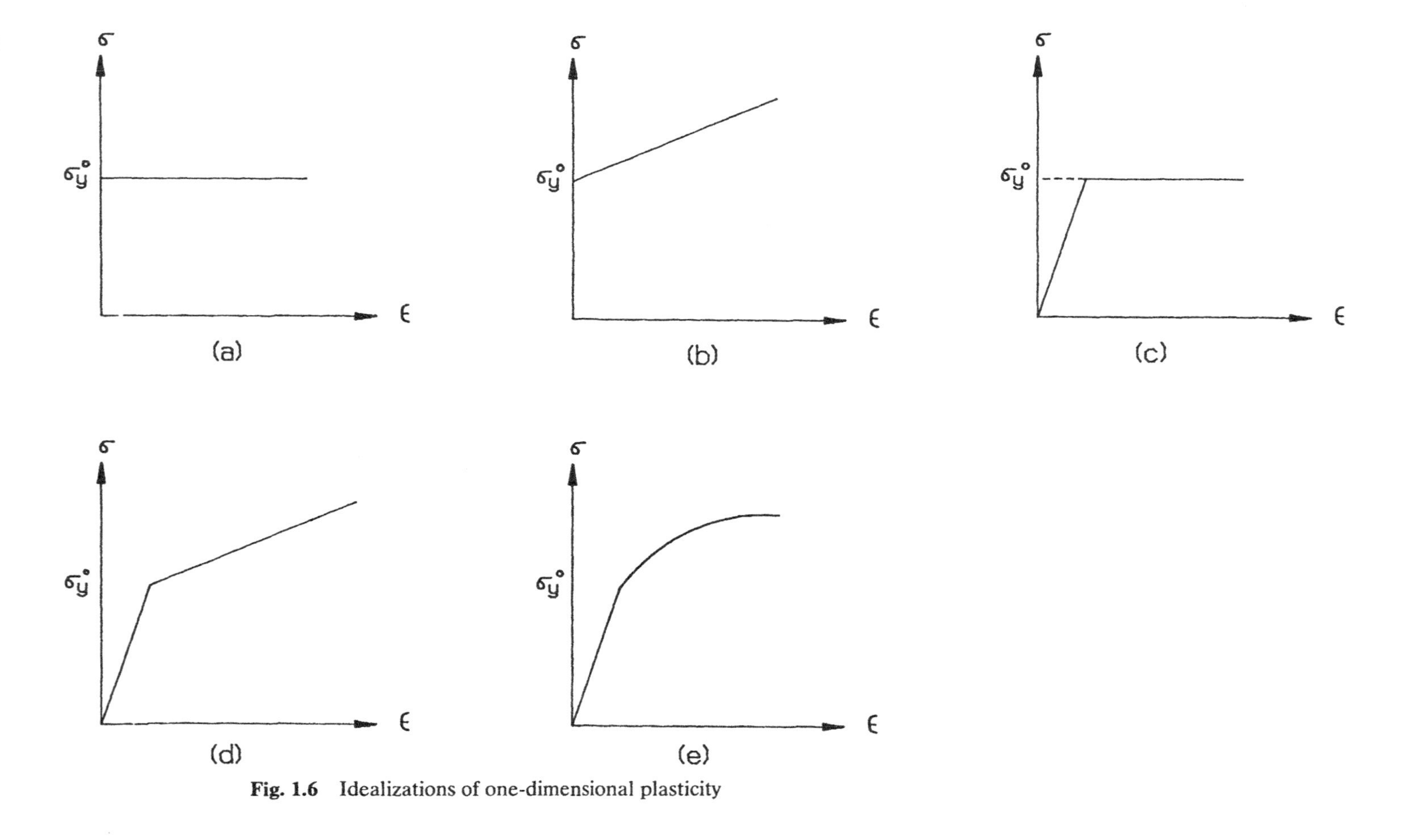

Fig. 1.6 Idealizations of one-dimensional plasticity

behavior shown in Fig. 1.6(e). In all the following cases σ and ϵ refer to true stress and true strain, respectively:

$$\sigma = \sigma_Y + H\epsilon^n \quad \text{(Ludwick 1909)} \tag{1.18}$$

$$\sigma = H\epsilon^n \quad \text{(Holloman 1944)} \tag{1.19}$$

$$\sigma = \sigma_Y + (\sigma_s - \sigma_Y)\{1 - \exp(-n\epsilon)\} \quad \text{(Voce 1948)} \tag{1.20}$$

$$\sigma = H(\epsilon_s + \epsilon)^n \quad \text{(Swift 1947)} \tag{1.21}$$

$$\sigma = \sigma_Y \tan\text{h}\left(\frac{E\epsilon}{\sigma_Y}\right) \quad \text{(Prager 1938)} \tag{1.22}$$

$$\epsilon = \frac{\sigma}{E} + H\left(\frac{\sigma}{E}\right)^n \quad \text{(Ramberg and Osgood 1943)} \tag{1.23}$$

where E, σ_Y, σ_s, ϵ_s, H, and n are material constants and must be determined experimentally.

1.3 SMALL AND LARGE DEFORMATIONS

It has been shown previously that only for small deformation are the incompressibility conditions written in terms of true strain and nominal strain the same. In general, one can expand the nominal strain in terms of the true strain using the Taylor series:

$$\hat{\epsilon} = e^{\epsilon} - 1 = \epsilon + \frac{\epsilon^2}{2!} + \frac{\epsilon^3}{3!} + \frac{\epsilon^4}{4!} + \cdots + \frac{\epsilon^n}{n!}1 + \cdots \tag{1.24}$$

It is obvious that $\hat{\epsilon} \approx \epsilon$ if and only if ϵ is small and all the higher-order terms in this expansion can be neglected. The difference between $\hat{\epsilon}$ and ϵ results from the difference of the reference in their definition. The former, $\hat{\epsilon}$, has reference in the initial or undeformed configuration. The latter, ϵ, is based on the current or deformed configuration. If the strain is small, the difference of the initial and the current dimensions can be ignored, as is usually done in the linear elasticity theory. As the strain becomes large, the difference between the initial and current dimensions becomes appreciable and can no longer be ignored. This simple geometrical consideration is the reason to introduce the concepts of initial and current configurations and the different strain and stress measures as defined in reference to these two different configurations. Moreover, since substantial rigid body rotation may occur in the case of large deformation, the so-called corotational rate concept, which is used to guarantee the objectivity and eliminate the effect of rigid body rotation on the constitutive equations describing plastic deformation, has been introduced. Thus the plasticity theory for finite strain is, by necessity, fairly complex.

In addition to the mathematical complexities discussed above, determination of the real material response at finite plastic strain is a problem. The experiments for finite deformation are much more difficult than those for small deformation. For example, to avoid torsional buckling of the tube specimen in simple torsion tests, one has to use short specimens that cause substantial edge effect and thus make the interpretation of the experimental results very difficult. Due to such experimental difficulties, most of the experimental research has been concentrated in the small strain range; the experiments at finite strain are very few. Therefore the finite plasticity investigations are far less conclusive than the ones for small strain, although several mathematical finite plasticity theories have been proposed. Actually a major portion of the research in plasticity during the last decade has aimed at explaining the material behavior during finite deformation by phenomenological macroscopic experimental observations, or by microscopic measurements and considerations.

In this book we will introduce the basic kinematics and stress measures for finite deformation in Chapter 2 and discuss the phenomenological finite plasticity theory in Chapters 7 and 8. Chapters 9, 10, and 11, will describe the micromechanisms of finite plastic deformation and the plasticity theories based on the microscale consideration.

1.4 MICROSCOPIC PHENOMENA DURING PLASTIC DEFORMATION

In explaining the nature of the plastic deformation on the macroscale we will view metals as a continuum as we have in the previous sections. Experimental observations have yielded several hypotheses and assumptions. We will look at these and discuss several established plasticity theories. These mathematical theories are made simple in order to be practically applicable. Obviously theories can predict the real features of material behavior only in a very limited range. We will not discuss the mechanisms of plastic deformation and their manifestation at the macroscopic level as observed experimentally. These phenomena can be understood and quantified if one looks into the microstructures of the materials and explores the mechanisms of the plastic deformation or flow at the microscopic level.

Suppose that a standard dog-bone type of metal specimen is subjected to uniaxial tension. If the test is carefully planned and performed so that perfect alignment is achieved during the test, there will be a fairly uniform stress distribution over a length in the central portion of the specimen where the edge effect can be ignored. From a macroscopic point of view, there is no reason to expect inhomogeneous deformation over the same length. Actually the deformation is *not* homogeneous when the specimen is stressed into the plastic region. In the early part of this century, Rosenhain and Ewing (1900) observed in a similar experiment that the plastic deformation of metal produced many parallel microscopic steps, called *slip bands*, on the surface. This suggested that the metals had somehow undergone shear along these bands, although they were subjected to macroscopically uniform tensile or compressive stress. In addition Rosenhain and Ewing found that the direction of the shear bands changed at the boundary of single-crystal grains and that the bands in different grains appeared to develop in different directions and to different extent. It was evident then that the investigation of the behavior of individual grains, or single crystals, was necessary. This gave motivation to Andrade, in 1910, to develop a technique for growing large single crystals, and this in turn facilitated many outstanding scientists at that time (e.g., Taylor) to obtain basic and fundamental experimental results on plastic deformation of single crystals. These early studies established that shear occurred along well-defined crystallographic directions and on certain crystallographic planes. Plastic deformation on the microscopic scale is thus inhomogeneous in nature. Although crystallographic slip is the mechanism of the plastic deformation, it cannot accurately predict the strength of the metals (which is the main concern of structural engineers). In fact the strength predicted using the slip mechanism was several orders higher than the experimentally observed one. This discrepancy was eliminated by the dislocation theory introduced independently by Orowan (1934), Polanyi (1934), and Taylor (1934). Also Taylor (1938), using the results of single-crystal plasticity and introducing several consistent assumptions, calculated the plastic response of polycrystals and obtained fairly reasonable results. Taylor's work will be discussed in more detail in Chapters 10 and 11.

Now, it is well-known that metals are crystalline solids that consist of atoms arranged in a pattern repeated periodically in three dimensions. The knowledge of this crystallography forms the basis for further study of physical plasticity, which will be covered briefly in Chapter 9. We mention these physical considerations here because any accurate and applicable mathematical plasticity theory should have its basis in the physics of plastic deformation. Even structural or mechanical engineers must have some basic facts about the physics of plastic deformation or flow.

REFERENCES

Bridgman, P. W. 1947. The effect of hydrostatic pressure on the fracture of brittle substances. *J. Appl. Phys.* **18**:246.

Bridgman, P. W. 1952. *Studies in Large Plastic Flow and Fracture with Special Emphasis on the Effects of Hydrostatic Pressure*. New York: McGraw-Hill.

Ewing, J. A., and Rosenhain, W. 1900. The crystalline structure of metals. *Phil. Trans. R. Soc. London* **193**:353.

Magnusson, A. W., and W. M. Baldwin 1957. Low temperature brittleness. *J. Mech. Phys. Solids* **5**:172.

Orowan, E. 1934. Zur Kristall plastizitat. III: Uber den Mechanisms des Gleituorganges. *Z. Phys.* **89**:634.

Polanyi, Von, M. 1934. Uber eine Ant Gitterstorung, die einen Kristall plastisch machen Konnte. *Z. Phys.* **89**:660.

Taylor, G. I. 1934. The mechanism of plastic deformation of crystals. Part I: Theoretical. *Proc. R. Soc. London* **A105**:362.

Taylor, G. I. 1938. Plastic strain in metals. *J. Inst. Metals* **62**:307.

2

STRAIN AND STRESS

The main idea behind any plasticity theory, or more generally of any purely mechanical constitutive theory, is to formulate the relationship between stress and strain. To explore this relationship, one must understand how stress and strain characterize the deformation of the material under applied forces. For one-dimensional cases, the situation is quite simple. One can use the nominal stress and strain $(\hat{\sigma}, \hat{e})$ or the true stress and strain (σ, ϵ), as discussed in Chapter 1. For infinitesimal strain, these two measures are approximately the same. However, for three-dimensional cases involving finite strain, the concepts and various measures of stress and strain are much more complex. This chapter contains the fundamental kinematics and various stress measures from the continuum mechanics point of view; these topics will be necessary for the understanding of the finite plasticity theory discussed in subsequent chapters. For the sake of clarity, only the rectangular cartesian coordinate system will be used here.

2.1 CONFIGURATIONS AND DISPLACEMENT

Physically any material consists of molecules, atoms, and even smaller particles. In other words, its structure is discrete in nature. If the scale or dimension of the material body under consideration is much larger than that of atoms or molecules, then the discrete nature can be ignored and the material body can be treated as a continuum consisting of particles continuously distributed in space. This idealization of the matter is called *continuum*, or more precisely, the *continuum model of the matter*. This model, or the whole set of all the particles of a matter, is referred to as a *body*. In a general sense kinematics is the study of the motion and deformation aspects of bodies which are isolated from external forces or any other sources that can cause motion or deformation.

The components of vectors and tensors are introduced in a fixed rectangular cartesian coordinate system. The origin of this system is denoted by O, and the base vectors are given by $\mathbf{i}_K$ ($K = 1, 2, 3$). Motion is defined relative to this fixed frame. Now suppose that at a given reference time, say, $t = t_0$ (usually taken as zero), a material body, denoted by $\mathscr{B}$, occupies a region of space denoted by $\mathscr{R}_0$, which may be finite or infinite depending on the extent of the body. Physically one can imagine that the particles of the body $\mathscr{B}$ are distributed continuously over the region $\mathscr{R}_0$. Mathematically one can say that there is a one-to-one correspondence between the material particles and the points in $\mathscr{R}_0$, and one can identify each of the particles denoted by P, by a position vector $\mathbf{X}$, as shown in Fig. 2.1. In the fixed coordinate frame, $\mathbf{X}$ is given by

$$\mathbf{X} = X_K \mathbf{i}_K \tag{2.1}$$

The summation convention for two repeated indices is used throughout this book.

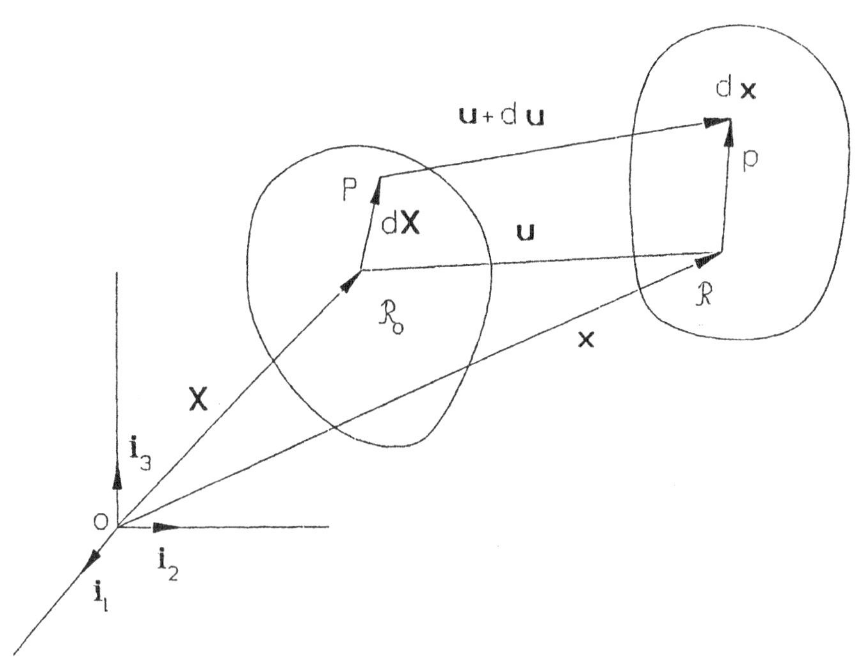

Fig. 2.1 Body and configurations

Imagine that at a typical subsequent time t, the body $\mathscr{B}$ moves to (and occupies) another region $\mathscr{R}$ in space. All particles of $\mathscr{B}$ are now continuously distributed over the region $\mathscr{R}$, and there is a one-to-one correspondence between the material particles of $\mathscr{B}$ and the points in $\mathscr{R}$. Similarly the new location of the particle P of body $\mathscr{B}$, denoted by p in the new configuration $\mathscr{R}$, can be specified by a position vector $\mathbf{x}$, as shown in Fig. 2.1:

$$\mathbf{x} = x_k \mathbf{i}_k \tag{2.2}$$

where $\mathbf{i}_k = \mathbf{i}_K$ since only one fixed rectangular cartesian coordinate system will be used in this book. Since the particle P in the reference configuration $\mathscr{R}_0$ with position vector $\mathbf{X}$ occupies the location p in the current configuration $\mathscr{R}$ with position vector $\mathbf{x}$, the motion can be mathematically described as

$$\mathbf{x} = \mathbf{x}(\mathbf{X}, t) \tag{2.3}$$

or in component form,

$$x_k = x_k(X_K, t) \tag{2.4}$$

The motion is assumed to be smooth. Equation (2.3) or (2.4) can therefore be inverted to give

$$\mathbf{X} = \mathbf{X}(\mathbf{x}, t) \tag{2.5}$$

or in component form

$$X_K = X_K(x_k, t) \tag{2.6}$$

To ensure that Eq. (2.3) or (2.4) is invertible, the function $\mathbf{x}(\mathbf{X}, t)$ should be single valued and continuously differentiable at a given time t. Thus the Jacobi determinant of $\mathbf{x}(\mathbf{X}, t)$ must be nonzero in the region $\mathscr{R}_0$:

$$\left| \frac{\partial x_k}{\partial X_K} \right| = |x_{k,K}| \neq 0 \tag{2.7}$$

Here a subscript preceded by a comma represents the partial derivative with respect to the coordinate denoted by that subscript. Equation (2.7) is called the *axiom of continuity* in continuum mechanics.

So far we have introduced two coordinates X_K and x_k, or two position vectors $\mathbf{X}$ and $\mathbf{x}$, to describe the motion of a typical particle P in a body $\mathscr{B}$. A typical particle that moves in space according to Eq. (2.3) or (2.4) must have its reference position $\mathbf{X}$ in the region $\mathscr{R}_0$; hence the point $\mathbf{X}$ is its starting point, or its reference point. At any subsequent time t, one can always find this particle by identifying its starting or reference point. Regardless of how it moves and the distance it travels, it can change only its current position and will always have the same original reference position. In this sense the coordinates X_K can be thought of as fixed labels of a typical particle and will not change for a given particle. Thus these coordinates are called *material coordinates* or *reference coordinates*, or even mistakenly *Lagrangian coordinates* (though, as pointed out by Truesdell 1952, they were introduced by Euler in 1762). Following the same terminology, the set of all particles of the body $\mathscr{B}$ at $t = t_0$ (i.e., the region $\mathscr{R}_0$ in space) is termed *reference* configuration. The "configuration" refers to the region in space occupied by the body at any given time. It should be noted that the reference configuration does not have to be the region that the material actually occupies. It can be an imaginary one. However, it is preferable to adopt any region actually occupied by the material at some previous time. Thus the position $\mathbf{X}$ is usually the starting point of the typical particle.

The position vector $\mathbf{x}$, on the other hand, specifies a spatial point that can be occupied by different particles at different times. The coordinates x_k are called *spatial coordinates* or *Eulerian coordinates*, even though they were introduced by d'Alembert in 1752 (see Truesdell 1952). According to the preceding definition of a configuration, the region $\mathscr{R}$ occupied by the body of material at time t is called the *current* configuration.

In describing the motion or deformation of materials, either the material coordinates X_K or the spatial coordinates x_k can be used as the independent variables to identify any particle of the material. This is true because the motion as denoted by Eq. (2.3) or (2.4) is invertible. One can determine the current position $\mathbf{x}$ of the particle at any time t if one knows its reference label $\mathbf{X}$. Inversely, one can specify its original or reference position $\mathbf{X}$ by knowing its current position $\mathbf{x}$ at time t. According to Eq. (2.5) or (2.6), the particle that currently occupies position $\mathbf{x}$ originated from reference position $\mathbf{X}$ at $t = t_0$. It takes a *material* or *Lagrangian* description if one chooses the material coordinates X_K as the independent variables. Alternatively, it takes the spatial or Eulerian description if the spatial coordinates are chosen as the independent variables. In the material description the typical particle or particles are followed. But in the spatial description the concern is with any phenomenon at a particular point in space as different particles occupy this point at different times. While both descriptions are useful, the choice of one or the other is to be dependent on the nature and interest of the problem under consideration.

As discussed above, the typical particle P with position $\mathbf{X}$ at $t = t_0$ moves to position $\mathbf{x}$ and is denoted by p at time t. As shown in Fig. 2.1, particle motion can be described by introducing a displacement vector $\mathbf{u}$ that relates $\mathbf{X}$ to $\mathbf{x}$ by

$$\mathbf{x} = \mathbf{X} + \mathbf{u} \tag{2.10}$$

In material descriptions the displacement $\mathbf{u}$ is represented as a function of $\mathbf{X}$

and t:

$$\mathbf{u}(\mathbf{X}, t) = \mathbf{x}(\mathbf{X}, t) - \mathbf{X} \tag{2.11}$$

Alternatively, one can employ the spatial description to write the displacement $\mathbf{u}$ as a function of $\mathbf{x}$ and t:

$$\mathbf{u}(\mathbf{x}, t) = \mathbf{x} - \mathbf{X}(\mathbf{x}, t) \tag{2.12}$$

It should be noted that Eqs. (2.11) and (2.12) have different physical meanings. Equation (2.11) expresses the displacement of the particle with position vector $\mathbf{X}$ in the reference configuration at the current time t. Equation (2.12) is the displacement of any particle which is currently at position $\mathbf{x}$.

Although for the sake of simplicity and clarity the material coordinates X_K and the spatial coordinates x_k are here both referred to the same cartesian coordinate system, it should be pointed out that they are in fact two different coordinate systems, as shown in Fig. 2.2. For some problems it is more convenient to use curvilinear coordinates for material and/or spatial descriptions. For example, to solve a problem of beam bending with initial rectangular configuration which is deformed to a circular shape (Fig. 2.3), it is simpler to choose the cartesian coordinate system to describe the initial configuration and the curvilinear polar coordinates to describe the current configuration.

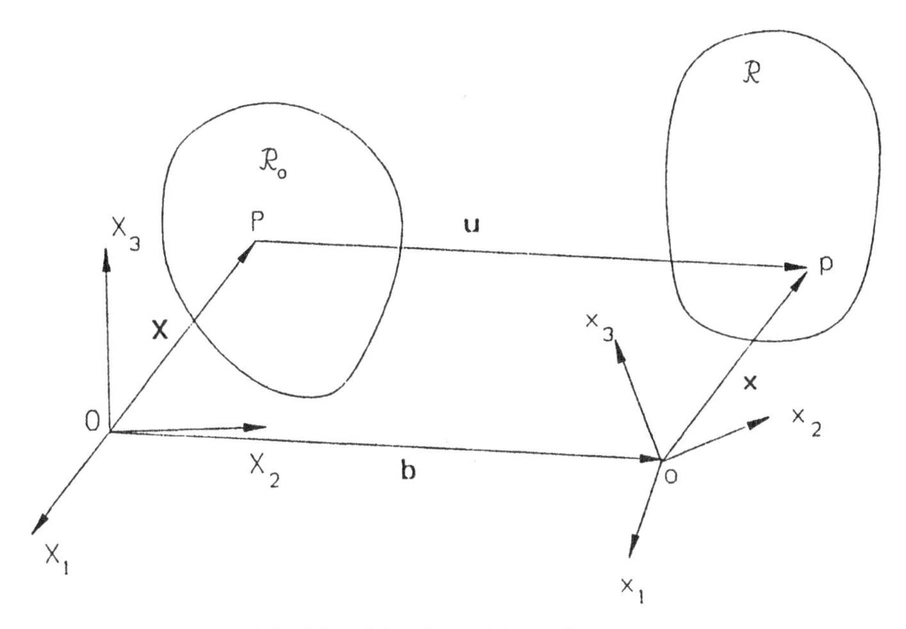

Fig. 2.2 Material and spatial coordinate systems

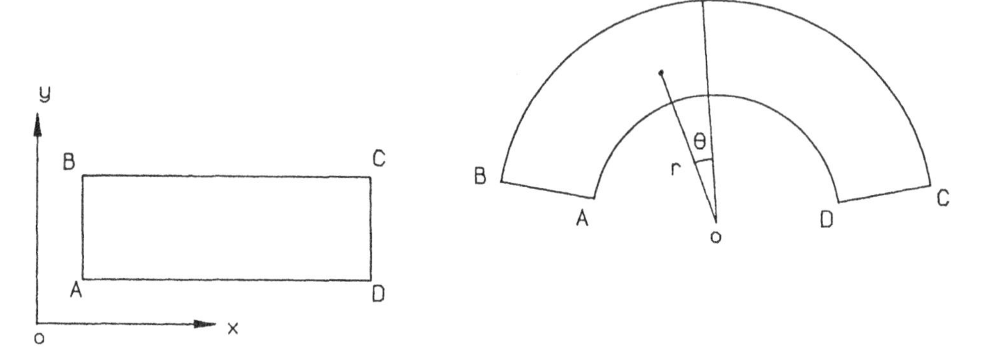

Fig. 2.3 Different material and spatial coordinates

Interested readers can refer to books by Malvern (1969) or Truesdell (1985) for more details about the material and spatial coordinates.

2.2 DEFORMATION GRADIENT AND DEFORMATION OF LINE, AREA, AND VOLUME ELEMENTS

The motion of a material body is given by Eq. (2.3) or (2.4). The configuration $\mathscr{R}$ at time t is deformed with respect to the initial configuration $\mathscr{R}_0$. Because of deformation the points, lines, areas, and volumes in $\mathscr{R}_0$ are transformed into the corresponding ones in $\mathscr{R}$. One of the main purposes of kinematic analysis is to determine the relationships between these geometrical elements in $\mathscr{R}_0$ and $\mathscr{R}$.

Obviously the relationship between points in $\mathscr{R}_0$ and $\mathscr{R}$ is given by Eq. (2.3) or (2.4) at any given time t. To find out the transformation equation between the line elements, let us consider a material line element at point **X**, denoted $d\mathbf{X}$ in Fig. 2.1. This line element becomes $d\mathbf{x}$ after deformation. From Eq. (2.3) or (2.4), one can easily obtain the following relation:

$$d\mathbf{x} = \frac{\partial \mathbf{x}}{\partial \mathbf{X}} \cdot d\mathbf{X} = \nabla \mathbf{x} \cdot d\mathbf{X} = \mathbf{F} \cdot d\mathbf{X}$$

$$dx_i = F_{iK}\, dX_K \tag{2.13}$$

The term **F** is called the *deformation gradient tensor* and has the following component form in the cartesian coordinate system:

$$F_{iK} = \frac{\partial x_i}{\partial X_K} = x_{i,K} = \delta_{iK} + u_{i,K} \tag{2.14}$$

It should be pointed out that $\mathbf{F} = F_{lK}\mathbf{i}_l\mathbf{i}_K$ is related to two coordinates (material and spatial) and is called the two-point tensor. The deformation gradient tensor $\mathbf{F}$ is a linear operator which relates every material line element $d\mathbf{X}$ in $\mathscr{R}_0$ to the corresponding $d\mathbf{x}$ in $\mathscr{R}$. According to the axiom of continuity of Eq. (2.7),

$$\det(\mathbf{F}) \neq 0 \tag{2.15}$$

Then Eq. (2.13) can be inverted to obtain

$$d\mathbf{X} = \mathbf{F}^{-1}\, d\mathbf{x} \quad \text{or} \quad dX_K = F_{lK}^{-1}\, dx_l \tag{2.16}$$

where $\mathbf{F}^{-1}$ is the inverse of $\mathbf{F}$ and has the component form

$$F_{iK}^{-1} = X_{K,i} \tag{2.17}$$

To determine the relationship between the area elements and volume elements, one needs to use permutation symbols and some identities (cf. Eringen 1965)

$$e_{ijk} = (\mathbf{i}_i \times \mathbf{i}_j) \cdot \mathbf{i}_k = \begin{cases} 1 & \text{if } (i, j, k) \text{ is an even permutation of } (1,2,3) \\ 0 & \text{if any two of } (i, j, k) \text{ are equal} \\ -1 & \text{if } (i, j, k) \text{ is an odd permutation of } (1,2,3) \end{cases} \tag{2.18}$$

$$\mathbf{i}_k \times \mathbf{i}_l = e_{klm}\mathbf{i}_m \tag{2.19}$$

$$e_{ijk}A_{im}A_{jn}A_{kl} = e_{mnl}\det(\mathbf{A}) \quad \text{or } \det(\mathbf{A}) = \tfrac{1}{6}e_{ijk}e_{mnl}A_{im}A_{jn}A_{kl} \tag{2.20}$$

where $\mathbf{i}_k$ and $\mathbf{A}$ are the base vectors and second-order tensor, respectively. Now consider an area element vector in $\mathscr{R}$, $d\mathbf{A} = d\mathbf{x}^{(1)} \times d\mathbf{x}^{(2)}$. According to Eqs. (2.13) and (2.14), one can write

$$\begin{aligned} d\mathbf{a} &= d\mathbf{x}^{(1)} \times d\mathbf{x}^{(2)} = \left(\mathbf{F} \cdot d\mathbf{X}^{(1)}\right) \times \left(\mathbf{F} \cdot d\mathbf{X}^{(2)}\right) \\ &= F_{iK}\, dX_K^{(1)} F_{jL}\, dX_L^{(2)}\, \mathbf{i}_i \times \mathbf{i}_j \\ &= e_{ijm}F_{iK}F_{jL}\, dX_K^{(1)}\, dX_L^{(2)}\, \mathbf{i}_m \\ &= e_{ijm}F_{iK}F_{jL}F_{mN}(F_{lN})^{-1}\, dX_K^{(1)}\, dX_L^{(2)}\, \mathbf{i}_l \\ &= \det(\mathbf{F})e_{KLN}(F_{lN})^{-1}\, dX_K^{(1)}\, dX_L^{(2)}\, \mathbf{i}_l \\ &= \det(\mathbf{F})\left(d\mathbf{X}^{(1)} \times d\mathbf{X}^{(2)}\right) \cdot \mathbf{F}^{-1} \\ &= \det(\mathbf{F})\, d\mathbf{A} \cdot \mathbf{F}^{-1} = \det(\mathbf{F})\left(\mathbf{F}^{-1}\right)^{\mathrm{T}} \cdot d\mathbf{A} \end{aligned} \tag{2.21}$$

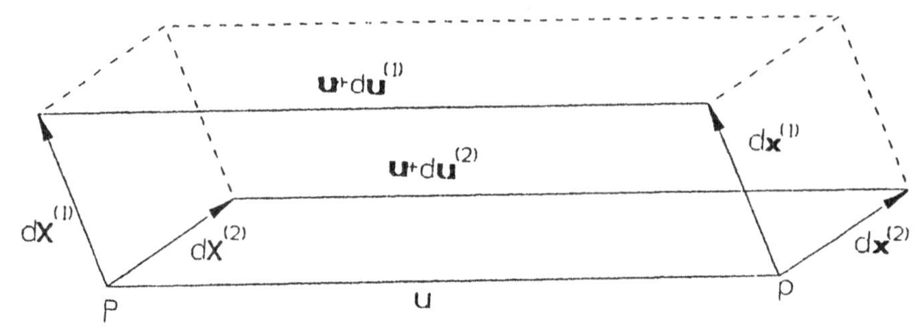

Fig. 2.4 Deformation of the area element

To obtain the final result as given above, Eqs. (2.16) and (2.17) have been used. As mentioned before, $\mathbf{i}_m = \mathbf{i}_M$ since both material and spatial coordinates refer to a single fixed cartesian coordinate system. After the deformation, the area $d\mathbf{A} = d\mathbf{X}^{(1)} \times d\mathbf{X}^{(2)}$ becomes $d\mathbf{a}$, as shown in Fig. 2.4.

From Eq. (2.21), it is easy to obtain an inverse relationship between $d\mathbf{A}$ and $d\mathbf{a}$:

$$d\mathbf{A} = [\det(\mathbf{F})]^{-1} d\mathbf{a} \cdot \mathbf{F} \tag{2.22}$$

In component form Eqs. (2.21) and (2.22) can be written, respectively, as

$$da_i = \det(\mathbf{F}) X_{K,i}\, dA_K \tag{2.23}$$

and

$$dA_K = [\det(\mathbf{F})]^{-1} x_{i,K}\, da_i \tag{2.24}$$

To consider the deformation of the volume element, let us express the volume of an elementary tetrahedron in the initial configuration $\mathscr{R}_0$ as

$$dV = \tfrac{1}{6}\left(d\mathbf{X}^{(1)} \times d\mathbf{X}^{(2)}\right) \cdot d\mathbf{X}^{(3)} \tag{2.25}$$

where $d\mathbf{X}^{(1)}$, $d\mathbf{X}^{(2)}$, and $d\mathbf{X}^{(3)}$ are three noncoplanar, infinitesimal line elements in the reference configuration and become $d\mathbf{x}^{(1)}$, $d\mathbf{x}^{(2)}$, and $d\mathbf{x}^{(3)}$, respectively, after deformation in the current configuration $\mathscr{R}$, as shown in Fig. 2.5. Then as in Eq. (2.25) one has the volume element dv as

$$dv = \tfrac{1}{6}\left(d\mathbf{x}^{(1)} \times d\mathbf{x}^{(2)}\right) \cdot d\mathbf{x}^{(3)} \tag{2.26}$$

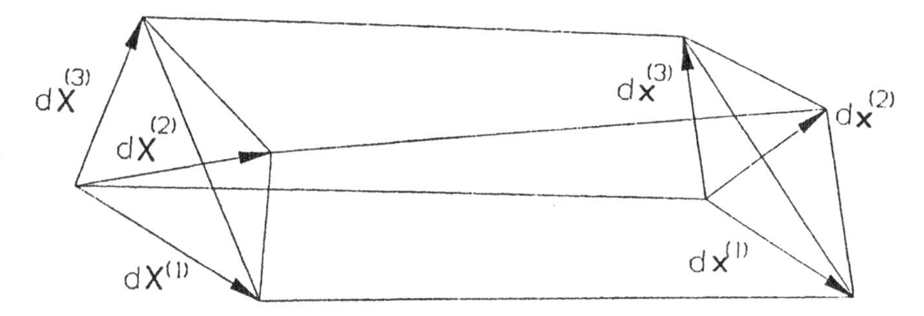

Fig. 2.5 Deformation of volume element

Substituting Eq. (2.13) into Eq. (2.26), we obtain

$$\begin{aligned} dv &= \tfrac{1}{6}\left(\mathbf{F}\cdot d\mathbf{X}^{(1)} \times \mathbf{F}\cdot d\mathbf{X}^{(2)}\right)\cdot \mathbf{F}\cdot d\mathbf{X}^{(3)} \\ &= \tfrac{1}{6}F_{iK}F_{jL}F_{lM}\, dX_K^{(1)}\, dX_L^{(2)}\, dX_M^{(3)}\,(\mathbf{i}_i \times \mathbf{i}_j)\cdot \mathbf{i}_l \\ &= \tfrac{1}{6}e_{ijl}F_{iK}F_{jL}F_{lM}\, dX_K^{(1)}\, dX_L^{(2)}\, dX_M^{(3)} \\ &= \det(\mathbf{F})\, dV \end{aligned} \tag{2.27}$$

Hence

$$\frac{dv}{dV} = \det(\mathbf{F}) \tag{2.28}$$

Equation (2.28) relates the volume elements in the configurations $\mathscr{R}_0$ and $\mathscr{R}$.

Using Eq. (2.14), we can write

$$\det(\mathbf{F}) = \det(\delta_{iK} + u_{i,K}) = 1 + u_{i,i} + O(u_{i,K})^2 \tag{2.29}$$

Here $u_{i,K} = (F_{iK} - \delta_{iK})$ is the displacement gradient. It is seen from Eq. (2.29) that if only an infinitesimal gradient is involved, terms of higher order than $u_{i,K}$ can be neglected, and Eq. (2.29) reduces to

$$\frac{dv}{dV} = \det(\mathbf{F}) \approx 1 + u_{i,i} \tag{2.30}$$

The volume strain in the classical infinitesimal elasticity and plasticity theory is given by

$$\frac{dv - dV}{dV} = \frac{dv}{dV} - 1 \approx \operatorname{tr}(u_{i,K}) = u_{i,i} \tag{2.31}$$

2.3 MEASURES OF FINITE DEFORMATIONS

The square of the length of any line element in $\mathscr{R}_0$ and $\mathscr{R}$ is given by

$$dS^2 = d\mathbf{X} \cdot d\mathbf{X} = dX_K \, dX_K \tag{2.32}$$

$$ds^2 = d\mathbf{x} \cdot d\mathbf{x} = dx_i \, dx_i \tag{2.33}$$

We can write dS^2 in terms of the deformation gradient $\mathbf{F}$ as follows:

$$\begin{aligned} dS^2 &= d\mathbf{X} \cdot d\mathbf{X} = \left(\mathbf{F}^{-1} \cdot d\mathbf{x}\right) \cdot \left(\mathbf{F}^{-1} \cdot d\mathbf{x}\right) \\ &= d\mathbf{x} \cdot \left(\mathbf{F}^{-1}\right)^{\mathrm{T}} \cdot \left(\mathbf{F}^{-1}\right) \cdot d\mathbf{x} \end{aligned} \tag{2.34}$$

Similarly ds^2 can be expressed as

$$\begin{aligned} ds^2 &= d\mathbf{x} \cdot d\mathbf{x} = (\mathbf{F} \cdot d\mathbf{X}) \cdot (\mathbf{F} \cdot d\mathbf{X}) \\ &= d\mathbf{X} \cdot \mathbf{F}^{\mathrm{T}} \cdot \mathbf{F} \cdot d\mathbf{X} \end{aligned} \tag{2.35}$$

In 1827 Cauchy introduced a deformation tensor $\mathbf{B}^{-1}$ given by

$$\begin{aligned} \mathbf{B}^{-1} &= \left(\mathbf{F}^{-1}\right)^{\mathrm{T}} \cdot \mathbf{F}^{-1} \\ B_{ij}^{-1} &= X_{K,i} X_{K,j} \end{aligned} \tag{2.36}$$

and in 1841 Green used the following tensor to describe finite deformation

$$\begin{aligned} \mathbf{C} &= \mathbf{F}^{\mathrm{T}} \cdot \mathbf{F} \\ C_{KL} &= x_{i,K} x_{i,L} \end{aligned} \tag{2.37}$$

These two tensors, **C** and **B**, are called *right* and *left Cauchy-Green tensors* by Truesdell and Noll (1965). **C** and **B** are also called, respectively, *referential* (Lagrangian) and *spatial* (Eulerian) descriptions of deformation.

Using Cauchy-Green tensors, one can write dS^2 and ds^2 as

$$dS^2 = d\mathbf{x} \cdot \mathbf{B}^{-1} \cdot d\mathbf{x} = dx_i \, B_{ij}^{-1} \, dx_j \tag{2.38}$$

and

$$ds^2 = d\mathbf{X} \cdot \mathbf{C} \cdot d\mathbf{X} = dX_K \, C_{KL} \, dX_L \tag{2.39}$$

The difference between ds^2 and dS^2 can be expressed by the equivalent forms

$$ds^2 - dS^2 = d\mathbf{X} \cdot (\mathbf{C} - \mathbf{I}) \cdot d\mathbf{X} = d\mathbf{x} \cdot \left(I - B^{-1}\right) \cdot d\mathbf{x}$$
$$= dX_K\left(C_{KL} - \delta_{KL}\right) dX_L = dx_i\left(\delta_{ij} - B_{ij}^{-1}\right) dx_j \qquad (2.40)$$

where δ_{ij} is the Kronecker delta.

From Eq. (2.40) we see that $\mathbf{C}$ or $\mathbf{B}^{-1}$ determines the difference $ds^2 - dS^2$. To further explore the physical implication of $\mathbf{C}$ and $\mathbf{B}^{-1}$, we need to consider the deformation ratios of a line element, an area element, and a volume element.

For a line element $d\mathbf{X}$ in $\mathscr{R}_0$, the unit vector along $d\mathbf{X}$ is defined by

$$\mathbf{N} = \frac{d\mathbf{X}}{|d\mathbf{X}|} \qquad (2.41)$$

The stretch ratio in $\mathbf{N}$ direction $\lambda_{\mathbf{N}}$ can be expressed

$$\lambda_{\mathbf{N}} = \frac{ds}{dS} = \frac{|d\mathbf{x}|}{|d\mathbf{X}|} = \frac{\sqrt{d\mathbf{X} \cdot \mathbf{C} \cdot d\mathbf{X}}}{|d\mathbf{X}|} = \sqrt{\mathbf{N} \cdot \mathbf{C} \cdot \mathbf{N}} \qquad (2.42)$$

Similarly we can define a unit vector $\mathbf{n}$ along an infinitesimal line element $d\mathbf{x}$ in $\mathscr{R}$ as

$$\mathbf{n} = \frac{d\mathbf{x}}{|d\mathbf{x}|} \qquad (2.43)$$

and the stretch ratio in $\mathbf{n}$ direction can be written

$$\lambda_{\mathbf{n}} = \frac{ds}{dS} = \frac{|d\mathbf{x}|}{|d\mathbf{X}|} = \frac{|d\mathbf{x}|}{\sqrt{d\mathbf{x} \cdot \mathbf{B}^{-1} \cdot d\mathbf{x}}} = \frac{1}{\sqrt{\mathbf{n} \cdot \mathbf{B}^{-1} \cdot \mathbf{n}}} \qquad (2.44)$$

If in Eqs. (2.42) and (2.44) $d\mathbf{x}$ and $d\mathbf{X}$ correspond to the same material element, then $\lambda_{\mathbf{N}} = \lambda_{\mathbf{n}}$, and Eqs. (2.42) and (2.44) represent the referential (Lagrangian) and the spatial (Eulerian) description, respectively, of the stretch of the same line element.

Now consider the area ratio (da/dA), where $da = |d\mathbf{a}|$ and $dA = |d\mathbf{A}|$. Similar to the derivation of Eqs. (2.42) and (2.44), the following relationships can be obtained:

$$\Lambda_N = \frac{da}{dA} = \det(\mathbf{F})\sqrt{\mathbf{N} \cdot \mathbf{C}^{-1} \cdot \mathbf{N}} \qquad (2.45)$$

and

$$\Lambda_n = \frac{da}{dA} = \det(\mathbf{F}) \frac{1}{\sqrt{\mathbf{n} \cdot \mathbf{B} \cdot \mathbf{n}}} \tag{2.46}$$

where $\mathbf{B} \cdot \mathbf{B}^{-1} = \mathbf{I}$ and $\mathbf{N}$ and $\mathbf{n}$ are given by

$$\mathbf{N} = \frac{d\mathbf{A}}{|d\mathbf{A}|}$$

$$\mathbf{n} = \frac{d\mathbf{a}}{|d\mathbf{a}|} \tag{2.47}$$

As for the volume ratio, it is obvious that from Eqs. (2.36) and (2.37), we get

$$\det(\mathbf{C}) = \det(\mathbf{F}^{\mathrm{T}} \cdot \mathbf{F}) = \det{}^2(\mathbf{F}) \tag{2.48}$$

$$\det(\mathbf{B}^{-1}) = \det\left[(\mathbf{F}^{-1})^{\mathrm{T}} \cdot \mathbf{F}^{-1}\right] = \left[\det{}^2(\mathbf{F})\right]^{-1} \tag{2.49}$$

Equation (2.28) can be rewritten, giving the volume ratio in terms of the tensor $\mathbf{C}$ or $\mathbf{B}^{-1}$ (or $\mathbf{B}$):

$$\frac{dv}{dV} = \sqrt{\det(\mathbf{C})} = \left[\sqrt{\det(\mathbf{B}^{-1})}\right]^{-1} \tag{2.50}$$

Besides length, area, and volume changes, there are angle changes to consider. Angle changes are an important aspect of the deformation resulting in *shear deformation*. Suppose that in $\mathscr{R}_0$ the angle between two unit vectors $\mathbf{N}^{(1)} = d\mathbf{X}^{(1)}/|d\mathbf{X}^{(1)}|$ and $\mathbf{N}^{(2)} = d\mathbf{X}^{(2)}/d\mathbf{X}^{(2)}$ is Θ. After deformation it becomes θ. The shear deformation is defined as the difference between Θ and θ (i.e., $\gamma = \Theta - \theta$). It is easy to show that

$$\cos\theta = \cos(\Theta - \gamma) = \frac{d\mathbf{x}^{(1)} \cdot d\mathbf{x}^{(2)}}{|d\mathbf{x}^{(1)}|\,|d\mathbf{x}^{(2)}|} = \frac{d\mathbf{X}^{(1)} \cdot \mathbf{F}^{\mathrm{T}} \cdot \mathbf{F} \cdot d\mathbf{X}^{(2)}}{\lambda_{\mathbf{N}^{(1)}}\lambda_{\mathbf{N}^{(2)}}|d\mathbf{X}^{(1)}|\,|d\mathbf{X}^{(2)}|}$$

$$= \frac{1}{\lambda_{\mathbf{N}^{(1)}}\lambda_{\mathbf{N}^{(2)}}}\mathbf{N}^{(1)} \cdot \mathbf{C} \cdot \mathbf{N}^{(2)}$$

or

$$\cos\theta = \cos(\Theta - \gamma) = \frac{1}{\lambda_{\mathbf{N}^{(1)}}\lambda_{\mathbf{N}^{(2)}}} C_{KL} N_K^{(1)} N_L^{(2)} \tag{2.51}$$

If $\Theta = \pi/2$, Eq. (2.51) can be written

$$\sin\gamma = \frac{1}{\lambda_{\mathbf{N}^{(1)}}\lambda_{\mathbf{N}^{(2)}}} C_{KL} N_K^{(1)} N_L^{(2)} \tag{2.52}$$

Equations (2.51) and (2.52) represent the referential or Lagrangian description of the shear deformation. In this case, since $\mathbf{N}^{(1)}$ and $\mathbf{N}^{(2)}$ are specified in advance, we can easily calculate the angle Θ by $\cos\Theta = \mathbf{N}^{(1)} \cdot \mathbf{N}^{(2)}$. Subtracting θ, determined by Eq. (2.51), from Θ gives the angle change (i.e., the shear deformation γ). Similarly, if one specifies two unit vectors in $\mathscr{R}$, $\mathbf{n}^{(1)}$ and $\mathbf{n}^{(2)}$, the angle θ can be calculated easily by $\cos\theta = \mathbf{n}^{(1)} \cdot \mathbf{n}^{(2)}$. The angle Θ can be calculated by the following equation using Eq. (2.44):

$$\cos\Theta = \cos(\theta + \gamma) = \frac{d\mathbf{X}^{(1)} \cdot d\mathbf{X}^{(2)}}{|d\mathbf{X}^{(1)}|\,|d\mathbf{X}^{(2)}|} = \lambda_{\mathbf{n}^{(1)}}\lambda_{\mathbf{n}^{(2)}}\mathbf{n}^{(1)} \cdot \mathbf{B}^{-1} \cdot \mathbf{n}^{(2)}$$

or

$$\cos\Theta = \cos(\theta + \gamma) = \lambda_{\mathbf{n}^{(1)}}\lambda_{\mathbf{n}^{(2)}} B_{ij}^{-1} n_i^{(1)} n_j^{(2)} \tag{2.53}$$

This is the spatial or Euler description of the shear. The shear deformation γ can be determined by subtracting θ from Θ determined by Eq. (2.53).

Various aspects of deformations can clearly be described by the Cauchy-Green tensors $\mathbf{C}$ and $\mathbf{B}^{-1}$ (or $\mathbf{B}$), as illustrated in Eqs. (2.32) through (2.53). Therefore $\mathbf{C}$ and $\mathbf{B}^{-1}$ can be used as the measures for finite deformation.

It should be noted that for rigid body motion,

$$\mathbf{C} = \mathbf{I}$$

or

$$\mathbf{B}^{-1} = \mathbf{I} \tag{2.54}$$

This can be proved easily by using Eqs. (2.42) and (2.44) if one notices that for rigid body motion, $\lambda_{\mathbf{N}} = \lambda_{\mathbf{n}} = 1$ in any direction. Usually the deformation measure is expected to be zero tensor for rigid body motion. This leads to the introduction of the following Almansi-Hemel strain tensors:

$$\mathbf{E} = \tfrac{1}{2}(\mathbf{C} - \mathbf{I})$$

$$E_{KL} = \tfrac{1}{2}(x_{i,K} x_{i,L} - \delta_{KL}) \tag{2.55}$$

and

$$\mathbf{e} = \tfrac{1}{2}(\mathbf{I} - \mathbf{B}^{-1})$$

$$e_{ij} = \tfrac{1}{2}(\delta_{ij} - X_{K,i} X_{K,j}) \tag{2.56}$$

In terms of the displacement vector **u**, one can write the above equations as

$$E_{KL} = \tfrac{1}{2}(u_{K,L} + u_{L,K} + u_{M,K}u_{M,L}) \tag{2.57}$$

and

$$e_{ij} = \tfrac{1}{2}(u_{i,j} + u_{j,i} - u_{m,i}u_{m,j}) \tag{2.58}$$

It is easily seen from Eq. (2.55) to (2.58) that for rigid body motion, $\mathbf{E} = \mathbf{e} = \mathbf{0}$.

For infinitesimal deformation, since $|u_{i,K}| \ll 1$, we obtain the following conventional infinitesimal strain tensor by neglecting the product term in Eq. (2.57) or (2.58):

$$\epsilon_{ij} = \tfrac{1}{2}(u_{i,j} + u_{j,i}) \tag{2.59}$$

In this case the difference between the referential or Lagrangian, and spatial or Eulerian, descriptions is not significant, and it is not necessary to distinguish between the two descriptions.

The definition of strain tensor is not unique. The basic requirement for a strain tensor is that it should completely determine the deformation state at any material point. The deformation state includes the stretches and shear in any direction. We have shown that the deformation tensors **C** and $\mathbf{B}^{-1}$ are sufficient for this purpose. In addition any single-valued, monotonic, invertible, and tensor-valued function of **C** can be defined as a Lagrangian-type strain tensor. Similarly any function of $\mathbf{B}^{-1}$ of this kind can be defined as the Eulerian-type strain tensor. In fact this is the equivalent theorem of strain tensors. On the basis of this theorem different types of strain tensors have been proposed. Here, as an example, we only introduce the logarithmic strain tensor used by Hencky in 1928, since we have already defined this kind of strain measure of the uniaxial case in Section 1.1. The definition of the logarithmic strain tensor (true or natural strain tensor) in referential or Lagrangian, and spatial or Eulerian, descriptions is given by Eqs. (2.60) and (2.61), respectively:

$$\overline{\mathbf{E}} = \tfrac{1}{2}\ln \mathbf{C} = \ln \mathbf{C}^{1/2} \tag{2.60}$$

$$\overline{\mathbf{e}} = -\tfrac{1}{2}\ln \mathbf{B}^{-1} = \ln\left(\mathbf{B}^{-1}\right)^{-1/2} \tag{2.61}$$

It is easy to see from Eq. (2.54) that for the rigid body deformation, both $\overline{\mathbf{E}}$ and $\overline{\mathbf{e}}$ are zero tensors. The tensor function $\ln \mathbf{A}$ means that it has the same principal directions as **A**, and its principal values are $\ln \lambda_i$, where λ_i are the corresponding principal values of **A** ($i = 1, 2, 3$).

2.4 DECOMPOSITION OF DEFORMATION GRADIENT AND PRINCIPAL STRETCHES

To further explain the deformation at any material point, consider the following Cauchy theorem of the polar decomposition. This theorem states that a nonsingular second-order tensor **A** can be decomposed uniquely into either of the following products:

$$\mathbf{A} = \mathbf{R} \cdot \mathbf{U} = \mathbf{V} \cdot \mathbf{R} \tag{2.62}$$

where **R** is an orthogonal tensor, and **U** and **V** are symmetric tensors. The proof of this theorem can be found in several books (e.g., Spencer 1980). As we discussed in Section 2.2, the deformation gradient tensor **F** is a nonsingular, second-order tensor. Using this polar deformation theorem,

$$\mathbf{F} = \mathbf{R} \cdot \mathbf{U} = \mathbf{V} \cdot \mathbf{R} \tag{2.63}$$

where **R** is an orthogonal rotation tensor, and **U** and **V** are symmetric, positive definite right and left Cauchy tensors, respectively.

From the definition of **B** and **C** and the decomposition of **F**, we can obtain

$$\mathbf{B} = \mathbf{F} \cdot \mathbf{F}^{\mathrm{T}} = \mathbf{V} \cdot \mathbf{R} \cdot \mathbf{R}^{\mathrm{T}} \cdot \mathbf{V} = \mathbf{V}^2 \tag{2.64}$$

$$\mathbf{C} = \mathbf{F}^{\mathrm{T}} \cdot \mathbf{F} = \mathbf{U} \cdot \mathbf{R}^{\mathrm{T}} \cdot \mathbf{R} \cdot \mathbf{U} = \mathbf{U}^2 \tag{2.65}$$

Now Eq. (2.13) can be rewritten in any one of the following forms:

$$d\mathbf{x} = \mathbf{F} \cdot d\mathbf{X} = \mathbf{R} \cdot \mathbf{U} \cdot d\mathbf{X} = \mathbf{V} \cdot \mathbf{R} \cdot d\mathbf{X} \tag{2.66a}$$

$$= \mathbf{I} \cdot \mathbf{R} \cdot \mathbf{U} \cdot d\mathbf{X} = \mathbf{R} \cdot \mathbf{I} \cdot \mathbf{U} \cdot d\mathbf{X} = \mathbf{R} \cdot \mathbf{U} \cdot \mathbf{I} \cdot d\mathbf{X} \tag{2.66b}$$

$$= \mathbf{I} \cdot \mathbf{V} \cdot \mathbf{R} \cdot d\mathbf{X} = \mathbf{V} \cdot \mathbf{I} \cdot \mathbf{R} \cdot d\mathbf{X} = \mathbf{V} \cdot \mathbf{R} \cdot \mathbf{I} \cdot d\mathbf{X} \tag{2.66c}$$

For the rigid body translation, $\mathbf{F} = \mathbf{I}$, we can make the following statements on the basis of the above expressions. The deformation at any material point may consist of three parts: rigid body translation, rotation, and pure stretching. The sequence of these three parts is arbitrary. The deformation given by **F** may be obtained by first stretching the material in three mutually orthogonal directions $\mathbf{N}_K$ $(K = 1, 2, 3)$ in $\mathscr{R}_0$ by amounts λ_K, followed by a rigid body rotation of these directions, followed by a rigid body translation. This deformation can also be obtained by first performing the same rigid body rotation, then stretching in the three mutually orthogonal directions $\mathbf{n}_k$ $(k = 1, 2, 3)$ in $\mathscr{R}$ by the same amounts λ_k $(\lambda_k = \lambda_K)$, followed by a rigid body translation. The quantities λ_K $(= \lambda_k)$ are the principal stretch ratios (or simply the principal stretches) along the principal axes of **U** and **V** in $\mathscr{R}_0$ and $\mathscr{R}$, respectively. Mathematically, λ_K and λ_k are the eigenvalues or principal values of **U** and **V**; namely they satisfy the following equations:

$$\mathbf{U} \cdot \mathbf{N}_K = \lambda_K \mathbf{N}_K \tag{2.67}$$

$$\mathbf{V} \cdot \mathbf{n}_k = \lambda_k \mathbf{n}_k \tag{2.68}$$

The identity $\lambda_K = \lambda_k$ is easy to understand because physically they correspond to the same stretching at a point of a deformed material. Mathematically one can also show, by noting $\mathbf{V} = \mathbf{R} \cdot \mathbf{U} \cdot \mathbf{R}^{\mathrm{T}}$ from Eq. (2.63), that

$$\mathbf{V} \cdot (\mathbf{R} \cdot \mathbf{N}_K) = \left(\mathbf{R} \cdot \mathbf{U} \cdot \mathbf{R}^{\mathrm{T}}\right) \cdot (\mathbf{R} \cdot \mathbf{N}_K) = \mathbf{R} \cdot \mathbf{U} \cdot \mathbf{N}_K = \lambda_K (\mathbf{R} \cdot \mathbf{N}_K) \tag{2.69}$$

Comparing Eq. (2.69) with Eq. (2.68) gives us the expression

$$\lambda_k = \lambda_K$$

$$\mathbf{n}_k = \mathbf{R} \cdot \mathbf{N}_K \tag{2.70}$$

This shows that the principal values of **U** and **V** are equal, and the principal axes of **V** can be obtained by rotating the corresponding ones of **U** through the tensor **R**. In addition one can calculate from Eqs. (2.64) and (2.65) that

$$\mathbf{C} \cdot \mathbf{N}_K = \lambda_K^2 \mathbf{N}_K \tag{2.71}$$

$$\mathbf{B} \cdot \mathbf{n}_k = \lambda_k^2 \cdot \mathbf{n}_k \tag{2.72}$$

So the deformation tensor **C** and **B** are coaxial with **U** and **V**, respectively. Further the principal values of **C** and **B** are squares of the respective principal stretches of **U** and **V**.

We close this section with the following equations for the logarithmic strain using Eqs. (2.60), (2.61), (2.64), and (2.65).

$$\bar{\mathbf{E}} = \ln \mathbf{U} \tag{2.73}$$

$$\bar{\mathbf{e}} = \ln \mathbf{V} \tag{2.74}$$

2.5 VELOCITY, ACCELERATION, AND MATERIAL DERIVATIVES

The velocity vector **v** of a particle is the rate of change of its displacement. This can be expressed mathematically as follows:

$$\mathbf{v} = \lim_{\Delta t \to 0} \frac{\mathbf{x}(\mathbf{X}, t + \Delta t) - x(\mathbf{X}, t)}{\Delta t} = \left(\frac{\partial \mathbf{x}}{\partial t}\right)_{\mathbf{X}} \tag{2.75}$$

where the subscript **X** denotes holding **X** constant while taking the partial derivative with respect to time. Thus, with **X** constant, a particular material particle's velocity is calculated according to the definition given above.

In terms of components v_i of the velocity vector **v**, one can write

$$v_i = \left[\frac{\partial x_i(X_K, t)}{\partial t} \right]_{\mathbf{X}} \tag{2.76}$$

Similarly the acceleration vector **a** can be defined as

$$\mathbf{a} = \lim_{\Delta t \to 0} \frac{\mathbf{v}(\mathbf{X}, t + \Delta t) - \mathbf{v}(\mathbf{X}, t)}{\Delta t} = \left(\frac{\partial \mathbf{v}}{\partial t} \right)_{\mathbf{X}} \tag{2.77}$$

$$a_i = \left(\frac{\partial v_i}{\partial t} \right)_{\mathbf{X}} \tag{2.78}$$

As discussed previously, either the referential (Lagrangian) or spatial (Eulerian) description can be used to describe the motion and deformation of a deformable material. This means that we can write the velocity vector **v** in either of the following forms:

$$\mathbf{v} \equiv \mathbf{v}(\mathbf{X}, t) \equiv \mathbf{v}(\mathbf{x}, t) \tag{2.79}$$

Expression of the acceleration of a particle **X** as given by Eq. (2.77) can be expressed by using the referential (Lagrangian) description or (2.79a). The time derivatives with the Lagrangian or material coordinate **X** held constant, as given by Eqs. (2.75) and (2.77), are called the material derivatives with respect to time t. If the velocity vector **v** is specified in terms of Eq. (2.79b), then Eq. (2.77) can no longer be used to calculate the acceleration vector **a**. However, as discussed earlier in Section 2.1, by assumption, there always is a sufficiently smooth function $\mathbf{x}(\mathbf{X}, t)$ so that one can rewrite Eq. (2.79b) in the form

$$\mathbf{v} = \mathbf{v}(\mathbf{x}, t) = \mathbf{v}[\mathbf{x}(\mathbf{X}, t), t] \tag{2.80}$$

Then we can calculate the material derivative in the following way:

$$\mathbf{a} = \left(\frac{\partial \mathbf{v}}{\partial t} \right)_{\mathbf{X}} = \left(\frac{\partial \mathbf{v}(\mathbf{x}, t)}{\partial t} \right)_{\mathbf{x}} + \frac{\partial \mathbf{v}(\mathbf{x}, t)}{\partial \mathbf{x}} \left(\frac{\partial \mathbf{x}}{\partial t} \right)_{\mathbf{X}}$$

$$a_i = \left(\frac{\partial v_i}{\partial t} \right)_{\mathbf{x}} + \frac{\partial v_i}{\partial x_j} \left(\frac{\partial x_j}{\partial t} \right)_{\mathbf{X}} \tag{2.81}$$

Since $(\partial \mathbf{x}/\partial t)_{\mathbf{X}} = \mathbf{v}$ or $(\partial x_j/\partial t)_{\mathbf{X}} = v_j$ according to the definition (2.75), we can rewrite Eq. (2.81b), as

$$a_i = \left(\frac{\partial v_i}{\partial t}\right)_{\mathbf{x}} + v_{i,j}v_j \tag{2.82}$$

In this case we use the notation D/Dt or a dot to represent the material derivative (e.g., $\mathbf{a} \equiv D\mathbf{v}/Dt \equiv \dot{\mathbf{v}}$). Since there are two coordinates, the material coordinates $\mathbf{X}$ and the spatial coordinates $\mathbf{x}$, it is also convenient to introduce different notations for the gradients with respect to $\mathbf{X}$ and $\mathbf{x}$. In this book the following notations are used

$$\nabla\mathbf{A} = \text{the gradient of } \mathbf{A} \text{ with respect to the material coordinates } \mathbf{X} \tag{2.83}$$

$$\operatorname{grad}\mathbf{A} = \text{the gradient of } \mathbf{A} \text{ with respect to the spatial coordinates } \mathbf{x} \tag{2.84}$$

Then

$$\mathbf{a} = \frac{D\mathbf{v}}{Dt} = \left(\frac{\partial \mathbf{v}}{\partial t}\right)_{\mathbf{x}} + \mathbf{v}\cdot\operatorname{grad}\mathbf{v} \tag{2.85}$$

In this expression $\mathbf{v}$ can be replaced by an arbitrary scalar or tensor-valued function $g(\mathbf{x}, t)$ or $\mathbf{G}(\mathbf{x}, t)$ to obtain a general definition of the material derivative of any scalar or tensor-valued function

$$\dot{g} \equiv \frac{Dg}{Dt} = \left(\frac{\partial \mathbf{g}}{\partial t}\right)_{\mathbf{x}} + \mathbf{v}\cdot\operatorname{grad} g \tag{2.86}$$

$$\mathbf{G} \equiv \frac{D\mathbf{G}}{Dt} = \left(\frac{D\mathbf{G}}{\partial t}\right)_{\mathbf{x}} + \mathbf{v}\cdot\operatorname{grad}\mathbf{G} \tag{2.86}$$

The material derivative is thus an operator defined by

$$\frac{D}{Dt} = \left[\left(\frac{\partial}{\partial t}\right)_{\mathbf{x}} + \mathbf{v}\cdot\operatorname{grad}\right] \tag{2.88}$$

2.6 VELOCITY GRADIENT AND THE MATERIAL DERIVATIVES OF THE LINE, AREA, AND VOLUME ELEMENTS

We will study the total deformation of a material subjected to finite straining by using the material and spatial coordinates $\mathbf{X}$ and $\mathbf{x}$ to calculate the deformation gradient $\mathbf{F}$. As we showed in Section 2.2, the infinitesimal line, area, and volume elements in the current configuration $\mathscr{R}$ can be related to the corresponding original elements in the referential or initial configuration $\mathscr{R}_0$ through $\mathbf{F}$. In the last section we introduced the concept of velocity or the rate of change of the displacement. This yielded a vector-valued velocity field defined either in $\mathscr{R}_0$ or $\mathscr{R}$, depending on whether material or spatial coordinates are used. An obvious question is, What will be the changing rates of the line, area, and volume elements in a finitely deformed material body subjected to a velocity field? We will determine in the subsequent discussion the changing rate (defined as the material derivative in Section 2.5) by a second-order tensor **L**, called the *velocity gradient*. The results obtained in this section may be compared with the ones obtained in Section 2.2.

Recall that the velocities at all points in the material $\mathscr{B}$ constitute a vector field $\mathbf{v}(\mathbf{x}, t)$. In general, this vector field varies with time. But for a fixed time moment t, the $\mathbf{v}(\mathbf{x}, t)$ varies only with the spatial coordinates $\mathbf{x}$. We can calculate the velocity increment $d\mathbf{v}$ caused by an infinitesimal variation of the position coordinates $d\mathbf{x}$ as

$$d\mathbf{v} = \frac{\partial \mathbf{v}}{\partial \mathbf{x}} \cdot d\mathbf{x} = \operatorname{grad} \mathbf{v} \cdot d\mathbf{x} = \mathbf{L} \cdot d\mathbf{x} \tag{2.89}$$

where

$$\mathbf{L} = \operatorname{grad} \mathbf{v} = \frac{\partial \mathbf{v}}{\partial \mathbf{x}}$$

$$L_{ij} = v_{i,j} = \dot{x}_{i,j} \tag{2.90}$$

is called the gradient of $\mathbf{v}$ with respect to the spatial coordinate $\mathbf{x}$; $d\mathbf{v}$ can be considered as the velocity at point $\mathbf{x} + d\mathbf{x}$ relative to point $\mathbf{x}$. Using the chain rule of the derivative, we calculate

$$\frac{D}{Dt}(F_{iK}) = \frac{D}{Dt}(x_{i,K}) = \dot{x}_{i,K} = \dot{x}_{i,j}x_{j,K} = L_{ij}F_{jK}$$

or

$$\dot{\mathbf{F}} = \mathbf{L} \cdot \mathbf{F} \tag{2.91}$$

which gives the material derivative of **F** in terms of **L** and **F**.

Since

$$\mathbf{F} \cdot \mathbf{F}^{-1} = \mathbf{I}$$

we take the material derivative of both sides of this equation, noting that $D\mathbf{I}/Dt = 0$, and derive

$$\dot{\mathbf{F}} \cdot \mathbf{F}^{-1} + \mathbf{F} \cdot \dot{\mathbf{F}}^{-1} = \mathbf{0}$$

Rearranging the above expression gives

$$\dot{\mathbf{F}}^{-1} = -\mathbf{F}^{-1} \cdot \dot{\mathbf{F}} \cdot \mathbf{F}^{-1}$$

Using Eq. (2.91), we obtain

$$\dot{\mathbf{F}}^{-1} = -\mathbf{F}^{-1} \cdot \mathbf{L} \tag{2.92}$$

To calculate the material derivatives of the line, area, and volume elements in the current configuration $\mathscr{R}$, first we observe that

$$\frac{D}{Dt}(d\mathbf{X}) = \frac{D}{Dt}(d\mathbf{A}) = \frac{D}{Dt}(dV) = 0 \tag{2.93}$$

This equation tells us that the material derivatives of these elements in the initial configuration $\mathscr{R}_0$ are zero. This is obvious because these elements are attached to the initial or reference configuration which, once chosen, will never change in the process of the subsequent motion.

Now, for $d\mathbf{x} = \mathbf{F} \cdot d\mathbf{X}$, we can calculate

$$\frac{D}{Dt}(d\mathbf{x}) = \frac{D}{Dt}[\mathbf{F} \cdot d\mathbf{X}] = \dot{\mathbf{F}} \cdot d\mathbf{X} = \mathbf{L} \cdot \mathbf{F} \cdot d\mathbf{X} = \mathbf{L} \cdot d\mathbf{x} \tag{2.94}$$

To calculate the material derivative of the volume element dv in $\mathscr{R}$, we need the following mathematical relations. According to Eq. (2.20b), we can write

$$\det(\mathbf{F}) = \tfrac{1}{6} e_{ijk} e_{LMN} x_{i,L} x_{j,M} x_{k,N} \tag{2.95}$$

Taking the derivative with respect to any element $x_{i,K}$, we obtain

$$\frac{\partial \det(\mathbf{F})}{\partial x_{i,K}} = \text{cofactor } x_{i,K} = \det(\mathbf{F}) X_{K,i} \tag{2.96}$$

Considering a volume element $dv = \det(\mathbf{F})\, dV$, we have

$$\begin{aligned} \frac{D}{Dt}(dv) &= \frac{D}{Dt}[\det(\mathbf{F})]\, dV = \frac{\partial \det(\mathbf{F})}{\partial x_{i,K}} \frac{Dx_{i,K}}{Dt}\, dV \\ &= \det(\mathbf{F}) X_{K,i} L_{ij} x_{j,K}\, dV \\ &= L_{ii} \det(\mathbf{F})\, dV = \text{tr}(\mathbf{L})\det(\mathbf{F})\, dV = \text{tr}(\mathbf{L})\, dv \end{aligned} \tag{2.97}$$

where Eqs. (2.91) and (2.96), and the relationship $X_{K,i} x_{j,K} = \delta_{ij}$ are used to derive Eq. (2.97).

Finally, using Eq. (2.21), we calculate

$$\frac{D}{Dt}(d\mathbf{a}) = \frac{D}{Dt}[\det(\mathbf{F})](\mathbf{F}^{-1})^{\mathrm{T}} \cdot d\mathbf{A} + \det(\mathbf{F})\frac{D}{Dt}\left[(\mathbf{F}^{-1})^{\mathrm{T}}\right] \cdot d\mathbf{A}$$

$$= \mathrm{tr}(\mathbf{L})\det(\mathbf{F})(\mathbf{F}^{-1})^{\mathrm{T}} \cdot d\mathbf{A} - \mathbf{L}^{\mathrm{T}} \det(\mathbf{F})(\mathbf{F}^{-1})^{\mathrm{T}} \cdot d\mathbf{A}$$

$$= \left[\mathrm{tr}(\mathbf{L})\mathbf{I} - \mathbf{L}^{\mathrm{T}}\right]\det(\mathbf{F})(\mathbf{F}^{-1})^{\mathrm{T}} \cdot d\mathbf{A} = \left[\mathrm{tr}(\mathbf{L})\mathbf{I} - \mathbf{L}^{\mathrm{T}}\right] \cdot d\mathbf{a} \quad (2.98)$$

Equations (2.92) and (2.97) are also used to obtain Eq. (2.98). According to Eqs. (2.94), (2.97), and (2.98), if configuration $\mathscr{R}$ is used as the reference, the velocity gradient **L** is sufficient to determine the rates of the infinitesimal elements $d\mathbf{x}$, $d\mathbf{a}$, and dv moving away from $\mathscr{R}$. That is, **L** completely determines the moving rates of these infinitesimal elements in reference to the current configuration $\mathscr{R}$. On the other hand, if we want to know how fast these elements move from the initial configuration $\mathscr{R}_0$ (i.e., the moving rates of the elements away from the initial configuration $\mathscr{R}_0$), we need to know the deformation gradient $\mathscr{F}$ in addition to the velocity gradient $\mathscr{L}$. As discussed in Section 2.2, the infinitesimal material elements in the current configuration $\mathscr{R}$ are related to the corresponding elements in $\mathscr{R}_0$ through the deformation gradient **F**. Because of the simplicity introduced by using the current configuration $\mathscr{R}$ as the reference, it is sometimes used as the reference configuration in studying the finite deformation of a material body; this is known as *relative description*. So far we have only used the initial configuration $\mathscr{R}_0$, which is fixed in time as the reference. However, we can use a configuration that varies with time as a reference. A natural choice is the current configuration $\mathscr{R}$, not only because it is a configuration changing with time and having physical meaning but also because it is a configuration changing with time and having physical meaning but also because its choice as reference introduces simplicity for the kinematic description above. The reader is directed for more details about the relative description to Truesdell (1985).

2.7 DEFORMATION RATE AND SPIN TENSORS

As we discussed in Section 2.3, the infinitesimal elements $d\mathbf{x}$, da, and dv in $\mathscr{R}$ are related to dX, dA, and dV in $\mathscr{R}_0$ only through **F**. Even so the deformation state of a material point (i.e., information about the stretch ratio, the shear deformation, and the ratio of those infinitesimal elements before and after deformation) is given by **C** and **B**, not by **F** itself. Similarly, although **L** plays a fundamental role in determining the material derivatives of those elements, as discussed in Section 2.6, we will see next that it is the symmetric part of **L**, not **L** itself, which completely determines their material derivatives. These derivatives are important in many problems, when the primary interest is in the *rate* at which the change of body shape takes place rather than in the change itself. In determining the changing rate of a material element, it is simpler to use the current configuration $\mathscr{R}$ as the reference and consider the rate relative to this configuration.

First, let us consider the stretch rate defined by $d_{\mathbf{n}} = |d\dot{\mathbf{x}}|/|d\mathbf{x}|$. Let $\mathbf{n} = d\mathbf{x}/|d\mathbf{x}|$ be the unit vector in $\mathscr{R}$. Then

$$
\begin{aligned}
d_{\mathbf{n}} &= \frac{|d\mathbf{x}|^{\cdot}}{|d\mathbf{x}|} = \frac{(D/Dt)\sqrt{d\mathbf{x}\cdot d\mathbf{x}}}{|d\mathbf{x}|} = \frac{d\dot{\mathbf{x}}\cdot d\mathbf{x} + d\mathbf{x}\cdot d\dot{\mathbf{x}}}{2|d\mathbf{x}|^2} \\
&= \frac{d\mathbf{x}\cdot \mathbf{L}^{\mathrm{T}}\cdot d\mathbf{x} + d\mathbf{x}\cdot \mathbf{L}\cdot d\mathbf{x}}{2|d\mathbf{x}|^2} \\
&= \mathbf{n}\frac{\mathbf{L}^{\mathrm{T}} + \mathbf{L}}{2}\mathbf{n} = \mathbf{n}\cdot\mathbf{D}\cdot\mathbf{n}
\end{aligned}
\tag{2.99}
$$

where $\mathbf{D}$, which is the symmetric part of $\mathbf{L}$, is the deformation rate tensor. The above expression means that the stretch rate in any direction $\mathbf{n}$ is given by the normal component of the deformation rate $\mathbf{D}$ in that direction. To further explore the physical meaning of $\mathbf{D}$, we calculate

$$
\begin{aligned}
\frac{D}{Dt}(ds^2) &= \frac{D}{Dt}(dx_i\, dx_i) = 2d\dot{x}_i\, dx_i = 2dx_i\, v_{i,j}\, dx_j \\
&= 2dx_i\, L_{ij}\, dx_j = 2d\mathbf{x}\cdot(\mathbf{D}+\mathbf{W})\cdot d\mathbf{x},
\end{aligned}
\tag{2.100}
$$

where Eq. (2.94) has been used and the velocity gradient has been decomposed into two parts:

$$
\mathbf{L} = \mathbf{D} + \mathbf{W} \tag{2.101}
$$

$$
\begin{aligned}
\mathbf{W} &= \tfrac{1}{2}(\mathbf{L} - \mathbf{L}^{\mathrm{T}}) \\
\mathbf{D} &= \tfrac{1}{2}(\mathbf{L} + \mathbf{L}^{\mathrm{T}})
\end{aligned}
\tag{2.102}
$$

Since $W_{ij} = -W_{ji}$,

$$
d\mathbf{x}\cdot\mathbf{W}\cdot d\mathbf{x} = 0 \tag{2.103}
$$

Eq. (2.99) becomes

$$
\frac{D}{Dt}(ds^2) = 2d\mathbf{x}\cdot\mathbf{D}\cdot d\mathbf{x} = 2(ds^2)d_{\mathbf{n}} \tag{2.104}
$$

Also tr(**L**) in Eqs. (2.97) and (2.98) can be replaced by tr(**D**), since

$$
\mathrm{tr}(\mathbf{L}) = \mathrm{tr}(\mathbf{D}) + \mathrm{tr}(\mathbf{W}) = \mathrm{tr}(\mathbf{D}) \tag{2.105}
$$

For the shear rate, we first calculate $\dot{\mathbf{n}}$:

$$\dot{\mathbf{n}} = \frac{D}{Dt}\left(\frac{d\mathbf{x}}{|d\mathbf{x}|}\right) = \frac{d\dot{\mathbf{x}}}{|d\mathbf{x}|} - \frac{D|d\mathbf{x}|/Dt}{|d\mathbf{x}|^2}\, d\mathbf{x}$$
$$= \mathbf{L}\cdot\mathbf{n} - (\mathbf{n}\cdot\mathbf{D}\cdot\mathbf{n})\mathbf{n} = (\mathbf{L} - d_{\mathbf{n}}\mathbf{I})\mathbf{n} \qquad (2.106)$$

Using the two unit vectors $\mathbf{n}^{(1)}$ and $\mathbf{n}^{(2)}$, according to Eq. (2.51), we can calculate

$$\frac{D}{Dt}(\cos\theta) = \dot{\gamma}\sin\theta = \frac{D}{Dt}\left(\mathbf{n}^{(1)}\cdot\mathbf{n}^{(2)}\right) = \dot{\mathbf{n}}^{(1)}\cdot\mathbf{n}^{(2)} + \mathbf{n}^{(1)}\cdot\dot{\mathbf{n}}^{(2)}$$
$$= \left(\mathbf{L}\cdot\mathbf{n}^{(1)}\right)\cdot\mathbf{n}^{(2)} - d_{\mathbf{n}^{(1)}}\mathbf{n}^{(1)}\cdot\mathbf{n}^{(2)} + \mathbf{n}^{(1)}\cdot\mathbf{L}\cdot\mathbf{n}^{(2)} - d_{\mathbf{n}^{(2)}}\mathbf{n}^{(1)}\cdot\mathbf{n}^{(2)}$$
$$= 2\mathbf{n}^{(1)}\cdot\mathbf{D}\cdot\mathbf{n}^{(2)} - \mathbf{n}^{(1)}\cdot\mathbf{n}^{(2)}\left(d_{\mathbf{n}^{(1)}} + d_{\mathbf{n}^{(2)}}\right) \qquad (2.107)$$

From this equation it can be seen that if $\mathbf{D}$ is known, the shear rate can be completely determined. If $\mathbf{n}^{(1)}\cdot\mathbf{n}^{(2)} = 0$, then $\sin\theta = 1$, and the following simplified formula for the shear rate can be obtained:

$$\dot{\gamma} = 2\mathbf{n}^{(1)}\cdot\mathbf{D}\cdot\mathbf{n}^{(2)} \qquad (2.108)$$

Thus the changing rate of the angle between two orthogonal line elements in the current configuration $\mathscr{R}$ is equal to twice that of the "shear component" of $\mathbf{D}$ in $\mathbf{n}^{(1)}$ and $\mathbf{n}^{(2)}$ directions.

To calculate the rate of a volume element dv, we note from Eq. (2.97) that

$$\frac{D}{Dt}[\det(\mathbf{F})] = \operatorname{tr}(\mathbf{D})\det(\mathbf{F}) \qquad (2.109)$$

Using Eqs. (2.27) and (2.109), the volume rate is given by

$$\frac{d\dot{v}}{dv} = \frac{(D/Dt)[\det(\mathbf{F})]\, dV}{dv} = \frac{\operatorname{tr}(\mathbf{D})\det(\mathbf{F})\, dV}{dv} = \operatorname{tr}(\mathbf{D}) \qquad (2.110)$$

We can derive an expression for the rate of change of an area element ($= |d\dot{\mathbf{a}}|/|d\mathbf{a}|$) by using Eqs. (2.98) and (2.99):

$$\frac{|d\mathbf{a}|^{\cdot}}{|d\mathbf{a}|} = \frac{D/Dt\sqrt{d\mathbf{a}\cdot d\mathbf{a}}}{|d\mathbf{a}|} = \frac{d\dot{\mathbf{a}}\cdot d\mathbf{a} + d\mathbf{a}\cdot d\dot{\mathbf{a}}}{2|d\mathbf{a}|^2}$$
$$= \frac{\left[\operatorname{tr}(\mathbf{D})\mathbf{I} - \mathbf{L}^{\mathrm{T}}\right]\cdot d\mathbf{a}\cdot d\mathbf{a} + d\mathbf{a}\cdot\left[\operatorname{tr}(\mathbf{D})\mathbf{I} - \mathbf{L}^{\mathrm{T}}\right]\cdot d\mathbf{a}}{2|d\mathbf{a}|^2}$$
$$= \frac{d\mathbf{a}\cdot\left[2\operatorname{tr}(\mathbf{D})\mathbf{I} - \left(\mathbf{L}^{\mathrm{T}} + \mathbf{L}\right)\right]\cdot d\mathbf{a}}{2|d\mathbf{a}|^2} = \mathbf{n}\cdot\left(\operatorname{tr}(\mathbf{D})\mathbf{I} - \mathbf{D}\right)\cdot\mathbf{n}$$
$$= \operatorname{tr}(\mathbf{D}) - d_{\mathbf{n}} \qquad (2.111)$$

where $\mathbf{n}$ is the unit normal vector of the area element $d\mathbf{a}$ $(= d\mathbf{a}/|d\mathbf{a}|)$ and $d_\mathbf{n}$ is the normal component of $\mathbf{D}$ in the direction of $\mathbf{n}$.

In Section 2.3, we defined $\lambda_\mathbf{n}$ and $\Lambda_\mathbf{n}$ by Eqs. (2.44) and (2.46), respectively. Using Eqs. (2.99) and (2.109), we can obtain $\dot{\lambda}_\mathbf{n}$ and $\dot{\Lambda}_\mathbf{n}$ relative to the current configuration $\mathscr{R}$, namely in the Eulerian description

$$\dot{\lambda}_\mathbf{n} = \frac{D|d\mathbf{x}|/Dt}{|dX|} = \frac{|d\mathbf{x}|}{|d\mathbf{X}|}\frac{D|d\mathbf{x}|/Dt}{|d\mathbf{x}|} = \lambda_\mathbf{n} d_\mathbf{n} \tag{2.112}$$

and

$$\dot{\Lambda}_\mathbf{n} = \frac{D|d\mathbf{a}|/Dt}{|d\mathbf{A}|} = \frac{|d\mathbf{a}|}{|d\mathbf{A}}\frac{D|d\mathbf{a}|/Dt}{|d\mathbf{a}|} = \sigma_\mathbf{n}(\mathrm{tr}(\mathbf{D}) - d_\mathbf{n}) \tag{2.113}$$

From Eqs. (2.109), (2.112), and (2.113), the following relationships are established:

$$\frac{D}{Dt}\{\ln[\det(\mathbf{F})]\} = \mathrm{tr}(\mathbf{D}) \tag{2.114}$$

$$\frac{D}{Dt}(\ln \lambda_\mathbf{n}) = d_\mathbf{n} \tag{2.115}$$

$$\frac{D}{Dt}(\ln \sigma_\mathbf{n}) = \mathrm{tr}(\mathbf{D}) - d_\mathbf{n} \tag{2.116}$$

Adding Eq. (2.115) and Eq. (2.116) and comparing the result with Eq. (2.114), we get

$$\frac{D}{Dt}\{\ln[\det(\mathbf{F})\} = \frac{D}{Dt}(\ln \lambda_\mathbf{n}) + \frac{D}{Dt}(\ln \sigma_\mathbf{n}) = \frac{D}{Dt}[\ln(\lambda_\mathbf{n}\sigma_\mathbf{n})] \tag{2.117}$$

It is implied from this expression that

$$\det(\mathbf{F}) = \lambda_\mathbf{n}\sigma_\mathbf{n} \tag{2.118}$$

Since det(**F**) represents the volume ratio, Eq. (2.118) indicates that the volume ratio equals the product of the stretch ratio and the area ratio.

So far we have discussed the role played by the symmetric part of **L** (or **D**), describing the deformation rate at any material point in the current configuration $\mathscr{R}$. It has been shown that as the deformation tensor **C** or **B**, or equivalently **E** or **e**, can completely determine the deformation state at any material point, the deformation rate state relative to $\mathscr{R}$ is totally determined by **D**. Thus **D** is known as the *deformation rate tensor*.

It can be shown that the antisymmetric part of **L**, the tensor **W** given by Eq. (2.102d), represents rotation just as the antisymmetric part of the displacement gradient, $\frac{1}{2}(u_{i,j} - u_{j,i})$, represents an infinitesimal rotation in case of infinitesimal deformation (cf. Eringen 1965). For this reason **W** is called the *spin tensor*. The reader may find more detailed discussion about this in Truesdell (1985) or Malvern (1969).

2.8 MATERIAL DERIVATIVES OF THE STRAIN TENSORS

This section discusses the material derivatives of the deformation tensors **C** and **B** and the strain tensors **E** and **e**. This is necessary because the deformation rate **D** can determine the changing rate of all the material elements relative to the current configuration $\mathscr{R}$; however, it is not sufficient to give all these rates of the elements in reference to the initial configuration $\mathscr{R}_0$. From the discussion in Section 2.3 we know that **C** and **E** are the material description of the deformation in $\mathscr{R}_0$. Thus it is expected that the material derivatives of **C** or **E** will play an important role in describing the rate of stretch, $\dot{\lambda}_N$, and the rate of area ratio, $\dot{\sigma}_N$. To complete our discussion about the kinematics of finite deformation, the material derivatives of **B** and **e** and the relationship between the deformation rate **D** and the strain tensors are also presented here.

Using Eq. (2.37) and the material derivative of **F** as given by Eq. (2.91), we get

$$\begin{aligned}\dot{\mathbf{C}} &= \dot{\mathbf{F}}^{\mathrm{T}} \cdot \mathbf{F} + \mathbf{F}^{\mathrm{T}} \cdot \dot{\mathbf{F}} = \mathbf{F}^{\mathrm{T}} \cdot \mathbf{L}^{\mathrm{T}} \cdot \mathbf{F} + \mathbf{F}^{\mathrm{T}} \cdot \mathbf{L} \cdot \mathbf{F} \\ &= \mathbf{F}^{\mathrm{T}} \cdot \left(\mathbf{L} + \mathbf{L}^{\mathrm{T}}\right) \cdot \mathbf{F} = 2\mathbf{F}^{\mathrm{T}} \cdot \mathbf{D} \cdot \mathbf{F}\end{aligned} \tag{2.119}$$

Since $\mathbf{C}^{-1} \cdot \mathbf{C} = \mathbf{I}$, taking the material derivative of both sides results in

$$\begin{aligned}\dot{\mathbf{C}}^{-1} &= -\mathbf{C}^{-1} \cdot \dot{\mathbf{C}} \cdot \mathbf{C}^{-1} = -\mathbf{F}^{-1} \cdot \left(\mathbf{F}^{-1}\right)^{\mathrm{T}} \cdot 2\mathbf{F}^{\mathrm{T}} \cdot \mathbf{D} \cdot \mathbf{F} \cdot \mathbf{F}^{-1} \cdot \left(\mathbf{F}^{-1}\right)^{\mathrm{T}} \\ &= -2\mathbf{F}^{-1} \cdot \mathbf{D} \cdot \left(\mathbf{F}^{-1}\right)^{\mathrm{T}}\end{aligned} \tag{2.120}$$

It is seen that $\dot{\mathbf{C}}$ and $\dot{\mathbf{C}}^{-1}$ are still symmetric.

Similarly, according to the definition of $\mathbf{B}^{-1}$ given by Eq. (2.36), and using Eq. (2.92), we obtain

$$\begin{aligned}\dot{\mathbf{B}}^{-1} &= \left(\dot{\mathbf{F}}^{-1}\right)^{\mathrm{T}} \cdot \mathbf{F}^{-1} + \left(\mathbf{F}^{-1}\right)^{\mathrm{T}} \cdot \dot{\mathbf{F}}^{-1} \\ &= -\mathbf{L}^{\mathrm{T}} \cdot \left(\mathbf{F}^{-1}\right)^{\mathrm{T}} \cdot \mathbf{F}^{-1} - \left(\mathbf{F}^{-1}\right)^{\mathrm{T}} \cdot \mathbf{F}^{-1} \cdot \mathbf{L} \\ &= -\left(\mathbf{L}^{\mathrm{T}} \cdot \mathbf{B}^{-1} + \mathbf{B}^{-1} \cdot \mathbf{L}\right)\end{aligned} \tag{2.121}$$

$$\begin{aligned}\dot{\mathbf{B}} &= -\mathbf{B}\dot{\mathbf{B}}^{-1} \cdot \mathbf{B} = \mathbf{B} \cdot \left(\mathbf{L}^{\mathrm{T}} \cdot \mathbf{B}^{-1} + \mathbf{B}^{-1} \cdot \mathbf{L}\right) \cdot \mathbf{B} \\ &= \mathbf{B} \cdot \mathbf{L}^{\mathrm{T}} + \mathbf{L} \cdot \mathbf{B}\end{aligned} \tag{2.122}$$

Again, $\dot{\mathbf{B}}$ and $\dot{\mathbf{B}}^{-1}$ are symmetric.

Once $\dot{\mathbf{C}}$ and $\dot{\mathbf{B}}^{-1}$ have been determined, $\mathbf{E}$ and $\mathbf{e}$ can be calculated using Eqs. (2.55) and (2.56):

$$\dot{\mathbf{E}} = \tfrac{1}{2}\dot{\mathbf{C}} = \mathbf{F}^{\mathrm{T}} \cdot \mathbf{D} \cdot \mathbf{F} \tag{2.123}$$

$$\begin{aligned}\dot{\mathbf{e}} &= -\tfrac{1}{2}\dot{\mathbf{B}}^{-1} = \tfrac{1}{2}\left(\mathbf{L}^{\mathrm{T}} \cdot \mathbf{B}^{-1} + \mathbf{B}^{-1} \cdot \mathbf{L}\right) \\ &= \tfrac{1}{2}\left[\mathbf{L}^{\mathrm{T}} \cdot (\mathbf{I} - 2\mathbf{e}) + (\mathbf{I} - 2\mathbf{e}) \cdot \mathbf{L}\right] = \mathbf{D} - \left(\mathbf{e} \cdot \mathbf{L} + \mathbf{L}^{\mathrm{T}} \cdot \mathbf{e}\right)\end{aligned} \tag{2.124}$$

Now we can obtain $\dot{\lambda}_{\mathbf{N}}$ and $\dot{\sigma}_{\mathbf{N}}$ relative to the initial configuration. From the Lagrangian description of $\lambda_{\mathbf{N}}$ given by Eq. (2.42),

$$\begin{aligned}\dot{\lambda}_{\mathbf{N}} &= \frac{D}{Dt}\sqrt{\mathbf{N} \cdot \mathbf{C} \cdot \mathbf{N}} = \frac{\mathbf{N} \cdot \dot{\mathbf{C}} \cdot \mathbf{N}}{2\sqrt{\mathbf{N} \cdot \mathbf{C} \cdot \mathbf{N}}} = \frac{2\mathbf{N} \cdot \mathbf{F}^{\mathrm{T}} \cdot \mathbf{D} \cdot \mathbf{F} \cdot \mathbf{N}}{2\lambda_{\mathbf{N}}} \\ &= \frac{(\mathbf{F} \cdot \mathbf{N}) \cdot \mathbf{D} \cdot (\mathbf{F} \cdot \mathbf{N})}{\lambda_{\mathbf{N}}}\end{aligned} \tag{2.125}$$

From Eq. (2.45), $\dot{\sigma}_{\mathbf{N}}$ can be evaluated as

$$\begin{aligned}\dot{\sigma}_{\mathbf{N}} &= \frac{D}{Dt}\left[\det(\mathbf{F})\sqrt{\mathbf{N} \cdot \mathbf{C}^{-1} \cdot \mathbf{N}}\right] \\ &= \frac{D}{Dt}[\det(\mathbf{F})]\sqrt{\mathbf{N} \cdot \mathbf{C}^{-1} \cdot \mathbf{N}} + \det(\mathbf{F})\frac{D}{Dt}\sqrt{\mathbf{N} \cdot \mathbf{C}^{-1} \cdot \mathbf{N}} \\ &= \operatorname{tr}(\mathbf{D})\sigma_{\mathbf{N}} + \det(\mathbf{F})\frac{\mathbf{N} \cdot \dot{\mathbf{C}}^{-1} \cdot \mathbf{N}}{2\sqrt{\mathbf{N} \cdot \mathbf{C}^{-1} \cdot \mathbf{N}}} \\ &= \operatorname{tr}(\mathbf{D})\sigma_{\mathbf{N}} - \det{}^2(\mathbf{F})\frac{\left[\left(\mathbf{F}^{-1}\right)^{\mathrm{T}} \cdot \mathbf{N}\right] \cdot \mathbf{D} \cdot \left[\left(\mathbf{F}^{-1}\right)^{\mathrm{T}} \cdot \mathbf{N}\right]}{\sigma_{\mathbf{N}}}\end{aligned} \tag{2.126}$$

A comparison of Eqs. (2.125) and (2.126) with Eqs. (2.112) and (2.113) shows that the rate of change of stretch and area ratios relative to the current configuration $\mathscr{R}$ depends only on $\mathbf{D}$. But when the *initial* configuration is used as the reference, the rate depends on the deformation gradient $\mathbf{F}$ in addition to the deformation rate $\mathbf{D}$.

It should be pointed out that $\mathbf{D}$ and $\dot{\mathbf{E}}$ are different, as can be inferred from Eq. (2.123). Since $\mathbf{x} = \mathbf{X} + \mathbf{u}$, hence $\mathbf{F} = \mathbf{I} + \nabla\mathbf{u}$. If the displacement gradient $\nabla\mathbf{u}$ is small compared to the identity tensor $\mathbf{I}$ (i.e., the classical infinitesimal displacement gradient case), the following approximations are possible:

$$\mathbf{F} \approx \mathbf{I} \tag{2.127}$$

$$\dot{\mathbf{E}} \approx \mathbf{I} \cdot \mathbf{D} \cdot \mathbf{I} = \mathbf{D} \tag{2.128}$$

When the displacement gradient is small, the difference between the material and spatial coordinates and the nonlinear terms in the expression for E_{KL}, given by Eq. (2.57), can be ignored. Thus

$$E_{KL} \approx \tfrac{1}{2}(u_{K,L} + u_{L,K}) \tag{2.129}$$

which is the infinitesimal strain tensor ϵ_{ij}, given by Eq. (2.59). The difference between X_K and x_i becomes insignificant in this case, and the lower case and upper case indices are interchangeable. Therefore $\dot{\mathbf{E}}$ and $\mathbf{D}$ are equal for the case of infinitesimal deformation *only*. In general,

$$\frac{d\mathbf{E}}{dt} \neq \mathbf{D} \tag{2.130}$$

For the logarithmic strain tensor introduced in Sections 2.3 and 2.4, the following relationships exists when the principal axes of the stretches do not rotate (i.e., if the principal axes of the strain are fixed in space):

$$\mathbf{D} = \dot{\mathbf{e}} = \frac{D}{Dt}[\ln \mathbf{V}] \tag{2.131}$$

In general, this relationship does not hold. For more details, the reader can refer to Gurtin et al. (1983).

2.9 FORCES, SURFACE TRACTION, AND CAUCHY STRESS TENSOR

Thus far, this chapter has dealt with the purely geometrical features of a material body, namely motion and deformation and their mathematical descriptions. The remainder of the chapter will discuss forces acting on a material body that cause motion and deformation of that body. These forces are classified as external or internal forces. External forces represent loads or actions applied on the boundary surface of the body or to the particles within the body by external sources. Internal forces denote the actions or loads applied by one part of the body to another part of the same body in response to external forces. We are considering only the "additional internal forces" caused by the deformation of the materials. Intrinsic bonding forces between particles of the body that exist without the action of external forces are not discussed here.

The external forces can be divided as body and surface forces. Body forces are actions distributed over each particle of the body. They are long-range actions or the loads applied by an external agent at a distance. The most commonly encountered body force is the gravity of earth. Surface forces act on the boundary of the body's surface; they are created by the contact action of two bodies. Therefore surface forces are short-range actions in comparison to body forces.

Consider a deformed body $\mathscr{B}$ that occupies the spatial region $\mathscr{R}$ with the boundary surface $\mathscr{S}$. Let the body force per unit mass acting on an infinitesimal volume element dv of the body be denoted by $\mathbf{b}$. The body force acting on the volume dv is then $\rho \mathbf{b} dv$, where ρ is the mass density in the current configuration $\mathscr{R}$. The total force on $\mathscr{B}$ can be calculated by the following integration:

$$\mathbf{B} = \int_{\mathscr{R}} \rho \mathbf{b} dv \tag{2.132}$$

Similarly consider an infinitesimal surface element da on $\mathscr{S}$, and let $\mathbf{t}$ represent the surface force per unit area on da. Then, the surface force on da is $\mathbf{t}\, da$ and the total surface force on $\mathscr{S}$ is

$$\mathbf{T} = \int_{\mathscr{S}} \mathbf{t}\, da \tag{2.133}$$

Surface force per unit area $\mathbf{t}$ is called *traction*.

Internal forces are one of the important variables used to characterize the behavior of materials in a constitutive equation. The existence of invisible internal forces is a plausible assumption, known as the *Euler-Cauchy stress principle*, which can be stated as follows (cf. Scipio 1967):

Postulate: Tractions exist across every internal surface element of a body, and, in terms of these tractions, the laws of motion apply for any interior region as well as to the body as a whole.

The significance of the Euler-Cauchy principle is that it introduces the concept of the Cauchy stress tensor at every point in the deformed body. We can then use classical field theory for continuum mechanics. According to this principle, we can specify the action between any two parts of the body using traction $\mathbf{t}^{(\mathbf{n})}$, as shown in Fig. 2.6. The body is sectioned with a plane with normal $\mathbf{n}$. Consider an infinitesimal element Δa on the section plane of the body, as shown in this figure. It is assumed that the action applied by the $\mathscr{R}^2$ part to the $\mathscr{R}^1$ part can be represented by the force

$$\Delta \mathbf{p} = \mathbf{t}^{(\mathbf{n})}\, \Delta a \tag{2.134}$$

where $\mathbf{n}$ and $\mathbf{t}^{(\mathbf{n})}$ are the unit normal to Δa and the traction vector (i.e., the surface force per unit area on Δa), respectively. Furthermore consider the ratio $\Delta \mathbf{p}/\Delta a$ as $\Delta a \to 0$. The Euler-Cauchy principle assumes that this limit

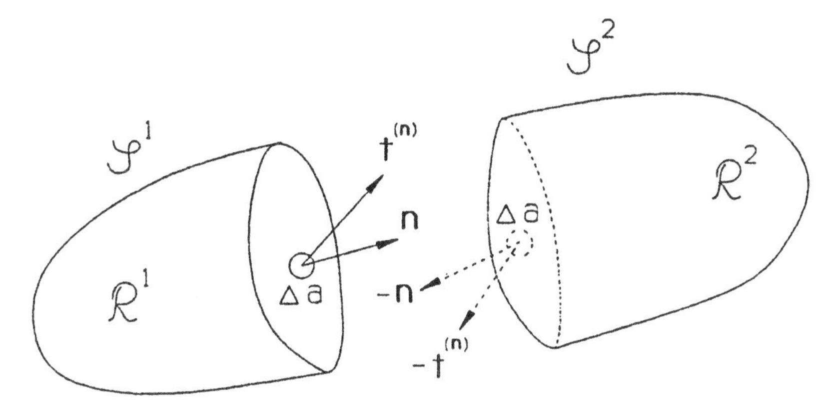

Fig. 2.6 Euler-Cauchy stress principle

exists as $\Delta a \to 0$ and is equal to $\mathbf{t}^{(\mathbf{n})}$:

$$\mathbf{t}^{(\mathbf{n})} = \lim_{\Delta a \to 0} \frac{\Delta \mathbf{p}}{\Delta a} \tag{2.135}$$

where the limit does not depend on the shape of Δa and the way in which $\Delta a \to 0$. This limit $\mathbf{t}^{(\mathbf{n})}$, the traction vector, is also called the *stress vector*. Clearly $\mathbf{t}^{(\mathbf{n})}$ is dependent on $\mathbf{n}$, but in general the direction of $\mathbf{t}^{(\mathbf{n})}$ does not coincide with $\mathbf{n}$. The justification of the Euler-Cauchy stress principle is that conclusions based on it agree with experimental observations. It is possible that in addition to the stress vector $\mathbf{t}^{(\mathbf{n})}$, there may be a couple vector, say, $\mathbf{m}^{(\mathbf{n})}$, acting on $\Delta \mathbf{a}$. The introduction of such a couple vector is beyond the scope of this book. The interested reader will find more information about couple vector and couple stress in Malvern (1969).

To put the Euler-Cauchy stress principle into practical use, Cauchy proposed the following fundamental Lemma (cf. Truesdell 1985): There is a tensor $\boldsymbol{\sigma}$, called the *Cauchy stress tensor*, that is related to the stress vector $\mathbf{t}^{(\mathbf{n})}$ in the following way:

$$\mathbf{t}^{(\mathbf{n})} = \mathbf{n} \cdot \boldsymbol{\sigma} \tag{2.136}$$

It can be seen from Eq. (2.136) that there exists a stress tensor $\boldsymbol{\sigma}$ that is the characteristic variable at each point in the deformed body, and its projection in any direction $\mathbf{n}$ gives the traction vector $\mathbf{t}^{(\mathbf{n})}$. In the cartesian coordinate system introduced in Section 2.1, the three base vectors are $\mathbf{i}_1$, $\mathbf{i}_2$, and $\mathbf{i}_3$. The traction vectors on surfaces with normals in $\mathbf{i}_l$ $(l = 1, 2, 3)$ directions are then given by

$$\mathbf{t}^{(\mathbf{i}_l)} = \mathbf{i}_l \cdot \boldsymbol{\sigma} \tag{2.137}$$

Since $\mathbf{t}^{(\mathbf{i}_l)}$ is a vector and generally does not point along the $\mathbf{i}_l$ direction, its components along $\mathbf{i}_1$, $\mathbf{i}_2$, and $\mathbf{i}_3$ can be calculated by

$$t_k^{(\mathbf{i}_l)} = \sigma_{lk} = \mathbf{i}_l \cdot \boldsymbol{\sigma} \cdot \mathbf{i}_k \qquad (k = 1, 2, 3) \tag{2.138}$$

The quantity σ_{lk} is called *stress component* and stands for the component of $\mathbf{t}^{(\mathbf{i}_l)}$ in the direction of i_k. Therefore in any orthogonal coordinate system σ can be represented by a matrix $[\sigma_{ij}]$:

$$[\sigma_{ij}] = \begin{bmatrix} \sigma_{11} & \sigma_{12} & \sigma_{13} \\ \sigma_{21} & \sigma_{22} & \sigma_{23} \\ \sigma_{31} & \sigma_{32} & \sigma_{33} \end{bmatrix} \tag{2.139}$$

and it can be shown that $\boldsymbol{\sigma} = \boldsymbol{\sigma}^{\mathrm{T}}$; that is, σ is symmetric. Since $\boldsymbol{\sigma}$ is a tensor, the transformation rule of its components σ_{ij} in the coordinate systems x_i, into components σ'_{ij} in a new coordinate system x'_j, is the following:

$$\sigma'_{ij} = \alpha_{ik}\alpha_{jl}\sigma_{kl} \tag{2.140}$$

where $\alpha_{ik} = \mathbf{i}_i \cdot \mathbf{i}'_k$ is the direction cosines between the axes of the old and new coordinates.

2.10 ALTERNATIVE STRESS MEASURES

The Cauchy stress tensor $\boldsymbol{\sigma}$ introduced in the previous section is defined in the current configuration $\mathscr{R}$. It varies with the spatial coordinate $\mathbf{x}$, and it is the true stress in a deformed body. As discussed at the beginning of this chapter, however, there are two kinds of descriptions for finite deformation depending on whether the material coordinate $\mathbf{X}$ or the spatial coordinate $\mathbf{x}$ is used. In some cases it is more convenient to use a stress tensor that is defined in the reference configuration (i.e., the Lagrangian or referential representation of the stress tensor).

Consider an element of a material surface ΔA in the reference configuration $\mathscr{R}_0$ that has a normal $\mathbf{N}$. This element $\Delta\mathbf{A} = \Delta A\mathbf{N}$ becomes, after the deformation, $\Delta\mathbf{a} = \mathbf{n}\,\Delta a$ in the current configuration. From Eq. (2.21) we have

$$\mathbf{n} = \det(\mathbf{F})\frac{\Delta A}{\Delta a}\mathbf{N} \cdot \mathbf{F}^{-1} \tag{2.141}$$

Following the definition of the Cauchy stress tensor σ, Eq. (2.136), the first Piola-Kirchhoff stress $\boldsymbol{\Sigma}^{\mathrm{I}}$ is introduced to give the actual force or traction $\Delta\mathbf{p}$ on the deformed Δa in the current deformed configuration $\mathscr{R}$. It refers to the area ΔA in the reference configuration (usually the undeformed configuration). Therefore

$$(\mathbf{N} \cdot \boldsymbol{\Sigma}^{\mathrm{I}}) \, \Delta A = \Delta \mathbf{p} = (\mathbf{n} \cdot \boldsymbol{\sigma}) \, \Delta a \tag{2.142}$$

Substituting Eq. (2.141) in the right-hand side of the equation above, we get

$$\mathbf{N} \cdot \left[\boldsymbol{\Sigma}^{\mathrm{I}} - \det(\mathbf{F}) \mathbf{F}^{-1} \cdot \boldsymbol{\sigma} \right] = 0$$

Hence

$$\boldsymbol{\Sigma}^{\mathrm{I}} = \det(\mathbf{F}) \mathbf{F}^{-1} \cdot \boldsymbol{\sigma} \tag{2.143}$$

since $\mathbf{N}$ is arbitrary. From its definition we conclude that the component Σ^{I}_{Ki} of $\boldsymbol{\Sigma}^{\mathrm{I}}$ represents the component in the x_i direction of the force per unit surface area in the reference configuration $\mathscr{R}_0$ on a surface that is normal to the X_K axis in that configuration. For this reason it is sometimes called the *nominal stress tensor*. It should be noted from Eq. (2.143) that $\boldsymbol{\Sigma}^{\mathrm{I}}$ is unsymmetric since $\boldsymbol{\sigma}$ is symmetric in the absence of couple stress.

To overcome the disadvantage of the unsymmetry of the first Piola-Kirchhoff stress tensor $\boldsymbol{\Sigma}^{\mathrm{I}}$, the following second Piola-Kirchhoff stress $\boldsymbol{\Sigma}^{\mathrm{II}}$ is usually used:

$$\boldsymbol{\Sigma}^{\mathrm{II}} = \boldsymbol{\Sigma}^{\mathrm{I}} \left(\mathbf{F}^{-1} \right)^{\mathrm{T}} = \det(\mathbf{F}) \mathbf{F}^{-1} \cdot \boldsymbol{\sigma} \cdot \left(\mathbf{F}^{-1} \right)^{\mathrm{T}} \tag{2.144}$$

It can be easily seen that $\boldsymbol{\Sigma}^{\mathrm{II}}$ is symmetric since σ is symmetric.

From Eqs. (2.143) and (2.144) we can write the inverse relationships

$$\boldsymbol{\sigma} = \frac{1}{\det(\mathbf{F})} \mathbf{F} \cdot \boldsymbol{\Sigma}^{\mathrm{I}} \tag{2.145}$$

$$\boldsymbol{\sigma} = \frac{1}{\det(\mathbf{F})} \mathbf{F} \cdot \boldsymbol{\Sigma}^{\mathrm{II}} \cdot \mathbf{F}^{\mathrm{T}} \tag{2.146}$$

and

$$\boldsymbol{\Sigma}^{\mathrm{I}} = \boldsymbol{\Sigma}^{\mathrm{II}} \cdot \mathbf{F}^{\mathrm{T}} \tag{2.147}$$

Similar to the definition of strain measures in finite deformation, the stress measure is not unique. In addition to the three stress measures given above, other different stress tensors have been used. For example, in developing a general theory for finite plastic deformation, Green and Naghdi (1965) found the following corotated Cauchy stress $\boldsymbol{\sigma}_u$ more suitable for their use:

$$\boldsymbol{\sigma}_u = \frac{1}{\det(\mathbf{F})} \mathbf{U} \cdot \boldsymbol{\Sigma}^{\mathrm{II}} \cdot \mathbf{U} = \mathbf{R}^{\mathrm{T}} \cdot \boldsymbol{\sigma} \cdot \mathbf{R} \tag{2.148}$$

Also in the literature of finite plasticity we can find the so-called Kirchhoff stress tensor $\boldsymbol{\tau}$:

$$\boldsymbol{\tau} = \det(\mathbf{F})\boldsymbol{\sigma} \tag{2.149}$$

The choice of a particular stress measure depends somewhat on whether or not some degree of simplicity can be achieved in an application but more critically on the ease of finding the related strain measure that is the work-conjugate of the adopted stress measure (Hill 1970).

2.11 EXAMPLES

In this section we will present several examples to calculate various kinds of strain and stress tensors. These examples are simple experimental configurations most often used to study elastic-plastic behaviors. We will discuss how to calculate strains from experimental measurements and how to relate different strain and stress measures. Some basic concepts and difficulties encountered in the analysis of finite deformation will be presented at the end of this section.

2.11.1 Uniaxial Tension or Compression Experiment

Figure 2.7 shows a cylindrical specimen under uniaxial tension (uniaxial compression can be treated in the same way). The length and radius at any deformed state (current configuration $\mathscr{R}$) are l and r, respectively. The corresponding length and radius are L and R in the undeformed state, which is taken as reference configuration $\mathscr{R}_0$. The deformation throughout the specimen is assumed homogeneous; therefore l in Fig. 2.7 represents the gauge length in the actual experiment.

Using the coordinate system shown in the figure, we can write

$$\begin{aligned} x &= \lambda_1 X \\ y &= \lambda_2 Y \\ z &= \lambda_2 Z \end{aligned} \tag{2.150}$$

where λ_1 and λ_2 are the principal stretch ratios and are given by

$$\begin{aligned} \lambda_1 &= \frac{l}{L} \\ \lambda_2 &= \frac{r}{R} \end{aligned} \tag{2.151}$$

Using Eq. (2.150) and equations in section 2.4, we obtain the relations

$$\mathbf{R} = \mathbf{I}, \quad \mathbf{F} = \mathbf{U} = \mathbf{V} = \begin{bmatrix} \lambda_1 & 0 & 0 \\ 0 & \lambda_2 & 0 \\ 0 & 0 & \lambda_2 \end{bmatrix}, \quad \mathbf{F}^{-1} = \begin{bmatrix} \lambda_1^{-1} & 0 & 0 \\ 0 & \lambda_2^{-1} & 0 \\ 0 & 0 & \lambda_2^{-1} \end{bmatrix} \tag{2.152}$$

$$\mathbf{C} = \mathbf{U}^2 = \mathbf{F}^{\mathrm{T}} \cdot \mathbf{F} = \begin{bmatrix} \lambda_1^2 & 0 & 0 \\ 0 & \lambda_2^2 & 0 \\ 0 & 0 & \lambda_2^2 \end{bmatrix},$$

$$\mathbf{B}^{-1} = \mathbf{V}^{-2} = (\mathbf{F}^{-1})^{\mathrm{T}}(\mathbf{F}^{-1}) \begin{bmatrix} \dfrac{1}{\lambda_1^2} & 0 & 0 \\ 0 & \dfrac{1}{\lambda_2^2} & 0 \\ 0 & 0 & \dfrac{1}{\lambda_2^2} \end{bmatrix} \tag{2.153}$$

$$\mathbf{E} = \tfrac{1}{2}(\mathbf{B} - \mathbf{I}) = \tfrac{1}{2} \begin{bmatrix} \lambda_1^2 - 1 & 0 & 0 \\ 0 & \lambda_2^2 - 1 & 0 \\ 0 & 0 & \lambda_2^2 - 1 \end{bmatrix},$$

$$\mathbf{e} = \tfrac{1}{2}(\mathbf{I} - \mathbf{B}^{-1}) = \tfrac{1}{2} \begin{bmatrix} 1 - \dfrac{1}{\lambda_1^2} & 0 & 0 \\ 0 & 1 - \dfrac{1}{\lambda_2^2} & 0 \\ 0 & 0 & 1 - \dfrac{1}{\lambda_2^2} \end{bmatrix} \tag{2.154}$$

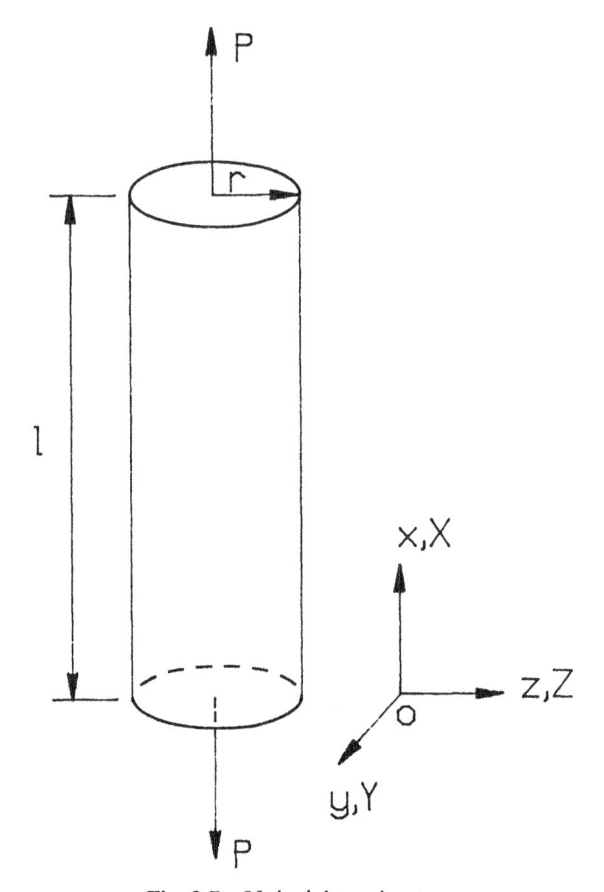

Fig. 2.7 Uniaxial tension test

According to the definition of logarithmic strain,

$$\bar{\mathbf{e}} = \ln \mathbf{V} = \begin{bmatrix} \ln \lambda_1 & 0 & 0 \\ 0 & \ln \lambda_2 & 0 \\ 0 & 0 & \ln \lambda_2 \end{bmatrix} = \begin{bmatrix} \ln \dfrac{l}{L} & 0 & 0 \\ 0 & \ln \dfrac{r}{R} & 0 \\ 0 & 0 & \ln \dfrac{r}{R} \end{bmatrix} \tag{2.155a}$$

We use Eq. (2.91) to obtain the velocity gradient **L**:

$$\mathbf{L} = \dot{\mathbf{F}} \cdot \mathbf{F}^{-1} = \begin{bmatrix} \dot{\lambda}_1 & 0 & 0 \\ 0 & \dot{\lambda}_2 & 0 \\ 0 & 0 & \dot{\lambda}_2 \end{bmatrix} \begin{bmatrix} \lambda_1^{-1} & 0 & 0 \\ 0 & \lambda_2^{-1} & 0 \\ 0 & 0 & \lambda_2^{-1} \end{bmatrix} \tag{2.155b}$$

$$= \begin{bmatrix} \dfrac{\dot{\lambda}_1}{\lambda_1} & 0 & 0 \\ 0 & \dfrac{\dot{\lambda}_2}{\lambda_2} & 0 \\ 0 & 0 & \dfrac{\dot{\lambda}_2}{\lambda_2} \end{bmatrix} = \begin{bmatrix} \dfrac{\dot{l}}{l} & 0 & 0 \\ 0 & \dfrac{\dot{r}}{r} & 0 \\ 0 & 0 & \dfrac{\dot{r}}{r} \end{bmatrix} \tag{2.156}$$

In this case it is obvious that

$$\mathbf{L} = \mathbf{D}$$

$$\mathbf{W} = 0 \tag{2.157}$$

From Eq. (2.155) we get

$$\dot{\mathbf{e}} = \begin{bmatrix} \dfrac{\dot{\lambda}_1}{\lambda_1} & 0 & 0 \\ 0 & \dfrac{\dot{\lambda}_2}{\lambda_2} & 0 \\ 0 & 0 & \dfrac{\dot{\lambda}_2}{\lambda_2} \end{bmatrix} = \mathbf{D} \tag{2.158}$$

The identity above holds because the principal axes of the strain are fixed in space. During an experiment the so-called nominal strain (e.g., the readings of the strain gauges) is usually measured. It is defined by

$$\hat{\mathbf{E}} = \begin{bmatrix} \dfrac{l-L}{L} & 0 & 0 \\ 0 & \dfrac{r-R}{R} & 0 \\ 0 & 0 & \dfrac{r-R}{R} \end{bmatrix} = \begin{bmatrix} \lambda_1 - 1 & 0 & 0 \\ 0 & \lambda_2 - 1 & 0 \\ 0 & 0 & \lambda_2 - 1 \end{bmatrix} = \mathbf{V} - \mathbf{I} \tag{2.159}$$

and

$$\dot{\hat{\mathbf{E}}} = \begin{bmatrix} \dfrac{\dot{l}}{L} & 0 & 0 \\ 0 & \dfrac{\dot{r}}{R} & 0 \\ 0 & 0 & \dfrac{\dot{r}}{R} \end{bmatrix} \tag{2.160}$$

$\hat{\mathbf{E}}$ should not be confused with $\mathbf{D}$. Using Eqs. (2.156), (2.159), and (2.160), we have

$$\mathbf{D} = \dot{\hat{\mathbf{E}}} \cdot \left(\hat{\mathbf{E}} + I\right)^{-1} \tag{2.161}$$

During an experiment, measurements are made after a successive finite time interval Δt; then

$$\dot{\hat{\mathbf{E}}}\,\Delta \mathbf{t} \approx \begin{bmatrix} \dfrac{l_2 - l_1}{L} & 0 & 0 \\ 0 & \dfrac{r_2 - r_1}{R} & 0 \\ 0 & 0 & \dfrac{r_2 - r_1}{R} \end{bmatrix},$$

$$\mathbf{D}\,\Delta t \approx \begin{bmatrix} \dfrac{l_2 - l_1}{l_1} & 0 & 0 \\ 0 & \dfrac{r_2 - r_1}{r_1} & 0 \\ 0 & 0 & \dfrac{r_2 - r_1}{r_1} \end{bmatrix} \tag{2.162}$$

If the axial load is P, then the Cauchy stress is given by the following matrices:

$$\boldsymbol{\sigma} = \begin{bmatrix} \frac{P}{\pi r^2} & 0 & 0 \\ 0 & 0 & 0 \\ 0 & 0 & 0 \end{bmatrix} = \begin{bmatrix} \frac{P}{\lambda_2^2 \pi R^2} & 0 & 0 \\ 0 & 0 & 0 \\ 0 & 0 & 0 \end{bmatrix} = \begin{bmatrix} \frac{P}{\lambda_2^2 A_0} & 0 & 0 \\ 0 & 0 & 0 \\ 0 & 0 & 0 \end{bmatrix} \tag{2.163}$$

Using Eq. (2.143), we obtain the first Piola-Kirchhoff stress $\boldsymbol{\Sigma}^{\mathrm{I}}$:

$$\boldsymbol{\Sigma}^{\mathrm{I}} = \det(\mathbf{F})\mathbf{F}^{-1} \cdot \boldsymbol{\sigma} = \lambda_1 \lambda_2^2 \begin{bmatrix} \frac{1}{\lambda_1} & 0 & 0 \\ 0 & \frac{1}{\lambda_2} & 0 \\ 0 & 0 & \frac{1}{\lambda_2} \end{bmatrix} \begin{bmatrix} \frac{P}{\lambda_2^2 A_0} & 0 & 0 \\ 0 & 0 & 0 \\ 0 & 0 & 0 \end{bmatrix}$$

$$= \begin{bmatrix} \frac{P}{A_0} & 0 & 0 \\ 0 & 0 & 0 \\ 0 & 0 & 0 \end{bmatrix} \tag{2.164}$$

This equation indicates that $\boldsymbol{\Sigma}^{\mathrm{I}}$ is the nominal stress. Similarly we can determine the second Piola-Kirchhoff stress:

$$\boldsymbol{\Sigma}^{\mathrm{II}} = \det(\mathbf{F})\mathbf{F}^{-1} \cdot \boldsymbol{\sigma} \cdot \left(\mathbf{F}^{-1}\right)^{\mathrm{T}} = \begin{bmatrix} \frac{P}{\lambda_1 A_0} & 0 & 0 \\ 0 & 0 & 0 \\ 0 & 0 & 0 \end{bmatrix} \tag{2.165}$$

If the deformation is assumed incompressible, then $\lambda_1 \lambda_2^2 = 1$, and hence $\lambda_2 = (\lambda_1)^{-1/2}$.

2.11.2 Simple Shear Experiment

In the simple shear experiment geometry (Fig. 2.8), the motion is described by

$$\begin{aligned} x &= X + \gamma Y \\ y &= Y \\ z &= Z \end{aligned} \tag{2.166}$$

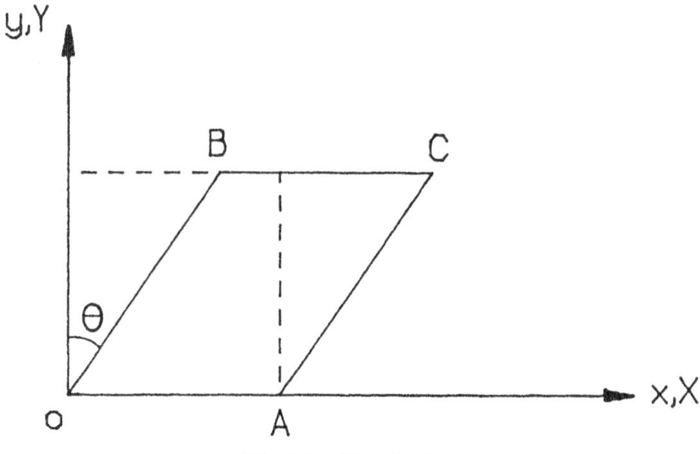

Fig. 2.8 Simple shear

where

$$\gamma = \tan\theta \tag{2.167}$$

Then, following the standard procedure, we obtain

$$\mathbf{F} = \begin{bmatrix} 1 & \gamma & 0 \\ 0 & 1 & 0 \\ 0 & 0 & 1 \end{bmatrix}, \quad \mathbf{F}^{-1} = \begin{bmatrix} 1 & -\gamma & 0 \\ 0 & 1 & 0 \\ 0 & 0 & 1 \end{bmatrix}, \quad \det(\mathbf{F}) = 1 \tag{2.168}$$

$$\mathbf{C} = \mathbf{U}^2 = \mathbf{F}^{\mathrm{T}} \cdot \mathbf{F} = \begin{bmatrix} 1 & \gamma & 0 \\ \gamma & 1+\gamma^2 & 0 \\ 0 & 0 & 1 \end{bmatrix},$$

$$\mathbf{B}^{-1} = \mathbf{V}^{-2} = (\mathbf{F}^{\mathrm{T}})^{-1} \cdot (\mathbf{F})^{-1} = \begin{bmatrix} 1 & -\gamma & 0 \\ -\gamma & 1+\gamma^2 & 0 \\ 0 & 0 & 1 \end{bmatrix} \tag{2.169}$$

$$\mathbf{E} = \tfrac{1}{2}(\mathbf{C} - \mathbf{I}) = \tfrac{1}{2}\begin{bmatrix} 0 & \gamma & 0 \\ \gamma & \gamma^2 & 0 \\ 0 & 0 & 0 \end{bmatrix}, \quad \mathbf{e} = \tfrac{1}{2}(\mathbf{I} - \mathbf{B}^{-1}) = \tfrac{1}{2}\begin{bmatrix} 0 & -\gamma & 0 \\ -\gamma & \gamma^2 & 0 \\ 0 & 0 & 0 \end{bmatrix}$$

To obtain **U** and **V**, note that the deformation is confined in x-y plane, so we can write

$$\mathbf{U} = \begin{bmatrix} U_{xx} & U_{xy} & 0 \\ U_{xy} & U_{yy} & 0 \\ 0 & 0 & 1 \end{bmatrix}, \quad \mathbf{V} = \begin{bmatrix} V_{xx} & V_{xy} & 0 \\ V_{xy} & V_{yy} & 0 \\ 0 & 0 & 1 \end{bmatrix} \tag{2.171}$$

and

$$\mathbf{R} = \begin{bmatrix} \cos\phi & \sin\phi & 0 \\ -\sin\phi & \cos\phi & 0 \\ 0 & 0 & 1 \end{bmatrix} \tag{2.172}$$

where $\mathbf{R}$ is the most general form of an orthogonal tensor representing a rotation in the x-y plane. We get the following components of $\mathbf{U}$, $\mathbf{V}$, and $\mathbf{R}$ by solving a system of algebraic equations and by comparing components on both sides of Eq. (2.63):

$$\cos\phi = \frac{2}{\sqrt{4+\gamma^2}}, \quad \sin\phi = \frac{\gamma}{\sqrt{4+\gamma^2}} \tag{2.173}$$

$$\mathbf{U} = \begin{bmatrix} \cos\phi & \sin\phi & 0 \\ \sin\phi & \gamma\sin\phi + \cos\phi & 0 \\ 0 & 0 & 1 \end{bmatrix}, \quad \mathbf{V} = \begin{bmatrix} \cos\phi + \gamma\sin\phi & \sin\phi & 0 \\ \sin\phi & \cos\phi & 0 \\ 0 & 0 & 1 \end{bmatrix} \tag{2.174}$$

The velocity gradient is calculated by

$$\mathbf{L} = \dot{\mathbf{F}} \cdot \mathbf{F}^{-1} = \begin{bmatrix} 0 & \dot{\gamma} & 0 \\ 0 & 0 & 0 \\ 0 & 0 & 0 \end{bmatrix} \begin{bmatrix} 0 & -\gamma & 0 \\ 0 & 1 & 0 \\ 0 & 0 & 1 \end{bmatrix} = \begin{bmatrix} 0 & \dot{\gamma} & 0 \\ 0 & 0 & 0 \\ 0 & 0 & 0 \end{bmatrix} \tag{2.175}$$

Thus

$$\mathbf{D} = \begin{bmatrix} 0 & \frac{\dot{\gamma}}{2} & 0 \\ \frac{\dot{\gamma}}{2} & 0 & 0 \\ 0 & 0 & 0 \end{bmatrix}, \quad \mathbf{W} = \begin{bmatrix} 0 & \frac{\dot{\gamma}}{2} & 0 \\ -\frac{\dot{\gamma}}{2} & 0 & 0 \\ 0 & 0 & 0 \end{bmatrix} \tag{2.176}$$

In this case, since the principal axes of stretch ratio rotate in the process of deformation, we cannot obtain a simple relationship between $\mathbf{D}$ and $\bar{\mathbf{e}}$. Also, because the definition of $\ln \mathbf{V}$ is based on the principal values, it is inconvenient in this case to introduce $\bar{\mathbf{e}}$. Note that $\mathbf{V}$ as given by Eq. (2.174) is not in diagonal form; in other words, the coordinate system used in Fig. 2.8 does not coincide with the principal axes of **V**.

The Cauchy stress σ and first and second Piola-Kirchhoff stresses are given by Eqs. (2.177), (2.178), and (2.179), respectively:

$$\sigma = \begin{bmatrix} \sigma_{xx} & \tau & 0 \\ \tau & \sigma_{yy} & 0 \\ 0 & 0 & 0 \end{bmatrix} \tag{2.177}$$

$$\Sigma^{I} = \det(\mathbf{F})\mathbf{F}^{-1} \cdot \boldsymbol{\sigma} = \begin{bmatrix} 1 & -\gamma & 0 \\ 0 & 1 & 0 \\ 0 & 0 & 1 \end{bmatrix} \begin{bmatrix} \sigma_{xx} & \tau & 0 \\ \tau & \sigma_{yy} & 0 \\ 0 & 0 & 0 \end{bmatrix}$$

$$= \begin{bmatrix} \sigma_{xx} - \gamma\tau & \tau - \gamma\sigma_{yy} & 0 \\ \tau & \sigma_{yy} & 0 \\ 0 & 0 & 0 \end{bmatrix} \tag{2.178}$$

$$\Sigma^{II} = \det(\mathbf{F})\mathbf{F}^{-1} \cdot \boldsymbol{\sigma} \cdot (\mathbf{F}^{-1})^{T}$$

$$= \begin{bmatrix} \sigma_{xx} - \gamma\tau & \tau - \gamma\sigma_{yy} & 0 \\ \tau & \sigma_{yy} & 0 \\ 0 & 0 & 0 \end{bmatrix} \begin{bmatrix} 1 & 0 & 0 \\ -\gamma & 1 & 0 \\ 0 & 0 & 1 \end{bmatrix}$$

$$= \begin{bmatrix} \sigma_{xx} - \gamma^2\sigma_{yy} - 2\gamma\tau & \tau - \gamma\sigma_{yy} & 0 \\ \tau - \gamma\sigma_{yy} & \sigma_{yy} & 0 \\ 0 & 0 & 0 \end{bmatrix} \tag{2.179}$$

where $\tau = \sigma_{yx} = \sigma_{xy}$ is the shear stress, and σ_{xx} and σ_{yy} are the normal stresses caused by the shear deformation.

2.11.3 Combined Tension and Torsion Experiment

Thin-walled hollow cylinders subjected to combined tension and torsion loads are widely used to study the material response to finite plastic deformation. The deformation for this configuration is given by

$$\begin{aligned} r &= \alpha R \\ \theta &= \Theta + \omega Z \\ z &= \lambda Z \end{aligned} \tag{2.180}$$

where the polar coordinates are used for both referential and spatial descriptions. In this case it is easier to describe the geometry and deformation in polar coordinates than cartesian coordinates. Any material particle at (R, Θ, z) before deformation moves to (r, θ, z) after deformation. For a thin-walled tube specimen the geometry is bounded as follows

$$R_i \le R \le R_o, \quad 0 \le \Theta \le 2\pi, \quad 0 \le Z \le L$$
$$r_i \le r \le r_o, \quad 0 \le \theta \le 2\pi, \quad 0 \le z \le l \tag{2.181}$$

Here R_o and R_i are the undeformed outer and inner radii, r_o and r_i are the deformed outer and inner radii, L and l are the undeformed and deformed lengths of the cylinder, respectively, and ω is the angle of twist per unit undeformed length,

$$\omega = \frac{\phi}{L} \tag{2.182}$$

where ϕ is the relative twist angle of the two end sections of the cylinder.

In the cylindrical polar coordinates, the deformation gradient is given by (cf. Spencer 1980)

$$\mathbf{F} = \begin{bmatrix} \dfrac{\partial r}{\partial R} & \dfrac{1}{R}\dfrac{\partial r}{\partial \Theta} & \dfrac{\partial r}{\partial Z} \\ r\dfrac{\partial \theta}{\partial R} & \dfrac{r}{R}\dfrac{\partial \theta}{\partial \Theta} & r\dfrac{\partial \theta}{\partial Z} \\ \dfrac{\partial z}{\partial R} & \dfrac{1}{R}\dfrac{\partial z}{\partial \Theta} & \dfrac{\partial z}{\partial Z} \end{bmatrix} = \begin{bmatrix} \alpha & 0 & 0 \\ 0 & \alpha & \omega\alpha R_m \\ 0 & 0 & \lambda \end{bmatrix}$$

$$\mathbf{F}^{-1} = \begin{bmatrix} \dfrac{1}{\alpha} & 0 & 0 \\ 0 & \dfrac{1}{\alpha} & -\dfrac{\omega R_m}{\lambda} \\ 0 & 0 & \dfrac{1}{\lambda} \end{bmatrix} \tag{2.183}$$

where the mean undeformed radius, $R_m \approx R$, is used to replace R; this replacement is a standard assumption for thin-walled tubes. Following the same procedures as introduced for simple shear, we obtain

$$\mathbf{V} = \begin{bmatrix} \alpha & 0 & 0 \\ 0 & \alpha\cos\phi + \alpha\omega R_m \sin\phi & \lambda\sin\phi \\ 0 & \lambda\sin\phi & \lambda\cos\phi \end{bmatrix} \tag{2.184}$$

$$\mathbf{U} = \begin{bmatrix} \alpha & 0 & 0 \\ 0 & \alpha\cos\phi & \alpha\sin\phi \\ 0 & \alpha\sin\phi & \alpha\omega R_m \sin\phi + \lambda\cos\phi \end{bmatrix} \tag{2.185}$$

where

$$\cos\phi = \frac{\lambda + \alpha}{\sqrt{(\alpha+\lambda)^2 + (\alpha\omega R_m)}}, \quad \sin\phi = \frac{\alpha\omega R_m}{\sqrt{(\alpha+\lambda)^2 + (\alpha\omega R_m)^2}}, \tag{2.186}$$

R is given by Eq. (2.172):

$$\mathbf{C} = \mathbf{F}^{\mathrm{T}} \cdot \mathbf{F} = \begin{bmatrix} \alpha^2 & 0 & 0 \\ 0 & \alpha^2 & \omega\alpha^2 R_m \\ 0 & \omega\alpha^2 R_m & \lambda^2 + \omega^2\alpha^2 R_m^2 \end{bmatrix}$$

$$\mathbf{B}^{-1} = (\mathbf{F}^{-1})^{\mathrm{T}} \cdot \mathbf{F}^{-1} = \begin{bmatrix} \dfrac{1}{\alpha^2} & 0 & 0 \\ 0 & \dfrac{1}{\alpha^2} & -\dfrac{\omega R_m}{\alpha\lambda} \\ 0 & -\dfrac{\omega R_m}{\alpha\lambda} & \dfrac{1}{\lambda^2} + \dfrac{\omega^2 R_m^2}{\lambda^2} \end{bmatrix} \tag{2.187}$$

The Lagrangian and Eulerian strains can be determined by using $\mathbf{E} = \frac{1}{2}(\mathbf{C} - \mathbf{I})$ and $\mathbf{e} = \frac{1}{2}(\mathbf{I} - \mathbf{B}^{-1})$. Equation (2.188) gives the velocity gradient

$$\mathbf{L} = \dot{\mathbf{F}} \cdot \mathbf{F}^{-1} = \begin{bmatrix} \dfrac{\dot{\alpha}}{\alpha} & -\dot{\omega}Z & 0 \\ \dot{\omega}Z & \dfrac{\dot{\alpha}}{\alpha} & \dfrac{\dot{\omega}\alpha R_m}{\lambda} \\ 0 & 0 & \dfrac{\dot{\lambda}}{\lambda} \end{bmatrix} \tag{2.188}$$

Hence

$$\mathbf{D} = \begin{bmatrix} \dfrac{\dot{\alpha}}{\alpha} & 0 & 0 \\ 0 & \dfrac{\dot{\alpha}}{\alpha} & \dfrac{\dot{\omega}\alpha R_m}{2\lambda} \\ 0 & \dfrac{\dot{\omega}\alpha R_m}{2\lambda} & \dfrac{\dot{\lambda}}{\lambda} \end{bmatrix}, \quad \mathbf{W} = \begin{bmatrix} 0 & -\dot{\omega}Z & 0 \\ \dot{\omega}Z & 0 & \dfrac{\dot{\omega}\alpha R_m}{2\lambda} \\ 0 & -\dfrac{\dot{\omega}\alpha R_m}{2\lambda} & 0 \end{bmatrix} \tag{2.189}$$

In the case of combined tension-torsion loading, the Cauchy, first Piola-

Kirchhoff and second Piola-Kirchhoff stresses are given by the following equations:

$$\boldsymbol{\sigma} = \begin{bmatrix} 0 & 0 & 0 \\ 0 & 0 & \tau \\ 0 & \tau & \sigma \end{bmatrix} \tag{2.190}$$

$$\boldsymbol{\Sigma}^{\mathrm{I}} = \det(\mathbf{F})\mathbf{F}^{-1} \cdot \boldsymbol{\sigma} = \alpha^2\lambda \begin{bmatrix} \dfrac{1}{\alpha} & 0 & 0 \\ 0 & \dfrac{1}{\alpha} & -\dfrac{\omega R_m}{\lambda} \\ 0 & 0 & \dfrac{1}{\lambda} \end{bmatrix} \begin{bmatrix} 0 & 0 & 0 \\ 0 & 0 & \tau \\ 0 & \tau & \sigma \end{bmatrix}$$

$$= \begin{bmatrix} 0 & 0 & 0 \\ 0 & -\tau\alpha^2\omega R_m & \alpha\lambda\tau - \alpha^2\omega\sigma R_m \\ 0 & \alpha^2\tau & \alpha^2\sigma \end{bmatrix} \tag{2.191}$$

$$\boldsymbol{\Sigma}^{\mathrm{II}} = \boldsymbol{\Sigma}^{\mathrm{I}} \cdot \left(\mathbf{F}^{-1}\right)^{\mathrm{T}} = \begin{bmatrix} 0 & 0 & 0 \\ 0 & -\tau\alpha^2\omega R_m & \tau\alpha\lambda - \sigma\alpha^2\omega R_m \\ 0 & \alpha^2\tau & \alpha^2\sigma \end{bmatrix}$$

$$\times \begin{bmatrix} \dfrac{1}{\alpha} & 0 & 0 \\ 0 & \dfrac{1}{\alpha} & 0 \\ 0 & -\dfrac{\omega R_m}{\lambda} & \dfrac{1}{\lambda} \end{bmatrix}$$

$$= \begin{bmatrix} 0 & 0 & 0 \\ 0 & -2\tau\alpha\omega R_m + \dfrac{\sigma\alpha^2\omega^2 R_m^2}{\lambda} & \tau\alpha - \dfrac{\sigma\alpha^2\omega R_m}{\lambda} \\ 0 & \tau\alpha - \dfrac{\sigma\alpha^2\omega R_m}{\lambda} & \dfrac{\alpha^2}{\lambda} \end{bmatrix} \tag{2.192}$$

Here τ is deduced from the applied torque and σ from the applied axial load.

We close the chapter with two important remarks. First, it is emphasized that the tensor function introduced in Section 2.3 is based on principal values. Supposing the principal values of a tensor **A** are A_1, A_2, and A_3, then in the principal axes coordinates **A** has the form

$$\mathbf{A} = \begin{bmatrix} A_1 & 0 & 0 \\ 0 & A_2 & 0 \\ 0 & 0 & A_3 \end{bmatrix} \tag{2.193}$$

and $\ln \mathbf{A}$ is defined as

$$\ln \mathbf{A} = \begin{bmatrix} \ln A_1 & 0 & 0 \\ 0 & \ln A_2 & 0 \\ 0 & 0 & \ln A_3 \end{bmatrix} \tag{2.194}$$

This means $\ln \mathbf{A}$ is coaxial with **A** and has the principal value $\ln A_i$ $(i = 1, 2, 3)$. One can define $e^{\mathbf{A}}$, $\sin \mathbf{A}$, and so on, in a similar manner. This is valid because this definition can keep the tensorial characteristics for $\ln \mathbf{A}$; it guarantees that $\ln \mathbf{A}$ is still a tensor. Otherwise, if we define the components of $\ln \mathbf{A}$ in any coordinates as $\ln A_{ij}$, then, in general, $\ln \mathbf{A}$ will *not* be a tensor and will *not* obey the tensorial transformation law.

Our second remark pertains to strain measures for finite deformation. As discussed in Section 2.7, the deformation rate **D** can determine the deformation rate state in reference to the current configuration $\mathscr{R}$. Thus, for $\mathscr{R}$ at time t as the reference and the infinitesimal deformation change in an infinitesimal time interval, we can define an infinitesimal strain tensor by

$$d\epsilon_{ij} = \tfrac{1}{2}(du_{i,j} + du_{j,i}) = D_{ij}\, dt \tag{2.195}$$

While the infinitesimal strain $d\epsilon_{ij}$ in $\mathscr{R}$ is well-defined above, there is no strain measure, in general, that is the integration of $d\epsilon_{ij}$. As discussed in Section 2.8, *only* in cases where the principal axes of the strain are fixed do we have $\dot{\mathbf{e}}^{\mathrm{L}} = \mathbf{D}$. Only in those cases it is

$$\mathbf{e}^{\mathrm{L}} = \int \mathbf{D}\, dt \tag{2.196}$$

This expression does not hold in general, and the integration on the right-hand side of Eq. (2.196) has no physical meaning; even it is not a tensor. For this reason the application of the logarithmic strain is limited to those cases where the principal axes of strain are fixed.

REFERENCES

Eringen, A. E. 1967. Mechanics of continua. New York: Wiley.

Green, A. E., and P. M. Naghdi. 1965. A general theory of an elastic-plastic continuum. *Arch. Rat. Mech. Anal.* **18**:251.

Gurtin, M. E., and K. Spear. 1983. On the relationship between the logarithmic strain rate and the stretching tensor. *Int. J. Solids Struct.* **19**(5):437.

Hill, R. 1970. Constitutive inequalities for isotropic elastic solids under finite strain. *Proc. R. Soc. London* **A314**:457.

Malvern, L. E. 1969. *Introduction to the Mechanics of a Continuous Medium.* Englewood Cliffs, NJ: Prentice-Hall.

Scipio, L. A. 1967. *Principles of Continua with Applications.* New York: Wiley.

Spencer, A. J. M. 1980. *Continuum Mechanics.* London: Longman Group.

Truesdell, C. 1952. The mechanical foundations of elasticity and fluid dynamics. *J. Rat. Mech. Anal.* **1**:125.

Truesdell, C. 1985. *The Elements of Continuum Mechanics.* Berlin: Springer-Verlag.

Truesdell, C., and W. Noll. 1965. The nonlinear field theories of mechanics. *Encyclopedia of Physics*, vol. III/3, ed. S. Flügge. Berlin: Springer-Verlag.

3

GENERAL PRINCIPLES

The preceding chapter discussed different strain and strain rate measures for finite deformation as well as several stress measures in different reference configurations. These are the mathematical descriptions of the three elements of classical mechanics: bodies, motions, and forces. Ultimately we are interested in determining the response of bodies or structures to the applied external forces or loads, and vice versa. To achieve this, one has to know the principles or laws that govern these elements and their relations.

There are two kinds of principles or laws that relate the forces to the motions of bodies. The first kind concerns the universal or general principles that govern the motions of bodies under applied external forces regardless of the material properties of the bodies. In other words, any material undergoing elastic, plastic, or viscoplastic deformation must obey these principles whenever it moves and deforms under the applied loads. There are five such universal principles for us to consider in plasticity theory:

1. Conservation of mass.
2. Conservation of momentum.
3. Conservation of moment of momentum.
4. Conservation of energy (the first law of thermodynamics).
5. Clausius-Duhem inequality (the second law of thermodynamics).

The second kind of laws relate the stress to strain and/or strain rate that characterize the behavior of a material under an application of forces or loads. These are called *constitutive equations*, and they vary for different materials. They can even differ for the same material in different regimes of deformation. For example, the response during elastic deformation will be different from that during plastic deformation. Also the constitutive laws for quasi-static and dynamic behavior can differ.

Since the general principles also govern plastic deformation which is the main concern of this book, and since some plasticity theories and concepts that will be presented and discussed later in this book are developed using thermodynamics, the general principles listed above will be briefly discussed in this chapter. Detailed mathematical treatments can be found in advanced continuum mechanics and physics books (Truesdell and Noll 1965; Malvern 1969).

A constitutive equation, commonly used to describe an objective physical phenomenon such as material behaviors under applied loads, must obey certain axioms or principles during the occurrence of the physical phenomenon. Among them is the axiom or principle of objectivity or material frame indifference. This principle is especially important in finite plastic deformation. In this chapter we will consider the physical meaning and the consequences of objectivity and introduce the objective material derivatives of stress. Other axioms that govern a constitutive equation, such as axioms of determinism and equipresence, can be found in the references mentioned above.

3.1 CONSERVATION OF MASS

Suppose that at any fixed time t the body $\mathscr{B}$ occupies space region $\mathscr{R}$ which defines the current configuration. Let v and a denote the volume of $\mathscr{R}$ and its boundary, respectively, as shown in Fig. 3.1. We further assume that the mass with a density ρ at time t is continuously distributed over v. Then the total mass of the volume v at the moment t is given by

$$M = \int_v \rho \, dv \tag{3.1}$$

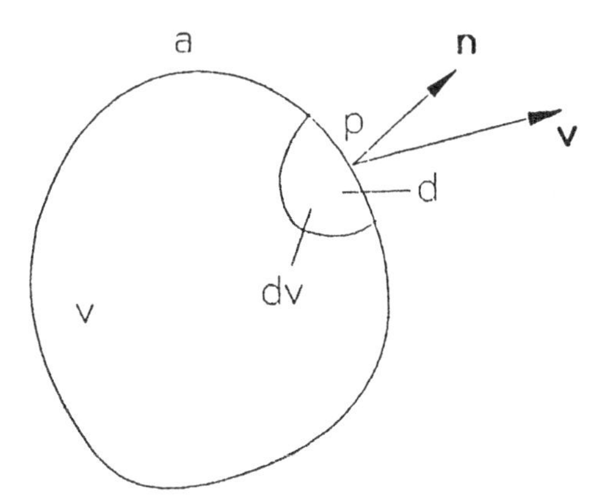

Fig. 3.1 A volume of mass in space

where

$$\rho = \rho(x, t) \tag{3.2}$$

is a continuous function of the spatial coordinates x and time t. Consider an infinitesimal surface element da at point p as shown in Fig. 3.1. If the mass moves across da outward with the speed v, the rate of the mass loss across da (i.e., the mass flowing out of v through the surface element da at point p per unit time) is $\rho\mathbf{n} \cdot \mathbf{v}\, da$, where $\mathbf{n}$ is the unit outward normal at point p. The total mass loss rate through the whole surface a is expressed by

$$\int_a \rho\mathbf{n} \cdot \mathbf{v}\, da = \int_v \nabla_{\mathbf{x}} \cdot (\rho\mathbf{v})\, dv \tag{3.3}$$

where the following divergence theorem

$$\int_a \mathbf{n} \cdot \mathbf{A}\, da = \int_v \nabla_{\mathbf{x}} \cdot \mathbf{A}\, dv \tag{3.4}$$

has been used to derive the right side of Eq. (3.3), and ∇_x is given by $i_l(\partial/\partial x_l)$. In the divergence theorem, $\mathbf{A}$ can be a vector, a tensor, or even a scalar. In the later case the dot product becomes simply a scalar product. Further we can calculate the rate of increase of the total mass in the volume v as

$$\frac{\partial M}{\partial t} = \int_v \frac{\partial \rho}{\partial t}\, dv \tag{3.5}$$

According to the conversion of mass, no mass can be destroyed or created in v. The term $\partial M/\partial t$ must be equal to the negative value given by Eq. (3.3).

$$\int_v \left[\frac{\partial \rho}{\partial t} + \nabla_{\mathbf{x}} \cdot (\rho\mathbf{v})\right] dv = 0 \tag{3.6}$$

Since the choice of the volume v is arbitrary, and the equation above holds for any point of v, it follows that the integrand must vanish everywhere inside v. That results in the following continuity or conservation of mass equation:

$$\frac{\partial \rho}{\partial t} + \nabla_{\mathbf{x}} \cdot (\rho\mathbf{v}) = 0 \tag{3.7}$$

In terms of components, we write

$$\frac{\partial \rho}{\partial t} + (\rho v_i)_{,i} = 0 \tag{3.8}$$

Since

$$(\rho v_i)_{,i} = \rho_{,i} v_{,i} + \rho v_{i,i}$$

and

$$\frac{d\rho}{dt} = \dot{\rho} = \frac{\partial \rho}{\partial t} + v_i \rho_{,i}$$

where $d\rho/dt = \dot{\rho}$ is the material derivative of ρ introduced in Section 2.6. We can write the continuity equation (3.7) in the following different form:

$$\frac{d\rho}{dt} + \rho v_{i,i} = 0$$

or

$$\dot{\rho} + \rho \nabla_{\mathbf{x}} \cdot \mathbf{v} = 0 \tag{3.9}$$

If the material is incompressible so that the mass density ρ remains the same when the material moves or deforms, we have $\dot{\rho} = 0$. Then the continuity equation will take the form

$$v_{i,i} = 0 \tag{3.10}$$

According to the definition for the deformation rate $\mathbf{D}$, Eq. (2.102), this will read

$$\text{tr}(\mathbf{D}) = 0 \tag{3.11}$$

for the incompressible material.

The conservation of mass discussed above is in the current configuration. The resulting continuity equation can also be derived in reference configuration. Suppose that material occupying the space volume v in the current configuration occupies the volume V in the initial reference configuration. Then the conservation of mass requires that

$$\int_V \rho_0(\mathbf{X}, t_0)\, dV = \int_v \rho(\mathbf{x}, t)\, dv \tag{3.12}$$

where ρ_0 is the mass density in the reference configuration, given in terms of the coordinates $\mathbf{X}$ in that configuration, and t_0 is the initial time or reference time. According to the motion, Eq. (2.3) and the relationship between the infinitesimal volume elements dv and dV in the current and reference configurations of Eq. (2.28), we have

$$\rho(\mathbf{x}, t)\, dv = \rho[\mathbf{x}(\mathbf{X}, t), t] \det(\mathbf{F})\, dV$$

Then we can write

$$\int_V \rho(\mathbf{x}, t)\, dv = \int_V \rho(\mathbf{X}, t) \det(\mathbf{F})\, dV$$

As a result Eq. (3.12) becomes

$$\int_v [\rho_0 - \rho \det(\mathbf{F})]\, dV = 0 \tag{3.13}$$

where ρ_0 and ρ are functions of the coordinates X in the reference configuration. Since V is arbitrary, it follows that the integrand of the equation above has to be zero:

$$\rho_0 = \rho \det(\mathbf{F}) \quad \text{or} \quad \frac{\rho_0}{\rho} = \det(\mathbf{F}) \tag{3.14}$$

This equation is called the *material description of the continuity* or *conservation of mass equation*. For incompressible materials, since $\rho_0 = \rho =$ constant, we can conclude that

$$\det(\mathbf{F}) = 1 \tag{3.15}$$

Almost all plasticity theories for metals assume that the plastic deformation is isochoric, namely that it takes place with negligible volume change, based on Bridgeman's experimental observations. In these cases, if we further assume that elastic deformation is small compared with plastic deformation, Eqs. (3.11) and (3.15) can be used as a first-order approximation in formulating the plasticity models.

3.2 CONSERVATION OF MOMENTUM

The conservation of momentum principle is in fact the generalization of Newton's second law of motion for continuum mechanics, provided that the internal action among particles of a continuum can be represented by internal tractions and that these tractions follow the same laws as the external forces. The conservation of momentum can be stated as follows: The rate of change of the total momentum of any given set of particles equals the vectorial sum of all the forces acting on this set of particles.

Consider any set of particles that currently occupies a spatial volume v with the boundary a, as shown in Fig. 3.2. The forces acting on v are the surface traction $t^{(\mathbf{n})}$ per unit area and the body force $\mathbf{b}$ per unit mass. The momentum per unit volume is $\rho \mathbf{v}\, dv$, where $\mathbf{v}$ is the velocity of the particles contained in volume dv. Then the total momentum of v can be calculated by

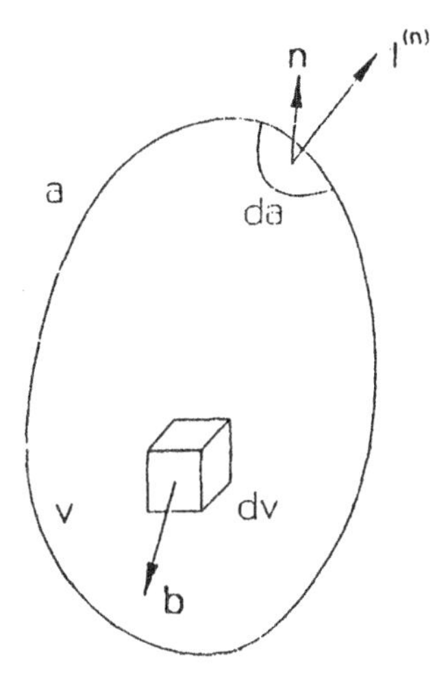

Fig. 3.2 Conservation of momentum.

the integral

$$\text{Total momentum} = \int_v \rho \mathbf{v} \, dv \tag{3.16}$$

The rate of change of the total momentum is given by

$$= \frac{d}{dt} \int_v \rho \mathbf{v} \, dv = \int_v \rho \frac{d\mathbf{v}}{dt} \, dv \tag{3.17}$$

The last result comes from the observation that

$$\rho \, dv = \rho_0 \, dV \tag{3.18}$$

and that

$$\frac{d}{dt} \int_v \rho \mathbf{A} \, dv = \frac{d}{dt} \int_V \rho_0 \mathbf{A} \, dV = \int_V \rho_0 \frac{d\mathbf{A}}{dt} \, dV = \int_v \rho \frac{d\mathbf{A}}{dt} \, dv \tag{3.19}$$

where Eq. (3.18) is the conservation of mass and **A** in Eq. (3.19) can be a scalar, vector, or a tensor of any order. Using Eq. (3.17), we can write the conservation of momentum in the form

$$\int_v \rho \frac{d\mathbf{v}}{dt} \, dv = \int_v \rho \mathbf{b} \, dv + \int_a \mathbf{t}^{(\mathbf{n})} \, da \tag{3.20}$$

Recall that $\mathbf{t}^{(\mathbf{n})} = \mathbf{n} \cdot \boldsymbol{\sigma}$ and using the divergence theorem, Eq. (3.4), the integral on surface a is transformed into

$$\int_a \mathbf{t}^{(\mathbf{n})} \, da = \int_a \mathbf{n} \cdot \boldsymbol{\sigma} \, da = \int_v \boldsymbol{\nabla}_{\mathbf{x}} \cdot \boldsymbol{\sigma} \, dv \tag{3.21}$$

where $\boldsymbol{\sigma}$ is the Cauchy stress tensor and $\mathbf{n}$ is the unit normal to the surface. Substitute Eq. (3.21) into Eq. (3.20) and rearrange it to obtain

$$\int_v \left[\nabla_x \cdot \boldsymbol{\sigma} + \rho \mathbf{b} - \rho \frac{dv}{dt} \right] dv = 0 \tag{3.22}$$

Since the volume v is arbitrary, the integrand must be zero everywhere. Thus

$$\nabla_{\mathbf{x}} \cdot \boldsymbol{\sigma} + \rho \mathbf{b} = \rho \frac{d\mathbf{v}}{dt} \tag{3.23}$$

This is the differential form of the conservation of momentum and is known as Cauchy's equations of motion. In terms of components, we can write

$$\sigma_{ji,j} + \rho b_i = \rho \frac{dv_i}{dt} \tag{3.24}$$

In case of static equilibrium, $d\mathbf{v}/dt = \mathbf{0}$, and we have the following equilibrium equations

$$\nabla_x \cdot \boldsymbol{\sigma} + \rho \mathbf{b} = \mathbf{0} \quad \text{or} \quad \sigma_{ji,j} + \rho b_i = 0 \tag{3.25}$$

These Cauchy's equations of motion are written in the current configuration, so $\mathbf{b} = \mathbf{b}(x, t)$ and $\mathbf{v} = \mathbf{v}(x, t)$. Sometimes it is more convenient to use the equations of motion in the reference configuration. To obtain these equations, we need to transform Eq. (3.25) into the reference configuration and express the variables in terms of coordinates in that configuration.

The transformation of the body force terms is simply

$$\int_v \rho \mathbf{b}(\mathbf{x}, t)\, dv = \int_V \rho_0 \mathbf{b}(\mathbf{X}, t)\, dV = \int_V \rho_0 \mathbf{b}_0\, dV \tag{3.26}$$

where the motion $\mathbf{x} = \mathbf{x}(\mathbf{x}, t)$ has been used to replace $\mathbf{x}$ by $\mathbf{X}$, and $\mathbf{b}_0 = \mathbf{b}((\mathbf{X}, t), t) = \mathbf{b}(\mathbf{X}, t)$ is the body force per mass in the reference configuration. Similarly we can write

$$\int_v \rho \frac{d\mathbf{v}(\mathbf{x}, t)}{dt}\, dv = \int_V \rho_0 \frac{d\mathbf{v}(\mathbf{X}, t)}{dt}\, dV \tag{3.27}$$

To transform the traction term in Eq. (3.25), Eq. (2.142) is used to derive

$$\int_a \mathbf{t}^{(\mathbf{n})}\, da = \int_a \mathbf{n} \cdot \boldsymbol{\sigma}\, da = \int_A \mathbf{N} \cdot \boldsymbol{\Sigma}^{\mathrm{I}}\, dA \tag{3.28}$$

Using the divergence theorem gives the following

$$\int_A \mathbf{N} \cdot \boldsymbol{\Sigma}^{\mathrm{I}}\, d\mathbf{A} = \int_V \boldsymbol{\nabla} \cdot \boldsymbol{\Sigma}^{\mathrm{I}}\, dV \tag{3.29}$$

where $\boldsymbol{\nabla}$ is the differential operator with respect to the referential coordinates X, and $\boldsymbol{\Sigma}^{I}$ is the first Piola-Kirchhoff stress. The equations of motion in the reference configuration can be obtained by substituting Eqs. (3.26), (3.27), and (3.28) into Eq. (3.25) to give

$$\int_V \left(\boldsymbol{\nabla} \cdot \boldsymbol{\Sigma}^{\mathrm{I}} + \rho_0 \mathbf{b}_0 - \rho_0 \frac{d\mathbf{v}}{dt} \right) dV = \mathbf{0} \tag{3.30}$$

Since the volume V is arbitrary, this implies

$$\boldsymbol{\nabla} \cdot \boldsymbol{\Sigma}^{\mathrm{I}} + \rho_0 \mathbf{b} = \rho_0 \frac{d\mathbf{v}}{dt} \quad \text{or} \quad \Sigma^{\mathrm{I}}_{Ki,K} + \rho_0 b_{0i} = \rho_0 \frac{dv_i}{dt} \tag{3.31}$$

Furthermore we can obtain the equations of motion in terms of the second Piola-Kirchhoff stress, $\boldsymbol{\Sigma}^{\mathrm{II}}$, by substituting Eq. (2.147) into the above equation

$$\boldsymbol{\nabla} \cdot \left(\boldsymbol{\Sigma}^{\mathrm{II}} \cdot \mathbf{F}^{\mathrm{T}} \right) + \rho_0 \mathbf{b}_0 = \rho_0 \frac{d\mathbf{v}}{dt} \quad \text{or} \quad \left[\Sigma^{\mathrm{II}}_{KL} F_{iL} \right]_{,K} + \rho_0 b_{0i} = \frac{dv_i}{dt} \tag{3.32}$$

Recall that $\mathbf{F} = \mathbf{x}\boldsymbol{\nabla} = \mathbf{I} + \mathbf{u}\boldsymbol{\nabla}$, where $\mathbf{u}$ is the displacement vector. Equation (3.32) can then be written

$$\boldsymbol{\nabla} \cdot \left[\boldsymbol{\Sigma}^{\mathrm{II}} \cdot (\mathbf{I} + \boldsymbol{\nabla}\mathbf{u}) \right] + \rho_0 \mathbf{b}_0 = \rho_0 \frac{d\mathbf{v}}{dt} \quad \text{or}$$

$$[\Sigma^{\mathrm{II}}_{KL}(\delta_{iL} + u_{i,L})]_{,K} + \rho_0 b_{0i} = \rho_0 \frac{dv_i}{dt} \tag{3.33}$$

3.3 CONSERVATION OF MOMENT OF MOMENTUM

The principle of conservation of moment of momentum is also a generalization of Newton's second law of motion for continuum mechanics. It states that the rate of change of the moment of momentum for any given set of particles equals the vectorial sum of the moments acting on it. If it is assumed that there are no distributed couples and the action between the internal particles of the body is the traction force only, we can write the conservation of moment of momentum as

$$\frac{d}{dt} \int_v (\mathbf{r} \times \rho \mathbf{v})\, dv = \int_v \left(\mathbf{r} \times \mathbf{t}^{(\mathbf{n})} \right) da + \int_v (\mathbf{r} \times \mathbf{b})\, dv \tag{3.34}$$

where $\mathbf{r}$ is the position vector of the particle under consideration. Using Eq. (2.18), the equation above can be changed into the component form

$$\frac{d}{dt}\int_v \rho e_{ijk} x_j v_k \, dv = \int_a e_{ijk} x_j t_k^{(n)} \, da + \int_v e_{ijk} x_j b_k \, dv \qquad (3.35)$$

From Eq. (3.19),

$$\frac{d}{dt}\int_v \rho e_{ijk} x_j v_k \, dv = \int_v \rho e_{ijk} \frac{d}{dt}(x_j v_k) \, dv \qquad (3.36)$$

Note that $\mathbf{t}^{(\mathbf{n})} = \mathbf{n} \cdot \boldsymbol{\sigma}$ and the following integral theorem

$$\int_a (\mathbf{r} \times \mathbf{n} \cdot \mathbf{A}) \, da = \int_a \mathbf{n}\left(\mathbf{r} \times \mathbf{A}^{\mathrm{T}}\right)^{\mathrm{T}} da = \int_v \nabla_{\mathbf{x}} \cdot \left(\mathbf{r} \times \mathbf{A}^{\mathrm{T}}\right)^{\mathrm{T}} dv \qquad (3.37)$$

where $\mathbf{A}$ is a vector or tensor of any order.

The proof of Eq. (3.37) can be found in any advance continuum mechanics book (e.g., Malvern 1969). The first term on the right-hand side of Eq. (3.34) is transformed to the volume integral

$$\int_a \left(\mathbf{r} \times \mathbf{t}^{(\mathbf{n})}\right) da = \int_a (\mathbf{r} \times \mathbf{n} \cdot \boldsymbol{\sigma}) \, da = \int_v \nabla_{\mathbf{x}} \cdot \left(\mathbf{r} \times \boldsymbol{\sigma}^{\mathrm{T}}\right)^{\mathrm{T}} dv \qquad (3.38)$$

or in the component form

$$\int_a e_{ijk} x_j t_k^{(\mathbf{n})} \, da = \int_v e_{ijk} (x_j \sigma_{lk})_{,l} \, dv = \int_v e_{ijk}\left[\delta_{jl}\sigma_{lk} + x_j \sigma_{lk \cdot l}\right] dv \qquad (3.39)$$

Substituting Eqs. (3.36) and (3.39) into Eq. (3.34) gives

$$\int_v \rho e_{ijk} \frac{d}{dt}(x_j v_k) \, dv = \int_v e_{ijk}\left[\delta_{jl}\sigma_{lk} + x_j \sigma_{lk,l}\right] dv + \int_v e_{ijk} x_j \cdot b_k \, dv \qquad (3.40)$$

We can further obtain

$$\int_v \rho e_{ijk} \frac{d}{dt}(x_j v_k) \, dv + \int_v \rho e_{ijk}\left[v_j v_k + x_j \frac{dv_k}{dt}\right] dv = \int_v \rho e_{ijk} x_j \frac{dv_k}{dt} \qquad (3.41)$$

since

$$e_{ijk} v_j v_k = \tfrac{1}{2}(e_{ijk} v_j v_k + e_{ikj} v_k v_j) = \tfrac{1}{2} v_j v_k (e_{ijk} - e_{ijk}) = 0$$

Using Cauchy's equations of motion, we can derive the following from Eq. (3.40)

$$\int_v \left[e_{ijk} x_j \left(\rho \frac{dv_k}{dt} - \sigma_{lk,k} - b_k \right) + e_{ijk}\sigma_{jk} \right] dv = \int_v e_{ijk}\sigma_{jk}\, dv = 0 \quad (3.42)$$

Since the volume v is arbitrary, the following results are obtained from the consideration of the conservation of moment of momentum

$$e_{ijk}\sigma_{jk} = 0 \qquad (3.43)$$

which should hold at any point in the body. From the definition of e_{ijk}, Eq. (2.18), the preceding equation actually establishes the symmetry of the Cauchy stress tensor σ:

$$\boldsymbol{\sigma} = \boldsymbol{\sigma}^{\mathrm{T}} \quad \text{or} \quad \sigma_{ij} = \sigma_{ji} \qquad (3.44)$$

Therefore the symmetry of the Cauchy stress tensor $\boldsymbol{\sigma}$ is due to the conservation of moment of momentum provided that the body couples and the distributed couples over any boundary surface are absent. This is called the *nonpolar case*. Throughout this book we consider only the nonpolar case so the Cauchy stress σ is always symmetric.

3.4 CONSERVATION OF ENERGY: THE FIRST LAW OF THERMODYNAMICS

The conservation of energy principle is one of the most important general principles in the universe. It comes from the extensive experimental observation that energy can never be created or destroyed in the universe; it can only be transformed from one form into another. For a closed system the total rate of work done on the system by all the external agencies must equal the rate of the increase of the total energy of the system. This principle is also called the *first law of thermodynamics*.

There are various types of energies—for example, mechanical, chemical, electrical, magnetic, and thermal energy. The principle of conservation of energy holds in the most general sense for all types of energies. This book considers only mechanical and thermal energies. The principle of conservation of energy provides us a relationship between the mechanical work done, the heat transferred to the system, and the change in the internal energy of the system. The plasticity theory is based on continuum mechanics and establishes that the closed system is a continuous media which will not interchange the mass with its surroundings. The mechanical work done on a system is duc to the surface tractions $\mathbf{t}^{(\mathbf{n})}$ and the body forces $\mathbf{b}$. The heat transfer into the system is by the heat flow through its surface and the internal heat sources distributed over each mass particle.

At a fixed time t, consider a body of continuous media with volume v and surface a, as shown in Fig. 3.2. The rate of mechanical work $\dot{W}_m$ done on this system by the external forces can be expressed as

$$\dot{W}_m = \int_a \mathbf{t}^{(\mathbf{n})} \cdot \mathbf{v}\, da + \int_v \rho \mathbf{b} \cdot \mathbf{v}\, dv \tag{3.45}$$

where $\mathbf{t}^{(\mathbf{n})}$ and $\mathbf{b}$ are the traction per unit area and the body force per unit mass, respectively. In the current configuration $\mathbf{v}$ is the velocity field and ρ is the mass density. Recall that $\mathbf{t}^{(\mathbf{n})} = \mathbf{n} \cdot \boldsymbol{\sigma}$. We can denote the surface integral by the volume integral:

$$\begin{aligned} \dot{W}_m &= \int_v \nabla_{\mathbf{x}} \cdot (\boldsymbol{\sigma} \cdot \mathbf{v})\, dv + \int_v \rho \mathbf{b} \cdot \mathbf{v}\, dv \\ &= \int_v \left[\mathbf{v} \cdot (\nabla_{\mathbf{x}} \cdot \boldsymbol{\sigma} + \rho \mathbf{b}) + \operatorname{tr}(\boldsymbol{\sigma} \cdot \operatorname{grad} \mathbf{v})\right] dv \\ &= \int_v \left[\rho \mathbf{v} \cdot \frac{d\mathbf{v}}{dt} + \operatorname{tr}(\boldsymbol{\sigma} \cdot \operatorname{grad} \mathbf{v})\right] dv \\ &= \frac{d}{dt} \int_v \frac{1}{2} \rho \mathbf{v} \cdot \mathbf{v}\, dv + \int_v \operatorname{tr}(\boldsymbol{\sigma} \cdot \operatorname{grad} \mathbf{v})\, dv \end{aligned} \tag{3.46}$$

where Cauchy's equations of motion were used to derive the third expression on the right-hand side.

Note from Section 2.6 that

$$\operatorname{tr}(\boldsymbol{\sigma} \cdot \operatorname{grad} \mathbf{v}) = \operatorname{tr}(\boldsymbol{\sigma} \cdot \mathbf{L}) = \operatorname{tr}[\boldsymbol{\sigma} \cdot (\mathbf{D} + W)] = \operatorname{tr}(\boldsymbol{\sigma} \cdot \mathbf{D}) \tag{3.47}$$

since $\boldsymbol{\sigma}$ is symmetric and $\mathbf{w}$ is antisymmetric.

Substitute Eq. (3.47) into Eq. (3.46) to give

$$\dot{W}_m = \frac{d}{dt} \int_v \frac{1}{2} \rho \mathbf{v} \cdot \mathbf{v}\, dv + \int_v \operatorname{tr}(\boldsymbol{\sigma} \cdot \mathbf{D})\, dv \quad \text{or}$$

$$\dot{W}_m = \frac{d}{dt} \int_v \frac{1}{2} \rho v_i v_i\, dv + \int_v \sigma_{ji} D_{ij}\, dv \tag{3.48}$$

The second term on the right-hand side of this equation, $\operatorname{tr}(\boldsymbol{\sigma} \cdot \mathbf{D})$, is called the *stress power*. From Eq. (3.48) it can be seen that the mechanical work rate done on the system by external forces equals the sum of the rate of kinetic energy and the stress power.

Since the stress power is a very important quantity, we will discuss its expression in the reference configuration. Using Eqs. (2.145), (2.91), and (2.28), we obtain

$$\int_v \operatorname{tr}(\boldsymbol{\sigma} \cdot \mathbf{D})\, dv = \int_v \operatorname{tr}\left[\frac{1}{\det(\mathbf{F})} \mathbf{F} \cdot \boldsymbol{\Sigma}^{\mathrm{I}} \cdot \mathbf{L}\right] dv$$

$$= \int_V \operatorname{tr}\left(\boldsymbol{\Sigma}^{\mathrm{I}} \cdot \dot{\mathbf{F}}\right) dV \tag{3.49}$$

Furthermore it can be shown that

$$\operatorname{tr}(\boldsymbol{\Sigma}^{\mathrm{I}} \cdot \dot{\mathbf{F}}) = \operatorname{tr}\left[\boldsymbol{\Sigma}^{\mathrm{II}} \cdot \mathbf{F}^{\mathrm{T}} \dot{\mathbf{F}}\right] = \operatorname{tr}\left[\boldsymbol{\Sigma}^{\mathrm{II}} \cdot \mathbf{F}^{\mathrm{T}} \cdot \frac{\mathbf{L} + \mathbf{L}^{\mathrm{T}}}{2} \cdot \mathbf{F}\right]$$

$$= \operatorname{tr}\left(\boldsymbol{\Sigma}^{\mathrm{II}} \cdot \dot{\mathbf{E}}\right) = \tfrac{1}{2} \operatorname{tr}\left(\boldsymbol{\Sigma}^{\mathrm{II}} \cdot \dot{\mathbf{C}}\right) \tag{3.50}$$

where Eqs. (2.147) and (2.123) have been used to obtain the final result. From Eqs. (2.49) and (2.50), the following result is derived:

$$\det(\mathbf{F}) \operatorname{tr}(\boldsymbol{\sigma} \cdot \mathbf{D}) = \operatorname{tr}\left(\boldsymbol{\Sigma}^{\mathrm{I}} \cdot \dot{\mathbf{F}}\right) = \operatorname{tr}\left(\boldsymbol{\Sigma}^{\mathrm{II}} \cdot \dot{\mathbf{E}}\right) \tag{3.51}$$

This means that the work conjugate variables to $\boldsymbol{\sigma}$, $\boldsymbol{\Sigma}^{\mathrm{I}}$, and $\boldsymbol{\Sigma}^{\mathrm{II}}$ are, respectively, $\mathbf{D}$, $\dot{\mathbf{F}}$, and $\dot{\mathbf{E}}$.

Denote the heat transfer into the system by $\dot{W}_h$. As discussed above, $\dot{W}_h$ consists of two parts: the heat flow through the boundary surface which is governed by the conduction law of heat and the heat generated by the internal sources. Mathematically we can express $\dot{W}_h$ as

$$\dot{W}_h = -\int_a \mathbf{q} \cdot \mathbf{n}\, da + \int_v \rho r\, dv \tag{3.52}$$

where $\mathbf{q}$ is the heat flux vector per unit time, $\mathbf{n}$ is the unit normal to the boundary surface, and r is heat source per unit mass. The negative sign in the equation is introduced because $\mathbf{q}$, by convention, is the outward heat flux.

The total energy of the system U is composed of two parts: the total kinetic energy and the total internal energy. Denote the specific internal energy per unit mass by u. We can write the rate of the total energy of the system in the following form:

$$\dot{U} = \frac{d}{dt} \int_v \frac{1}{2} \rho \mathbf{v} \cdot \mathbf{v}\, dv + \frac{d}{dt} \int_v \rho u\, dv \tag{3.53}$$

Note that in the equation above the kinetic energy term is obtained by simply generalizing the result of Newtonian mechanics to the continuum mechanics.

Then, according to the first law of thermodynamics, we get

$$\dot{\mathbf{U}} = \dot{\mathbf{W}}_m + \dot{W}_h \tag{3.54}$$

With the use of Eqs. (3.48), (3.52), and (3.53), this equation takes the form

$$\frac{d}{dt}\int_v \rho u \, dv = \int_v \operatorname{tr}(\boldsymbol{\sigma} \cdot \mathbf{D}) \, dv + \int_v \rho r \, dv - \int_a \mathbf{q} \cdot \mathbf{n} \, da \tag{3.55}$$

Note that the kinetic energy terms have been eliminated. After the surface integral is transformed to volume integral by using the divergence theorem, we can rearrange the equation to give

$$\int_V \left[\rho \frac{du}{dt} - \operatorname{tr}(\boldsymbol{\sigma} \cdot \mathbf{D}) - \rho r + \nabla_{\mathbf{x}} \mathbf{q} \right] dv = 0 \tag{3.56}$$

by using Eq. (3.19). Since the volume **v** is arbitrary, we get the following differential equation expressing the conservation of energy:

$$\rho \frac{du}{dt} = \operatorname{tr}(\boldsymbol{\sigma} \cdot \mathbf{D}) + \rho r - \nabla_{\mathbf{x}} \mathbf{q} \tag{3.57}$$

or, in terms of components,

$$\rho \frac{du}{dt} = \sigma_{ij} D_{ji} + \rho r - q_{i,i} \tag{3.58}$$

Equation (3.57) or (3.58) shows that the increase of the internal energy $\rho(du/dt)$ per unit volume consists of three parts: the stress power $\operatorname{tr}(\boldsymbol{\sigma} \cdot \mathbf{D})$, which represents the mechanical work done by the external forces not converted into kinetic energy, the heat ρr supplied by the internally distributed sources, and the heat $-\nabla_{\mathbf{x}} q$ provided by the flow of thermal energy through the boundary into the system or continuous body.

3.5 CLAUSIUS-DUHEM INEQUALITY: THE SECOND LAW OF THERMODYNAMICS

The second law of thermodynamics limits the direction of the energy transformation. As noted previously, we have known for a long time that energy can be neither created nor destroyed; it can only be transformed from one form into another. Whenever this transformation happens, the energies involved in the process must obey the first law of thermodynamics. It has been experimentally observed that while some energies transform from one type to another, there are other types of transformations that are impossible.

For example, heat can flow from a warmer system to a colder system, but the reversed heat flow can never occur. The kinetic energy of a moving body can be converted into heat by friction and the body will stop due to the frictional resistance. However, the heat caused by friction is never converted into the kinetic energy of the body causing it to move. These observations cannot be described by the first law of thermodynamics, since it simply relates the quantities of energies involved in a transformation process. It is the second law of thermodynamics that governs this directional phenomenon observed in the energy transformation processes.

The second law of thermodynamics postulates that there exists a state function called *entropy*; the change in entropy (S) is given by

$$\Delta S = S_2 - S_1 = \int_1^2 \frac{dQ}{\theta} \qquad \text{for a reversible process} \tag{3.59}$$

and

$$\Delta S = S_2 - S_1 > \int_1^2 \frac{dQ}{\theta} \qquad \text{for an irreversible process} \tag{3.60}$$

where dQ is the heat input during the process, θ the temperature, and indices 1 and 2 denote the starting and ending points of the process. From Eqs. (3.59) and (3.60) it is seen that the change in entropy of the system for any process can never be negative. It is zero for a reversible process if there is no heat inflow to the system, and positive for an irreversible process.

The second law of thermodynamics can be rewritten in the notation of continuum mechanics. It has been pointed out that the system in continuum mechanics is a set of continuous particles (i.e., a body of continuous media). Using the notations defined in the last section, we obtain

$$dQ = \rho r \, dt \, dv - \mathbf{q} \cdot \mathbf{n} \, dt \, da \tag{3.61}$$

where r is the internal heat supply per unit mass and unit time, and $\mathbf{q}$ is the outward heat flux vector. We can define an entropy input rate as

$$\int_v \frac{\rho r}{\theta} \, dv - \int_a \frac{\mathbf{q} \cdot \mathbf{n}}{\theta} \, da \tag{3.62}$$

Then, according to the second law of thermodynamics,

$$\frac{d}{dt} \int_v \rho s \, dv \geq \int_v \frac{\rho r}{\theta} \, dv - \int_a \frac{\mathbf{q} \cdot \mathbf{n}}{\theta} \, da \tag{3.63}$$

This is called the *Clausius-Duhem inequality*. The greater than and equal to signs in Eq. (3.63) are for irreversible and reversible processes, respectively.

The surface integral on the right-hand side can be converted into a volume integral by using the divergence theorem to give

$$\frac{d}{dt}\int_{v} \rho s \, dv \geq \int_{v} \left[\frac{\rho r}{\theta} - \nabla_x \cdot \left(\frac{\mathbf{q}}{\theta} \right) \right] dv \tag{3.64}$$

Since the choice of the volume v is arbitrary, we derive the following differential version of the Clausius-Duhem inequality

$$\frac{ds}{dt} \geq \frac{r}{\theta} - \frac{1}{\rho} \nabla_x \cdot \left(\frac{\mathbf{q}}{\theta} \right) \tag{3.65}$$

or

$$\frac{ds}{dt} - \frac{r}{\theta} + \frac{1}{\rho\theta} \nabla_x \cdot \mathbf{q} - \frac{\mathbf{q}}{\rho\theta^2} \cdot \operatorname{grad} \theta \geq 0 \tag{3.66}$$

In terms of components, we have

$$\frac{ds}{dt} - \frac{r}{\theta} + \frac{1}{\rho\theta} q_{i,i} - \frac{1}{\rho\theta^2} q_i \theta_{,i} \geq 0 \tag{3.67}$$

or, elimination of r by using Eq. (3.58) gives

$$\rho\theta\dot{s} - \rho\dot{u} + \sigma_{ij} D_{ji} - \frac{1}{\theta} q_i \theta_{,i} \geq 0 \tag{3.68}$$

Again, the inequality implies an irreversible process and the equality represents a reversible process. By distinguishing the irreversible and reversible processes, the directional characteristics of the energy transformation can be clearly described by the second law of thermodynamics.

In thermodynamics the internal energy u, entropy s, heat flux $\mathbf{q}$, and the Cauchy stress $\boldsymbol{\sigma}$ are all considered state functions that can be determined by the state variables. The formulas that relate the state functions to the state variables are called *state equations* or *constitutive equations*. From a purely mechanical consideration, the only constitutive equation is that for the Cauchy stress σ or Piola-Kirchhoff stress $\boldsymbol{\Sigma}$.

For thermoelasticity one can expect that the state variables would be the deformation gradient $\mathbf{F}$ and the temperature θ since here u, s, $\mathbf{q}$, and $\boldsymbol{\sigma}$ are determined completely by their current values. Thus, for an ideal thermoelastic behavior,

$$\begin{aligned} u &= u(\mathbf{F}, \theta), \quad s = s(\mathbf{F}, \theta) \\ \mathbf{q} &= \mathbf{q}(\mathbf{F}, \theta), \quad \boldsymbol{\sigma} = \boldsymbol{\sigma}(\mathbf{F}, \theta) \end{aligned} \tag{3.69}$$

The situation becomes much more complex if the deformation is inelastic. For example, the stress of a plastically deformed solid cannot be determined by the current value of deformation **F** only. The history of the deformation is also necessary. Simple state equations or constitutive equations, such as Eq. (3.69), cannot describe correctly the plastic deformation of solids. If u, x, $\mathbf{q}$, and $\boldsymbol{\sigma}$ are assumed to be state functions, we face two fundamental problems. First, we have to find or specify the set of all state variables. For example, it is necessary to know for a plastically deformed solid what other variables are needed to uniquely describe the current state. As pointed out previously, knowing only **F** and θ will be insufficient to characterize a state in the case of plastic deformation of solids, although they are adequate for thermoelastic materials. It is very hard to enumerate all the relevant state variables from macroscopic observations alone; microscopic deformation considerations and some assumptions have to be made to use certain macroscopic observable variable(s) as the representative(s) of the microscopic phenomena. Second the mathematical forms of the constitutive equations should be determined after the state variables are chosen. This involves the experimental evaluation and mathematical formalization. At times severe experimental or mathematical difficulties may arise during this process. In recent years great efforts have been made to study the constitutive equation for various materials under different loading conditions, as described in this book.

To solve the problem of specifying state variables, two different methods are usually adopted. The first method ignores the problem of state variables and assumes that u, s, $\mathbf{q}$, and $\boldsymbol{\sigma}$ are determined by the histories of **F** and θ, and not by their current values only (Coleman 1964). Therefore these quantities should be expressed as the functional, not functions, of **F** and θ. The second method introduces the concept of internal (or hidden) variables. In this method it is postulated that the current state of an inelastically deformed solid can be determined by the current values of **F**, θ as well as a set of internal variables. The history of the deformation is indirectly included in the evolution of these internal variables. The material response will be different if the values of the internal variables are different even though **F**, θ are the same. We state this mathematically as

$$\begin{aligned} u &= u(\mathbf{F}, \theta, \alpha_i) \\ s &= s(\mathbf{F}, \theta, \alpha_i) \\ \mathbf{q} &= \mathbf{q}(\mathbf{F}, \theta, \alpha_i) \\ \boldsymbol{\sigma} &= \boldsymbol{\sigma}(\mathbf{F}, \theta, \alpha_i) \end{aligned} \tag{3.70}$$

where α_i, $i = 1, 2, \ldots, n$, are a set of n internal variables including mechanical, or thermal, or even electrical state variables. These variables can be scalars, vectors, or tensors, although they are all denoted by scalar symbols here. The specific physical meaning for each internal variable and the actual number n need to be chosen and identified for different materials and different conditions. Different choices result in different models, which will be discussed in the subsequent chapters. It should be pointed out here that, instead of using θ as the state variable and s as the state function, it is quite common in the literature to find s used as the state variable and θ as the state function. For example,

$$u = u(\mathbf{F}, s, \alpha_i) \tag{3.71}$$

Similarly one can write θ, $\mathbf{q}$, and $\boldsymbol{\alpha}$ as functions of $\mathbf{F}$, s, and α_i.

In closing this section, we introduce another thermodynamic potential ψ called Helmholtz free energy:

$$\psi = u - s\theta \tag{3.72}$$

Note that $\dot{u} = \dot{\psi} + \dot{s}\theta + \dot{s}\theta$; thus Eq. (3.68) is rewritten as

$$-\rho\dot{\psi} - \rho\eta\dot{\theta} + \sigma_{ij}D_{ji} + \frac{1}{\theta}q_i\theta_{,i} \geq 0 \tag{3.73}$$

3.6 PRINCIPLE OF MATERIAL OBJECTIVITY OR FRAME INDIFFERENCE

Truesdell and Noll (1965) postulated three fundamental principles for a purely mechanical constitutive equation. These are (1) principle of determination for stress, (2) principle of local action, and (3) principle of material frame indifference. The first principle simply states that the stress in a body is determined by the history of the motion of that body. The principle of local action is that in determining the stress at a given particle, the motion outside an arbitrarily small neighborhood of it may be disregarded. These two principles are self-explanatory. We are more interested in the third principle, since it introduces the concept of objectivity that is significant when we consider the finite plasticity theory.

The detailed and rigorous treatment of the frame indifference principle can be found in any advance continuum mechanics book. Here we only present this principle and discuss various quantities under the change of frame of reference.

According to the principle of material frame indifference constitutive equations must be invariant under changes of reference frame. The change of a time-space reference is given by

$$x^* = \mathbf{c}(t) + \mathbf{Q}(t)\mathbf{x} \tag{3.74}$$

$$t^* = t - a \tag{3.75}$$

where $\mathbf{c}(t)$ is a vector and $\mathbf{Q}(t)$ is an orthogonal tensor representing the translation and rotation of the spatial coordinate system, respectively, and a is a constant denoting the shift of time. This principle can then be stated mathematically as follows (Malvern 1969): If a constitutive equation is satisfied by the motion and stress

$$\mathbf{x} = \mathbf{x}(\mathbf{X}, t)$$
$$\boldsymbol{\sigma} = \boldsymbol{\sigma}(\mathbf{x}, t) \tag{3.76}$$

then it must be also satisfied by the following motion and stress

$$\mathbf{x}^* = \mathbf{c}(t) + \mathbf{Q}(t)\mathbf{x}$$
$$\boldsymbol{\sigma}^* = \mathbf{Q}(t) \cdot \boldsymbol{\sigma} \cdot \mathbf{Q}^{\mathrm{T}}(t) \tag{3.77}$$

It should be noted that the coordinates involved in this principle are the spatial coordinates x. The principle of material frame indifference is sometimes called the *principle of material objectivity*.

Under the change of reference (3.74) and (3.75), a vector and tensor are called *frame indifference* or *objective* if they transform according to the following:

$$\mathbf{v}^* = \mathbf{Q}(t) \cdot \mathbf{v} \qquad \text{(vector } \mathbf{v}) \tag{3.78}$$

$$\mathbf{A}^* = \mathbf{Q} \cdot \mathbf{A} \cdot \mathbf{Q}^{\mathrm{T}} \qquad \text{(tensor } \mathbf{A}) \tag{3.79}$$

Here a symbol with a star represents the quantity in the new reference frame; without a star the quantity is in the old reference frame. It is obvious that scalar is frame indifference, so any scalar a remains unchanged under the change of reference:

$$a^* = a \tag{3.80}$$

Following is a discussion of the transformation laws of various quantities introduced in Chapter 2 by Eq. (2.74) and by Eq. (3.75) above.

Consider the transformation law of the deformation gradient $\mathbf{F}$. Since the reference configuration can be fictitious, we can assume, without loss of generality, that the two frames $\mathbf{x}$ and $\mathbf{x}^*$ coincide at time t_0 in that $d\mathbf{X}^* = d\mathbf{X}$ at t_0. Then it follows from the definition that

$$\mathbf{dx}^* = \mathbf{F}^* \cdot d\mathbf{X}^* = \mathbf{F}^* \cdot d\mathbf{X}$$
$$\mathbf{dx} = \mathbf{F} \cdot \mathbf{dX} \tag{3.81}$$

On the other hand, from Eq. (3.74), we have

$$\mathbf{dx}^* = \mathbf{Q}(t)\,\mathbf{dx} = \mathbf{Q}(t) \cdot \mathbf{F} \cdot \mathbf{dX} \tag{3.82}$$

Substituting Eq. (3.82) into the first expression of Eq. (3.81) results in

$$\mathbf{F}^* \cdot \mathbf{dX} = \mathbf{Q} \cdot \mathbf{F} \cdot \mathbf{dX} \tag{3.83}$$

From Eq. (3.83) we obtain the transformation law for $\mathbf{F}$:

$$\mathbf{F}^* = \mathbf{Q} \cdot \mathbf{F} \tag{3.84}$$

It should be pointed out that $\mathbf{F}$ is objective even though Eq. (3.84) is different from Eq. (3.79). The reason is that, as mentioned in Section 2.2 $\mathbf{F}$ is a two-point tensor in that only one of its two indices is in the spatial coordinate $\mathbf{x}$. This two-point tensor transforms like a vector under change of frame if it is objective.

As explained in Section 2.4,

$$\begin{aligned} \mathbf{F} &= \mathbf{R} \cdot \mathbf{U} = \mathbf{V} \cdot \mathbf{R} \\ \mathbf{F}^* &= \mathbf{R}^* \cdot \mathbf{U}^* = \mathbf{V}^* \cdot \mathbf{R}^* \end{aligned} \tag{3.85}$$

Using Eq. (3.84), we write

$$\mathbf{Q} \cdot \mathbf{R} \cdot \mathbf{U} = \mathbf{R}^* \cdot \mathbf{U}^* \tag{3.86}$$

$$\mathbf{Q} \cdot \mathbf{V} \cdot \mathbf{R} = \mathbf{V}^* \cdot \mathbf{R}^* \tag{3.87}$$

Since $\mathbf{Q} \cdot \mathbf{R}$ is still an orthogonal tensor and the polar decomposition is unique, Eq. (3.86) leads to

$$\begin{aligned} \mathbf{R}^* &= \mathbf{Q} \cdot \mathbf{R} \\ \mathbf{U}^* &= \mathbf{U} \end{aligned} \tag{3.88}$$

From Eq. (3.87) we derived

$$\mathbf{V}^* = \mathbf{Q} \cdot \mathbf{V} \cdot \mathbf{R} \cdot \mathbf{R}^{*\mathrm{T}} = \mathbf{Q} \cdot \mathbf{V} \cdot \mathbf{Q}^{\mathrm{T}} \tag{3.89}$$

By definition and using Eqs. (3.88) and (3.89), we can obtain the transformation law for $\mathbf{C}$ and $\mathbf{B}$:

$$\mathbf{C}^* = \mathbf{U}^{*2} = \mathbf{C} \tag{3.90}$$

$$\mathbf{B}^* = \mathbf{V}^{*2} = \mathbf{Q} \cdot \mathbf{B} \cdot \mathbf{Q}^{\mathrm{T}} \tag{3.91}$$

It follows immediately that

$$\begin{aligned} \mathbf{E}^* &= \mathbf{E} \\ \mathbf{e}^* &= \mathbf{Q} \cdot \mathbf{e} \cdot \mathbf{Q}^{\mathrm{T}} \end{aligned} \tag{3.92}$$

Thus $\mathbf{U}$, $\mathbf{C}$, and $\mathbf{E}$ are not objective tensors, while $\mathbf{V}$, $\mathbf{B}$, and $\mathbf{e}$ are objective quantities.

Next we consider the quantities related to the velocity field. From Eq. (3.84), we calculate

$$\dot{\mathbf{F}}^* = \dot{\mathbf{Q}} \cdot \mathbf{F} + \mathbf{Q} \cdot \dot{\mathbf{F}} \tag{3.93}$$

Then we have

$$\begin{aligned}\mathbf{L}^* = \dot{\mathbf{F}}^* \mathbf{F}^{*-1} &= \left(\dot{\mathbf{Q}} \cdot \mathbf{F} + \mathbf{Q} \cdot \dot{\mathbf{F}}\right) \cdot \mathbf{F}^{-1} \cdot \mathbf{Q}^{\mathrm{T}} \\ &= \dot{\mathbf{Q}} \cdot \mathbf{Q}^{\mathrm{T}} + \mathbf{Q} \cdot \dot{\mathbf{F}} \cdot \mathbf{F}^{-1} \cdot \mathbf{Q}^{\mathrm{T}} \\ &= \dot{\mathbf{Q}} \cdot \mathbf{Q}^{\mathrm{T}} + \mathbf{Q} \cdot \mathbf{L} \cdot \mathbf{Q}^{\mathrm{T}} \end{aligned} \tag{3.94}$$

It can be seen from the above that the velocity gradient $\mathbf{L}$ is not objective. Since $\mathbf{Q} \cdot \mathbf{Q}^{\mathrm{T}} = \mathbf{I}$, it follows that

$$\dot{\mathbf{Q}} \cdot \mathbf{Q}^{\mathrm{T}} + \mathbf{Q} \cdot \dot{\mathbf{Q}}^{\mathrm{T}} = \mathbf{0}$$

which gives

$$\dot{\mathbf{Q}} \cdot \mathbf{Q}^{\mathrm{T}} = -\left(\dot{\mathbf{Q}} \cdot \mathbf{Q}^{\mathrm{T}}\right)^{\mathrm{T}} \tag{3.95}$$

We conclude that $\dot{\mathbf{Q}} \cdot \mathbf{Q}^{\mathrm{T}}$ is an antisymmetric tensor, and thus we can decompose $\mathbf{L}^*$ into symmetric part $\mathbf{D}^*$ and antisymmetric part $\mathbf{W}^*$ in the form

$$\mathbf{L}^* = \mathbf{Q} \cdot \frac{\mathbf{L} + \mathbf{L}^{\mathrm{T}}}{2} \cdot \mathbf{Q}^{\mathrm{T}} + \mathbf{Q} \cdot \frac{\mathbf{L} - \mathbf{L}^{\mathrm{T}}}{2} \cdot \mathbf{Q}^{\mathrm{T}} + \dot{\mathbf{Q}} \cdot \mathbf{Q}^{\mathrm{T}} \tag{3.96}$$

This leads to the following transformation laws:

$$\begin{aligned}\mathbf{D}^* &= \mathbf{Q} \cdot \mathbf{D} \cdot \mathbf{Q}^{\mathrm{T}} \\ \mathbf{W}^* &= \mathbf{Q} \cdot \mathbf{W} \cdot \mathbf{Q}^{\mathrm{T}} + \dot{\mathbf{Q}} \cdot \mathbf{Q}^{\mathrm{T}} \end{aligned} \tag{3.97}$$

The deformation rate $\mathbf{D}$ is objective, but the spin $\mathbf{W}$ is not objective.

Finally, we discuss the transformation laws for stress measures and their material derivatives. In continuum mechanics, the forces (internal and external) are assumed to be objective (Malvern 1969). They and the unit normal vector $\mathbf{n}$ transform according to Eq. (3.78). Since $\mathbf{t}^{(\mathbf{n})} = \mathbf{n} \cdot \boldsymbol{\sigma}$ and $\mathbf{t}^{(\mathbf{n})*} = \mathbf{n}^* \cdot \boldsymbol{\sigma}^*$, it is easy to show that

$$\boldsymbol{\sigma}^* = \mathbf{Q} \cdot \boldsymbol{\sigma} \cdot \mathbf{Q}^{\mathrm{T}} \tag{3.98}$$

That is, the Cauchy stress $\boldsymbol{\sigma}$ is objective. From Eq. (2.143), the first Piola-

Kirchhoff stress in x^* is

$$\boldsymbol{\Sigma}^{\mathrm{I}*} = \det(\mathbf{F}^*)\mathbf{F}^{*-1} \cdot \boldsymbol{\sigma}^* = \det(\mathbf{F})\mathbf{F}^{-1} \cdot \boldsymbol{\sigma} \cdot \mathbf{Q}^{\mathrm{T}} = \boldsymbol{\Sigma}^{\mathrm{I}} \cdot \mathbf{Q}^{\mathrm{T}} \tag{3.99}$$

where $\det(\mathbf{F}^*) = \det(\mathbf{Q} \cdot \mathbf{F}) = \det(\mathbf{Q}) \cdot \det(\mathbf{F})$ has been used.

Similarly the following transformation law is for the second Piola-Kirchhoff stress $\boldsymbol{\Sigma}^{\mathrm{II}}$:

$$\boldsymbol{\Sigma}^{\mathrm{II}*} = \det(\mathbf{F}^*)\mathbf{F}^{*-1} \cdot \boldsymbol{\sigma}^* \cdot \mathbf{F}^* = \det(\mathbf{F})\mathbf{F}^{-1} \cdot \boldsymbol{\sigma} \cdot \mathbf{F} = \boldsymbol{\Sigma}^{\mathrm{II}} \tag{3.100}$$

Therefore $\boldsymbol{\Sigma}^{\mathrm{I}}$ and $\boldsymbol{\Sigma}^{\mathrm{II}}$ are not frame indifferent.

In plasticity theory the constitutive equations are usually written in an incremental form, since the stress is dependent on history of the plastic deformation. Thus the material derivative of the stress tensor, rather than stress tensor itself, is used in formulating the constitutive equations. To obtain the transformation law of the Cauchy stress $\boldsymbol{\sigma}$ under change of frame, we take the material derivative on both sides of Eq. (3.98):

$$\dot{\boldsymbol{\sigma}}^* = \dot{\mathbf{Q}} \cdot \boldsymbol{\sigma} \cdot \mathbf{Q}^{\mathrm{T}} + \mathbf{Q} \cdot \dot{\boldsymbol{\sigma}} \cdot \mathbf{Q}^{\mathrm{T}} + \mathbf{Q} \cdot \boldsymbol{\sigma} \cdot \dot{\mathbf{Q}}^{\mathrm{T}} \tag{3.101}$$

Unfortunately, as can be seen from this equation, the material rate of $\boldsymbol{\sigma}$ is not objective because of the first and last terms in Eq. (3.101) even though σ itself is objective. Therefore $\dot{\sigma}^*$ is inappropriate to use in formulating constitutive equations in plasticity theory. To solve this difficulty, Jaumann introduced a corotational stress rate, which can be derived as follows: From the second expression of Eqs. (3.97) take

$$\begin{aligned} \dot{\mathbf{Q}} &= \mathbf{W}^* \cdot \mathbf{Q} - \mathbf{Q} \cdot \mathbf{W} \\ \dot{\mathbf{Q}}^{\mathrm{T}} &= -\mathbf{Q}^{\mathrm{T}} \cdot \mathbf{W}^* + \mathbf{W} \cdot \mathbf{Q}^{\mathrm{T}} \end{aligned} \tag{3.102}$$

Substitute this $\dot{\mathbf{Q}}$ into Eq. (3.101) to get

$$\begin{aligned} \dot{\boldsymbol{\sigma}}^* &= \mathbf{Q} \cdot \boldsymbol{\sigma} \cdot \mathbf{Q}^{\mathrm{T}} + \mathbf{W}^* \cdot \mathbf{Q} \cdot \sigma \cdot \mathbf{Q}^{\mathrm{T}} - \mathbf{Q} \cdot \mathbf{W} \cdot \boldsymbol{\sigma} \cdot \mathbf{Q}^{\mathrm{T}} - \mathbf{Q} \cdot \boldsymbol{\sigma} \cdot \mathbf{Q}^{\mathrm{T}} \cdot W^* \\ &\quad + \mathbf{Q} \cdot \boldsymbol{\sigma} \cdot \mathbf{W} \cdot \mathbf{Q}^{\mathrm{T}} \cdot \mathbf{W} \cdot \mathbf{Q}^{\mathrm{T}} \\ &= \mathbf{Q} \cdot (\dot{\boldsymbol{\sigma}} - \mathbf{W} \cdot \boldsymbol{\sigma} + \boldsymbol{\sigma} \cdot \mathbf{W}) \cdot \mathbf{Q}^{\mathrm{T}} + \mathbf{W}^* \cdot \sigma^* - \boldsymbol{\sigma}^* \cdot \mathbf{W}^* \end{aligned} \tag{3.103}$$

Moving the last two terms on the right-hand side of this equation to the left results in

$$\dot{\boldsymbol{\sigma}}^* - \mathbf{W}^* \cdot \boldsymbol{\sigma}^* + \boldsymbol{\sigma}^* \cdot \mathbf{W}^* = \mathbf{Q} \cdot (\dot{\boldsymbol{\sigma}} - \mathbf{W} \cdot \boldsymbol{\sigma} + \boldsymbol{\sigma} \cdot \mathbf{W}) \cdot \mathbf{Q}^{\mathrm{T}} \tag{3.104}$$

A comparison of both sides of Eq. (3.104) reveals that the quantity $\dot{\boldsymbol{\sigma}} - \mathbf{W} \cdot \boldsymbol{\sigma} + \boldsymbol{\sigma} \cdot \mathbf{W}$ is objective under change of frame of reference. Denote it by $\overset{\circ}{\boldsymbol{\sigma}}$,

$$\overset{\circ}{\boldsymbol{\sigma}} = \dot{\boldsymbol{\sigma}} - \mathbf{W} \cdot \boldsymbol{\sigma} + \boldsymbol{\sigma} \cdot \mathbf{W} \tag{3.105}$$

This corotational stress rate is called the *Jaumann rate*. It is corotational in that it represents a rate relative to a rotating frame of reference with the rate of rotation given by the spin tensor **W**.

It should be pointed out that $\overset{\circ}{\boldsymbol{\sigma}}$ is not the only stress rate that is objective. For example, the following two stress rates are also frame indifferent:

$$\overset{\Delta}{\boldsymbol{\sigma}} = \dot{\boldsymbol{\sigma}} + \mathbf{L}^{\mathrm{T}} \cdot \boldsymbol{\sigma} + \boldsymbol{\sigma} \cdot \mathbf{L} \tag{3.106}$$

$$\overset{\nabla}{\boldsymbol{\sigma}} = \dot{\boldsymbol{\sigma}} + \left(\dot{\mathbf{R}} \cdot \mathbf{R}^{\mathrm{T}}\right) \cdot \boldsymbol{\sigma} - \boldsymbol{\sigma} \cdot \left(\dot{\mathbf{R}} \cdot \mathbf{R}^{\mathrm{T}}\right)^{\mathrm{T}} \tag{3.107}$$

It is easy to verify that $\overset{\Delta}{\boldsymbol{\sigma}}$ and $\overset{\nabla}{\boldsymbol{\sigma}}$ satisfy the transformation law (3.79) under change of frame of reference given by Eqs. (3.88), (3.94), and (3.98). This task is left to readers as an exercise.

Similarly we obtain the following results:

$$\dot{\boldsymbol{\Sigma}}^{\mathrm{I}*} + \boldsymbol{\Sigma}^{\mathrm{I}*} \cdot \mathbf{W}^* = \left(\dot{\boldsymbol{\Sigma}}^{\mathrm{I}} + \boldsymbol{\Sigma}^{\mathrm{I}} \cdot \mathbf{W}\right) \cdot \mathbf{Q}^{\mathrm{T}} \tag{3.108}$$

$$\dot{\boldsymbol{\Sigma}}^{\mathrm{II}*} = \dot{\boldsymbol{\Sigma}}^{\mathrm{II}} \tag{3.109}$$

$$\overset{\nabla}{\boldsymbol{\sigma}} = \mathbf{R} \cdot \dot{\boldsymbol{\sigma}}_{\mathrm{u}} \cdot \mathbf{R}^{\mathrm{T}} \tag{3.110}$$

$$\overset{\square}{\boldsymbol{\sigma}} = \dot{\boldsymbol{\sigma}} - \mathbf{L} \cdot \boldsymbol{\sigma} - \boldsymbol{\sigma} \cdot \mathbf{L}^{\mathrm{T}} + \boldsymbol{\sigma}\, \mathrm{tr}(\mathbf{D}) = \frac{1}{\det(\mathbf{F})} \mathbf{F} \cdot \dot{\boldsymbol{\Sigma}}^{\mathrm{II}} \cdot \mathbf{F}^{\mathrm{T}} \tag{3.111}$$

where $\boldsymbol{\sigma}_{\mathrm{u}}$ is the unrotated Cauchy stress defined by Eq. (2.148) and $\overset{\square}{\boldsymbol{\sigma}}$ is the objective stress rate introduced by Truesdell for problems of nonlinear elasticity. Equations (3.110) and (3.111) can be obtained by differentiating Eqs. (2.148) and (2.146), respectively, and rearranging the resulting expressions. $\overset{\nabla}{\boldsymbol{\sigma}}$ is sometimes called the *Green-McInnis rate*.

REFERENCES

Coleman, B. D. 1964. Thermodynamics of materials with memory. *Arch. Rat. Mech. Anal.* **17**: 1.

Malvern, L. E. 1969. *Introduction to the Mechanics of a Continuous Medium*. Prentice Hall.

Truesdell, C., and W. Noll 1965. The nonlinear field theories of mechanics. *Encyclopedia of Physics*, vol. 3, ed. S. Flügge. Berlin: Springer-Verlag.

4

YIELD CRITERIA

As we explained in Chapter 1, some materials like mild steel exhibit a fairly sharp yield point, while others may not have pronounced yield stress (see Fig. 1.1). For materials that have a pronounced yield point, it is natural to make a distinction between the elastic region and the plastic region of deformation and to define the yield stress σ_Y for a one-dimensional case as the boundary point that separates the two regions. But for materials that do not have the pronounced yield point, it is hard to conceptualize a threshold below which there is only pure elastic deformation. In fact, as shown in Fig. 1.1(*b*), the elastic-plastic transition is not sharp, and there may be some plastic deformation right from the beginning of the application of the load. The stress-strain curve, in uniaxial tension or compression, may be slightly nonlinear starting from the origin. We can classify all constitutive models for plastic deformation to two categories. The first assumes the existence of yield point, and the second does not. In this book we will mostly deal with the first category, though in Chapter 6 (Section 6.5) a constitutive model is presented that does not use the concept of a yield point.

Suppose that there exists a yield stress σ_Y, such as noted in Chapter 1.1. In general, for the three-dimensional cases, if we use the six independent stress components as independent coordinate axes and define a stress state as a point in this six-dimensional space, all the stress states that cause yielding can be imagined to constitute a continuous surface, called *yield surface*, that divides the stress space into the elastic and the plastic domains. The yield

In this and the next two chapters we consider only the infinitesimal displacement gradient; therefore we do not distinguish between the initial reference and the current configurations, and ignore the different between different stress tensors.

surface is the boundary of these two domains. Because the stress-strain relationship is different for elastic and plastic deformation, we need to determine the shape and position of the yield surface in the stress space in order to know if the stress state is in the elastic or plastic deformation domain. In this chapter we present the experimental results and the mathematical descriptions of yield surfaces. The general features of a yield surface, especially for isotropic materials, are also discussed.

The yield surface or yield function can be represented mathematically in the stress space by a surface separating the elastic and plastic domains. To get a clear picture of this surface—for example, its shape, size, and orientation—we need to know the geometrical representation of any stress state in the stress space as well as some other characteristics. In Section 4.1 the geometrical representation of a stress state is described, then in Section 4.2 we will resume the discussion of the yield function.

4.1 STRESS STATE AND STRESS SPACE

The stress state at any material point may be characterized by the Cauchy stress tensor $\boldsymbol{\sigma}$. Since the deformation is infinitesimal, the difference between the Cauchy and the Piola-Kirchoff stress tensors is negligible. We will simply call it *stress tensor* $\boldsymbol{\sigma}$ in Chapters 4 through 6. Its components σ_{ij} in the cartesian coordinate system (x, y, z) can be represented by a matrix of the second order

$$[\sigma_{ij}] = \begin{bmatrix} \sigma_{xx} & \sigma_{xy} & \sigma_{xz} \\ \sigma_{yx} & \sigma_{yy} & \sigma_{yz} \\ \sigma_{zx} & \sigma_{zy} & \sigma_{zz} \end{bmatrix} \tag{4.1}$$

Since the stress tensor is symmetric, only six stress components σ_{xx}, σ_{yy}, σ_{zz}, σ_{xy}, σ_{yz}, and σ_{xz} are independent. Thus six independent stress components determine uniquely a stress state, and vice versa. From the matrix $[\sigma_{ij}]$ given by Eq. (4.1), we can calculate the three principal stresses using the equation

$$|\sigma_{ij} - \sigma\delta_{ij}| = 0$$

or

$$\begin{vmatrix} \sigma_{xx} - \sigma & \sigma_{xy} & \sigma_{xz} \\ \sigma_{xy} & \sigma_{yy} - \sigma & \sigma_{yz} \\ \sigma_{xz} & \sigma_{yz} & \sigma_{zz} - \sigma \end{vmatrix} = 0 \tag{4.2}$$

Equation (4.2) can be expanded to

$$\sigma^3 - J_1\sigma^2 + J_2\sigma - J_3 = 0 \tag{4.3}$$

where

$$J_1 = \mathrm{tr}(\boldsymbol{\sigma}) = \sigma_{ii} = \sigma_{xx} + \sigma_{yy} + \sigma_{zz} \tag{4.4}$$

$$J_2 = \tfrac{1}{2}(\sigma_{ii}\sigma_{kk} - \sigma_{ij}\sigma_{ij}) = \sigma_{xx}\sigma_{yy} + \sigma_{yy}\sigma_{zz} + \sigma_{xx}\sigma_{zz} - \sigma_{xy}^2 - \sigma_{xz}^2 - \sigma_{yz}^2 \tag{4.5}$$

$$J_3 = (\tfrac{1}{6})e_{ijk}e_{mnl}\sigma_{im}\sigma_{jn}\sigma_{kl} = \det[\sigma_{ij}] = \begin{vmatrix} \sigma_{xx} & \sigma_{xy} & \sigma_{xz} \\ \sigma_{yx} & \sigma_{yy} & \sigma_{yz} \\ \sigma_{zx} & \sigma_{zy} & \sigma_{zz} \end{vmatrix} \tag{4.6}$$

are invariants of the stress tensor. In terms of the principal stresses σ_1, σ_2, and σ_3 they can also be written

$$J_1 = \sigma_1 + \sigma_2 + \sigma_3 \tag{4.7}$$

$$J_2 = \sigma_1\sigma_2 + \sigma_2\sigma_3 + \sigma_3\sigma_1 \tag{4.8}$$

$$J_3 = \sigma_1\sigma_2\sigma_3$$

For each principal stress σ_i we can obtain the associated principal direction $\mathbf{n}^{(i)}$ $(i = 1, 2, 3)$ by

$$\boldsymbol{\sigma} \cdot \mathbf{n}^{(i)} = \sigma_i \mathbf{n}^{(i)} \tag{4.9}$$

or

$$\begin{bmatrix} \sigma_{xx} & \sigma_{xy} & \sigma_{xz} \\ \sigma_{yx} & \sigma_{yy} & \sigma_{yz} \\ \sigma_{xz} & \sigma_{yz} & \sigma_{zz} \end{bmatrix} \begin{pmatrix} n_1^{(i)} \\ n_2^{(i)} \\ n_3^{(i)} \end{pmatrix} = \sigma_i \begin{pmatrix} n_1^{(i)} \\ n_2^{(i)} \\ n_3^{(i)} \end{pmatrix} \qquad (i = 1, 2, 3) \tag{4.10}$$

where $\mathbf{n}^{(i)}$ is the unit vector in the ith principal direction and the following relation holds for a pair of $\mathbf{n}^{(i)}$ $(i = 1, 2, 3)$:

$$\mathbf{n}^{(i)} \cdot \mathbf{n}^{(j)} = \begin{cases} 1 & \text{for } i = j \\ 0 & \text{for } i \neq j \end{cases} \tag{4.11}$$

Equation (4.10) can also be written

$$\begin{bmatrix} \sigma_{xx} & \sigma_{xy} & \sigma_{xz} \\ \sigma_{yx} & \sigma_{yy} & \sigma_{yz} \\ \sigma_{xz} & \sigma_{yz} & \sigma_{zz} \end{bmatrix} \begin{bmatrix} n_1^{(1)} & n_1^{(2)} & n_1^{(3)} \\ n_2^{(1)} & n_2^{(2)} & n_2^{(3)} \\ n_3^{(1)} & n_3^{(2)} & n_3^{(3)} \end{bmatrix} = \begin{bmatrix} n_1^{(1)} & n_1^{(2)} & n_1^{(3)} \\ n_2^{(1)} & n_2^{(2)} & n_2^{(3)} \\ n_3^{(1)} & n_3^{(2)} & n_3^{(3)} \end{bmatrix} \begin{bmatrix} \sigma_1 & 0 & 0 \\ 0 & \sigma_2 & 0 \\ 0 & 0 & \sigma_3 \end{bmatrix} \tag{4.12}$$

or, if we introduce

$$[R_{ij}] = [n_j^{(i)}] = \begin{bmatrix} n_1^{(1)} & n_1^{(2)} & n_1^{(3)} \\ n_2^{(1)} & n_2^{(2)} & n_2^{(3)} \\ n_3^{(1)} & n_3^{(2)} & n_3^{(3)} \end{bmatrix} \tag{4.13}$$

With the help of Eq. (4.11) it is easy to prove that

$$[R_{ij}]^{-1} = [R_{ij}]^{\mathrm{T}} \tag{4.14}$$

Thus we can express $[\sigma_{ij}]$ in terms of the principal stresses using Eqs. (4.12) and (4.14)

$$\begin{bmatrix} \sigma_{xx} & \sigma_{xy} & \sigma_{xz} \\ \sigma_{yx} & \sigma_{yy} & \sigma_{yz} \\ \sigma_{xz} & \sigma_{yz} & \sigma_{zz} \end{bmatrix} = [R_{ij}] \begin{bmatrix} \sigma_1 & 0 & 0 \\ 0 & \sigma_2 & 0 \\ 0 & 0 & \sigma_3 \end{bmatrix} [R_{ij}]^{\mathrm{T}} \tag{4.15}$$

According to this equation, if the principal stresses and their directions are known, the stress state can be determined.

In the plasticity theory it is customary to decompose the stress tensor into two parts:

$$\sigma_{ij} = p\delta_{ij} + S_{ij} \tag{4.16}$$

where p is the hydrostatic stress or pressure given by

$$\begin{aligned} p &= \tfrac{1}{3}\sigma_{ii} = \tfrac{1}{3}(\sigma_{xx} + \sigma_{yy} + \sigma_{zz}) \\ &= \tfrac{1}{3}(\sigma_1 + \sigma_2 + \sigma_3) \end{aligned} \tag{4.17}$$

and $p\delta_{ij}$ is called the *spherical* or *hydrostatic stress tensor*. The second part S_{ij} is calculated from

$$S_{ij} = \sigma_{ij} - p\delta_{ij} \tag{4.18}$$

and is called the deviatoric stress tensor. It is easy to see from Eq. (4.18) that

$$J_1' = \mathrm{tr}(\mathrm{S}) = S_{ii} = S_1 + S_2 + S_3 = 0 \tag{4.19}$$

$$\begin{aligned} J_2' &= \tfrac{1}{2}S_{ij}S_{ij} = \tfrac{1}{6}\left[(\sigma_{xx} - \sigma_{yy})^2 + (\sigma_{yy} - \sigma_{zz})^2 + (\sigma_{zz} - \sigma_{xx})^2\right] \\ &\quad + \sigma_{xy}^2 + \sigma_{yz}^2 + \sigma_{xz}^2 \\ &= \tfrac{1}{6}\left[(\sigma_1 - \sigma_2)^2 + (\sigma_2 - \sigma_3)^2 + (\sigma_3 - \sigma_1)^2\right] = \tfrac{1}{2}(S_1^2 + S_2^2 + S_3^2) \end{aligned} \tag{4.20}$$

$$J_3' = \det(S_{ij}) = (\tfrac{1}{6})e_{ijk}e_{mnl}S_{im}S_{jn}S_{kl} = S_1S_2S_3 \tag{4.21}$$

where S_i $(i = 1, 2, 3)$ are the principal values of the deviatoric stress tensor **S**. Noting Eqs. (4.9) and (4.18), we can derive

$$\mathbf{S} \cdot \mathbf{n}^{(i)} = (\boldsymbol{\sigma} - p\mathbf{I}) \cdot \mathbf{n}^{(i)} = (\sigma_i - p)n^{(i)} \qquad (i = 1, 2, 3) \tag{4.22}$$

which shows that the principal directions of the deviatoric stress tensor **S** are the same as those of the stress tensor $\boldsymbol{\sigma}$, and the principal deviatoric stresses S_i $(i = 1, 2, 3)$ are related to the principal stresses σ_i $(i = 1, 2, 3)$ by

$$S_1 = \frac{2\sigma_1 - \sigma_2 - \sigma_3}{3} \tag{4.23}$$

$$S_2 = \frac{2\sigma_2 - \sigma_1 - \sigma_3}{3} \tag{4.24}$$

$$S_3 = \frac{2\sigma_3 - \sigma_1 - \sigma_2}{3} \tag{4.25}$$

or

$$S_i = \sigma_i - p \qquad (i = 1, 2, 3) \tag{4.26}$$

Furthermore, except for $J_1' = 0$, the second and the third invariants of the deviatoric stress tensor **S** are related to the invariants J_i $(i = 1, 2, 3)$ of the stress tensor σ by the following equations:

$$J_2' = \tfrac{1}{3}(J_1^2 - 3J_2) \tag{4.27}$$

$$J_3' = \tfrac{1}{27}(2J_1^3 - 9J_1J_2 + 27J_3) \tag{4.28}$$

It is helpful for later use to introduce the octahedral stresses here. First, we need to define the octahedral plane. An octahedral plane is the plane whose normal makes equal angles with each of the principal directions (axes) of the stress tensor $\boldsymbol{\sigma}$. Thus, for each of these planes, the unit normal $\mathbf{n} = (n_1, n_2, n_3)$ must satisfy

$$n_1^2 = n_2^2 = n_3^2 \tag{4.29}$$

and

$$n_1^2 + n_2^2 + n_3^2 = 1$$

which gives

$$n_1 = \pm\frac{1}{\sqrt{3}}, \quad n_2 = \pm\frac{1}{\sqrt{3}}, \quad n_3 = \pm\frac{1}{\sqrt{3}} \tag{4.30}$$

Thus there are a total of eight families of octahedral planes whose equations are

$$\pm\sigma_1 \pm \sigma_2 \pm \sigma_3 = \sqrt{3}\,C \tag{4.31}$$

where C is a constant. With octahedral planes we can introduce octahedral stresses. The octahedral normal stress σ_{oct} is defined as the normal stress on an octahedral plane and is calculated by

$$\sigma_{\text{oct}} = \mathbf{n}\cdot\boldsymbol{\sigma}\cdot\mathbf{n} = \sigma_1 n_1^2 + \sigma_2 n_2^2 + \sigma_3 n_3^2 = \tfrac{1}{3}(\sigma_1 + \sigma_2 + \sigma_3) = \tfrac{1}{3}J_1 \tag{4.32}$$

This is obtained using the fact that $\boldsymbol{\sigma}$ takes the diagonal form in the principal axis coordinate system.

The octahedral shear stress τ_{oct} is defined as the shear stress on an octahedral plane and is given by

$$\begin{aligned}\tau_{\text{oct}}^2 &= |\boldsymbol{\sigma}\cdot\mathbf{n}|^2 - \sigma_{\text{oct}}^2\\ &= \tfrac{1}{9}\left[(\sigma_1-\sigma_2)^2 + (\sigma_2-\sigma_3)^2 + (\sigma_3-\sigma_1)^2\right]\\ &= \tfrac{2}{3}J_2'\end{aligned} \tag{4.33}$$

then

$$\begin{aligned}\tau_{\text{oct}} &= \sqrt{\tfrac{2}{3}J_2'} = \tfrac{1}{3}\left[(\sigma_1-\sigma_2)^2 + (\sigma_2-\sigma_3)^2 + (\sigma_3-\sigma_1)^2\right]^{1/2}\\ &= \tfrac{1}{3}\Big[(\sigma_{xx}-\sigma_{yy})^2 + (\sigma_{yy}-\sigma_{zz})^2 + (\sigma_{zz}-\sigma_{xx})^2\\ &\qquad + 6\sigma_{xy}^2 + 6\sigma_{yz}^2 + 6\sigma_{zx}^2\Big]^{1/2}\end{aligned} \tag{4.34}$$

As discussed previously, there are six independent stress components for a definite stress state. Hence, to represent a stress state geometrically, we need a six-dimensional space with the six independent stress components as its coordinate axes. With very few exceptions, to describe the yield surface or failure surface for anisotropic materials, we have no choice other than working with this six-dimensional stress space. For anisotropic material the orientation of the principal stresses is as important as the magnitude of the principal stresses. When the material is isotropic, the situation becomes much simpler. Since the material properties are the same in any direction, only the magnitude of the principal stresses plays a role in describing the yield or failure behavior. Therefore we only need a three-dimensional stress space using the three principal stresses as the coordinate axes. This principal stress space is called the Haigh-Westergaard stress space. In this space every point with coordinates σ_1, σ_2, and σ_3 represents a stress state with these principal stresses. Two stress states having the same principal stresses

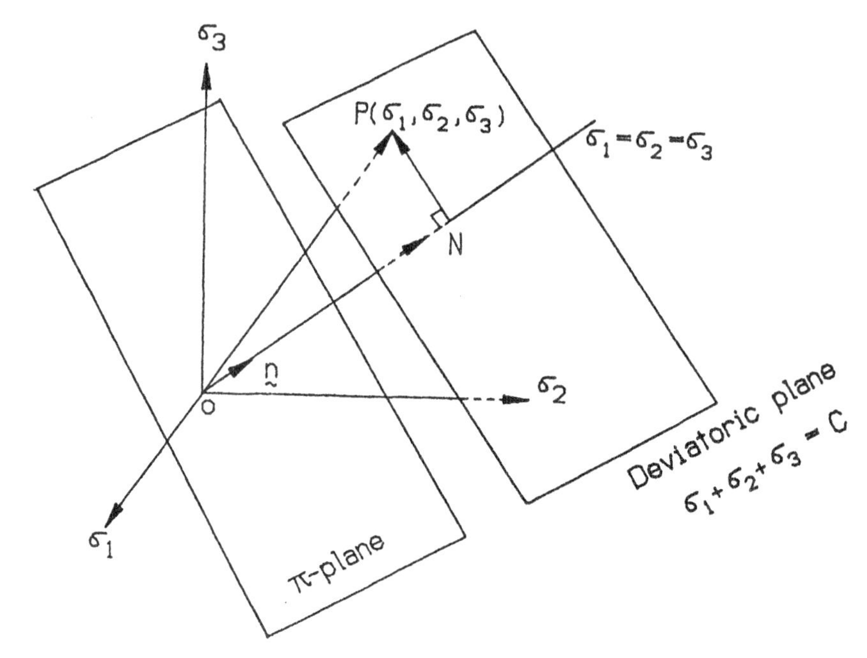

Fig. 4.1 Principal stress space and decomposition of $\boldsymbol{\sigma}$

$\sigma_1, \sigma_2, \sigma_3$, but different principal directions are not distinguishable in this space.

As we mentioned earlier, in the theory of plasticity we can split the stress tensor $\boldsymbol{\sigma}$ into a hydrostatic stress tensor $p\mathbf{I}$ and a deviatoric stress tensor $\mathbf{S}$. Therefore, it is useful to discuss here the geometric representation of this decomposition in the principal stress space.

Consider a diagonal $\overline{ON}$, as shown in Fig. 4.1, that passes through the origin and makes equal angles with the coordinate axes. We have

$$\sigma_1 = \sigma_2 = \sigma_3 = \tfrac{1}{3}J_1 \tag{4.35}$$

and

$$S_1 = S_2 = S_3 = 0 \tag{4.36}$$

at every point along this line. The points on the space diagonal represent the states of hydrostatic pressure, and this line is called hydrostatic axis. Any plane perpendicular to the diagonal is an octahedral plane and, from Eq. (4.31), is represented by the equation

$$\sigma_1 + \sigma_2 + \sigma_3 = C = \sqrt{3}\,d \tag{4.37}$$

where d is distance from the origin to the plane. If $d = 0$, the plane is called a π-plane.

Any stress state $(\sigma_1, \sigma_2, \sigma_3)$ can be represented in the principal stress space by a position vector $\overrightarrow{OP}$, and from Fig. 4.1, we can write

$$\overrightarrow{OP} = \overrightarrow{ON} + \overrightarrow{NP} \tag{4.38}$$

where $\overrightarrow{NP}$ is perpendicular to $\overrightarrow{ON}$ and thus lies on the octahedral plane. Since $\overrightarrow{ON} = |\overrightarrow{ON}|\mathbf{n}$, $\mathbf{n}$ is the unit vector along ON and is equal to $(1/\sqrt{3}, 1/\sqrt{3}, 1/\sqrt{3})$, and

$$\begin{aligned} |\overrightarrow{ON}| &= \overrightarrow{OP} \cdot \mathbf{n} = (\sigma_1, \sigma_2, \sigma_3) \cdot \frac{1}{\sqrt{3}}(1,1,1) \\ &= \frac{1}{\sqrt{3}}(\sigma_1 + \sigma_2 + \sigma_3) = \frac{1}{\sqrt{3}} J_1 = \sqrt{3}\, p \end{aligned} \tag{4.39}$$

Thus

$$\overrightarrow{ON} = (p, p, p) \tag{4.40}$$

and

$$\begin{aligned} \overrightarrow{NP} &= \overrightarrow{OP} - \overrightarrow{ON} = (\sigma_1 - p, \sigma_2 - p, \sigma_3 - p) \\ &= (S_1, S_2, S_3) \end{aligned} \tag{4.41}$$

Therefore for an arbitrary stress state the hydrostatic pressure part is represented by a vector along the space diagonal and the deviatoric part by a vector on the octahedral plane, which is perpendicular to the space diagonal. For this reason this plane is sometimes called *deviatoric plane*.

From Eqs. (4.41), (4.20), and (4.34) we can calculate the length of $\overrightarrow{NP}$

$$r = |\overrightarrow{NP}| = \left(S_1^2 + S_2^2 + S_3^2\right)^{1/2} = \sqrt{2J_2'} = \sqrt{3}\,\tau_{\text{oct}} \tag{4.42}$$

To determine the orientation of $\overrightarrow{NP}$ on the deviatoric plane, let us consider the projection of $\overrightarrow{NP}$ on this plane in Fig. 4.2. In Fig. 4.2(a), σ_1', σ_2', and σ_3' are the projections of the axes σ_1, σ_2, and σ_3 on the deviatoric plane. Let $\mathbf{n}_1'$ be the unit vector in σ_1' direction on the deviatoric plane, and its components with respect to σ_1, σ_2, and σ_3 axes are $[\cos(90° - \alpha), m, n]$, as shown in Fig. 4.2($b$). Since $\cos\alpha = 1/\sqrt{3}$, $m = n$ and should be negative, we have

$$\mathbf{n}_1' = \frac{1}{\sqrt{6}}(2, -1, -1) \tag{4.43}$$

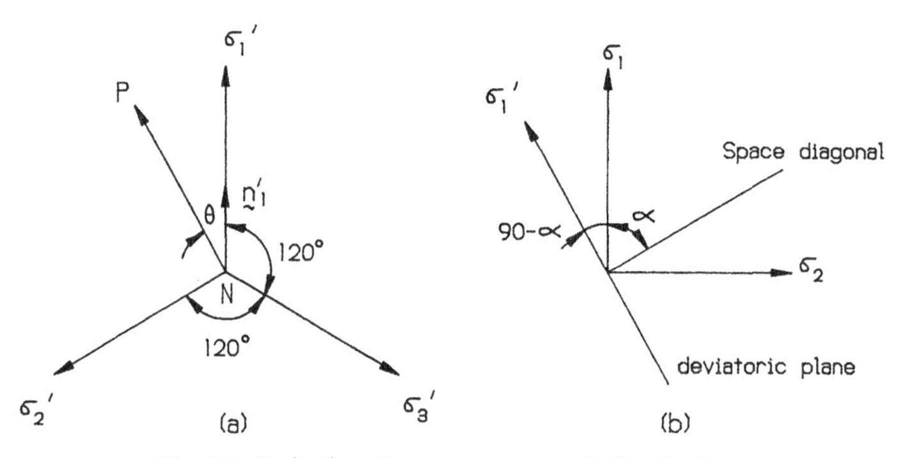

Fig. 4.2 Projection of a stress state on a deviatoric plane

Therefore we can determine the projection of $\overrightarrow{NP}$ on σ_1' as

$$\overrightarrow{NP} \cdot \mathbf{n}_1' = r \cos\theta = (S_1, S_2, S_3) \cdot \frac{1}{\sqrt{6}}(2, -1, -1)$$

or

$$\cos\theta = \frac{\sqrt{3}}{2} \frac{S_1}{\sqrt{J_2'}} \tag{4.44}$$

where Eqs. (4.19) and (4.42) have been used to eliminate r, S_2, and S_3. To represent θ in terms of the invariants of the deviatoric stress tensor **S**, we use the trigonometric identity $\cos 3\theta = 4\cos^3\theta - 3\cos\theta$ and Eq. (4.44) to get

$$\begin{aligned}\cos 3\theta &= 4\left(\frac{\sqrt{3}}{2} \frac{S_1}{\sqrt{J_2'}}\right)^3 - 3\left(\frac{\sqrt{3}}{2} \frac{S_1}{\sqrt{J_2'}}\right) \\ &= \frac{3\sqrt{3}}{2J_2'^{3/2}}(S_1^3 - S_1 J_2')\end{aligned} \tag{4.45}$$

Noting that $J_2' = -(S_1S_2 + S_2S_3 + S_3S_1)$, $S_2 + S_3 = -S_1$, and $J_3' = S_1S_2S_3$, we finally obtain

$$\cos 3\theta = \frac{3\sqrt{3}}{2} \frac{J_3'}{J_2'^{3/2}} \qquad (0 \le \theta \le 60°) \tag{4.46}$$

From Eq. (4.44) and Fig. 4.2(*a*), it is easy to show that

$$S_1 = \frac{2}{\sqrt{3}}\sqrt{J_2'}\cos\theta \tag{4.47}$$

$$S_2 = \frac{2}{\sqrt{3}}\sqrt{J_2'}\cos(120^\circ - \theta) \tag{4.48}$$

$$S_3 = \frac{2}{\sqrt{3}}\sqrt{J_2'}\cos(120^\circ + \theta) \tag{4.49}$$

In the principal stress space Eq. (4.16) can be written

$$\begin{pmatrix}\sigma_1\\ \sigma_2\\ \sigma_3\end{pmatrix} = \frac{J_1}{3}\begin{pmatrix}1\\1\\1\end{pmatrix} + \frac{2}{\sqrt{3}}\sqrt{J_2'}\begin{pmatrix}\cos\theta\\ \cos(120^\circ - \theta)\\ \cos(120^\circ + \theta)\end{pmatrix} \tag{4.50}$$

We can see from the following consideration that the value of θ is different for different stress states:

1. Uniaxial tension in the presence of the hydrostatic pressure $\sigma_1 > \sigma_2 = \sigma_3$ (uniaxial tension if $\sigma_2 = \sigma_3 = 0$). This stress state is sometimes referred to as *triaxial tension*, especially when dealing with geomaterials. From Eq. (4.44) we get

$$\theta = 0^\circ \tag{4.51}$$

2. Pure shear in the presence of a hydrostatic pressure $\sigma_1 - \sigma_2 = \sigma_2 - \sigma_3$, where $\sigma_1 > \sigma_2 > \sigma_3$. Substitution in Eq. (4.44) results in

$$\theta = 30^\circ \tag{4.52}$$

3. Uniaxial compression in the presence of the hydrostatic pressure $\sigma_1 = \sigma_2 > \sigma_3$ (uniaxial compression for $\sigma_1 = \sigma_2 = 0$). This stress state is also called triaxial compression in geomaterials applications. Using again Eq. (4.44), we obtain

$$\theta = 60^\circ \tag{4.53}$$

Clearly from Eq. (4.50) we can choose J_1, $\sqrt{J_2'}$, and θ as the coordinate axes in the principal stress space instead of σ_1, σ_2, and σ_3. In fact this choice is more convenient to use in describing the yield surface in the plasticity theory.

4.2 YIELD SURFACE

As we mentioned earlier, a yield criterion is a basic assumption about a material for the purpose of determining the onset of the plastic deformation. If a stress state at a point satisfies the yield criterion, then this point deforms plastically; otherwise, it undergoes elastic deformation. All the possible stress states corresponding to yielding constitute a closed hypersurface in the six-dimensional stress space, as noted in the introduction to this chapter. For clarity let us consider the initial yield surface that consists of all the yield points corresponding to initial yielding. Furthermore let us assume that the material is homogeneous, we will ignore any thermal effects. The yield function can be written mathematically in the general form

$$F(\sigma_{ij}) = 0 \tag{4.54}$$

with

$$\begin{aligned} F(\sigma_{ij}) &< 0 \quad \text{for elastic deformation domain} \\ F(\sigma_{ij}) &= 0 \quad \text{for plastic deformation domain} \end{aligned} \tag{4.55}$$

If the material is isotropic, the yielding depends only on the magnitudes of the principal stresses. For such materials the yield criterion is given by

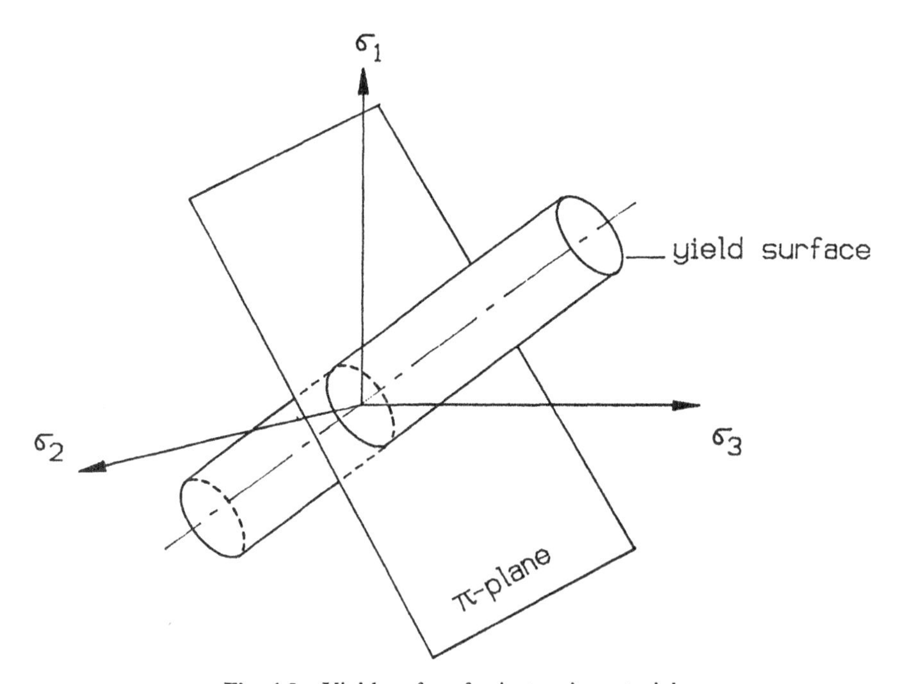

Fig. 4.3 Yield surface for isotropic materials

Eq. (4.56) or (4.57).

$$F(\sigma_1, \sigma_2, \sigma_3) = 0 \tag{4.56}$$

$$F(J_1, J_2, J_3) = 0 \tag{4.57}$$

Equation (4.56) or (4.57) represents a surface in the principal stress space. For metallic materials experimental evidence allows us to assume that the hydrostatic pressure has no effect on plastic yielding. The plastic yielding is only related to the deviatoric stress tensor **S**. Then the yield criterion can be written in terms of its invariants J_2' and J_3' for isotropic dense or nonporous materials:

$$F(J_2', J_3') = 0 \tag{4.58}$$

This represents a cylinder in the principal stress space with the generator parallel to the hydrostatic pressure axis since the hydrostatic pressure $p = J_1/3$ is not an argument in the yield function (4.58), as shown in Fig. 4.3. It suffices to consider the shape of yield surface on any one of the deviatoric planes, particularly the π-plane.

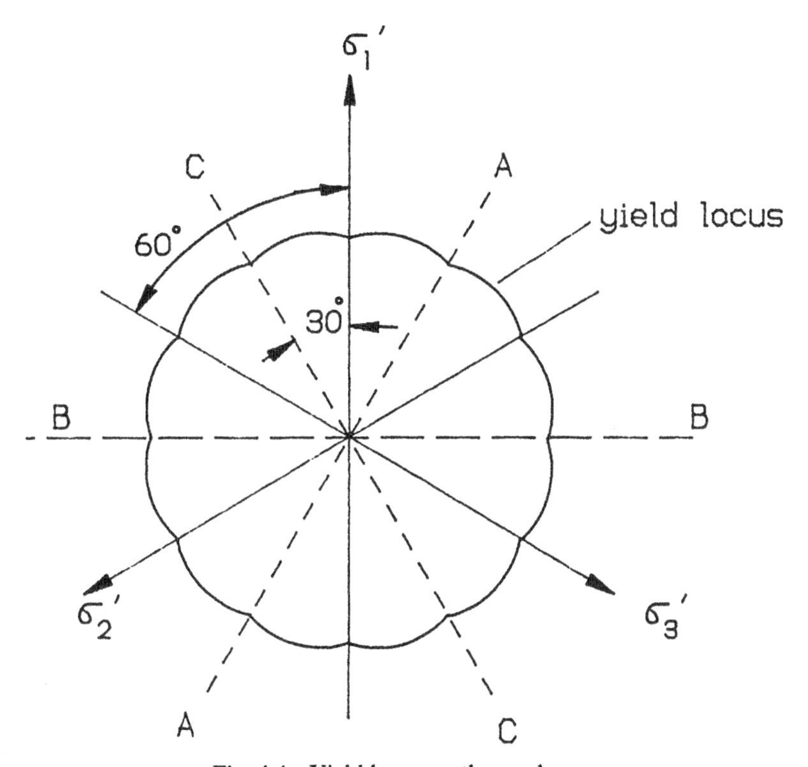

Fig. 4.4 Yield locus on the π-plane

Consider the locus of the yield surface on the π-plane, as shown in Fig. 4.4. First, according to the isotropy assumption, if point $(\sigma_1, \sigma_2, \sigma_3)$ is on the yield surface, so is the point $(\sigma_1, \sigma_3, \sigma_2)$. This means the yield locus must be symmetric about the σ_1' axis. For the same reason the axes σ_2' and σ_3' are also lines of symmetry. This symmetry about σ_1', σ_2', and σ_3' axes divides the yield locus on the π-plane into 6 identical segments. Therefore only any one of the 6 segments for a particular material needs to be determined experimentally. Symmetry can then be used to obtain the whole yield locus. This is the case for porous soil and rocklike materials.

Second, if the yield strength of the material for tension is the same as that for compression, then, when the point $(\sigma_1, \sigma_2, \sigma_3)$ is on the yield surface, so is the point $(-\sigma_1, \sigma_2, \sigma_3)$. Therefore the yield locus must be symmetric about the line B–B, which is perpendicular to the σ_1' axis. Similarly C–C and A–A, which are perpendicular to the σ_2' and σ_3' axes, respectively, must be lines of symmetry. This divides any segment, as mentioned above, into two identical sectors. For these dense or nonporous materials, such as metals, it is sufficient to experimentally determine any one of 12 segments, say, $0° \leq \theta \leq 30°$. A schematic is given in Fig. 4.4.

4.3 YIELD CRITERIA FOR METALS

In this section two commonly used yield criteria—von Mises and Tresca criteria—are presented for metal plasticity. Bridgeman (1952) established that at even 25,000 bars there was no yielding in metals. Thus, as discussed in the last section, the yield criteria for metals does not depend on the first invariant J_1 of the stress tensor.

4.3.1 Maxwell-Huber-von Mises Criterion

It is usually believed that this yield criterion was suggested by von Mises (1913). But in fact it was first published by Huber (1904), and there is evidence that even earlier, in 1856, Maxwell came up with this criterion in a letter to Kelvin (Bell 1973). Later Hencky (1924) interpreted this criterion as the deviatoric strain energy.

The criterion assumes that the plastic yielding will occur only when the second invariant $\mathbf{J}_2'$ of the deviatoric stress tensor $\mathbf{S}$ reaches a critical value k^2, which is a material property as given by Eq. (4.59):

$$
\begin{aligned}
J_2' - \kappa^2 &= 0 \qquad \text{for yielding or plastic deformation} \\
J_2' &< \kappa^2 \qquad \text{for elastic deformation}
\end{aligned}
\tag{4.59}
$$

In terms of stress components the yield criterion takes the form

$$\frac{1}{6}\left[(\sigma_{xx} - \sigma_{yy})^2 + (\sigma_{yy} - \sigma_{zz})^2 + (\sigma_{zz} - \sigma_{xx})^2\right] + \sigma_{xy}^2 + \sigma_{yz}^2 + \sigma_{zx}^2 = \kappa^2 \tag{4.60}$$

or

$$\frac{1}{6}\left[(\sigma_1 - \sigma_2)^2 + (\sigma_2 - \sigma_3)^2 + (\sigma_3 - \sigma_1)^2\right] = \kappa^2 \tag{4.61}$$

Obviously the von Mises yield surface is a cylinder surface parallel to the hydrostatic stress axis, its locus on the π-plane is a circle since in Eq. (4.59), there is no dependence on J_1 or on θ. From Eq. (4.42) the radius of the circle is $r = \sqrt{2}\,\kappa$, as shown in Fig. 4.5.

In case of plane stress ($\sigma_3 = 0$), Eq. (4.61) reduces to

$$\sigma_1^2 - \sigma_1\sigma_2 + \sigma_2^2 = 3\kappa^2 \tag{4.62}$$

This represents an ellipse in $\sigma_1 - \sigma_2$ space, as shown in Fig. 4.6.

To determine the constant κ, we will conduct a simple tension test:

$$\sigma_1 = \sigma_Y \quad \text{and} \quad \sigma_2 = \sigma_3 = 0 \tag{4.63}$$

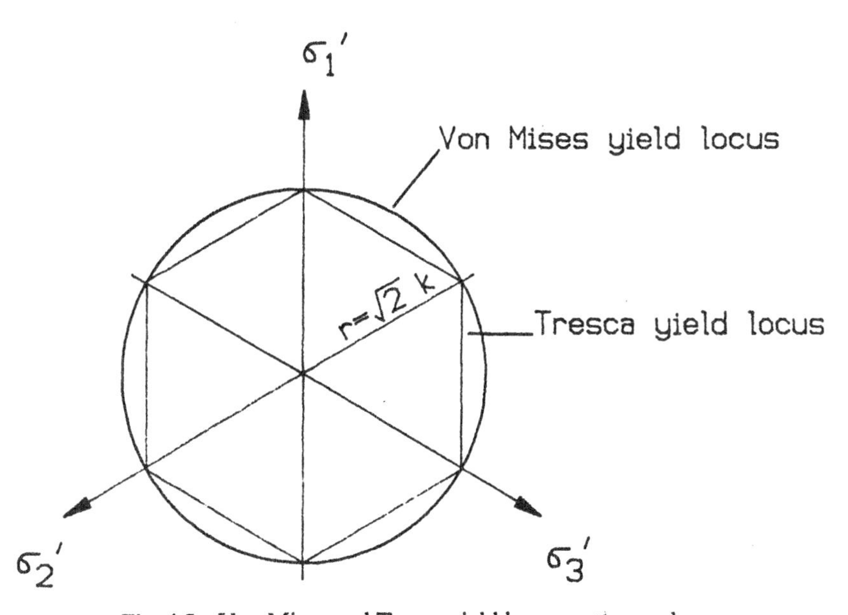

Fig. 4.5 Von Mises and Tresca yield locus on the π-plane

where σ_Y is the yield stress in simple tension. Substituting Eq. (4.63) into (4.61) gives

$$\frac{1}{3}\sigma_Y^2 = \kappa^2 \tag{4.64}$$

Therefore

$$\kappa = \frac{\sigma_Y}{\sqrt{3}} \tag{4.65}$$

If we determine κ using the yield stress in pure shear test τ_Y, we have $\sigma_1 = -\sigma_3 = \tau_Y$ and $\sigma_2 = 0$. Then from Eq. (4.61) we have

$$\tau_Y^2 = \kappa^2 \tag{4.66}$$

and hence

$$\kappa = \tau_Y \tag{4.67}$$

Comparing Eqs. (4.67) and (4.65), we obtain

$$\kappa = \tau_Y = \frac{\sigma_Y}{\sqrt{3}} \tag{4.68}$$

For a Maxwell-Huber-von Mises material, the yield strength in uniaxial tension is $\sqrt{3}$ times the yield strength in pure shear.

4.3.2 Tresca Criterion

Tresca (1864), based on Coulomb's results on soil mechanics and his own experiments on metal extrusion, proposed a yield criterion for metallic solids, which is now well known as the Tresca yield criterion.

This criterion assumes that the plastic yielding will occur only when the maximum shear stress reaches the critical value k of a material. Mathematically it is written

$$\tau_{max} = \kappa \quad \text{for yielding or plastic deformation} \tag{4.69}$$

$$\tau_{max} < \kappa \quad \text{for elastic deformation} \tag{4.70}$$

In terms of principal stresses, Eq. (4.69) can be written

$$\max\left[\tfrac{1}{2}|\sigma_1 - \sigma_2|, \tfrac{1}{2}|\sigma_2 - \sigma_3|, \tfrac{1}{2}|\sigma_3 - \sigma_1|\right] = \kappa \tag{4.71}$$

or

$$\tfrac{1}{2}(\sigma_{max} - \sigma_{min}) = \kappa \tag{4.72}$$

Using Eq. (4.50) we can also write the equation above in terms of $\sqrt{J_2'}$ and θ:

$$\sqrt{J_2'}\,\sin(\theta + 60°) = \kappa \qquad (0° \leq \theta \leq 60°) \tag{4.73}$$

It is seen from Eq. (4.73) that Tresca yield surface depends on θ but not on J_1. It is easy to show that in the principal stress space, the Tresca yield surface is a hexagonal cylinder parallel to the J_1-axis; its locus on the π-plane is a regular hexagon, as shown in Fig. 4.5.

In case of plane stress ($\sigma_3 = 0$), Eq. (4.71) reduces to

$$\begin{aligned} \sigma_1 - \sigma_2 &= \pm 2\kappa \\ \sigma_1 &= \pm 2\kappa \\ \sigma_2 &= \pm 2\kappa \end{aligned} \tag{4.74}$$

This represents the Tresca hexagon in $\sigma_1 - \sigma_2$ space shown in Fig. 4.6.

The material constants can be determined by considering the uniaxial test (4.63). Applying Eq. (4.72) to this case gives

$$\kappa = \frac{\sigma_Y}{2} \tag{4.75}$$

Alternatively, we can determine the material constant κ using pure shear test

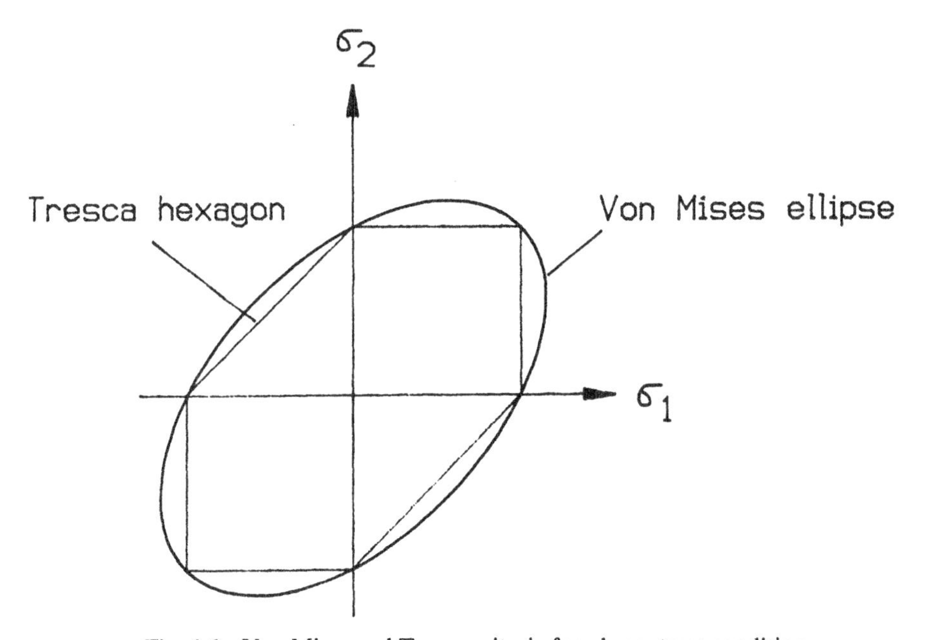

Fig. 4.6 Von Mises and Tresca criteria for plane stress condition

$\sigma_1 = -\sigma_3 = \tau_Y$, $\sigma_2 = 0$. Now we have

$$\kappa = \tau_Y \tag{4.76}$$

From Eqs. (4.75) and (4.76) we derive

$$\kappa = \tau_Y = \frac{\sigma_Y}{2} \tag{4.77}$$

Thus, for Tresca materials, the shear strength in pure shear is half of that in uniaxial tension.

4.3.3 Experimental Investigations

In 1925 Lode used thin-walled tubes of steel, copper, and nickel subjected to the combined uniaxial tension and internal pressure loads in an effort to determine the influence of the intermediate principal stress on yielding. He assumed $\sigma_1 > \sigma_2 > \sigma_3$ and used the following Lode parameter to characterize the stress state:

$$\mu_\sigma = -\frac{2\sigma_2 - \sigma_1 - \sigma_3}{\sigma_1 - \sigma_3} \tag{4.78}$$

It is seen that

$$-1 \leq \mu_\sigma \leq 1 \tag{4.79}$$

we have $\mu_\sigma = 1$ when $\sigma_2 = \sigma_1$, $\mu_\sigma = -1$ and $\sigma_2 = \sigma_3$.

Suppose that the Tresca and von Mises criteria give the same prediction for uniaxial tension (and compression) test, and the yield stress in the uniaxial tension and compression test is the same. Then we have

$$\frac{\sigma_1 - \sigma_3}{\sigma_Y} = 1 \tag{4.80}$$

$$\frac{1}{\sqrt{2}}\left[(\sigma_1 - \sigma_2)^2 + (\sigma_2 - \sigma_3)^2 + (\sigma_3 - \sigma_1)^2\right]^{1/2} = \sigma_Y \tag{4.81}$$

Solving for σ_2 from Eq. (4.78), substituting it into Eq. (4.81) for the von Mises case, and rearranging the equation results in

$$\frac{\sigma_1 - \sigma_3}{\sigma_Y} = \frac{2}{\sqrt{3 + \mu_\sigma^2}} \tag{4.82}$$

For Lode's test we have

$$\sigma_\theta = \frac{pR}{t}$$
$$\sigma_z = \frac{F}{2\pi R t} \tag{4.83}$$
$$\sigma_r \approx 0$$

where F and p are the axial tensile force and the internal pressure, respectively. R is the mean radius of the thin-walled tube and t is wall thickness. Therefore we have

$$\mu_\sigma = \frac{F - \pi R^2 p}{\pi R^2 p} \tag{4.84}$$

and

$$\begin{aligned} \mu_\sigma &= 0 \qquad \text{when } F = \pi R^2 p \\ \mu_\sigma &= -1 \qquad \text{when } F = 0 \end{aligned} \tag{4.85}$$

It is possible to obtain the stress state $-1 \le \mu_\sigma \le 0$ by controlling F and p so that $0 \le F \le \pi R^2 p$. Note that both the Tresca and von Mises criteria are even functions of μ_σ; hence the entire range $-1 \le \mu_\sigma \le +1$ is covered. Lode's results of this experimental study are shown in Fig. 4.7. Lode's experimental results favor von Mises yield criterion. Taylor and Quinney (1931) performed another extensive study of yield criterion for metals. They

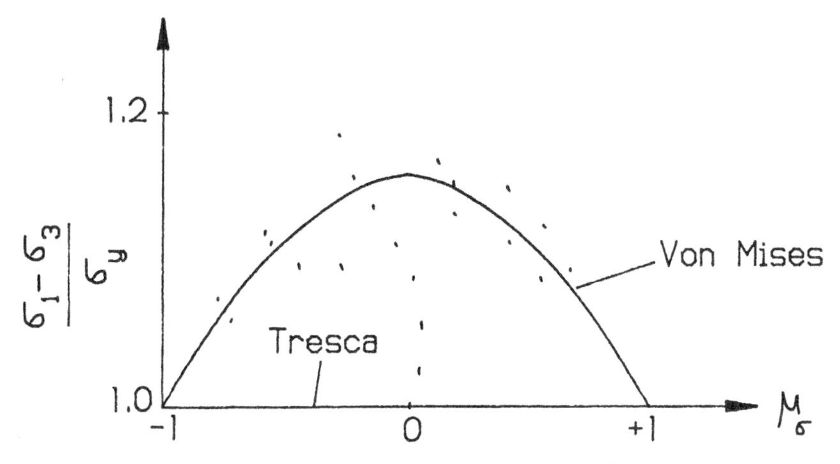

Fig. 4.7 Lode's verification of the yield criteria

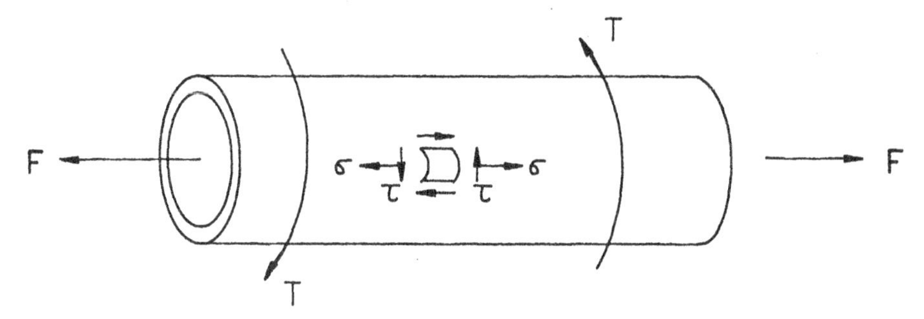

Fig. 4.8 Thin-walled tube under combined torque and tensile loading

used copper, mild steel, and aluminum thin-walled tubes and loaded them with the combined uniaxial tension F and torque T, as shown in Fig. 4.8.

For this type of test only normal stress σ and shear stress τ are present, as shown in Fig. 4.8. The Tresca and von Mises criteria can be simplified as

$$\left(\frac{\sigma}{\sigma_Y}\right)^2 + 4\left(\frac{\tau}{\sigma_Y}\right)^2 = 1 \qquad \text{(Tresca criterion)} \tag{4.86}$$

$$\left(\frac{\sigma}{\sigma_Y}\right)^2 + 3\left(\frac{\tau}{\sigma_Y}\right)^2 = 1 \qquad \text{(von Mises criterion)} \tag{4.87}$$

On σ-τ plane the criteria are represented by ellipses and are shown in Fig. 4.9. Also included in the figure are the experimental results obtained by

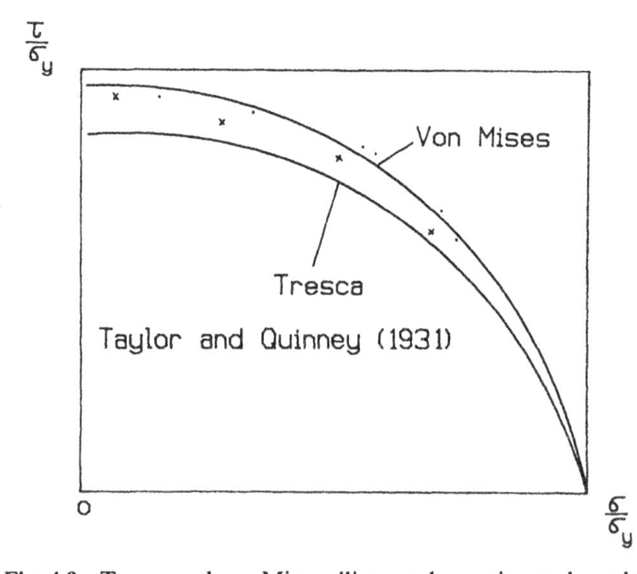

Fig. 4.9 Tresca and von Mises ellipse and experimental results

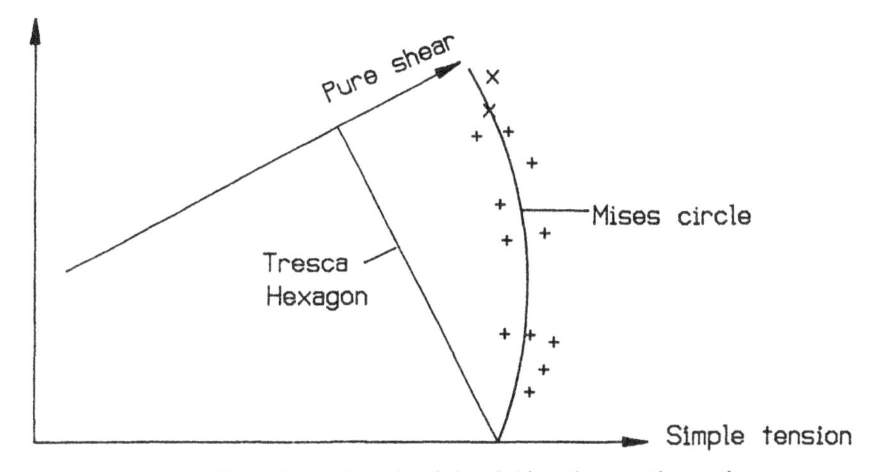

Fig. 4.10 Experimental study of the yield surface on the π-plane

Taylor and Quinney (1931). It is seen from this figure that although the experimental data fall between the two ellipses, they are closer to the von Mises ellipse.

The preceding experiments with thin-walled tubes as the specimen have required rather complicated apparatus. In 1940 Biljaard suggested the idea of using a notched strip in place of the thin-walled tube. The idea was further investigated by Hill (1953) through his theory of localized necking. Hill's method was used by Lianis and Ford (1957) and later on by Parker and Bassett (1964). Lianis and Ford's experimental results using commercially pure aluminum are plotted on the π-plane and compared with Tresca and von Mises criteria in Fig. 4.10. Obviously consistent with Lode and Taylor and Quinney's results discussed earlier, Lianis and Ford's study provides convincing evidence that the von Mises criterion is better for predicting yielding of metals than the Tresca criterion.

4.3.4 Yield Criterion for Anisotropic Materials

As mentioned earlier, for fully anisotropic materials the initial yield criterion should be expressed in terms of six independent components of the stress tensor $\boldsymbol{\sigma}$:

$$F(\sigma_{ij}) = 0$$

or

$$F(\sigma_{xx}, \sigma_{yy}, \sigma_{zz}, \sigma_{xy}, \sigma_{yx}, \sigma_{xz}) = 0 \tag{4.88}$$

This can be geometrically interpreted as a limiting envelope of the elastic domain in the general six-dimensional stress space. Alternatively, in terms of the principal stresses σ_i and the associated principal directions $\mathbf{n}_i$ ($i = 1, 2, 3$), we can equivalently write Eq. (4.88) as

$$F(\sigma_1, \sigma_2, \sigma_3, \mathbf{n}_1, \mathbf{n}_2, \mathbf{n}_3) = 0 \tag{4.89}$$

Equation (4.89) strongly suggests that the yielding of an anisotropic material depends not only on the "intensity" of the stress tensor $\boldsymbol{\sigma}$ (magnitudes of the principal stresses) but also on its "orientation" (the principal directions). As can be expected, the determination of the yield criterion for anisotropic materials is much more difficult than for isotropic materials. Also many more material constants are needed to characterize yield behavior for anisotropic materials. For example, the uniaxial tensile yield strength in one direction may differ from that in another direction. Therefore in an effort to predict the uniaxial tensile yield strength in any direction for an anisotropic material, a single constant σ_Y obtained in only one direction is no longer sufficient.

The search for yield criterion for anisotropic materials was started by Jackson, Smith, and Lankford (1948), Dorn (1949), and Hill (1948). Basically they tried to modify the Tresca or von Mises criterion for isotropic materials to describe the yielding of anisotropic materials by using additional material constants. Here we only present Hill's yield criterion (1948) for orthotropic materials.

Hill's yield criterion is based on the following assumptions.

1. The material is orthotropic. That is, there exist three mutually orthogonal planes of symmetry at each material point. The intersections of these planes are the principal axes of anisotropy.
2. Hydrostatic stress does not affect yielding.
3. There is no Bauschinger effect.

Mathematically Hill's yield criterion takes the following form according to the principal axes of anisotropy x, y, and z:

$$\begin{aligned} 2f(\sigma_{ij}) = F(\sigma_{yy} - \sigma_{zz})^2 + G(\sigma_{zz} - \sigma_{xx})^2 + H(\sigma_{xx} - \sigma_{yy})^2 + 2L\sigma_{yz}^2 \\ +2M\sigma_{zx}^2 + 2N\sigma_{xy}^2 - 1 = 0 \end{aligned} \tag{4.90}$$

where F, G, H, L, M, and N are material constants characterizing the current state of anisotropic yield behavior.

To determine these six material constants, it is necessary to measure three tensile yield stresses X, Y, and Z in three principal directions of anisotropy. From Eq. (4.90) it is easy to arrive at

$$\frac{1}{X^2} = G + H, \quad \frac{1}{Y^2} = H + F, \quad \frac{1}{Z^2} = F + G \tag{4.91}$$

Then F, G, and H can be solved to be

$$F = \frac{1}{2}\left(\frac{1}{Y^2} + \frac{1}{Z^2} - \frac{1}{X^2}\right) \tag{4.92}$$

$$G = \frac{1}{2}\left(\frac{1}{Z^2} + \frac{1}{X^2} - \frac{1}{Y^2}\right) \tag{4.93}$$

$$H = \frac{1}{2}\left(\frac{1}{Y^2} + \frac{1}{X^2} - \frac{1}{Z^2}\right) \tag{4.94}$$

In addition to uniaxial yield stresses X, Y, and Z, we also need to determine the yield stresses for pure shear on each of the orthogonal planes of anisotropy. Denoting these stresses by R, S, and T, we obtain the following by applying Eq. (4.90) to these cases:

$$L = \frac{1}{2R^2}, \quad M = \frac{1}{2S^2}, \quad N = \frac{1}{2T^2} \tag{4.95}$$

It is easy to verify that if $L = M = N = 3F = 3G = 3H$, Eq. (4.90) reduces to the von Mises criterion for isotropic materials.

There is a more systematic way to deal with the yield criterion for anisotropic materials using tensor functions and their analysis. For example, one can introduce a material tensor of fourth-order **M** and write the following more general form of yield criterion for anisotropic materials

$$F(\boldsymbol{\sigma}, \mathbf{M}) = F(\sigma_{ij}, M_{ijkl}) = 0 \tag{4.96}$$

Further, assuming the quadratic form and hydrostatic pressure independence, Eq. (4.96) can be reduced to

$$F(\boldsymbol{\sigma}, \mathbf{M}) = \mathrm{tr}(\mathbf{S}\colon \mathbf{M}\colon \mathbf{S}) - 1 = 0 \tag{4.97}$$

It should be pointed out that Eq. (4.97) is a generalization of Hill's criterion Eq. (4.90). For example, Hill's criterion is obtained if

$$M_{xxxx} = G + H, \quad M_{yyyy} = H + F, \quad M_{zzzz} = F + G \tag{4.98}$$

$$M_{xxyy} = -H, \quad M_{yyzz} = -F, \quad M_{zzxx} = -G \tag{4.99}$$

$$M_{xyxy} = \frac{N}{2}, \quad M_{yzyz} = \frac{L}{2}, \quad M_{zxzx} = \frac{M}{2} \tag{4.100}$$

$$\text{other } M_{ijkl} = 0 \tag{4.101}$$

and if M_{ijkl} has the symmetry

$$M_{ijkl} = M_{klij} = M_{ijlk} = M_{jikl} \tag{4.102}$$

For isotropic material

$$M_{ijkl} = a_1 \delta_{ij} \delta_{kl} + a_2 (\delta_{ik} \delta_{jl} + \delta_{il} \delta_{jk}) \tag{4.103}$$

where a_1, a_2 are constants. Then Eq. (4.97) reduces to

$$J_2' = C \tag{4.104}$$

which is the von Mises criterion, where C is a constant in terms of a_2.

From the preceding discussion it is clear that tensor **M** plays a role in the measure of the anisotropy of the materials; hence it is called the *anisotropy tensor* of the material. If the effect of hydrostatic pressure on the yielding is desired, then **S** in Eq. (4.97) can be replaced by the stress tensor $\boldsymbol{\sigma}$. For example, Betten (1982) obtained the following form of yield criterion for pressure sensitive materials

$$F(\boldsymbol{\sigma}, \mathbf{M}) = \operatorname{tr}[(\boldsymbol{\sigma} - b\mathbf{I}) : \mathbf{M} : (\boldsymbol{\sigma} - b\mathbf{I})] - 1 = 0 \tag{4.105}$$

where b is a constant. For isotropic material we substitute Eq. (4.103) into this equation to get

$$J_2' + C_1 J_1^2 + C_2 J_1 = C_3 \tag{4.106}$$

where C_1, C_2, and C_3 are material constants, which can be experimentally determined. Furthermore, if $C_1 = -\alpha^2$, $C_2 = 2\alpha\beta$, and $C_3 = \beta$, then Eq. (4.106) can be reduced to the Drucker-Prager yield criterion.

$$\sqrt{J_2'} + \alpha_1 J_1 = \beta \tag{4.107}$$

This criterion will be introduced in the next section.

Here, for demonstration only, we have discussed the use of a fourth-order material tensor as a measure of anisotropy. In fact additional material tensors of first and second orders can be used to characterize the material anisotropy. An extensive and systematic study of this problem has been carried out by Betten (1982, 1984, 1985, 1988). Readers are directed to these papers for more details.

4.4 YIELD CRITERIA FOR PRESSURE-SENSITIVE MATERIALS

The yielding of porous materials is pressure sensitive. The yield criterion for these materials should include the influence of the hydrostatic pressure

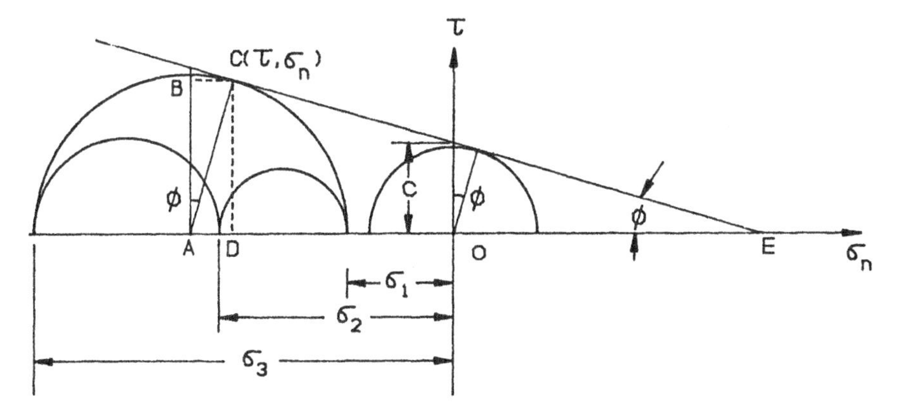

Fig. 4.11 Coulomb-Mohr yield criterion

(i.e., dependence on J_1). In this section we will discuss two classical yield criteria for pressure-dependent materials, the Coulomb-Mohr and Drucker-Prager criteria.

4.4.1 Coulomb-Mohr Criterion

This criterion is a generalization of the Coulomb (1773) friction failure law, which can be expressed by

$$\tau = C - \sigma_n \tan \phi \tag{4.108}$$

where τ is the magnitude of the shear stress on the failure plane, σ_n is the normal stress on that plane (tensile stress is assumed positive), C is the cohesion, and ϕ is the angle of the internal friction.

Based on the above idea and also in an effort to improve the Tresca criterion to include the effect of the mean stress (hydrostatic stress) on yielding, Mohr assumed that the critical shear stress is not only related to the maximum shear stress but also depends on the normal stress applied on the shearing plane. It was first demonstrated by Mohr (1882) that graphically Eq. (4.108) represents a straight-line tangent to the largest principal stress circle, as shown in Fig. 4.11.

From Fig. 4.11 it is easy to see that if $\sigma_1 \geq \sigma_2 \geq \sigma_3$,

$$\tau = BA = CA \cos \phi = \frac{\sigma_1 - \sigma_3}{2} \cos \phi \tag{4.109}$$

$$\sigma_n = DO = AO - AD = \frac{\sigma_1 + \sigma_3}{2} + \frac{\sigma_1 - \sigma_3}{2} \sin \phi \tag{4.110}$$

Therefore Eq. (4.108) can be rewritten as

$$\frac{1}{2}(\sigma_1 - \sigma_3)\cos\phi = C - \left(\frac{\sigma_1 + \sigma_3}{2} + \frac{\sigma_1 - \sigma_3}{2}\sin\phi\right)\tan\phi \quad (4.111)$$

After rearranging, we obtain

$$(\sigma_1 - \sigma_3) = 2C\cos\phi - (\sigma_1 + \sigma_3)\sin\phi \quad (4.112)$$

But from Eq. (4.50) we have

$$\sigma_1 = \frac{1}{3}J_1 + \frac{2}{\sqrt{3}}\sqrt{J_2'}\cos\theta \quad (4.113)$$

$$\sigma_3 = \frac{1}{3}J_1 + \frac{2}{\sqrt{3}}\sqrt{J_2'}\cos(\theta + 120°) \quad (4.114)$$

Substituting σ_1, σ_3 in terms of J_1, $\sqrt{J_2'}$, and θ into Eq. (4.112), we obtain

$$\frac{1}{3}J_1\sin\phi + \sqrt{J_2'}\left[\sin(\theta + 60°) + \frac{1}{\sqrt{3}}\sin\phi\cos(\theta + 60°)\right] = C\cos\phi \quad (4.115)$$

In principal stress space this gives a conical yield surface whose locus on a deviatoric plane ($J_1 = C$) is an irregular hexagon, as shown in Figs. 4.12 and 4.13.

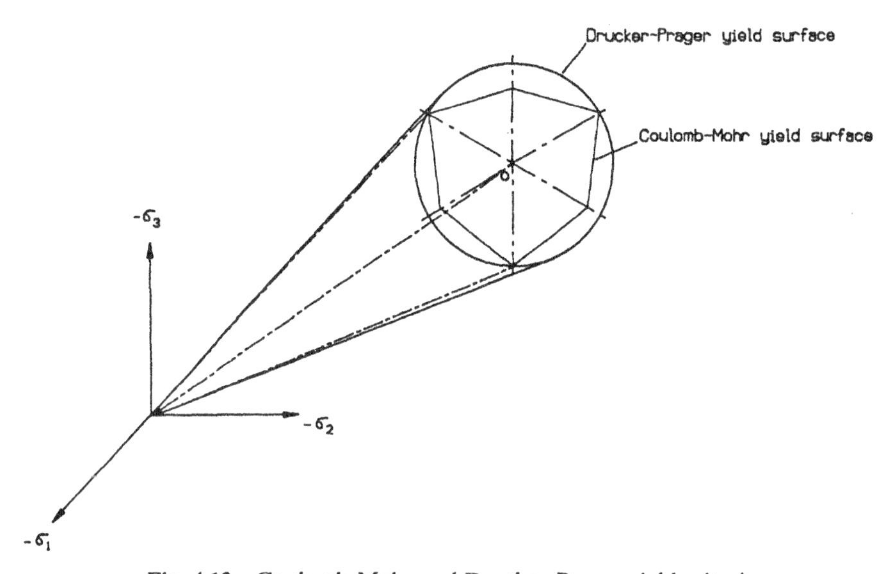

Fig. 4.12 Coulomb-Mohr and Drucker-Prager yield criteria

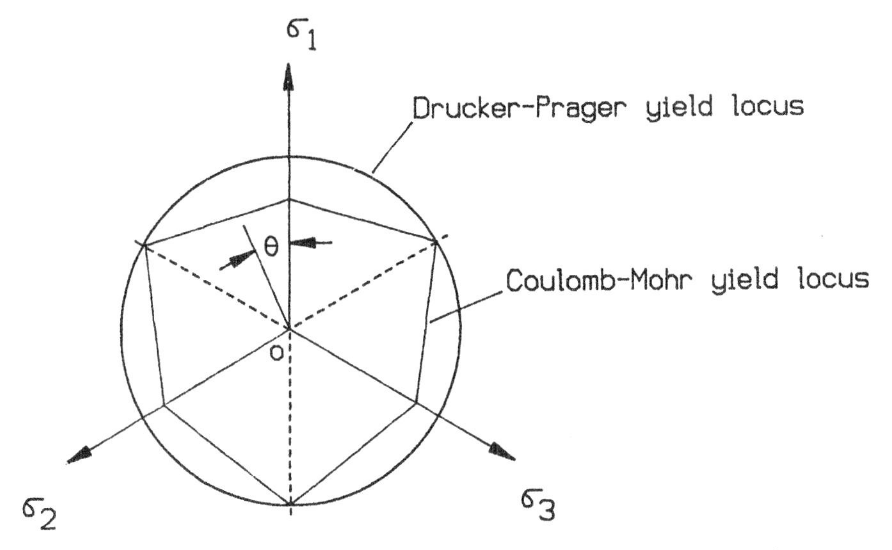

Fig. 4.13 Coulomb-Mohr and Drucker-Prager yield loci on the π-plane

Compared to the cylindrical shape for the von Mises yield surface, the conical shape is due to the fact that the hydrostatic pressure J_1 does influence yielding, as seen from Eq. (4.115). In addition the cross section is irregular, instead of regular circular or hexagonal, because for pressure-sensitive materials, such as rock, soil, and concrete, the yield stress in tension is smaller than the yield stress in compression at the same hydrostatic stress.

For two-dimensional stress ($\sigma_3 = 0$) the Coulomb-Mohr criterion can be written

$$\sigma_1 = \sigma_t \qquad \text{if } \sigma_1 > 0, \sigma_2 > 0 \tag{4.116}$$

$$\sigma_1 = \sigma_c \qquad \text{if } \sigma_1 < 0, \sigma_2 < 0 \tag{4.117}$$

and

$$\frac{\sigma_1}{\sigma_t} - \frac{\sigma_2}{\sigma_c} = 1 \qquad \text{if } \sigma_1 > 0 > \sigma_2 \tag{4.118}$$

where σ_t and σ_c are the yield strength in tension and the yield strength in compression, respectively. Because of the assumption of isotropy, σ_1 and σ_2 can be interchanged, giving another set of equations. On the $\sigma_1 - \sigma_2$ plane these equations represent an irregular hexagon, as shown in Fig. 4.14. To determine the material constant C and ϕ, two experiments are needed. Assuming that the yield stresses σ_t in uniaxial tension and σ_c in uniaxial

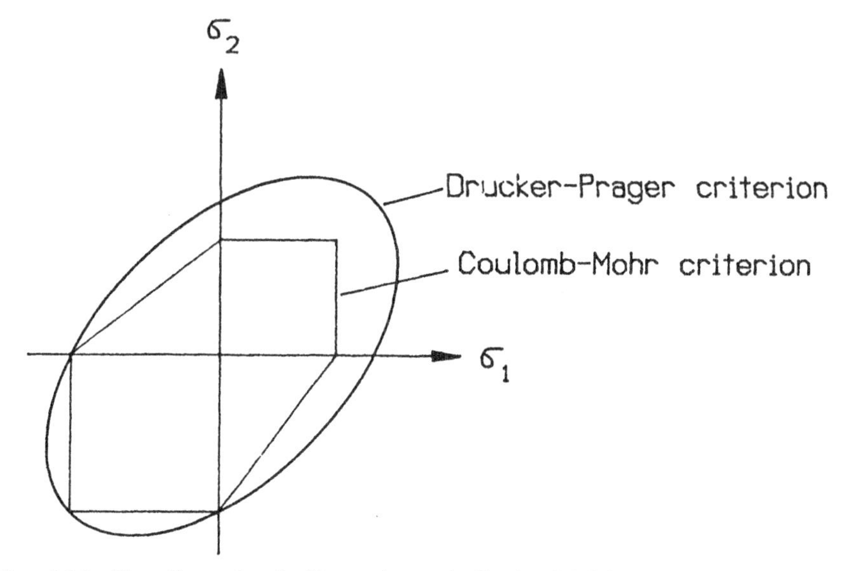

Fig. 4.14 Two-dimensional illustration of Coulomb-Mohr and Drucker-Prager criteria

compression are known, we have two known stress states:

$$\sigma_1 = \sigma_t, \quad \sigma_2 = \sigma_3 = 0 \qquad \text{(tension)} \tag{4.119}$$

$$\sigma_1 = \sigma_2 = 0, \quad \sigma_3 = -\sigma_c \qquad \text{(compression)} \tag{4.120}$$

Substituting Eqs. (4.119) and (4.120) into Eq. (4.112), we obtain, respectively,

$$\sigma_t = 2C\cos\phi - \sigma_t \sin\phi \tag{4.121}$$

$$\sigma_c = 2C\cos\phi + \sigma_c \sin\phi \tag{4.122}$$

which can be solved for C and ϕ

$$\phi = \sin^{-1}\frac{\sigma_c - \sigma_t}{\sigma_c + \sigma_t} \tag{4.123}$$

$$C = \frac{\sigma_t(1 + \sin\phi)}{2\cos\phi} \tag{4.124}$$

4.4.2 Drucker-Prager Criterion

Coulomb-Mohr and Tresca yield surfaces have hexagonal-shaped corners. These corners present difficulties when the classical associated plasticity theory is used to determine the plastic strain increment. Drucker and Prager (1950) suggested treating the surface as a smooth circular cone by adding a hydrostatic stress term αJ_1 to the von Mises criterion:

$$F(J_1, J_2') = \sqrt{J_2'} - \alpha J_1 - \kappa = 0 \tag{4.125}$$

Since this criterion does not include θ as a variable and its dependence on J_1 is linear, graphically Eq. (4.125) represents a conical surface with a circular cross section on any deviatoric plane, as shown in Figs. 4.12 and 4.13.

To determine the material constants α and k, at least two experimental data are needed. There are two options if one has already determined the material constants C and ϕ for Coulomb-Mohr material and wishes to use a Drucker-Prager yield surface in place of the Coulomb-Mohr yield surface so that corners are eliminated. By coinciding a Drucker-Prager circular cone with the outer apexes of the Coulomb-Mohr hexagon at any section, the following relations are obtained:

$$\alpha = \frac{2\sin\phi}{\sqrt{3}\,(3 - \sin\phi)} \qquad \kappa = \frac{6C\cos\phi}{\sqrt{3}\,(3 - \sin\phi)} \tag{4.126}$$

Coincidence of circular cross section with the inner apexes of the Coulomb-Mohr hexagon requires that

$$\alpha = \frac{2\sin\phi}{\sqrt{3}\,(3 + \sin\phi)} \qquad \kappa = \frac{6C\cos\phi}{\sqrt{3}\,(3 + \sin\phi)} \tag{4.127}$$

The Drucker-Prager yield surface coincident with the outer apexes of the Coulomb-Mohr yield surface are shown in Figs. 4.12 and 4.13. As in the von Mises yield surface, the yield locus of the Drucker-Prager yield cone on σ_1-σ_2-plane is an ellipse, though in Fig. 4.14 only the case coincident with outer apexes is shown.

4.5 SUBSEQUENT YIELD SURFACE

Thus far we have only discussed the initial yield surface at which the material yields for the first time on initial loading. Points on the initial yield surface correspond to the point A in Fig. 1.1, which is the initial yield point on the uniaxial stress-strain curve. As Fig. 1.1 shows, after initial yielding further deformation requires an increase of the applied force. As we noted in Section 1.1, if we unload the material after the deformation in the plastic region, say, at point D in Fig. 1.1 and then reload it, the material will deform elastically until the stress reaches the subsequent yield point G, at a stress level slightly lower than that at D. At point G the material yields again, in the same way as it yielded initially at point A. Point G is called the *subsequent yield point*. For metals the subsequent yield point is higher than the initial yield point. This means that the yield stress is raised by the previous plastic deformation.

Just as the initial yield surface is the generalization of the initial yield point for three-dimensional loading, corresponding to the subsequent yield point G, there exists a subsequent yield surface for three dimensional loading in the stress space. Again, it is the boundary between the elastic deformation region and the plastic deformation region at that stage of loading. During the plastic deformation the subsequent yield surface will expand, translate, and distort in the stress space. For some materials such as rock, concrete, and soil the stress-strain curve is lower in the plastic region, a phenomenon called *strain softening*. The subsequent yield surface for those materials contracts in size due to the softening phenomenon.

4.5.1 Experimental Studies

Naghdi, Essenberg, and Koff (1958) examined the subsequent yield surface by carrying out tension-torsion tests using aluminum alloy tubes. They loaded the tubes initially with axial tension and then with various ratios of torsion to axial tension to obtain the initial yield locus in σ-τ-plane. They systematically unloaded and reloaded the tubes to obtain the subsequent yield loci. Their results are shown in Fig. 4.15.

In this figure the initial von Mises ellipse, which is symmetric about the σ axis, becomes unsymmetric because of previous loading. The initial yield locus ellipse moves up, in the previous loading direction. This is due to the Bauschinger effect as discussed in Chapter 1. Also the shape of the subsequent yield locus has been changed or distorted.

Phillips and coworkers have conducted a more extensive and systematic experimental study of the subsequent yield surface. Their results are documented in many papers (1965–1986), but here we present results taken only from the studies reported in 1971, 1985, and 1986.

As reported by Phillips and Das (1985), during these experiments the yield point was obtained by a well-defined operational procedure whereby small excursions were made at the stress point into the plastic region, limiting the plastic strain from 2×10^{-6} to 5×10^{-6}; the yield point was then obtained by backward extrapolation to the elastic line.

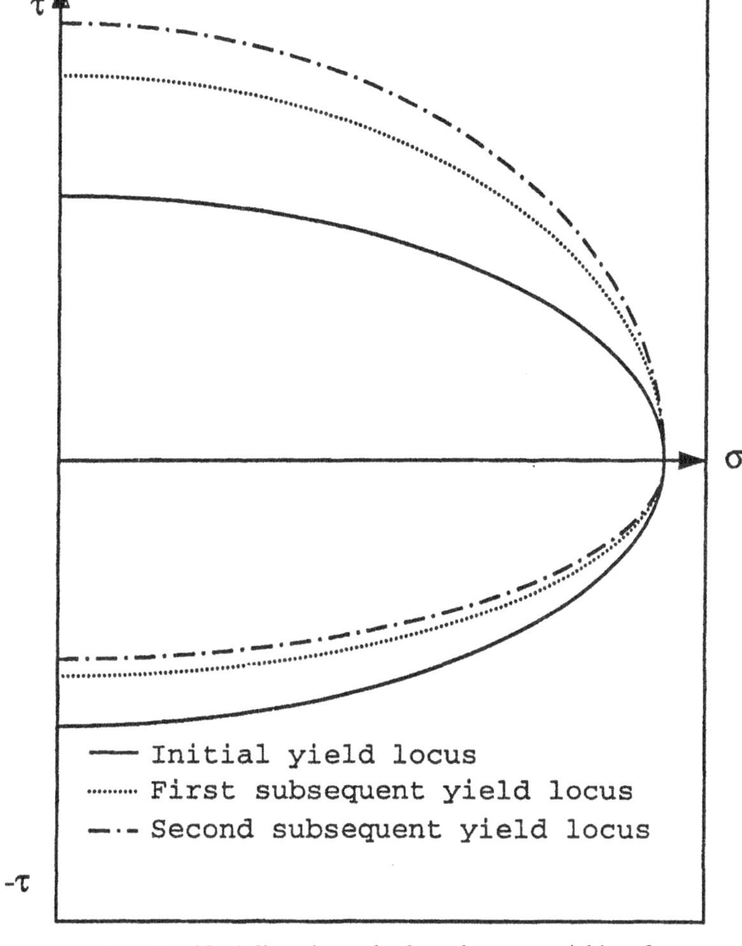

Fig. 4.15 Naghdi et al. results for subsequent yield surface

As Fig. 4.16 shows, during the plastic deformation the yield surface underwent a combination of three changes: expansion or contraction, translation, and distortion. Besides the plastic deformation, temperature had a definite effect on the shape of the final yield surface.

4.5.2 Isotropic and Kinematic Hardening

It is difficult to determine the hardening behavior—or the expansion (contraction), translation, and distortion of the yield surface—quantitatively

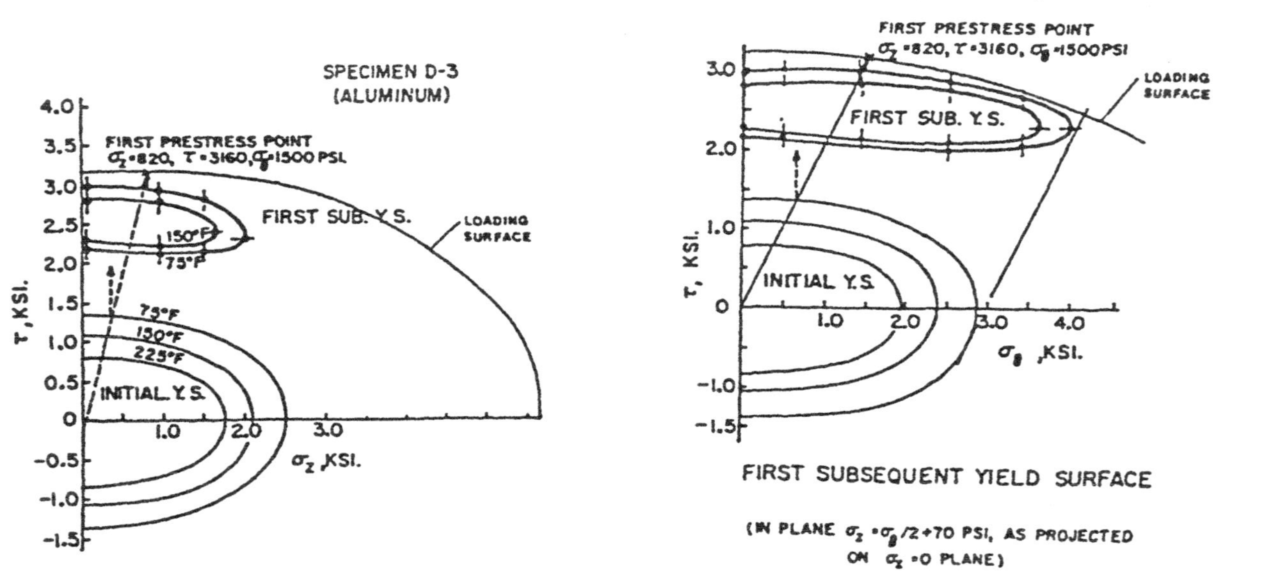

Fig. 4.16 Subsequent yield surface and the influence of the temperature

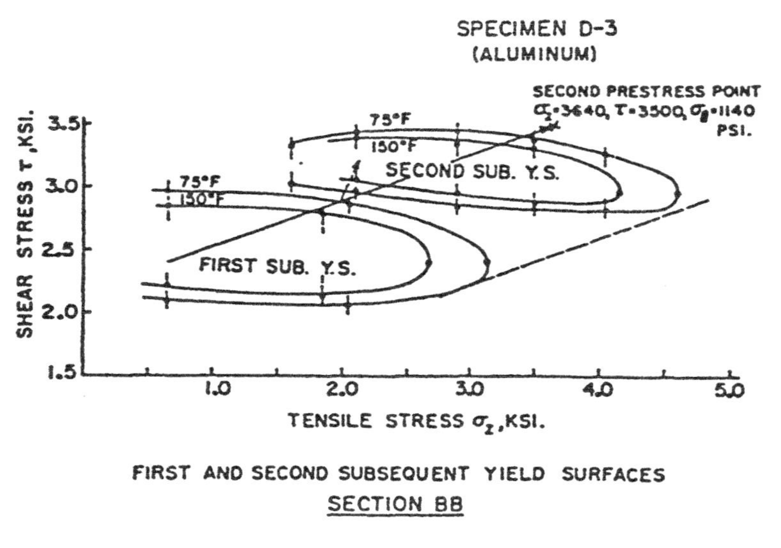

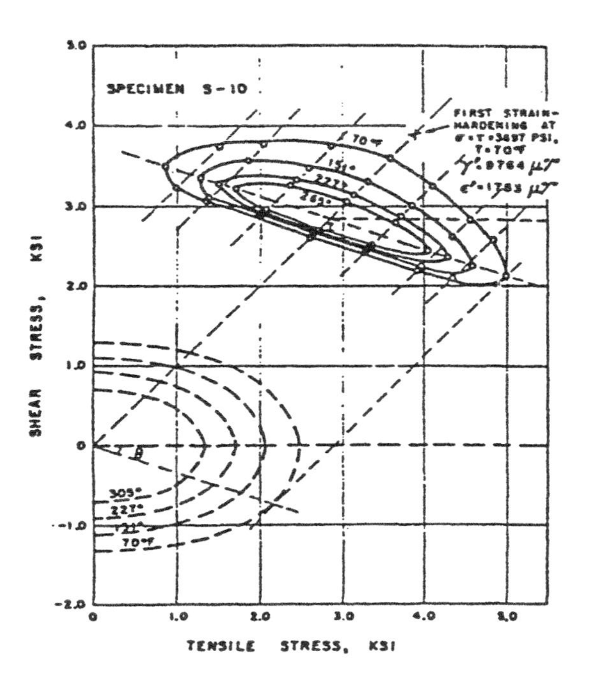

Fig. 4.16 *Continued.*

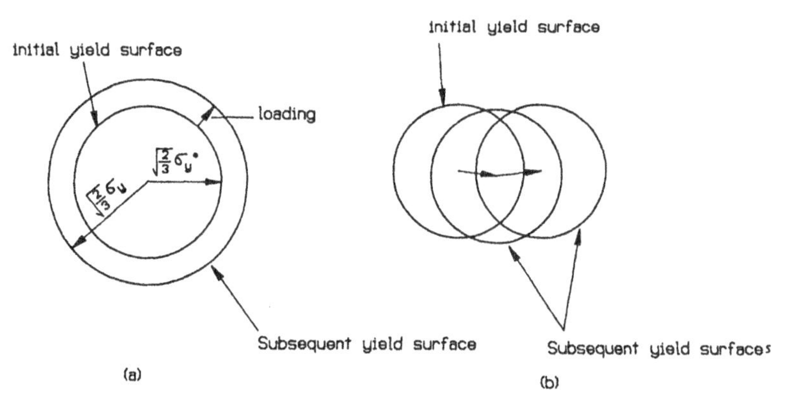

Fig. 4.17 Graphic representation of isotropic and kinematic hardening

without knowledge of the stress-strain relation for plastic deformation. We will cover this problem later in this text. Here we will deal with two simple, classical models: the isotropic hardening and kinematic hardening models.

Isotropic hardening assumes that the subsequent yield surface is a uniform expansion of the initial yield surface, as shown in Fig. 4.17(*a*), and that the material's isotropic response to yielding remains unchanged during plastic deformation. Thus the center of initial and subsequent yield surfaces is the same. It neglects the anisotropic effect on the subsequent yielding induced by the deformation.

If we ignore the effect of hydrostatic pressure on yield,

$$F(J_2', J_3', \kappa) = f(J_2', J_3') - \kappa = 0 \tag{4.128}$$

describes mathematically the isotropic hardening behavior, where κ is a material constant characterizing the isotropic hardening effect. For the von Mises yield surface $\kappa = \sqrt{2/3}\,\sigma_Y$ is the radius of the von Mises circle on the π-plane. Initially $k = \kappa_0 = \sqrt{2/3}\,\sigma_Y^0$. In Eq. (4.128), κ is the only parameter dependent on plastic deformation. For perfectly plastic materials κ remains a constant during plastic deformation, so the yield surface is fixed in size and position in the stress space.

The isotropic hardening model is the simplest one to use, but it cannot predict the Bauschinger effect which is observed experimentally. To calculate the Bauschinger effect, Prager (1956) suggested the kinematic hardening model. This model assumes that the yield surface translates as a rigid body in the stress space during the plastic deformation. As a result the shape of the subsequent yield surface during plastic deformation remains unchanged. This is shown in Fig. 4.17(*b*). This model can be written

$$F(\mathbf{S}, \boldsymbol{\alpha}) = f(\mathbf{S} - \boldsymbol{\alpha}) - \kappa_0 = 0 \tag{4.129}$$

where $\boldsymbol{\alpha}$ is a second-order tensor, known as *back stress*. Geometrically it represents the center of the yield surface in the stress space.

It should be noted that neither the isotropic nor the kinematic hardening model is truly representative of the real material hardening behavior, which can be quite complicated as indicated by Fig. 4.16. Nevertheless, in some cases, such as in proportional loading, these models can provide satisfactory results because the complexity in the subsequent yield surface is more pronounced when the loading path or direction is reversed or drastically changed.

Generally, the yield surface, including the hardening effect, can be written

$$F(\boldsymbol{\sigma}, \alpha_i) = 0 \qquad (i = 1, 2, \ldots, n) \tag{4.130}$$

where α_i are scalar or tensor hardening parameters.

4.5.3 Loading and Unloading in Stress Space

We know from our previous discussion that under mechanical loading a material will deform elastically or plastically depending on the applied loading level. Because the stress-strain relations for elastic deformation and plastic deformation are different, it is important to distinguish between elastic and plastic regions and to identify whether the deformation is elastic or plastic. We can use the yield surface to identify elastic and plastic regions and loading and unloading criteria to identify the characteristics of the deformation.

For perfectly plastic materials the subsequent yield surface at any instant is exactly the same as the initial yield surface. Loading occurs when the stress state point is on the yield surface and remains there, and the plastic deformation can increase indefinitely. If the stress point moves to the inside of the yield surface, it causes unloading. Let $F(\boldsymbol{\sigma}) = f(\boldsymbol{\sigma}) - \kappa_0 = 0$ be the yield function. We have

$$F(\boldsymbol{\sigma}) < 0 \qquad \text{(elastic deformation)}$$

$$F(\boldsymbol{\sigma}) = 0, \quad dF = f(\boldsymbol{\sigma} + d\boldsymbol{\sigma}) - f(\boldsymbol{\sigma}) = \frac{\partial f}{\partial \boldsymbol{\sigma}} : d\boldsymbol{\sigma} = 0 \qquad \text{(loading)}$$

$$\tag{4.131}$$

$$F(\boldsymbol{\sigma}) = 0, \quad dF = f(\boldsymbol{\sigma} + d\boldsymbol{\sigma}) - f(\boldsymbol{\sigma}) = \frac{\partial f}{\partial \boldsymbol{\sigma}} : d\boldsymbol{\sigma} < 0 \qquad \text{(unloading)}$$

In the stress space the direction of vector $\partial f / \partial \boldsymbol{\sigma}$ is in the direction of outer normal $\mathbf{n}$ to the yield surface. So $\partial f / \partial \boldsymbol{\sigma} : d\boldsymbol{\sigma} < 0$ represents a stress increment $d\boldsymbol{\sigma}$ which points inward from the yield surface, while $\partial f / \partial \boldsymbol{\sigma} : d\boldsymbol{\sigma} = 0$ represents a stress increment $d\boldsymbol{\sigma}$ which is on the tangential plane at

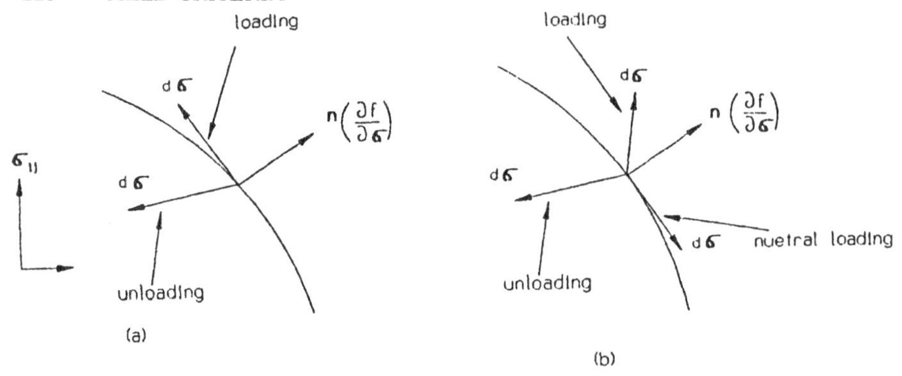

Fig. 4.18 Loading and unloading for (*a*) perfect plastic and (*b*) work-hardening materials

the stress point to the yield surface. Because the yield surface is fixed and can not expand, $d\boldsymbol{\sigma}$ cannot point outward. This is shown in Fig. 4.18(*a*).

Similarly for work-hardening materials

$$\begin{aligned}
&F(\boldsymbol{\sigma}) < 0 && \text{(elastic deformation)} \\
&F(\boldsymbol{\sigma}) = 0, \quad \frac{\partial f}{\partial \boldsymbol{\sigma}} : d\boldsymbol{\sigma} > 0 && \text{(loading)} \\
&F(\boldsymbol{\sigma}) = 0, \quad \frac{\partial f}{\partial \boldsymbol{\sigma}} : d\boldsymbol{\sigma} = 0 && \text{(neutral loading)} \\
&F(\boldsymbol{\sigma}) = 0, \quad \frac{\partial f}{\partial \boldsymbol{\sigma}} : d\boldsymbol{\sigma} < 0 && \text{(unloading)}
\end{aligned} \tag{4.132}$$

It should be emphasized that during plastic deformation, the yield criterion should be satisfied at all times. Thus

$$F(\boldsymbol{\sigma}) > 0 \tag{4.133}$$

is meaningless for time-independent classical plasticity. If a stress point is on the intersecting line of two smooth yield surfaces $F_l(\boldsymbol{\sigma}) = 0$ and $F_m(\boldsymbol{\sigma}) = 0$, such as the corner of the Tresca yield surface, Eq. (4.132) can be modified to

$$\begin{aligned}
&F_l(\boldsymbol{\sigma}) = 0, \quad F_m(\boldsymbol{\sigma}) = 0, \quad \max\left(\frac{\partial f_l}{\partial \boldsymbol{\sigma}} : d\boldsymbol{\sigma}; \frac{\partial f_m}{\partial \boldsymbol{\sigma}} : d\boldsymbol{\sigma}\right) > 0 && \text{(loading)} \\
&F_l(\boldsymbol{\sigma}) = 0, \quad F_m(\boldsymbol{\sigma}) = 0, \quad \max\left(\frac{\partial f_l}{\partial \boldsymbol{\sigma}} : d\boldsymbol{\sigma}; \frac{\partial f_m}{\partial \boldsymbol{\sigma}} : d\boldsymbol{\sigma}\right) = 0 && \text{(neutral loading)} \\
&F_l(\boldsymbol{\sigma}) = 0, \quad F_m(\boldsymbol{\sigma}) = 0, \quad \max\left(\frac{\partial f_l}{\partial \boldsymbol{\sigma}} : d\boldsymbol{\sigma}; \frac{\partial f_m}{\partial \boldsymbol{\sigma}} : d\boldsymbol{\sigma}\right) < 0 && \text{(unloading)}
\end{aligned} \tag{4.134}$$

REFERENCES

Bell, J. F. 1973. The experimental foundations of solid mechanics. In *Handbuck der Physik*, Vol. VIa/1. Berlin: Springer-Verlag.

Betten, J. 1988. Applications of tensor functions to the formulation of yield criteria for anisotropic materials. *Int. J. Plast.* **4**:29.

Betten, J. 1985. Irreducible invariants of fourth-order tensors. In *Mathematical Modeling in Science and Technology, Proc. Fifth Int. Conf. Math. Modeling.* X. J. R. Avula, G. Leitmann, C. D. Mote, and E. Y. Rodin (eds.).

Betten, J. 1984. CISM lecture notes. In *Applications of Tensor Function in Solids Mechanics*, J. P. Boehler (ed.). New York: Springer-Verlag.

Betten, J. 1982. Pressure-dependent yield behaviour of isotropic and anisotropic materials. In *Deformation and Failure of Granular Materials*, P. V. Vermeer and H. J. Luger (eds.). Rotterdam: A. R. Balkema, p. 81.

Bridgeman, P. W. 1952. Studies in large plastic flow and fracture. Metallurgy and Metallurgical Engineering Series. New York: McGraw-Hill.

Coulomb, C. A. 1773. Sur une application des regles de maxims et minims a qulques problems de statique relatifs à l'architecture. *Mémoires de Mathématique et de Physique, Acad. Sci.*, Paris **7**:343.

Dorn, J. E. 1949. Stress-strain relations for anisotropic plastic flow. *J. Appl. Phys.* **20**:15.

Drucker, D. C., and W. Prager. 1952. Soil mechanics and plastic analysis or limit design. *Quart. Appl. Math.* **10**:157.

Hencky, H. Z. 1924. Zur Theorie plasticher deformationen und der hierdurch im Material hervorgerufenen Nachspannungen. *Z. Angew. Math. Mech.* **4**:323.

Hill, R. 1953. On discontinuous plastic states with special reference to localized necking in thin sheets. *J. Mech. Phys. Solids* **1**:19.

Hill, R. 1948. Theory of yielding and plastic flow of anisotropic metals. *Proc. R. Soc., London* **A193**:281.

Huber, M. T. 1904. *Czasopismo Techniczne*, Lemberg, Austria, **22**:181.

Jackson, L. R., K. F. Smith, and W. T. Lankford. 1948. Plastic flow in anisotropic sheet metal. Metals Technology Technical Publication.

Lianis, G., and H. Ford. 1957. An experimental investigation of the yield criterion and the stress-strain law. *J. Mech. Phys. Solids* **5**:215.

Lode, W. 1925. Versuche über den Einfluss der mittleren Hauptspannung auf die Fliessgrenze. *Z. Angew. Math. Mech.* **5**:142.

Naghdi, P. M., F. Essenberg, and W. Koff. 1958. An experimental study of initial and subsequent yield surfaces in plasticity. *J. Appl. Mech.* **25**:201.

Phillips, A. 1986. A review of quasistatic experimental plasticity and viscoplasticity. *Int. J. Plast.* **2**:315.

Phillips, A., and P. K. Das. 1985. Yield surfaces and loading surfaces of aluminum and brass: An experimental investigation at room and elevated temperatures. *Int. J. Plast.* **1**:87.

Phillips, A., and J. Tang. 1972. The effect of loading path on the yield surface at elevated temperatures. *Int. J. Solids Struct.* **8**:463.

Prager, W. 1986. A new method of analyzing stresses and strains in work-hardening plastic solids. *J. Appl. Mech.* **23**:493.

Taylor, G. I., and H. Quinney. 1931. The plastic distortion of metals. *Phil. Trans. R. Soc., London* **A230**:323.

Tresca, H. 1864. Sur l'écoulement des corps solides soumis á de fortes pressions. *C. R. Acad. Sci., Paris* **59**:754.

von Mises, R. 1913. Mechanik der festen Körper im plastisch deformablen Zustand. *Nachr. Ges. Wiss. Göttingen*:582.

5

CLASSICAL THEORY OF PLASTICITY

The remaining chapters of this book will focus on plasticity theory: the mathematical relationships between stress and strain for plastically deformed materials, namely the constitutive equations for plastic deformation of metals and selected other engineering materials. Since the first proposed mathematical model for plastic strain rate by Levy and Saint-Venant in 1870, enormous progress has been made in developing the plasticity theory for infinitesimal deformation. This chapter presents the main elements of the classical theory of plasticity. We will discuss the classical plasticity equations after considering the basic principles of plasticity theory. Included are the classical plastic potential theory based on the Drucker's postulate for stable materials and detailed explanations of the isotropic and kinematic hardening rules. We will derive the general plasticity equations for isotropic or kinematic hardening and combined hardening. Then we will briefly cover the plasticity theories for perfectly plastic materials and the nonassociated flow rules. Also we briefly consider the deformation theory of plasticity, which may be useful under some circumstances. We should emphasize that all these classical aspects are concerned only with the infinitesimal deformation.

5.1 BASIC CONSIDERATIONS OF PLASTICITY THEORY

In Chapter 1 the basic characteristics of plastic deformation were introduced by presenting some simple and fundamental experimental observations. It can be concluded from those experiments that plastic deformation has the

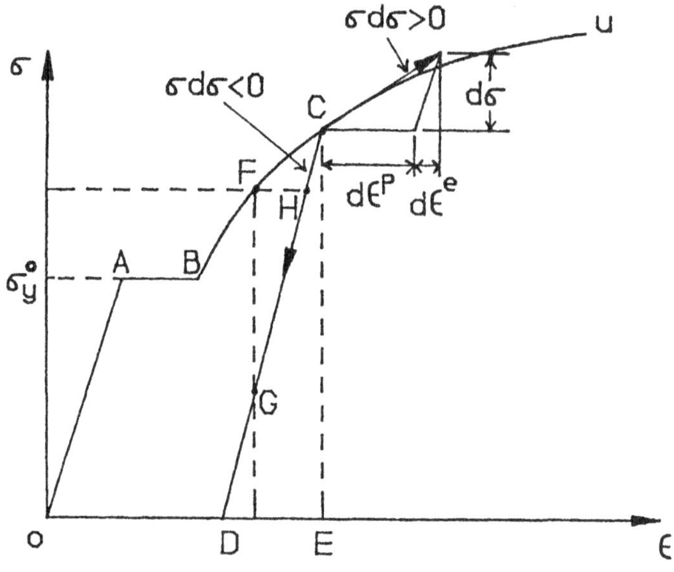

Fig. 5.1 Uniaxial stress-strain curve

following features:

1. Plastic deformation is associated with the dissipation of energy so that it is irreversible. This can be seen from Fig. 5.1 which is a schematic representation of the σ–ϵ diagram in an uniaxial tension test. When unloading occurs at point C, only part of the strain (DE) can be recovered while another part (OD) remains after the load is removed. The energy represented by the triangle DCE is the recovered elastic energy, while the area $OABCD$ represents the energy dissipated during the process producing plastic strain OD. Hence the plastic strain is the permanent strain after unloading. From the figure it can be seen that the total strain OE at an arbitrary point C in the plastic region is the sum of the plastic strain OD and elastic strain DE:

$$\epsilon = \epsilon^{\mathrm{e}} + \epsilon^{\mathrm{p}} \tag{5.1}$$

This additive law can be generalized for a three-dimensional case as

$$\epsilon_{ij} = \epsilon_{ij}^{\mathrm{e}} + \epsilon_{ij}^{\mathrm{p}} \tag{5.2}$$

where ϵ_{ij} is the infinitesimal strain tensor defined by Eq. (2.59). The preceding equation represents a basic assumption used in the classical theory of plasticity and is called the *additive decomposition* of the strain tensor. It should be pointed out that this decomposition is correct for cases of in-

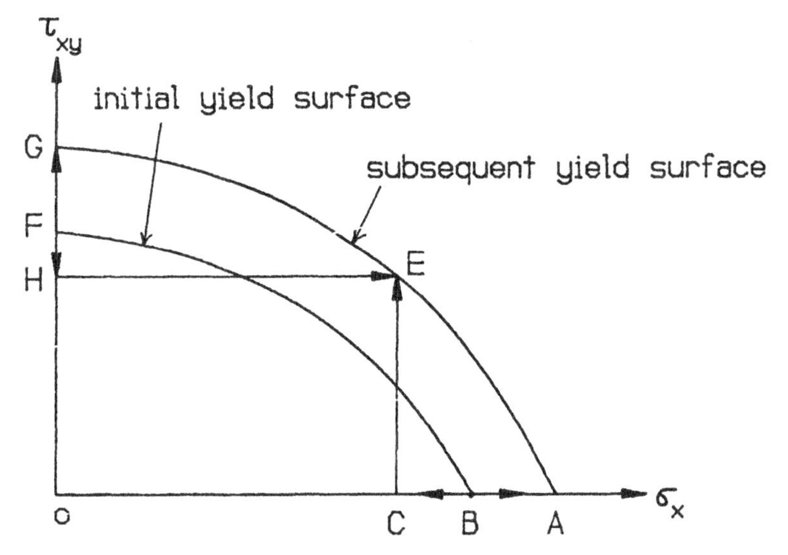

Fig. 5.2 History dependence of plastic deformation

finitesimal strain only. In the case of finite strain, which will be discussed in Chapter 7, it can be shown that there will be geometrical elastic-plastic coupling between the elastic and plastic strain measures so that Eq. (5.2) will lose its conventional physical meaning.

2. Due to its dissipation feature the plastic deformation process is history or path dependent. In other words, there will not be a one-to-one correspondence between stress-strain during plastic deformation. For example, in Fig. 5.1 points F and H have the same stress level but different strains. On the other hand, the strains at points F and G are the same, but stresses are different. It is obvious that these differences result from different deformation histories or stress paths between points F and H, and F and G.

To further explore the history- or path-dependent nature of plastic deformation, consider the following example: Suppose that a specimen is loaded with uniaxial tension in the x direction. It deforms elastically until the stress reaches point B on the initial yield surface shown in Fig. 5.2. If the stress is increased to point A, the material of the specimen will deform elastoplastically, and the yield surface will expand to become the subsequent yield surface due to the hardening behavior of the materials. Here, for simplicity, only isotropic hardening is assumed. At point A, since the stress state is uniaxial,

$$\epsilon_{xx}^{p} = \epsilon^{p} \tag{5.3}$$

where ϵ^{p} is the plastic strain along the x axis at point A. Due to the assumption of incompressibility of plastic deformation, it follows that

$$\epsilon^{p}_{yy} = \epsilon^{p}_{zz} = -\tfrac{1}{2}\epsilon^{p} \tag{5.4}$$

$$\gamma^{p}_{xy} = \gamma^{p}_{yz} = \gamma^{p}_{zx} = 0 \tag{5.5}$$

where γ_{ij} represents twice the shear component of the infinitesimal strain tensor ϵ.

Now unload the specimen from point A to point C, keep the axial force constant, and apply the shear stress τ_{xy} at the same time until point E is reached on the subsequent yield surface. During this process following the path A–C–E, the material behaves elastically because this path is inside the subsequent yield surface which is also the current yield surface. Therefore no further plastic deformation occurs when the stress state of the specimen changes from A to E following A–C–E. The plastic strains at point E can still be represented by Eqs. (5.3), (5.4), and (5.5).

Now consider another loading history. Let the specimen be loaded by pure shear with shear stress τ_{xy}. The elastic deformation continues until the initial yielding at point F. Then plastic deformation will result with increasing τ_{xy}. The yield surface will expand at the same time. Suppose that the shear stress τ_{xy} is increased until point G on the subsequent yield surface, which also contains points A and E from the previous loading history. Denote the plastic shear strain at point G by γ^{p}; then the plastic strains at point G are given by

$$\gamma^{p}_{xy} = \gamma^{p} \tag{5.6}$$

$$\gamma^{p}_{yz} = \gamma^{p}_{zx} = 0 \tag{5.7}$$

$$\epsilon^{p}_{xx} = \epsilon^{p}_{yy} = \epsilon^{p}_{zz} = 0 \tag{5.8}$$

The specimen is then unloaded from point G to point H, and the shear stress τ_{xy} is kept constant while the axial load is applied and increased to bring the stress state of the specimen to point E on the subsequent yield surface. Since the path G–H–E is inside the current yield surface AEG, the material deforms elastically, and no additional plastic deformation occurs following the stress path G–H–E. The plastic strains at point E are also given by Eqs. (5.6), (5.7), and (5.8). Obviously the plastic strains represented by Eqs. (5.6)–(5.8), which are the results of the stress path O–G–H–E, are completely different from those given by Eqs. (5.3)–(5.5), which are obtained by a different stress path O–A–C–E. This example clearly shows that the plastic deformation is a history- or path-dependent process. To obtain the correct final state of deformation, the deformation path or history of deformation must be traced. Recall that, in general, to determine a curve in space mathematically, the differential equation(s) of this curve and the starting and ending points of this curve must be given. Thus the path-dependent nature of plastic deformation requires that the constitutive equations for plastic deformation be in differential equation or incremental form. The incremental plastic strains throughout the deformation history have to be calculated and accumulated to determine the total plastic strains. This is the reason most plasticity theories are normally rate or incremental types in contrast to elastic deformation where constitutive equations have a one-to-one correspondence between the final strain and stress states.

3. Throughout this book, plastic deformation is assumed to be rate insensitive. This means that the constitutive equations for plastic deformation must be homogeneous in time (i.e., invariant with respect to the time scale) and that the rate forms and incremental forms must be equivalent to each other. The viscous effect, and hence the viscous resistance during plastic deformation, are neglected. Thus the stress is linearly related to the elastic strain and can be expressed by

$$\sigma = E\epsilon^{e} = E(\epsilon - \epsilon^{p}) \tag{5.9}$$

Assuming infinitesimal strain and linear elastic response, this equation can be generalized to the three-dimensional case:

$$\boldsymbol{\sigma} = \mathbf{C} : \boldsymbol{\epsilon}^{e} = \mathbf{C} : (\boldsymbol{\epsilon} - \boldsymbol{\epsilon}^{p}) \quad \text{or} \quad \sigma_{ij} = C_{ijkl}(\epsilon_{kl} - \epsilon_{kl}^{p}) \tag{5.10}$$

where **C** is the fourth-order elastic stiffness tensor.

The main point of this discussion is that plastic deformations are governed by nonlinear, homogeneous (in time), differential or rate form, constitutive equations. This notion is further elaborated in the description of simple elastic plastic deformation given in Fig. 5.1 for one-dimensional loading. A constitutive theory for plastic deformation should include the following elements:

1. The initial yield point σ_Y^0 should be known, since for stress values lower than this value deformation is linear and the correspondence between σ and ϵ is one to one. For stress values higher than σ_Y^0, the deformation is nonlinear and history dependent. Generalized to the three-dimensional case, an initial yield surface should be known and can be written in the following form:

$$F(\boldsymbol{\sigma}, \sigma_Y^0) = 0 \tag{5.11}$$

where σ_Y^0 is the initial yield stress.

2. The growth of subsequent yield surfaces should be known. In general, the yield strength is not a constant except for perfectly plastic materials, or in the small region of plastic flow of mild steel, denoted by AB in Fig. 5.1. Thus the subsequent yield strength (i.e., the entire curve BU) should be

known. Generally,

$$\sigma_Y = \sigma_Y(\alpha_i) \qquad (i = 1, 2, \ldots, n) \tag{5.12}$$

where α_i, $i = 1, 2, 3, \ldots, n$, are all possible hardening parameters such as the equivalent plastic strain and internal variables. Of course in the one-dimensional case it is not necessary to include more than one hardening parameter to determine the curve BU. Equations (1.18)–(1.23) are particular forms of Eq. (5.12). For instance, in Eq. (1.18) only one hardening parameter, $\alpha = \epsilon^p$, is used for the monotonically uniaxial loading case. According to the history-dependent nature of the total ϵ^p, it should be calculated by

$$\epsilon^p = \int d\epsilon^p = \int \dot{\epsilon}\, dt \tag{5.13}$$

From Fig. 5.1 it can be shown that

$$d\alpha = d\epsilon^p = d\epsilon - d\epsilon^e = \frac{d\sigma}{E^t} - \frac{d\sigma}{E} = \left(\frac{1}{E^t} - \frac{1}{E}\right) d\sigma \tag{5.14}$$

and equivalently, in rate form

$$\dot{\alpha} = \left(\frac{1}{E^t} - \frac{1}{E}\right)\dot{\sigma} \tag{5.15}$$

where E is the Young's modulus and E^t is the tangential stiffness at a given stress level. In the case of linear hardening E^t is a constant, but in the general nonlinear case shown in Fig. 5.1, E^t is considered to be a function of stress and plastic strain ϵ^p:

$$E^t = E^t(\sigma, \epsilon^p) \tag{5.16}$$

Then the general form of (5.15) becomes

$$\dot{\alpha} = \dot{\alpha}(\sigma, \dot{\sigma}, \alpha) \tag{5.17}$$

Note that $\alpha = \epsilon^p$ in uniaxial loading. As mentioned above, $\dot{\alpha}$ needs to be a linear, homogeneous function of $\dot{\sigma}$ in order to describe the rate-independent behavior. Therefore Eq. (5.17) should be rewritten as

$$\dot{\alpha} = \alpha_1(\sigma, \alpha)\dot{\sigma}$$

or equivalently

$$d\alpha = \alpha_1(\sigma, \alpha)\, d\sigma \tag{5.18}$$

where α_1 is a nonlinear function of σ and α and represents the experimental observations.

Equation (5.18) can be generalized for the three-dimensional case as

$$\dot{\alpha}_i = \boldsymbol{\phi}_i(\boldsymbol{\sigma}, \alpha_i) : \dot{\boldsymbol{\sigma}} \cdot$$

or equivalently

$$d\alpha_i = \boldsymbol{\phi}_i(\sigma, \alpha^{\mathrm{p}}) : \mathbf{d}\sigma \tag{5.19}$$

where $\boldsymbol{\phi}_i$, $i = 1, 2, \ldots, n$, are second-, third- or fourth-order tensor-valued functions of $\boldsymbol{\sigma}$, and α_i, depending on whether α_i is a scalar, vector, or second-order tensor-valued hardening parameter. Equation (5.19) is the evolution or constitutive equation for the hardening parameters α_i. Once the current values of these hardening parameters are determined from the evolution Eq. (5.19) by the stress path or history, they can be substituted into the expression of the yield surface to obtain the current or subsequent yield surface

$$F(\boldsymbol{\sigma}, \alpha_i) = 0 \qquad (i = 1, 2, \ldots, n) \tag{5.20}$$

where the general form of yield surface Eq. (4.130) is used.

Recall that for α_i, $i = 1, 2, \ldots, n$, can be scalar(s), vector(s), or tensor(s), thus representing the expansion, translation, and distortion of the yield surface in the stress space. Note that the yield surface will change only when plastic deformation occurs. This implies that Eqs. (5.18) and (5.19) are valid only during the process of elastic-plastic loading. During elastic deformation or unloading, no plastic deformation will occur, and the yield surface will remain unchanged, with

$$\dot{\alpha}_i = 0 \tag{5.21}$$

for these processes.

3. The constitutive equations for plastic deformation should be formulated. This process is central to plasticity theory. As discussed earlier, these equations should be in rate or increment form because of path dependence and rate insensitivity. For the one-dimensional case the plastic strain rate is easily given by Eq. (5.14), rewritten as

$$\dot{\epsilon}^{\mathrm{p}} = \left(\frac{1}{E^t} - \frac{1}{E}\right)\dot{\sigma} \quad \text{or} \quad d\epsilon^{\mathrm{p}} = \left(\frac{1}{E^t} - \frac{1}{E}\right) d\sigma \tag{5.22}$$

Using Eqs. (5.18) and (5.15), we obtain the general form of $\dot{\epsilon}^{\mathrm{p}}$ in the one-dimensional case:

$$\dot{\epsilon}^{\mathrm{p}} = \alpha^{\mathrm{p}}(\sigma, \alpha)\dot{\sigma} \quad \text{or} \quad d\epsilon^{\mathrm{p}} = \alpha^{\mathrm{p}}(\sigma, \alpha)\, d\sigma \tag{5.23}$$

where α^{p} is a nonlinear function of σ and ϵ^{p}. For the three-dimensional case this equation is generalized to

$$\dot{\boldsymbol{\epsilon}}^{\mathrm{p}} = \boldsymbol{\alpha}^{\mathrm{p}}(\boldsymbol{\sigma}, \alpha_i) : \dot{\boldsymbol{\sigma}} \quad \text{or} \quad d\dot{\boldsymbol{\epsilon}}^{\mathrm{p}} = \boldsymbol{\alpha}^{\mathrm{p}}(\boldsymbol{\sigma}, \alpha_i) : d\boldsymbol{\sigma} \qquad (i = 1, 2, \ldots, n) \quad (5.24)$$

where $\boldsymbol{\alpha}^{\mathrm{p}}$ is a fourth-order tensor valued function of $\boldsymbol{\sigma}$ and α_i, and $\dot{\boldsymbol{\sigma}}$ and $d\boldsymbol{\sigma}$ are the rate and increment of the stress tensor, respectively.

The elastic strain rate or increment is given by the linear law

$$\dot{\epsilon}^{\mathrm{e}} = \mathbf{C}^{-1} : \dot{\boldsymbol{\sigma}} \quad \text{or} \quad d\epsilon^{\mathrm{e}} = \mathbf{C}^{-1} : \mathbf{d}\boldsymbol{\sigma} \qquad (5.25)$$

which is obtained by differentiating Eq. (5.19) and assuming that $\mathbf{C}$ is a constant tensor during the deformation. $\mathbf{C}^{-1}$ is the inverse of $\mathbf{C}$, and represents the elastic compliance tensor. The total strain rate is then obtained by combining Eqs. (5.24) and (5.25) to obtain

$$\dot{\boldsymbol{\epsilon}} = \left(\boldsymbol{\alpha}^{\mathrm{p}} + \mathbf{C}^{-1}\right) : \dot{\boldsymbol{\sigma}} \quad \text{or} \quad d\boldsymbol{\epsilon} = \left(\boldsymbol{\alpha}^{\mathrm{p}} + \mathbf{C}^{-1}\right) : d\boldsymbol{\sigma} \qquad (5.26)$$

Inverting this relationship, we get

$$\dot{\boldsymbol{\sigma}} = \left[\boldsymbol{\alpha}^{\mathrm{p}} + \mathbf{C}^{-1}\right]^{-1} : \dot{\boldsymbol{\epsilon}} \quad \text{or} \quad d\boldsymbol{\sigma} = \left[\boldsymbol{\alpha}^{\mathrm{p}} + \mathbf{C}^{-1}\right]^{-1} : d\boldsymbol{\epsilon} \qquad (5.27)$$

It should be emphasized here that Eqs. (5.23) and (5.24) are valid for elastic-plastic deformation processes in which additional plastic deformation occurs. Otherwise, there will be only elastic deformation, and the stress-strain relationship will be governed by the elastic deformation law:

$$\dot{\boldsymbol{\sigma}} = \mathbf{C} : \dot{\boldsymbol{\epsilon}}^{\mathrm{e}} = \mathbf{C} : \dot{\boldsymbol{\epsilon}} \quad \text{or} \quad d\boldsymbol{\sigma} = \mathbf{C} : d\boldsymbol{\epsilon} \qquad (5.28)$$

and

$$\dot{\boldsymbol{\epsilon}}^{\mathrm{P}} = 0 \qquad (5.29)$$

In the one-dimensional case this corresponds to the deformation following lines OA or DC.

4. The loading-unloading criterion must be specified. As mentioned earlier, the constitutive equations for α_i and ϵ^{p} differ for plastic loading and elastic loading or unloading. This is actually the main difference between nonlinear elasticity theory and plasticity theory. Therefore, to describe elastic-plastic behavior by the plasticity theory, it is necessary to identify the process as belonging to plastic loading or elastic unloading. The criterion for loading or unloading should become a part of the theory. Loading and unloading represent a deformation process starting from a plastic state and continuing to deform plastically and then returning to the elastic region. A previous plastic state is always implied. For a hardening material under

uniaxial loading, shown in Fig. 5.1, this means that the stress point is on the curve BU, say, at point C, and satisfies the yield condition

$$\sigma_c - \sigma_Y(\epsilon^p) = 0 \tag{5.30}$$

where σ_c is the stress at point C. The following equations represent loading and unloading criterion:

$$\sigma d\sigma \geq 0 \qquad \text{for loading} \tag{5.31}$$

$$\sigma d\sigma < 0 \qquad \text{for unloading} \tag{5.32}$$

For the uniaxial tension in Fig. 5.1, either $d\sigma > 0$ or $d\sigma < 0$ is enough to represent loading or unloading. However, $\sigma d\sigma$ should be used as an index for loading and unloading if the reverse loading in compression in which $\sigma < 0$ is being considered. The equality sign in Eq. (5.31) holds for perfectly plastic deformations, such as those in region AB in Fig. 5.1.

The loading and unloading criteria in the three-dimensional case were given in Section 4.5 by Eq. (4.131) for perfectly plastic materials and by Eq. (4.132) for hardening materials. It can be seen that the index governing loading or unloading is

$$l = \frac{\partial f}{\partial \boldsymbol{\sigma}} : d\boldsymbol{\sigma} \tag{5.33}$$

For hardening materials $l > 0$ means loading, $l < 0$ represents unloading, and $l = 0$ signifies neutral loading.

The four fundamental elements of plastic deformation discussed above constitute a theory of plasticity. The first element—the initial yield surface—was the subject of the last chapter, where we discussed in detail the basic concepts of the subsequent yield surface and its evolution—namely its change of position and shape. The evolution or constitutive equations for the hardening parameters α_i and the plastic strain ϵ^p are the second and third elements of our plasticity theory, and they are our main concern in the remainder of this book. This chapter focuses on the classical constitutive equations for ϵ^p and α_i. The fourth element—that is, the loading and unloading criteria—was briefly presented in Section 4.5 for perfectly plastic and strain hardening materials. For strain-softening materials, such as rocks, concrete, or granular metals, it is necessary to formulate loading and unloading criteria in the strain space. These concepts will be discussed in Chapter 8. In stress space, or more precisely in terms of l defined by Eq. (5.33), the plastic loading of unstable materials showing strain softening behavior, cannot be distinguished from the elastic unloading. This topic will also be covered in Chapter 8.

We now move to perfectly plastic and hardening materials. We will explain what is meant by the term "hardening materials" after we introduce Drucker's postulate in Section 5.6. For hardening materials modified forms of ϕ_i and α^p were defined in Eqs. (5.19) and (5.24) and given in terms of the index l by Eq. (5.33). We begin by summarizing the elements 2 through 4 of plastic deformation, $\dot{\alpha}_i$ and $\dot{\epsilon}^p$, as

$$\dot{\alpha}_i = \begin{cases} 0 & \text{for } F < 0 & (5.34a) \\ 0 & \text{for } F = 0, l < 0 & (5.34b) \\ 0 & \text{for } F = 0, l = 0 & (5.34c) \\ \boldsymbol{\phi}_i : \dot{\boldsymbol{\sigma}} & \text{for } F = 0, l > 0 & (5.34d) \end{cases}$$

$$\dot{\boldsymbol{\epsilon}}^p = \begin{cases} 0 & \text{for } F < 0 & (5.35a) \\ 0 & \text{for } F = 0, l < 0 & (5.35b) \\ 0 & \text{for } F = 0, l = 0 & (5.35c) \\ \boldsymbol{\alpha}^p : \dot{\boldsymbol{\sigma}} & \text{for } F = 0, l > 0 & (5.35d) \end{cases}$$

These equations represent, respectively, the four cases of elastic deformation, elastic unloading, neutral loading, and plastic loading. The reason why $\dot{\alpha}_i = 0$, $\dot{\boldsymbol{\epsilon}}^p = 0$, during neutral loading is that the continuity condition of plastic deformation has to be satisfied when the stress state changes from unloading to loading. The plastic strain rate $\dot{\boldsymbol{\epsilon}}^p$ is nonzero if $d\boldsymbol{\sigma}$ points outward from the yield surface and is zero if $d\boldsymbol{\sigma}$ is directed inward to the interior of the yield surface. To avoid the discontinuities in the stress-strain relations, $\dot{\boldsymbol{\epsilon}}^p$ has to be zero when $d\boldsymbol{\sigma}$ is in a tangential direction at any point on the yield surface. This is called the *continuity condition* or *postulate* for plastic hardening. Using the definition of l, Eqs. (5.34d) and (5.35d) can be modified to

$$\dot{\alpha}_i = \hat{\boldsymbol{\phi}}_i l \qquad \text{for } F = 0, l > 0 \tag{5.36}$$

$$\dot{\boldsymbol{\epsilon}}^p = \hat{\boldsymbol{\alpha}}^p l \qquad \text{for } F = 0, l > 0 \tag{5.37}$$

where $\hat{\phi}_i = \hat{\phi}_i(\sigma, \alpha_i)$ is a scalar-, vector-, or tensor-valued function of σ and α_i depending on whether α_i is a scalar, vector, or tensor respectively; $\hat{\boldsymbol{\alpha}}^p$ is a second-order tensor-valued functions of $\boldsymbol{\sigma}$ and α_i. Note that the constitutive equations of $\dot{\boldsymbol{\epsilon}}^p$ for perfectly plastic materials cannot be given by Eq. (5.37), since $l = 0$ for these materials even for loading.

5.2 STRESS-STRAIN RELATIONS FOR ELASTIC DEFORMATION

The stress-strain relations for elastic deformation are briefly reviewed here. The simplest constitutive model for elastic deformation was introduced by Hooke (1678). He proposed a linear relation between stress and strain for uniaxial loading. This relationship was extended to the three-dimensional case to give the generalized Hooke's law in the cartesian coordinate system for isotropic materials:

$$\begin{aligned}
\epsilon_{xx} &= \frac{1}{E}\left[\sigma_{xx} - \nu(\sigma_{yy} + \sigma_{zz})\right] & \gamma_{xy} &= \frac{1}{G}\tau_{xy} \\
\epsilon_{yy} &= \frac{1}{E}\left[\sigma_{yy} - \nu(\sigma_{xx} + \sigma_{zz})\right] & \gamma_{yz} &= \frac{1}{G}\tau_{yz} \\
\epsilon_{zz} &= \frac{1}{E}\left[\sigma_{zz} - \nu(\sigma_{xx} + \sigma_{yy})\right] & \gamma_{xz} &= \frac{1}{G}\tau_{xz}
\end{aligned} \tag{5.38}$$

where E, ν, and G are Young's modulus, Poisson's ratio, and the shear modulus, respectively. G is related to E and ν by the equation

$$G = \frac{E}{2(1 + \nu)} \tag{5.39}$$

Equations (5.38) can be written in tensorial notations as

$$\epsilon_{ij} = \frac{\sigma_{ij}}{2G} - \frac{\nu}{E}J_1\delta_{ij} \tag{5.40}$$

where δ_{ij} is the Kronecker delta and J_1 is the first invariant of the stress tensor $\boldsymbol{\sigma}$:

$$J_1 = \sigma_1 + \sigma_2 + \sigma_3 = \sigma_{xx} + \sigma_{yy} + \sigma_{zz} \tag{5.41}$$

From Eq. (5.40) the elastic deformation relationship for volume changes can be derived as

$$\epsilon_{ii} = \frac{J_1}{2G} - \frac{3\nu}{E}J_1 = J_1\left(\frac{1 + \nu}{E} - \frac{3\nu}{E}\right) = \frac{J_1}{3}\frac{3(1 - 2\nu)}{E} \tag{5.42}$$

Upon introducing the notations,

$$p = \frac{J_1}{3}, \quad I_1 = \epsilon_{xx} + \epsilon_{yy} + \epsilon_{zz} = \epsilon_1 + \epsilon_2 + \epsilon_3 = \epsilon_{kk} \tag{5.43}$$

where p is the mean stress or hydrostatic pressure and I_1 is the volume strain or the first invariant of the strain tensor $\boldsymbol{\epsilon}$. Bulk modulus K is given by

$$K = \frac{E}{3(1 - 2\nu)} \tag{5.44}$$

Using Eqs. (5.43) and (5.44), we can rewrite Eq. (5.42) as

$$p = K\epsilon_{kk} = 3K\epsilon_m \tag{5.45}$$

where $\epsilon_m = I_1/3$ is the mean strain. This equation represents the elastic deformation law for dilatation or volume change.

Subtracting $\epsilon_m\delta_{ij}$ from both sides of Eq. (5.40) and making use of Eq. (5.45) result in the equation

$$\epsilon_{ij} - \epsilon_m\delta_{ij} = \frac{1}{2G}(\sigma_{ij} - \delta_{ij}p) \tag{5.46}$$

From the definition for the deviatoric stress tensor S_{ij} given in Eq. (4.18) and by a similar definition for the deviatoric strain tensor which we express as

$$\epsilon'_{ij} = \epsilon_{ij} - (\epsilon_m)\delta_{ij} \tag{5.47}$$

we can rewrite Eq. (5.46) as

$$\epsilon'_{ij} = \frac{1}{2G}S_{ij} \tag{5.48}$$

This equation is called the *elastic deformation law of distortion or shape change*. It can be seen from Eqs. (5.45) and (5.48) that in the elastic region, the mean stress and mean strain are related by the elastic constant K, the bulk modulus, while the deviatoric stress and strain are related to each other through another elastic constant G, the shear modulus.

Sometimes it is more convenient to write the generalized Hooke's law, Eq. (5.40), in the inverse form (i.e., express stress σ_{ij} in terms of strain ϵ_{ij}). It is easy to show that the inverse form of Eq. (5.40) is given by

$$\sigma_{ij} = 2G\epsilon_{ij} + \lambda\epsilon_{kk}\delta_{ij} \tag{5.49}$$

where λ is Lame elastic constant and can be expressed in terms of E and ν by

$$\lambda = \frac{E\nu}{(1+\nu)(1-2\nu)} \tag{5.50}$$

or in terms of ν and K by

$$\lambda = \frac{3\nu K}{1+\nu} \tag{5.51}$$

If the following equivalent or effective stress σ_e and strain ϵ_e are introduced

$$\sigma_e = \sqrt{3J_2'} = \frac{1}{\sqrt{2}}\left[(\sigma_1 - \sigma_2)^2 + (\sigma_2 - \sigma_3)^2 + (\sigma_3 - \sigma_1)^2\right]^{1/2} = \sqrt{\frac{3}{2}S_{ij}S_{ij}} \tag{5.52}$$

$$\epsilon_e = \frac{2}{\sqrt{3}}\left[(\epsilon_1 - \epsilon_2)^2 + (\epsilon_2 - \epsilon_3)^2 + (\epsilon_1 - \epsilon_3)^2\right]^{1/2} = \sqrt{\frac{2}{3}\epsilon'_{ij}\epsilon'_{ij}} \tag{5.53}$$

For uniaxial loading along the x axis, from the definition above $\sigma_e = \sigma_x$ and $\epsilon_e = \epsilon_x$ if $\nu = \frac{1}{2}$ is assumed. It is easy to derive the following equation from Eq. (5.48):

$$\sigma_e = 3G\epsilon_e \tag{5.54}$$

Thus Eq. (5.48) can be written in the form

$$S_{ij} = \frac{2\sigma_e}{3\epsilon_e}\epsilon'_{ij} \tag{5.55}$$

This equation can be generalized and used in the deformation theory of plasticity which we will discuss later. Now we calculate the specific elastic energy, which is equal to the work done by the external forces per unit volume during the elastic deformation:

$$\begin{aligned} W^e &= \tfrac{1}{2}\sigma_{ij}\epsilon_{ij} = \tfrac{1}{2}(S_{ij} + p\delta_{ij})(\epsilon'_{ij} + \epsilon_m\delta_{ij}) \\ &= \tfrac{1}{2}p\epsilon_m + \tfrac{1}{2}S_{ij}\epsilon'_{ij} = W_v^e + W_s^e \end{aligned} \tag{5.56}$$

where the first term W_v^e represents the specific energy of volume strain, and the second term W_s^e is the specific energy of distortion or shape change. Using Eqs. (5.52)–(5.54), we can further show that

$$\begin{aligned} W_s^e &= \frac{1}{2}S_{ij}\epsilon'_{ij} = \frac{1}{2}S_{ij}\cdot\frac{1}{2G}S_{ij} = \frac{1}{2G}J_2' \\ &= \frac{1}{2}\sigma_e\epsilon_e \end{aligned} \tag{5.57}$$

Finally, it should be pointed out that if the material is anisotropic, then the generalized Hooke's law will read

$$\sigma_{ij} = C_{ijkl}\epsilon_{kl} \tag{5.58}$$

with

$$C_{ijkl} = C_{klij} = C_{jikl} = C_{ijlk} \tag{5.59}$$

where C_{ijkl} is the stiffness tensor of the fourth order and should have 21 independent components. Moreover the components C_{ijkl} do not have to be constant during the deformation. They can be functions of plastic deformation. For example, for materials showing elastic-plastic coupling during plastic deformation, the elastic moduli are the functions of the plastic strain. This concept will be discussed later when dealing with unstable materials.

5.3 VOLUME CHANGE AND POISSON'S RATIO FOR PLASTIC DEFORMATION

In the theory of plasticity the constitutive equation for the deviatoric plastic strain, which represents the shape change of the materials, is usually proposed separately from the one for the mean plastic strain, which is a measure of the volume change. This is because the responses of the volume change and shape change to the load are completely different so that they have to be described by entirely different constitutive equations.

As we noted in Chapter 1, one fundamental assumption used to establish the plasticity theory is that plastic deformation is isochoric or volume preserving. This assumption is based on Bridgman's experimental observations (1923, 1949a, 1949b) which established that even at very high hydrostatic pressure, volume change is only reversible thus indicating only elastic volume change as given by Eq. (5.45). In the case of infinitesimal strain this is expressed by

$$\epsilon_{kk}^{p} = \epsilon_{xx}^{p} + \epsilon_{yy}^{p} + \epsilon_{zz}^{p} = \epsilon_{1}^{p} + \epsilon_{2}^{p} + \epsilon_{3}^{p} = 0 \tag{5.60}$$

so that the plastic strain tensor is merely a deviatoric strain tensor

$$\boldsymbol{\epsilon}^{p} = \boldsymbol{\epsilon}'^{p} + \epsilon_{m}^{p}\mathbf{I} = \boldsymbol{\epsilon}'^{p} \tag{5.61}$$

Based on this assumption, only the constitutive equation for the plastic deviatoric strain tensor (i.e., the rule for plastic distortion) is needed to establish a plasticity theory.

A significant feature of this volume change rule is that it provides information about the transverse deformation under uniaxial loading conditions. In other words, Poisson's ratio can be derived by assuming no plastic volume change. The determination of Poisson's ratio in the plastic deformation range for elastically isotropic materials is defined by

$$\nu^{p} = -\frac{\epsilon_{yy}}{\epsilon_{xx}} = -\frac{\epsilon_{zz}}{\epsilon_{xx}} \tag{5.62}$$

where the uniaxial loading is assumed to be in the x direction. The superscript p implies that deformation is in the plastic regime. Thus

$$\epsilon_{yy} = -\nu^{\mathrm{p}}\epsilon_{xx}$$

$$\epsilon_{zz} = -\nu^{\mathrm{p}}\epsilon_{xx} \tag{5.63}$$

Since $\epsilon^{\mathrm{p}}_{kk} = 0$, the volume strain ϵ_{kk} has only an elastic component which is governed by Eq. (5.45). Using Eqs. (5.45) and (5.63), we obtain

$$(1 - 2\nu^{\mathrm{p}})\epsilon_{xx} = \frac{\sigma_{xx}}{3K} = \frac{1 - 2\nu}{E}\sigma_{xx} \tag{5.64}$$

In the equation above, ν should not be confused with ν^{p}. ν is the Poisson's ratio in the elastic region and is defined by

$$\nu = -\frac{\epsilon^{\mathrm{e}}_{yy}}{\epsilon^{\mathrm{e}}_{xx}} = -\frac{\epsilon^{\mathrm{e}}_{zz}}{\epsilon^{\mathrm{e}}_{xx}} \tag{5.65}$$

and is a constant in most cases. However, as shown by Eq. (5.62), ν^{p} is defined in terms of the total strain in the plastic deformation region and in general will vary with strain ϵ. This will become more clear if ν^{p} is determined from Eq. (5.64):

$$\nu^{\mathrm{p}} = \frac{1}{2} - \left(\frac{1}{2} - \nu\right)\frac{\sigma_{xx}}{E\epsilon_{xx}} \tag{5.66}$$

It can be seen that if the deformation is elastic, $\sigma_{xx} = E\epsilon_{xx}$ (according to Hooke's law), this equation gives $\nu^{\mathrm{p}} = \nu$. Generally, $\nu^{\mathrm{p}} = \nu^{\mathrm{p}}(\epsilon)$ is a function of strain and varies during the process of deformation. Consider the following three particular cases:

1. For perfectly plastic materials, $\sigma_{xx} = \sigma_{\mathrm{Y}} =$ constant, the dependence of ν^{p} on ϵ_{xx} is hyperbolic. Equation (5.66) shows that ν^{p} approaches $\frac{1}{2}$ asymptotically as ϵ_{xx} increases.
2. For linear hardening materials $\sigma_{xx} = \sigma^0_{\mathrm{Y}} + E'(\epsilon_{xx} - \epsilon^0_{\mathrm{Y}})$, where σ^0_{Y} is the initial yield stress, ϵ^0_{Y} is the corresponding strain, and E' is the slope of the linear hardening region. Then Eq. (5.66) becomes

$$\nu^{\mathrm{p}} = \frac{1}{2} - \left(\frac{1}{2} - \nu\right)\left(\frac{\sigma^0_{\mathrm{Y}} - E'\epsilon^0_{\mathrm{Y}}}{E\epsilon_{xx}} + \frac{E'}{E}\right) \tag{5.67}$$

The dependence of ν^{p} on ϵ_{xx} is still hyperbolic in this case. Note that the asymptotic value of ν^{p} when ϵ_{xx} becomes large is

$$\nu^{p} = \frac{1}{2} - \left(\frac{1}{2} - \nu\right)\frac{E'}{E} \tag{5.68}$$

which is smaller than $\frac{1}{2}$. However, for most of the materials, $E' \ll E$, ν^{p} is very close to $\frac{1}{2}$.

3. If the elastic strain is very small compared with the plastic strain, the volume change caused by elastic deformation can be neglected, and Eq. (5.60) should reduce to $\epsilon_{kk} = 0$. Then from Eq. (5.64) follows the Poisson ratio for incompressible materials

$$\nu^{p} = \tfrac{1}{2} \tag{5.69}$$

As indicated by these three cases, ν^{p} is always smaller than $\frac{1}{2}$ and approaches it from the lower side. Only for incompressible materials does $\nu^{p} = \frac{1}{2}$ exactly. However, for most of the materials, when the strain becomes finite, the hardening effect is negligible compared with Young's modulus. Elastic strain is always a negligible part of the total strain, and $\nu^{p} = \frac{1}{2}$ provides a good approximation for these cases. Moreover $\nu^{p} = \frac{1}{2}$ simplifies the solution of the problem of elastic-plastic deformation. Therefore it is commonly accepted in the plasticity theory. However, it should be kept in mind that this is only an approximation that is reasonably accurate in the case of large deformation or in the region where the plastic deformation increases without an increase in the stress (i.e., region *AB* in Fig. 5.1). If the magnitude of the elastic strain is comparable with that of the plastic strain, then ν^{p} should be in the range $\nu \le \nu^{p} \le \frac{1}{2}$, and it can be calculated from Eq. (5.65) if the volume change is only the elastic as represented by Eq. (5.45). The preceding discussion assumes that plastic incompressibility and linear elastic volume change, Eq. (5.45). Generally, if the elastic volume change is governed by a nonlinear relation, then

$$\epsilon_{m} = \epsilon_{m}(p) \tag{5.70}$$

where $\epsilon_{m}(p)$ is a nonlinear function of hydrostatic or volume stress. Equation (5.66) becomes

$$\nu^{p} = \frac{1}{2} - \frac{2}{3}\frac{\epsilon_{m}(\sigma_{xx}/3)}{\epsilon_{xx}} \tag{5.71}$$

where uniaxial loading in the x direction is implied and $p = \sigma_{xx}/3$.

5.4 LEVY-MISES EQUATIONS

The first attempt to formulate the stress-strain relationship for plastic deformation was made by Saint-Venant (1870). He worked on the plane plastic strain problem using Tresca's criterion and assuming the work hardening to be zero. For the first time he proposed that the principal axes of the strain increment coincided with the axes of principal stress. The elastic strain $\boldsymbol{\epsilon}^e$ was neglected so that the plastic strain $\boldsymbol{\epsilon}^p$ was equal to the total strain $\boldsymbol{\epsilon}$. The generalization of Saint-Venant's idea for the three-dimensional case was done by Levy (1870, 1871) and independently by Mises (1913), since Levy's work remained unknown outside of his country at that time. The theory, called the *Levy-Mises theory of plasticity*, can be stated as follows:

1. The elastic strain $\boldsymbol{\epsilon}^e$ is so small as to be negligible.
2. The increment of strain $d\boldsymbol{\epsilon}$, or equivalently the rate $\dot{\boldsymbol{\epsilon}}$, is coaxial with stress $\boldsymbol{\sigma}$.

As Eq. (4.22) shows, the principal axes of the stress $\boldsymbol{\sigma}$ are the same as those of the deviatoric stress $\mathbf{S}$, so $\dot{\boldsymbol{\epsilon}}$ or $d\boldsymbol{\epsilon}$ is coaxial with $\mathbf{S}$. This follows from the Levy-Mises equation

$$\dot{\boldsymbol{\epsilon}} = \dot{\lambda}\mathbf{S} \quad \text{or} \quad \dot{\epsilon}_{ij} = \dot{\lambda} S_{ij} \tag{5.72}$$

where $\dot{\lambda}$ is a proportional parameter which will be determined later. This equation can be rewritten in components of the cartesian coordinate system as

$$\frac{\dot{\epsilon}_{xx}}{S_{xx}} = \frac{\dot{\epsilon}_{yy}}{S_{yy}} = \frac{\dot{\epsilon}_{zz}}{S_{zz}} = \frac{\dot{\epsilon}_{xy}}{S_{xy}} = \frac{\dot{\epsilon}_{yz}}{S_{yz}} = \frac{\dot{\epsilon}_{zx}}{S_{zx}} = \dot{\lambda} \tag{5.73}$$

Parameter $\dot{\lambda}$ is determined from the yield criterion. Mises suggested that the von Mises criterion was more convenient for this theory, even though earlier Saint-Venant used Tresca criterion. Recalling Eqs. (4.20), (4.61), and (4.64), we can express the von Mises criterion by

$$S_{ij} S_{ij} = \tfrac{2}{3}\sigma_Y^2 \tag{5.74}$$

where σ_Y is the yield stress, which we assume to be constant since we ignore the strain-hardening effect here. The generalization to include the strain-hardening effect is straightforward as briefly outlined in the next section. We will discuss strain-hardening behavior after we introduce Drucker's postulate in Section 5.6. By squaring Eq. (5.72), we can derive

$$\dot{\epsilon}_{ij}\dot{\epsilon}_{ij} = \dot{\lambda}^2 S_{ij}S_{ij} = \dot{\lambda}^2 \frac{2}{3}\sigma_Y^2 \tag{5.75}$$

in terms of Eq. (5.74). Thus

$$\dot{\lambda} = \sqrt{\frac{3\dot{\epsilon}_{ij}\dot{\epsilon}_{ij}/2}{\sigma_Y}} = \frac{3}{2}\frac{\dot{\epsilon}_e}{\sigma_Y} \tag{5.76}$$

where $\dot{\epsilon}_e$, the equivalent or effective increment of strain, is defined by

$$\dot{\epsilon}_e = \frac{2}{\sqrt{3}}\left[\left(\dot{\epsilon}_1 - \dot{\epsilon}_2\right)^2 + \left(\dot{\epsilon}_2 - \dot{\epsilon}_3\right)^2 + \left(\dot{\epsilon}_3 - \dot{\epsilon}_1\right)^2\right]^{1/2} = \frac{2}{\sqrt{3}}\left(\frac{1}{2}\dot{\epsilon}_{ij}\dot{\epsilon}_{ij}\right)^{1/2} \tag{5.77}$$

For uniaxial loading in the x direction, $\dot{\epsilon}_e = \dot{\epsilon}_x$, since the elastic deformation is ignored and the material is incompressible; hence $\nu = \frac{1}{2}$. Substituting $\dot{\lambda}$, obtained by Eq. (5.76), into Eq. (5.72) results in the following expression for $\dot{\boldsymbol{\epsilon}}$:

$$\dot{\boldsymbol{\epsilon}} = \frac{3}{2}\frac{\dot{\epsilon}_e}{\sigma_Y}\mathbf{S} \quad \text{or} \quad \dot{\epsilon}_{ij} = \frac{3}{2}\frac{\dot{\epsilon}_e}{\sigma_Y}S_{ij} \tag{5.78}$$

An alternate form for the parameter $\dot{\lambda}$ is derived below.

Multiply both sides of Eq. (5.72) by S_{ij}, and use von Mises yield criterion of Eq. (5.74) to obtain

$$S_{ij}\dot{\epsilon}_{ij} = \dot{\lambda}S_{ij}S_{ij} = \tfrac{2}{3}\dot{\lambda}\sigma_Y^2 \tag{5.79}$$

Then $\dot{\lambda}$ can be determined, and it is different from Eq. (5.76):

$$\dot{\lambda} = \frac{3}{2}\frac{S_{ij}\dot{\epsilon}_{ij}}{\sigma_Y^2} = \frac{3\,\mathrm{tr}(\mathbf{S}\cdot\dot{\boldsymbol{\epsilon}})}{2\sigma_Y^2} \tag{5.80}$$

Substituting this into Eq. (5.72) yields

$$\dot{\boldsymbol{\epsilon}} = \frac{3\,\mathrm{tr}(\mathbf{S}\cdot\dot{\boldsymbol{\epsilon}})}{2\sigma_Y^2}\mathbf{S} \quad \text{or} \quad \dot{\epsilon}_{ij} = \frac{3S_{lm}\dot{\epsilon}_{lm}}{2\sigma_Y^2}S_{ij} \tag{5.81}$$

From Eqs. (5.76) and (5.80) we see that $\dot{\lambda}$ is the first-order homogeneous function of strain rate and that the constitutive equations (5.78) and (5.81) are homogeneous in time and are equivalent to the incremental forms

$$d\boldsymbol{\epsilon} = \frac{3d\epsilon_e}{2\sigma_Y}\mathbf{S} \quad \text{or} \quad d\epsilon_{ij} = \frac{3d\epsilon_e}{2\sigma_Y}S_{ij} \tag{5.82}$$

$$d\boldsymbol{\epsilon} = \frac{3\,\mathrm{tr}(\mathbf{S}\cdot d\boldsymbol{\epsilon})}{2\sigma_Y^2}\mathbf{S} \quad \text{or} \quad d\epsilon_{ij} = \frac{3S_{lm}d\epsilon_{lm}}{2\sigma_Y^2}S_{ij} \tag{5.83}$$

where $d\epsilon_e$ can be expressed by Eq. (5.77) with $\dot{\epsilon}_{ij}$ replaced by $d\epsilon_{ij}$.

Careful readers may find that Eqs. (5.78) and (5.81) cannot fit into the general frame of the plasticity theory. For example, if Eqs. (5.78) and (5.81) are compared with Eq. (5.37), it is easy to identify $\hat{\boldsymbol{\alpha}}^p$ with $\mathbf{S}$, but λ determined by either Eq. (5.76) or Eq. (5.80) is certainly not $l = \partial f/\partial\boldsymbol{\sigma} : \dot{\boldsymbol{\sigma}}$. The reason for this is that Eqs. (5.36) and (5.37) are valid for strain-hardening materials and cannot be applied to the perfectly plastic materials considered here. In fact, since σ_Y is assumed to be a constant and $\sigma = \sigma_Y$, $l = \partial f/\partial\boldsymbol{\sigma} : \dot{\boldsymbol{\sigma}}$ has to be zero during the process of plastic deformation. This is one of the disadvantages of formulating plasticity theory in stress space: A unified plasticity theory for both perfectly plastic and hardening materials cannot be obtained. This issue will be discussed later.

It can be seen from the Levy-Mises Eqs. (5.78) and (5.81), or (5.82) and (5.83), that for a given strain rate $\dot{\boldsymbol{\epsilon}}$, or strain increment $d\boldsymbol{\epsilon}$, $\mathbf{S}$ can be uniquely determined. On the other hand, if $\mathbf{S}$ is given, it is impossible to calculate $\dot{\boldsymbol{\epsilon}}$ or $d\boldsymbol{\epsilon}$ using these equations for perfectly plastic material. Only the ratio between the components of $\dot{\boldsymbol{\epsilon}}$ or $d\boldsymbol{\epsilon}$ can be determined for a given $\mathbf{S}$. In other words, the directions of the principal axes $\dot{\boldsymbol{\epsilon}}$ or $d\boldsymbol{\epsilon}$ can be calculated, but their magnitudes remain unknown. Again, the reason has to do with the nature of perfectly plastic materials. As can be seen from Fig. 5.1, points in the region AB have different strains corresponding to the same constant stress equal to σ_Y. The Levy-Mises Eq. (5.73) is usually called the *flow rule of plasticity* for this reason and because it resembles the constitutive equation of a viscous fluid. The flow rule is even generalized to imply any constitutive equation for $\dot{\boldsymbol{\epsilon}}^p$, and the plastic deformation is thus called *plastic flow*.

Finally, we can verify from Eq. (5.73) that

$$\dot{\epsilon}_{kk} = \lambda S_{kk} = 0 \tag{5.84}$$

which is the incompressibility condition (3.11) for infinitesimal deformation. By setting the plastic strain rate $\dot{\boldsymbol{\epsilon}}^p$ proportional to the deviatoric stress $\mathbf{S}$, and not to the total stress $\boldsymbol{\sigma}$, we can ensure that the plastic incompressibility condition is present. If the elastic deformation is neglected, as is the case here, the incompressibility condition of the material is retrieved. Finally, we note that the use of the Levy-Mises Eq. (5.72) is not limited to perfectly plastic materials (see Sections 5.5, 5.10–5.12).

5.5 PRANDTL-REUSS EQUATIONS

The Levy-Mises equations presented in the preceding section are useful in obtaining plastic deformation for most metallic materials, where the plastic strain ϵ^p is much larger than the elastic strain ϵ^e (i.e., $\dot{\boldsymbol{\epsilon}} = \dot{\boldsymbol{\epsilon}}^p$ holds approximately). When the magnitude of the elastic strain $\boldsymbol{\epsilon}^e$ is comparable to that of the plastic strain $\boldsymbol{\epsilon}^p$, we can expect a neglect of $\dot{\boldsymbol{\epsilon}}^e$ to result in significant error. Prandtl (1927) and Reuss (1930) proposed relationships similar to Levy-Mises equations for the plastic strain rate. Plastic deformation is isochoric, while elastic deformation causes volume change as well as shape change. The rate of the deviatoric elastic strain $\dot{\epsilon}'^e$, representing shape change, is obtained by differentiating Eq. (5.48) with respect to time:

$$\dot{\boldsymbol{\epsilon}}'^e = \frac{1}{2G}\dot{\mathbf{S}} \quad \text{or} \quad \dot{\epsilon}'^e_{ij} = \frac{1}{2G}\dot{S}_{ij} \tag{5.85}$$

The incremental form is given by

$$d\boldsymbol{\epsilon}'^e = \frac{1}{2G}\dot{\mathbf{S}} \quad \text{or} \quad d\epsilon'^e_{ij} = \frac{1}{2G}\,dS_{ij} \tag{5.86}$$

The rate of plastic strain is still assumed to be coaxial with the deviatoric stress $\mathbf{S}$ and thus is proportional to it:

$$\dot{\boldsymbol{\epsilon}}^p = \dot{\lambda}\mathbf{S} \quad \text{or} \quad \dot{\epsilon}^p_{ij} = \dot{\lambda}S_{ij} \tag{5.87}$$

Again, $\dot{\lambda}$ is a parameter and can be determined by using the yield criterion. Furthermore $\dot{\lambda}$ should be the first-order homogeneous function of strain rate $\dot{\boldsymbol{\epsilon}}$ or ($\dot{\boldsymbol{\epsilon}}^p$ or $\dot{\boldsymbol{\sigma}} \cdot \dot{\mathbf{S}}$) to ensure that the constitutive equation is time or rate independent. Noting that $\dot{\boldsymbol{\epsilon}}^p$ given by Eq. (5.87) is deviatoric in nature, since plastic deformation is isochoric, the total rate of shape change can be obtained by adding Eqs. (5.85) and (5.87) to get

$$\dot{\boldsymbol{\epsilon}}' = \frac{1}{2G}\dot{\mathbf{S}} + \dot{\lambda}\mathbf{S} \quad \text{or} \quad \epsilon'_{ij} = \frac{1}{2G}\dot{S}_{ij} + \dot{\lambda}S_{ij} \tag{5.88}$$

On the other hand, since $\dot{\epsilon}^p_{kk} = 0$ from Eq. (5.88) (i.e., the rate of volume change caused by plastic deformation is zero), the total rate of volume change is due to the elastic deformation, the governing equation for which can be obtained by differentiating Eq. (5.45):

$$\dot{\epsilon}_{kk} = \dot{\epsilon}^e_{kk} = \frac{\dot{P}}{K} \tag{5.89}$$

where $P = \frac{1}{3}(\sigma_1 + \sigma_2 + \sigma_3)$ is the mean or hydrostatic stress.

The elastic-plastic constitutive equations consist of two separate parts: the shape change as given by Eq. (5.88) and the volume change given by Eq. (5.89). Now, as can be seen from Eq. (5.88), the principal axes of the total strain rate $\dot{\boldsymbol{\epsilon}}$ are no longer coaxial with the principal axes of the total deviatoric stress $\mathbf{S}$ (hence the total stress $\boldsymbol{\sigma}$) due to the presence of the first term on the right-hand side of the equation.

To determine the parameter $\dot{\lambda}$, we use the yield criterion as follows: First, we multiply Eq. (5.88) by S_{ij} to get

$$
\begin{aligned}
S_{ij}\dot{\epsilon}'_{ij} &= \frac{1}{2G}S_{ij}\dot{S}_{ij} + \dot{\lambda}S_{ij}S_{ij} \\
&= \frac{1}{2G}S_{ij}\dot{S}_{ij} + \dot{\lambda}\frac{2}{3}\sigma_{\mathrm{Y}}^2
\end{aligned}
\tag{5.90}
$$

using the von Mises criterion (5.74). Then, by differentiating von Mises criterion (5.74) and assuming that σ_{Y} is a constant, we can show that

$$
S_{ij}\dot{S}_{ij} = 0 \tag{5.91}
$$

Next, we substitute Eq. (5.91) into Eq. (5.90) to get

$$
\dot{\lambda} = \frac{3S_{ij}\dot{\epsilon}'_{ij}}{2\sigma_{\mathrm{Y}}^2} = \frac{3\,\mathrm{tr}(\mathbf{S}\cdot\dot{\boldsymbol{\epsilon}}')}{2\sigma_{\mathrm{Y}}^2} \tag{5.92}
$$

which is the same as Eq. (5.80) for these Levy-Mises equations. It should be pointed out that Eq. (5.92) does not hold for strain hardening case. Note that $\dot{\lambda}$ is indeed a first-order homogeneous function of $\dot{\epsilon}$.

We take $\dot{\lambda}$ as expressed by Eq. (5.92) and rewrite Eq. (*5.88) as

$$
\dot{\boldsymbol{\epsilon}}' = \frac{1}{2G}\dot{\mathbf{S}} + \frac{3\,\mathrm{tr}(\mathbf{S}\cdot\dot{\boldsymbol{\epsilon}})}{2\sigma_{\mathrm{Y}}^2}\mathbf{S} \quad \text{or} \quad \dot{\epsilon}'_{ij} = \frac{1}{2G}\dot{S}_{ij} + \frac{3S_{lm}\dot{\epsilon}_{lm}}{2\sigma_{\mathrm{Y}}^2}S_{ij} \tag{5.93}
$$

By adopting the procedure used to derive Eq. (5.76), we can show that $\dot{\lambda}$ can also be expressed by

$$
\dot{\lambda} = \frac{3\dot{\epsilon}_{\mathrm{e}}^{\mathrm{p}}}{2\sigma_{\mathrm{Y}}} \tag{5.94}
$$

where $\dot{\epsilon}_{\mathrm{e}}^{\mathrm{p}}$ is the equivalent or effective rate of the plastic strain and defined by

$$
\dot{\epsilon}_{\mathrm{e}}^{\mathrm{p}} = \frac{2}{\sqrt{3}}\left(\frac{1}{2}\dot{\epsilon}_{ij}^{\mathrm{p}}\dot{\epsilon}_{ij}^{\mathrm{p}}\right)^{1/2} = \frac{\sqrt{2}}{\sqrt{3}}\left[(\dot{\epsilon}_1^{\mathrm{p}} - \dot{\epsilon}_2^{\mathrm{p}})^2 + (\dot{\epsilon}_2^{\mathrm{p}} - \dot{\epsilon}_3^{\mathrm{p}})^2 + (\dot{\epsilon}_3^{\mathrm{p}} - \dot{\epsilon}_1^{\mathrm{p}})^2\right]^{1/2} \tag{5.95}
$$

Note that Eq. (5.95) differs from Eq. (5.76) in that $\dot{\epsilon}_e$ is replaced by $\dot{\epsilon}_e^p$. If expression (5.94) is used, Eq. (5.88) becomes

$$\dot{\boldsymbol{\epsilon}}' = \frac{1}{2G}\dot{\mathbf{S}} + \frac{3\dot{\epsilon}_e^p}{2\sigma_Y}\mathbf{S} \quad \text{or} \quad \dot{\epsilon}'_{ij} = \frac{1}{2G}\dot{S}_{ij} + \frac{3\dot{\epsilon}_e^p}{2\sigma_Y}S_{ij} \tag{5.96}$$

Since the equation is independent of time, we can express it equivalently in incremental form

$$d\boldsymbol{\epsilon}' = \frac{1}{2G}\,d\mathbf{S} + \frac{3d\epsilon_e^p}{2\sigma_Y}\mathbf{S} \quad \text{or} \quad d\epsilon'_{ij} = \frac{1}{2G}\,dS_{ij} + \frac{3d\epsilon_e^p}{2\sigma_Y}S_{ij} \tag{5.97}$$

where $d\epsilon_e^p$ is defined in the same way as $\dot{\epsilon}_e^p$ with $\dot{\epsilon}_{ij}^p$ replaced by $d\epsilon_{ij}^p$ in Eq. (5.95). Similarly Eq. (5.93) can be written in the incremental form, which is simply the sum of Eqs. (5.85) and (5.86). Should it be necessary to express the rates of volume change and shape change to obtain the total rate or increment of $\boldsymbol{\epsilon}$, we take

$$\dot{\epsilon}_{ij} = \frac{\dot{\sigma}_{kk}}{9K}\delta_{ij} + \frac{1}{2G}\dot{S}_{ij} + \dot{\lambda}S_{ij} \tag{5.98}$$

with $\dot{\lambda}$ given by either Eq. (5.92) or (5.94).

The Prandtl-Reuss equations used in most engineering applications are expressed using the cartesian coordinate system:

$$\begin{aligned}
d\epsilon_{xx}^p &= \frac{d\epsilon_e^p}{\sigma_Y}\left[\sigma_{xx} - \frac{1}{2}(\sigma_{yy} + \sigma_{zz})\right], \quad & d\gamma_{xy} &= \frac{3d\epsilon_e^p}{\sigma_Y}\sigma_{xy} \\
d\epsilon_{yy}^p &= \frac{d\epsilon_e^p}{\sigma_Y}\left[\sigma_{yy} - \frac{1}{2}(\sigma_{xx} + \sigma_{zz})\right], \quad & d\gamma_{yz} &= \frac{3d\epsilon_e^p}{\sigma_Y}\sigma_{yz} \\
d\epsilon_{zz}^p &= \frac{d\epsilon_e^p}{\sigma_Y}\left[\sigma_{zz} - \frac{1}{2}(\sigma_{xx} + \sigma_{yy})\right], \quad & d\gamma_{zx} &= \frac{3d\epsilon_e^p}{\sigma_Y}\sigma_{zx}
\end{aligned} \tag{5.99}$$

In the equations above the engineering components of shear strain γ_{xy}, γ_{yz}, and γ_{zx} are used instead of the tensorial components ϵ_{xy}, ϵ_{yz}, and ϵ_{zx}. Note that $\gamma_{xy} = 2\epsilon_{xy}$, and so on. Also, in deriving these equations we assume that $\nu = \frac{1}{2}$, as is standard practice in plasticity theory (see Section 5.3).

The application of Prandtl-Reuss equations is not limited to perfectly plastic materials. The simplest way to generalize them to include strain-hardening behavior is to consider σ_Y as a function of a certain hardening parameter(s), such as the effective plastic strain ϵ_e^p:

$$\sigma_Y = \sigma_Y(\epsilon_e^p) \tag{5.100}$$

The function form $\sigma_Y(\epsilon_e^p)$ is determined from the uniaxial loading experiment by using the assumption of isotropic hardening. This will be discussed in detail later. Note that Eq. (5.92) does not hold for strain-hardening materials. Only Eq. (5.94) can be used.

The experimental verification of the Prandtl-Reuss equations (5.87), and hence the Levy-Mises equations (5.72) for negligible elastic strain, was first done by Lode (1926). His experiments were discussed earlier in Section 4.3. Experiments were performed on steel, copper, and nickel tubes under combined tension and internal pressure. To verify the Prandtl-Reuss equation, he introduced the Lode stress parameter μ_σ defined by Eq. (4.77) as well as the Lode plastic strain rate parameter $\mu_{\dot{\epsilon}^p}$ defined by

$$\mu_{\dot{\epsilon}^p} = \frac{2d\epsilon_2^p - d\epsilon_1^p - d\epsilon_3^p}{d\epsilon_3^p - d\epsilon_1^p} = \frac{d\epsilon_2^p - \frac{1}{2}(d\epsilon_3^p + d\epsilon_1^p)}{\frac{1}{2}(d\epsilon_3^p - d\epsilon_1^p)} \tag{5.101}$$

From the Prandtl-Reuss equation, we can derive

$$\frac{2d\epsilon_2^p - d\epsilon_1^p - d\epsilon_3^p}{d\epsilon_3^p - d\epsilon_1^p} = \frac{2S_2 - S_1 - S_3}{S_3 - S_1} = \frac{2\sigma_2 - \sigma_1 - \sigma_3}{\sigma_3 - \sigma_1} \tag{5.102}$$

or

$$\mu_{\dot{\epsilon}^p} = \mu_\sigma \tag{5.103}$$

This equation is obtained based on the assumption that Prandtl-Reuss equations hold. It follows that if the Prandtl-Reuss equations are valid, Eq. (5.103) should hold during the experiments. Figure 5.3 shows the experimental results obtained by Lode in 1926. As is clear from the figure, the Prandtl-Reuss equation is satisfied only approximately. The deviation has a definite pattern and is not due to random scattering in the experimental data.

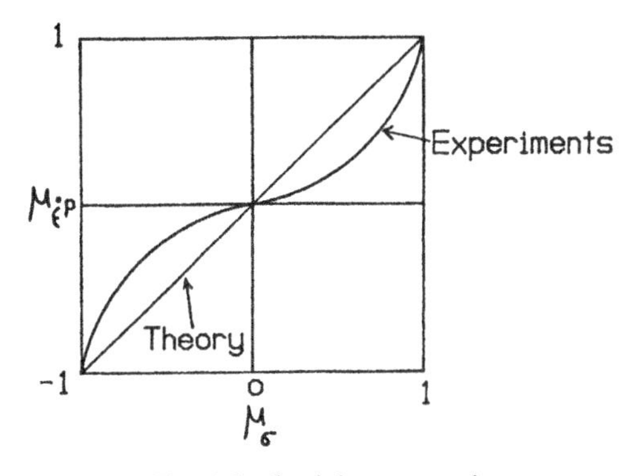

Fig. 5.3 Lode's test results

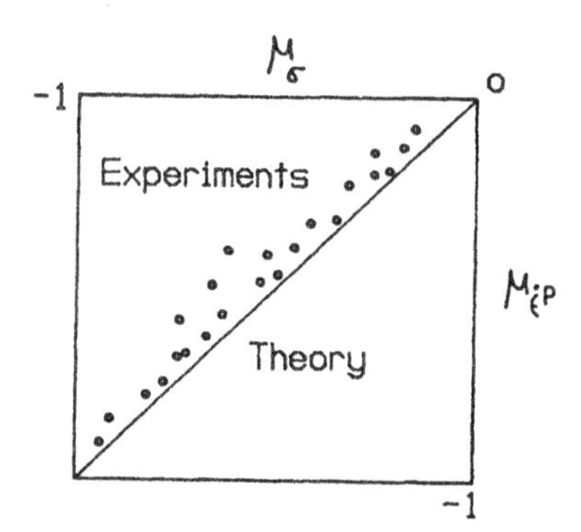

Fig. 5.4 Taylor and Quinney's test results

Taylor and Quinney (1921) later verified this deviation, and their results are shown in Fig. 5.4.

Although there exist deviations in the Prandtl-Reuss equations from the experimental observations, as shown in Figs. 5.3 and 5.4, these equations are still believed to be sufficiently accurate for engineering applications. Taylor and Quinney's tests using copper, aluminum, and mild steel tubes showed that the principal axes of the increment of plastic strain were basically coaxial with those of the total stress and the deviations did not exceed 2°.

The Prandtl-Reuss equations have been improved, for example, by Lianis and Ford (1957) through an elimination of the anisotropy effects, and by Prager (1945) who suggested including J_3' in the yield criterion. But these attempts have added considerable complexity without significant improvements in results.

5.6 PLASTIC POTENTIAL THEORY AND PLASTIC WORK

The Levy-Mises and the Prandtl-Reuss equations (5.72) and (5.87) were empirically postulated based on experimental observations in metals. The general mathematical treatment of the constitutive equation for plastic deformation or flow was proposed by von Mises in 1928. He noticed that in elasticity theory the strain tensor was related to the stress $\boldsymbol{\sigma}$ through an elastic potential function, the complementary strain energy U_c such that

$$\epsilon_{ij} = \frac{\partial U_c}{\partial \sigma_{ij}} \tag{5.104}$$

By generalizing and applying this idea to plasticity theory, Mises proposed that there existed a plastic potential function $Q(\sigma)$ or $Q(\sigma_{ij})$, and the plastic strain rate $\dot{\epsilon}^p$ could be derived similar to Eq. (5.104):

$$\dot{\boldsymbol{\epsilon}}^p = \dot{\lambda}\frac{\partial Q(\boldsymbol{\sigma})}{\partial \boldsymbol{\sigma}}, \quad \dot{\epsilon}_{ij}^p = \dot{\lambda}\frac{\partial Q(\sigma_{ij})}{\partial \sigma_{ij}} \tag{5.105}$$

where $\dot{\lambda}$ is a proportional positive scalar factor. To determine $\dot{\lambda}$, the yield criterion was used. This will be discussed in detail later. The plasticity theory based on flow rule (5.105) is *called plastic potential theory*. The following remarks should be noted about Eq. (5.105):

1. As we noted in Section 4.1, any stress state $\boldsymbol{\sigma}$ corresponds to a point in the six-dimensional stress space. The plastic potential $Q(\sigma_{ij}) = C$ represents a surface, and the plastic strain $\dot{\boldsymbol{\epsilon}}^p$ is a vector in this space. Geometrically Eq. (5.105) means that the plastic strain rate vector $\dot{\boldsymbol{\epsilon}}^p$ is perpendicular to the surface $Q(\sigma_{ij}) = C$. Therefore Eq. (5.105) is also referred to as the *normality rule* in plasticity theory. With this geometrical explanation, we can rewrite the flow rule (5.105) as

$$\dot{\boldsymbol{\epsilon}}^p = \dot{\lambda}\,\text{grad}\,Q\,(\boldsymbol{\sigma}) \tag{5.106}$$

2. For isotropic materials, $Q(\sigma_{ij})$ is a function of the invariants of the stress tensor $\boldsymbol{\sigma}$:

$$Q = Q(J_1, J_2, J_3) \tag{5.107}$$

where the invariants J_1, J_2, and J_3 are defined in Section 4.1. If the material is anisotropic, a certain number of vectors or tensors β_k, $k = 1, 2, \ldots, n$, characterizing the anisotropy of the material should be added as the arguments of Q, as noted for the yield function of anisotropic materials in Section 4.3.4:

$$Q = Q(\sigma_{ij}, \beta_k) \qquad (k = 1, 2, \ldots, n) \tag{5.108}$$

The number k can be determined from observed anisotropic properties of the material.

3. If incompressibility of the plastic flow is assumed, then the corresponding plastic potential surface must be cylindrical (not necessarily with circular cross section) with axis $\sigma_{xx} = \sigma_{yy} = \sigma_{zz}$ in the general six-dimensional stress space. For isotropic materials the three-dimensional principal stress space can be used; in this case Q is cylindrical with axis $\sigma_1 = \sigma_2 = \sigma_3$. $\dot{\boldsymbol{\epsilon}}^p$ is then perpendicular to this axis, representing shape changes only. For some materials such as rock, concrete, and soil, the plastic deformation includes the volume change as well, and the plastic potential surface is not parallel to this hydrostatic axis.

The main point in using the plastic potential theory is to determine the plastic potential Q. In fact the effective form of the function $Q(\sigma_{ij})$ is an open question. A common approach in plasticity theory is to assume that the plastic potential function $Q(\sigma_{ij})$ is the same as the yield function $F(\sigma_{ij})$:

$$Q(\sigma_{ij}) = F(\sigma_{ij}) \tag{5.109}$$

Equation (5.104) can then be rewritten as

$$\dot{\boldsymbol{\epsilon}}^{\mathrm{p}} = \dot{\lambda}\frac{\partial F}{\partial \boldsymbol{\sigma}} \tag{5.110}$$

and the plastic strain rate $\dot{\boldsymbol{\epsilon}}^{\mathrm{p}}$ is normal to the yield surface. This is called the *associated flow rule*. On the other hand, if $Q \neq F$, the flow rule is called *nonassociated*.

The association of Q with F is based on an assumption whose validity can be verified empirically. Experimental observations show that the plastic deformation of metals can be characterized quite well by the associated flow rule, but for some porous materials such as rocks, concrete, and soils, the nonassociated flow rule provides a better representation of their plastic deformation. Mathematically it can be proved by using Drucker's stability postulate that if the material is stable in Drucker's sense, then the flow rate must be associated. A full discussion of this, including the limitations of Drucker's stability postulate is presented in Section 5.7.

The Prandtl-Reuss equation (5.86) is a special case of the associated flow rule. Indeed applying the von Mises yield criterion (4.59) yields

$$\frac{\partial J_2'}{\partial \sigma_{ij}} = S_{ij} \tag{5.111}$$

and Eq. (5.110) gives

$$\dot{\boldsymbol{\epsilon}}^{\mathrm{p}} = \dot{\lambda}\frac{\partial f}{\partial \boldsymbol{\sigma}} = \dot{\lambda}\frac{\partial J_2'}{\partial \boldsymbol{\sigma}} = \dot{\lambda}\mathbf{S} \tag{5.112}$$

which is the Prandtl-Reuss equation, or the Levy-Mises equation if the elastic strain rate is ignored. Thus, within the general frame of the plastic potential theory, the Prandtl-Reuss or the Levy-Mises equation implies the von Mises yield condition and the associated flow rule.

Plastic work is an important concept that is often used in plasticity theory. During the process of deformation, the rate of work done per unit volume, as defined by Eq. (3.47), is

$$\begin{aligned} \dot{W} &= \sigma_{ij}\dot{\epsilon}_{ij} = \sigma_{ij}\left(\dot{\epsilon}_{ij}^{\mathrm{e}} + \dot{\epsilon}_{ij}^{\mathrm{p}}\right) \\ &= \dot{W}^{\mathrm{e}} + \dot{W}^{\mathrm{p}} \end{aligned} \tag{5.113}$$

where $\dot{W}^{\mathrm{e}} = \sigma_{ij}\dot{\epsilon}_{ij}^{\mathrm{e}}$ is the rate of elastic strain energy, as mentioned in Section 5.2, and $\dot{W}^{\mathrm{p}} = \sigma_{ij}\dot{\epsilon}_{ij}^{\mathrm{p}}$ is the rate of plastic work. $\dot{W}^{\mathrm{e}}$ is the recoverable part of the total rate of work $\dot{W}$, whereas $\dot{W}^{\mathrm{p}}$ is the irrecoverable part, since plastic deformation is an irreversible process. From the incompressibility of plastic deformations, it follows that

$$\dot{W}^{\mathrm{p}} = \sigma_{ij}\dot{\epsilon}_{ij}^{\mathrm{p}} = (S_{ij} + p\delta_{ij})\dot{\epsilon}_{ij}^{\mathrm{p}} = S_{ij}\dot{\epsilon}_{ij}^{\mathrm{p}} \tag{5.114}$$

since $p\delta_{ij}\dot{\epsilon}_{ij}^{\mathrm{p}} = p\dot{\epsilon}_{ii}^{\mathrm{p}} = 0$. In the stress space $\dot{W}^{\mathrm{p}}$ can be considered as the inner product of the vectors $\mathbf{S}$ and $\dot{\boldsymbol{\epsilon}}^{\mathrm{p}}$. To simplify the discussion, only the principal stress space for isotropic materials will be considered, but the results are the same whenever the general stress space is used. Equations (5.114) can be rewritten as

$$\dot{W}^{\mathrm{p}} = |\mathbf{S}|\,|\dot{\boldsymbol{\epsilon}}^{\mathrm{p}}|\cos\phi \tag{5.115}$$

where $|\mathbf{S}|$ and $|\dot{\boldsymbol{\epsilon}}^{\mathrm{p}}|$ are norms or lengths of $\mathbf{S}$ and $\dot{\boldsymbol{\epsilon}}^{\mathrm{p}}$, respectively, and ϕ is the angle between them. As stated in Section 4.1, the length of $\mathbf{S}$ given by Eq. (4.42) and a similar expression can be derived for $|\dot{\boldsymbol{\epsilon}}^{\mathrm{p}}|$, since $\dot{\boldsymbol{\epsilon}}^{\mathrm{p}}$ is also a deviatoric tensor:

$$|\mathbf{S}| = \left(S_1^2 + S_2^2 + S_3^2\right)^{1/2} = \sqrt{2J_2'} = \sqrt{\frac{2}{3}}\,\sigma_{\mathrm{e}} \tag{5.116}$$

$$|\dot{\boldsymbol{\epsilon}}^{\mathrm{p}}| = \left[\left(\dot{\epsilon}_1^{\mathrm{p}}\right)^2 + \left(\dot{\epsilon}_2^{\mathrm{p}}\right)^2 + \left(\dot{\epsilon}_3^{\mathrm{p}}\right)^2\right]^{1/2} = \sqrt{\frac{3}{2}}\,\dot{\epsilon}_{\mathrm{e}}^{\mathrm{p}} \tag{5.117}$$

where σ_{e} and $\dot{\epsilon}_{\mathrm{e}}^{\mathrm{p}}$ are the effective stress and effective plastic strain rate defined by Eqs. (5.52) and (5.77), respectively. Equation (5.115) then becomes

$$\dot{W}^{\mathrm{p}} = \sigma_{\mathrm{e}}\dot{\epsilon}_{\mathrm{e}}^{\mathrm{p}}\cos\phi \tag{5.118}$$

If the Prandtl-Reuss equation (5.85) is used to determine the plastic strain rate, $\dot{\boldsymbol{\epsilon}}^{\mathrm{p}}$ should be coaxial with $\mathbf{S}$ so that $\phi = 0$, and Eq. (5.118) reduces to

$$\dot{W}^{\mathrm{p}} = \sigma_{\mathrm{e}}\dot{\epsilon}_{\mathrm{e}}^{\mathrm{p}} \tag{5.119}$$

Using this equation, we can rewrite the Prandtl-Reuss equation as

$$\dot{\boldsymbol{\epsilon}}^{\mathrm{p}} = \frac{3}{2}\frac{\dot{W}^{\mathrm{p}}}{\sigma_{\mathrm{Y}}^2}\mathbf{S} \tag{5.120}$$

where the yield condition has been used to identify σ_{e} with σ_{Y}.

5.7 DRUCKER'S STABILITY POSTULATE AND ITS CONSEQUENCES

So far only the Levy-Mises equation and the Prandtl-Reuss equation have been cited as specific forms of the plastic constitutive equation or flow rule. These equations were formulated on the assumption that the rate or increment of the (plastic) strain is coaxial with the deviatoric part of the total stress, as observed experimentally. In the previous section it was shown that this flow rule incorporates the von Mises yield criterion and that it can be derived from the plastic potential theory using the associated flow rule. The plastic potential theory, although it is more rational and general, has the drawback of including the unspecified plastic potential Q.

In an effort to establish the general plastic stress-strain relations for any yield criterion, Drucker proposed a unified approach based on his stability postulate (1951). One major consequence of Drucker's postulate is that the flow rule for stable materials is associated (i.e., $Q \equiv F$). First, it is necessary to define a stress cycle or a closed loading-unloading path in stress space. Consider a material element in equilibrium with a given state of stress $\boldsymbol{\sigma}^0$. The stress is increased by $\Delta\boldsymbol{\sigma} = \boldsymbol{\sigma} - \boldsymbol{\sigma}^0$, which causes a change in the displacements, strains, and stresses of the element. Removing the additional stresses $\Delta\boldsymbol{\sigma}$ enables the stress of the body to return to the original state $\boldsymbol{\sigma}^0$. This process of application and removal of the additional stress is called a *stress cycle* or *closed loading-unloading path* in the stress space. Note that at the beginning and at the end of the stress cycle, the stress state of the element must be equal, but the strain state can be equal or different depending on whether plastic deformation occurred during the loading and unloading. The one-dimensional case is shown in Fig. 5.5, where the elastic

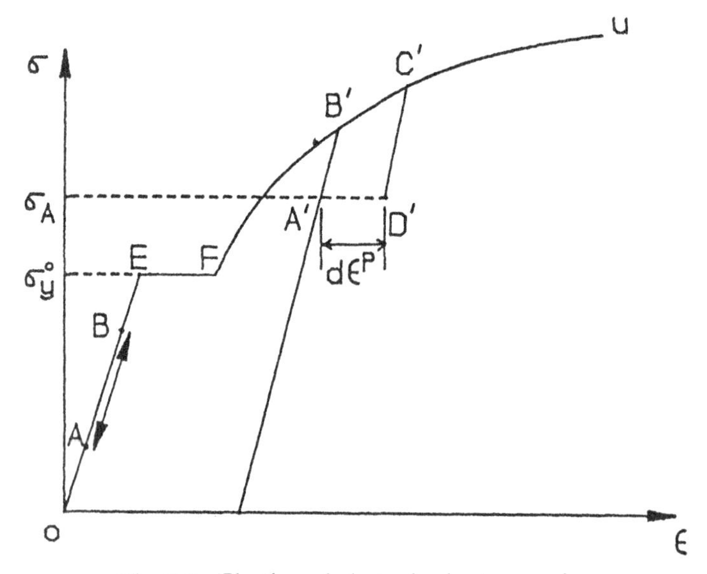

Fig. 5.5 Elastic and elastoplastic stress cycles

stress cycle is represented by ABA and the elastic-plastic stress cycle is denoted by $A'B'C'D'$. For the elastic-plastic stress cycle, it is obvious that the strain at the end of the cycle differs from that at the beginning by an amount $d\epsilon^p$.

Drucker's stability postulate asserts that if a material is stable, the following inequality must be satisfied:

$$W = \int_{C_\sigma} \Delta\boldsymbol{\sigma} : d\boldsymbol{\epsilon} = \int_{C_\sigma} \left(\boldsymbol{\sigma} - \boldsymbol{\sigma}^0\right) : d\boldsymbol{\epsilon} \geq 0 \tag{5.121}$$

where the integral is done over the closed stress cycle C_σ, a loading-unloading path in the stress space. The integral represents the work done by the external agency over C_σ. It was shown by Drucker (1956) that any material that does not obey this inequality is unstable.

Since a stress cycle consists of loading and unloading parts, and since during the process of unloading, the work done by the external agency is negative, it follows, from inequality (5.121), that

1. During the application of the additional stress $\Delta\boldsymbol{\sigma}$, the work done by the external agency must be positive,

$$d\boldsymbol{\sigma} : d\boldsymbol{\epsilon} > 0 \tag{5.122}$$

2. Over the cycle of the application and removal of the additional stress $\Delta\boldsymbol{\sigma}$, the work done by the external agency must be nonnegative. It is zero if only elastic deformation occurs over the cycle; it is positive if there is plastic deformation over the cycle.

The Drucker's stability postulate is a fairly strong requirement that can be satisfied only by hardening materials whose subsequent yield strength increases with the deformation, as shown by the curve FU in Fig. 5.5. This postulate is sometimes used as a definition of plastic hardening. Any material on which an external agency does positive work during an elastic-plastic stress cycle is considered as hardening material. Otherwise, it is called *nonhardening* or *softening* material. A stress-strain curve with a softening portion in simple tension is depicted in Fig. 5.6. Drucker's postulate does not apply to softening materials, and it needs to be modified for perfectly plastic materials (to be discussed later). For this reason this postulate is sometimes called *Drucker's hardening postulate.*

To explore the consequences of the Drucker's postulate, some mathematical manipulation is necessary. A simple calculation is presented here; the interested reader can find a more rigorous treatment in Iliushin (1960).

Assume that at $t = t_0$ the original stress state is $\boldsymbol{\sigma}^0$. This stress state can be on the current yield surface or inside this surface. Suppose that at $t = t_1$ the stress state reaches the current yield surface. From t_0 to t_1 the deformation is elastic. From t_1 on the elastic-plastic loading occurs until $t = t_2$.

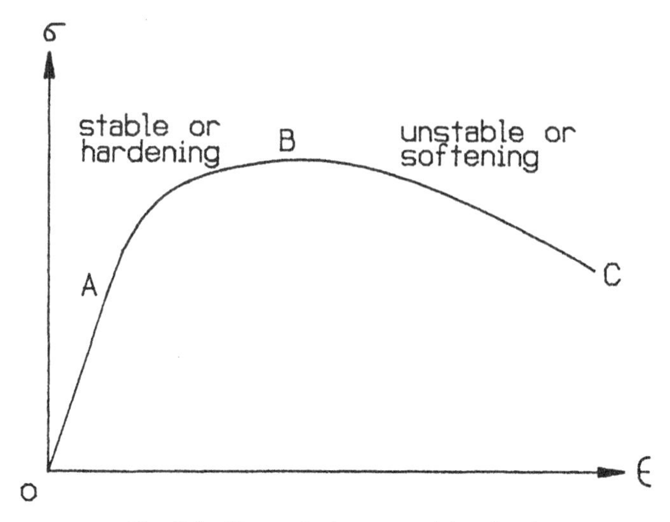

Fig. 5.6 Stress-strain curve with softening

During this process the stress state will stay on the current yield surface and will move outward with it. The unloading occurs from $t = t_2$ until $t = t_3$, which brings the stress state back to the original value $\boldsymbol{\sigma} = \boldsymbol{\sigma}^0$, hence completing the stress cycle as required by the Drucker's postulate. From t_2 to t_3 the deformation is elastic. For such a stress cycle with elastic-plastic deformation, the Drucker's postulate (5.121) can be written

$$\begin{aligned} W &= \int_{t_0}^{t_1} (\boldsymbol{\sigma} - \boldsymbol{\sigma}^0) : \dot{\boldsymbol{\epsilon}}^{\mathrm{e}} dt + \int_{t_1}^{t_2} (\boldsymbol{\sigma} - \boldsymbol{\sigma}^0) : (\dot{\boldsymbol{\epsilon}}^{\mathrm{e}} + \dot{\boldsymbol{\epsilon}}^{\mathrm{p}})\, dt + \int_{t_2}^{t_3} (\boldsymbol{\sigma} - \boldsymbol{\sigma}^0) : \dot{\boldsymbol{\epsilon}}^{\mathrm{e}}\, dt \\ &= \int_{t_0}^{t_2} (\boldsymbol{\sigma} - \boldsymbol{\sigma}^0) : \dot{\boldsymbol{\epsilon}}^{\mathrm{e}}\, dt + \int_{t_2}^{t_3} (\boldsymbol{\sigma} - \boldsymbol{\sigma}^0) : \dot{\boldsymbol{\epsilon}}^{\mathrm{e}}\, dt + \int_{t_1}^{t_2} (\boldsymbol{\sigma} - \boldsymbol{\sigma}^0) : \dot{\boldsymbol{\epsilon}}^{\mathrm{p}}\, dt > 0 \end{aligned} \tag{5.123}$$

Since the elastic deformation is reversible and the change of the elastic strain energy over a closed loading-unloading path is zero,

$$\int_{t_0}^{t_2} (\boldsymbol{\sigma} - \boldsymbol{\sigma}^0) : \dot{\boldsymbol{\epsilon}}^{\mathrm{e}}\, dt + \int_{t_2}^{t_3} (\boldsymbol{\sigma} - \boldsymbol{\sigma}^0) : \dot{\boldsymbol{\epsilon}}^{\mathrm{e}}\, dt = \int_{C_\sigma} (\boldsymbol{\sigma} - \boldsymbol{\sigma}^0) : \dot{\boldsymbol{\epsilon}}^{\mathrm{e}}\, dt = 0 \tag{5.124}$$

the Drucker's postulate requires that

$$W = \int_{t_1}^{t_2} (\boldsymbol{\sigma} - \boldsymbol{\sigma}^0) : \dot{\boldsymbol{\epsilon}}^{\mathrm{p}}\, dt > 0 \tag{5.125}$$

This should hold for any $\delta t = t_2 - t_1$. Assuming that $\delta t \to 0$, the integrand in this integral can be expanded using Taylor's series as

$$(\boldsymbol{\sigma} - \boldsymbol{\sigma}^0) : \dot{\boldsymbol{\epsilon}}^{\mathrm{p}} = \left[(\boldsymbol{\sigma} - \boldsymbol{\sigma}^0) : \dot{\boldsymbol{\epsilon}}^{\mathrm{p}}\right]_{t_1} + \tfrac{1}{2}\left[\dot{\boldsymbol{\sigma}} : \dot{\boldsymbol{\epsilon}}^{\mathrm{p}} + (\boldsymbol{\sigma} - \boldsymbol{\sigma}^0) : \ddot{\boldsymbol{\epsilon}}^{\mathrm{p}}\right]_{t_1} \delta t + 0(\delta t^2) \tag{5.126}$$

where $0(\delta t^2)$ represents second- or higher-order terms involving δt. Neglecting second-order terms, the inequality (5.126) can be integrated as

$$W = \left[(\boldsymbol{\sigma} - \boldsymbol{\sigma}^0) : \dot{\boldsymbol{\epsilon}}^{\mathrm{p}}\right]_{t_1} \delta t + \tfrac{1}{2}\left[\dot{\boldsymbol{\sigma}} : \dot{\boldsymbol{\epsilon}}^{\mathrm{p}} + (\boldsymbol{\sigma} - \boldsymbol{\sigma}^0) : \ddot{\boldsymbol{\epsilon}}^{\mathrm{p}}\right]_{t_1} \delta t^2 \tag{5.127}$$

or

$$W = \left[(\boldsymbol{\sigma} - \boldsymbol{\sigma}^0) : d\dot{\boldsymbol{\epsilon}}^{\mathrm{p}}\right]_{t_1} + \tfrac{1}{2}\left[d\boldsymbol{\sigma} : d\boldsymbol{\epsilon}^{\mathrm{p}} + (\boldsymbol{\sigma} - \boldsymbol{\sigma}^0) : \ddot{\boldsymbol{\epsilon}}^{\mathrm{p}} \delta t^2\right]_{t_1} \tag{5.128}$$

Two possibilities then exist:

1. If $\boldsymbol{\sigma}^0$ is not on the yield surface at $t = t_0$, then $\boldsymbol{\sigma} - \boldsymbol{\sigma}^0 \neq \mathbf{0}$ at t_1. As a result the second term in Eq. (5.127) can be neglected, since it is one order higher than the first term. From $W > 0$ and Eq. (5.127), it follows that

$$(\boldsymbol{\sigma} - \boldsymbol{\sigma}^0) : \dot{\boldsymbol{\epsilon}}^{\mathrm{p}} > 0 \tag{5.129}$$

   where the subscript t_1 has been dropped for simplicity.
2. If $\boldsymbol{\sigma}^0$ is on the yield surface at t_0, then $\boldsymbol{\sigma} - \boldsymbol{\sigma}^0 = 0$, so the Drucker's postulate reduces to

$$d\boldsymbol{\sigma} : d\boldsymbol{\epsilon}^{\mathrm{p}} > 0 \tag{5.130}$$

   This relationship and Eq. (5.123) refer to the stability or hardening condition. In fact Eq. (5.122) can be obtained from Eq. (5.130), since $d\boldsymbol{\sigma} : d\boldsymbol{\epsilon}^{\mathrm{e}}$ is always positive and $d\boldsymbol{\sigma} : d\boldsymbol{\epsilon} = d\boldsymbol{\sigma}(d\boldsymbol{\epsilon}^{\mathrm{e}} + d\boldsymbol{\epsilon}^{\mathrm{p}}) > 0$ whenever Eq. (5.130) is valid.

To explore the consequences of his postulate, Eqs. (5.129) and (5.130), Drucker introduced the following fundamental assumptions (1950, 1951):

1. There exists a yield surface that separates the plastic region from the elastic region. The change of the stress state within or along the surface will cause only elastic deformation, and change of the stress state from the yield surface toward its interior will cause elastic unloading. Plastic loading will occur only if the increment of the stress is directed toward the outside of the yield surface. (Obviously this assumption is the statement of the loading and unloading criteria presented earlier and based on experimental observations.)
2. The relationship between the infinitesimal increments of the stress $d\boldsymbol{\sigma}$ and the plastic strain $d\boldsymbol{\epsilon}^{\mathrm{p}}$ is linear.

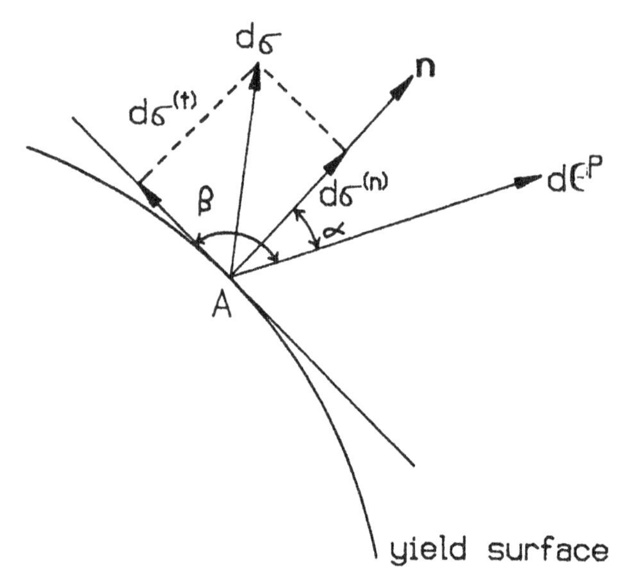

Fig. 5.7 Normality of the plastic strain increment

With the help of these two basic assumptions and Eqs. (5.130) and (5.131), two important conclusions about the shape of the yield surface and the direction of the plastic strain increment vector $d\boldsymbol{\epsilon}^{\mathrm{p}}$ in the general stress space can be derived.

First, the infinitesimal increment of the stress $d\boldsymbol{\sigma}$ for elastic-plastic loading from any point on the yield surface is directed toward the outside of the yield surface and can thus be decomposed uniquely to the tangential and normal components:

$$d\boldsymbol{\sigma} = d\boldsymbol{\sigma}^{(\mathrm{t})} + d\boldsymbol{\sigma}^{(\mathrm{n})} \tag{5.131}$$

as shown in Fig. 5.7. The second assumption assures that the plastic strain caused by $d\boldsymbol{\sigma}$ equals the sum of those caused by $d\boldsymbol{\sigma}^{(\mathrm{t})}$ and $d\boldsymbol{\sigma}^{(\mathrm{n})}$ separately. However, according to the first assumption or the continuity condition presented after Eq. (5.35), the stress increment along the tangential plane of the yield surface $d\boldsymbol{\sigma}^{(\mathrm{t})}$ will only cause elastic deformation (neutral loading), so the plastic strain increment $d\boldsymbol{\epsilon}^{\mathrm{p}}$ due to $d\boldsymbol{\sigma}$ is dependent only on $d\boldsymbol{\sigma}^{(\mathrm{n})}$. Thus Eq. (5.130) can be written

$$d\boldsymbol{\sigma} : d\boldsymbol{\epsilon}^{\mathrm{p}} = \left(d\boldsymbol{\sigma}^{(\mathrm{t})} + d\boldsymbol{\sigma}^{(\mathrm{n})}\right) : d\boldsymbol{\epsilon}^{\mathrm{p}} = d\boldsymbol{\sigma}^{(\mathrm{n})} : d\boldsymbol{\epsilon}^{\mathrm{p}} > 0 \tag{5.132}$$

or

$$|d\boldsymbol{\sigma}^{(\mathrm{n})}|\ |d\boldsymbol{\epsilon}^{\mathrm{p}}| \cos\alpha > 0 \tag{5.133}$$

and

$$d\boldsymbol{\sigma}^{(t)} : d\boldsymbol{\epsilon}^{p} = 0 \tag{5.134}$$

or

$$|d\boldsymbol{\sigma}^{(t)}|\ |d\boldsymbol{\epsilon}^{p}| \cos \beta = 0 \tag{5.135}$$

where α and β are respective angles between the vectors $d\boldsymbol{\sigma}^{(n)}$ and $d\boldsymbol{\epsilon}^{p}$, $d\boldsymbol{\sigma}^{(t)}$ and $d\boldsymbol{\epsilon}^{p}$ in the stress space, shown in Fig. 5.7.

Since Eq. (5.130) holds for any $d\boldsymbol{\sigma}$, so Eq. (5.135) must be true for $d\boldsymbol{\sigma}^{(t)}$ along any direction in the tangential plane. This is possible if and only if $\beta = \pi/2$, that is, the vector $d\boldsymbol{\epsilon}^{p}$ is directed along the normal $\mathbf{n}$ to the yield surface at the point **A**:

$$d\boldsymbol{\epsilon}^{p} = \dot{\lambda} \frac{\partial F}{\partial \boldsymbol{\sigma}} \tag{5.136}$$

Then $\alpha = 0$, and hence Eqs. (5.134) and Eq. (5.130) can be satisfied. Equation (5.136) is the well-known associated flow rule or normalty rule in the theory of plasticity, and it is the result of the Drucker's stability or hardening postulate.

Now the implication of Eq. (5.129) can be considered. Recall that stress state $\boldsymbol{\sigma}^{0}$ is inside the current yield surface, while stress state $\boldsymbol{\sigma}$ is on this surface, as shown in Fig. 5.8.

Consider $\boldsymbol{\sigma} - \boldsymbol{\sigma}^{0}$ and $d\boldsymbol{\epsilon}^{p}$ as the vectors in the stress space. Equation (5.129) can be written

$$|\boldsymbol{\sigma} - \boldsymbol{\sigma}^{0}|\ |d\boldsymbol{\epsilon}^{p}| \cos \phi > 0 \tag{5.137}$$

From this relationship,

$$-\frac{\pi}{2} \leq \phi \leq \frac{\pi}{2} \tag{5.138}$$

Since $d\boldsymbol{\epsilon}^{p}$ is in the normal $\mathbf{n}$ direction, as noted above, all stress states $\boldsymbol{\sigma}^{0}$ must lie on one side of the tangential plane of the yield surface at point $\boldsymbol{\sigma}$ which is perpendicular to $d\boldsymbol{\epsilon}^{p}$ (or $\mathbf{n}$), as shown in Fig. 5.8. Therefore the vector $\boldsymbol{\sigma} - \boldsymbol{\sigma}^{0}$ should not intersect the yield surface twice. This suggests that the yield surface must be convex. If the yield surface were concave at any point A, shown in Fig. 5.9, then it might be possible to find a point $\boldsymbol{\sigma}^{0}$ inside the yield surface, with an obtuse angle between $d\boldsymbol{\epsilon}^{p}$ at point A and $\boldsymbol{\sigma} - \boldsymbol{\sigma}^{0}$ (i.e., $\phi > \pi/2$, as shown in Fig. 5.9). In that case the inner product $(\boldsymbol{\sigma} - \boldsymbol{\sigma}^{0}) : d\boldsymbol{\epsilon}^{p}$ would be negative, as can be seen from Eq. (5.137), and this is contrary to Drucker's postulate. From Fig. 5.9 it can also be seen that the vector $\boldsymbol{\sigma} - \boldsymbol{\sigma}^{0}$ would intersect the yield surface twice, at point B as well as at

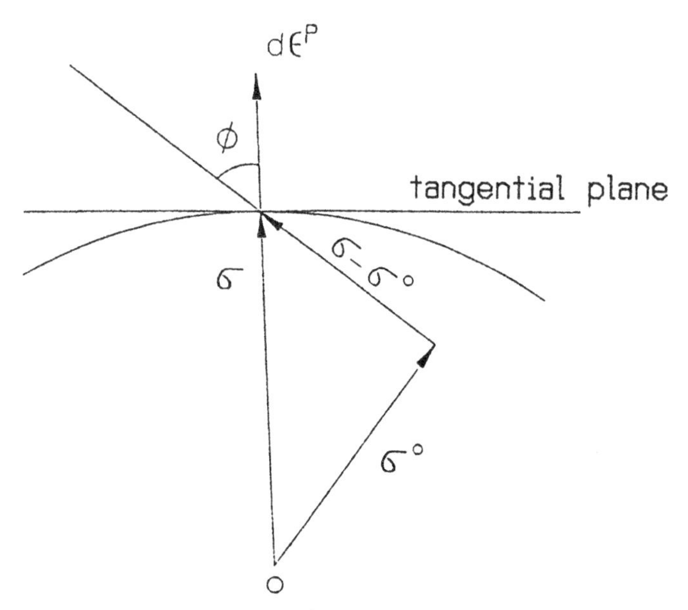

Fig. 5.8 Convexity of the yield surface

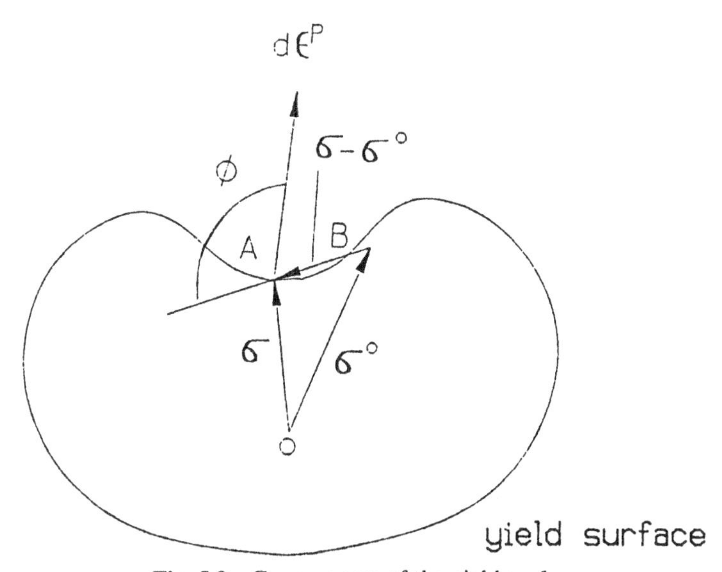

Fig. 5.9 Concaveness of the yield surface

point A. Thus, for the hardening or stable materials to satisfy the Drucker's postulate, the yield surface must be convex, and the plastic strain increment $d\boldsymbol{\epsilon}^p$, or equivalently the plastic strain rate $\dot{\boldsymbol{\epsilon}}^p$, must be directed along the normal $\mathbf{n}$ to the yield surface, resulting in the associated flow rule (5.136). However, as mentioned before, some limitations exist. The following three exceptions, or limitations of the postulate, illustrate this:

1. Drucker's postulate does not hold true for the strain-softening behavior, as shown by the portion BC in Fig. 5.6. Some engineering materials such as rocks, concrete, and soils may exhibit the softening phenomena for which

$$d\boldsymbol{\sigma} : d\boldsymbol{\epsilon} < 0 \quad \text{and} \quad d\boldsymbol{\sigma} : d\boldsymbol{\epsilon}^p < 0 \tag{5.139}$$

Obviously this is contrary to Drucker's postulate. Hence the preceding discussion about convexity of the yield surface and the normality of the plastic strain increment $d\boldsymbol{\epsilon}^p$ does not apply for softening materials in the stress space. A formulation of the plasticity theory in the strain space is needed to describe this softening behavior and will be discussed in detail in Chapter 8. Note here, however, that the convexity of the yield surface and the normality of $d\boldsymbol{\epsilon}^p$ still hold for softening materials when the elastic response of the material does not affect the plastic deformation; that is, when C_{ijkl} in Eq. (5.59) is constant.

2. A very important, implicit assumption used in the previous discussion is that the elastic response is constant during the process of the plastic deformation, that is, C_{ikjl} is constant and not a function of the plastic strain $\boldsymbol{\epsilon}^p$. If, instead, C_{ijkl} were a function of $\boldsymbol{\epsilon}^p$—that is, the elastic property is coupled with the plastic deformation

$$C_{ijkl} = C_{ijkl}(\boldsymbol{\epsilon}^p) \tag{5.140}$$

—it may be proved that, instead of Eq. (5.136), Drucker's postulate results in

$$d\boldsymbol{\epsilon}^p = \dot{\lambda}\frac{\partial F}{\partial \boldsymbol{\sigma}} - d\mathbf{C}^{-1}(\boldsymbol{\epsilon}^p) : \boldsymbol{\sigma} \tag{5.141}$$

where $\mathbf{C}^{-1}$ is the inverse of $\mathbf{C}$, the stiffness tensor with components C_{ikjl}. This identity was first derived by Iliushin (1960). A simplified derivation is given below.

The elastic response (5.58) can be inverted to give

$$\boldsymbol{\epsilon}^e = \mathbf{C}^{-1}(\boldsymbol{\epsilon}^p) : \boldsymbol{\sigma} \tag{5.142}$$

Then

$$d\boldsymbol{\epsilon}^{\mathrm{e}} = d\mathbf{C}^{-1}(\boldsymbol{\epsilon}^{\mathrm{p}}):\boldsymbol{\sigma} + \mathbf{C}^{-1}(\boldsymbol{\epsilon}^{\mathrm{p}}):d\boldsymbol{\sigma} \tag{5.143}$$

Hence

$$d\boldsymbol{\epsilon} = d\boldsymbol{\epsilon}^{\mathrm{e}} + d\boldsymbol{\epsilon}^{\mathrm{p}} = \mathbf{C}^{-1}(\boldsymbol{\epsilon}^{\mathrm{p}}):d\boldsymbol{\sigma} + d\mathbf{C}^{-1}(\boldsymbol{\epsilon}^{\mathrm{p}}):\boldsymbol{\sigma} + d\boldsymbol{\epsilon}^{\mathrm{p}} \tag{5.144}$$

Since $\mathbf{C}^{-1}(\boldsymbol{\epsilon}^{\mathrm{p}})$ is dependent only on the previous plastic deformation, it will be constant for the stress path inside the current yield surface as well as the unloading path starting from the yield surface and towards its interior. Thus Eq. (5.123) can be written

$$\begin{aligned} W = &\int_{t_0}^{t_1}(\boldsymbol{\sigma} - \boldsymbol{\sigma}^0):\mathbf{C}^{-1}(\boldsymbol{\epsilon}^{\mathrm{p}})_{t_1}:\dot{\boldsymbol{\sigma}}\,dt + \int_{t_1}^{t_2}(\boldsymbol{\sigma} - \boldsymbol{\sigma}^0): \\ &\times\left[\mathbf{C}^{-1}(\boldsymbol{\epsilon}^{\mathrm{p}})_t:\dot{\boldsymbol{\sigma}} + \dot{\mathbf{C}}^{-1}(\boldsymbol{\epsilon}^{\mathrm{p}})_t:\boldsymbol{\sigma} + \dot{\boldsymbol{\epsilon}}^{\mathrm{p}}\right]dt \\ &+\int_{t_2}^{t_3}(\boldsymbol{\sigma} - \boldsymbol{\sigma}^0):\mathbf{C}^{-1}(\boldsymbol{\epsilon}^{\mathrm{p}})_{t_2}:\dot{\boldsymbol{\sigma}}\,dt \geq 0 \end{aligned} \tag{5.145}$$

where $\dot{\mathbf{C}}^{-1}(\boldsymbol{\epsilon}^{\mathrm{p}})$ is the rate of change of $\mathbf{C}^{-1}(\boldsymbol{\epsilon}^{\mathrm{p}})$ due to plastic deformation and $\mathbf{C}^{-1}(\boldsymbol{\epsilon}^{\mathrm{p}})_{t_1}$ and $\mathbf{C}^{-1}(\boldsymbol{\epsilon}^{\mathrm{p}})_{t_2}$ are the elastic compliance calculated for the plastic strain $\boldsymbol{\epsilon}^{\mathrm{p}}$ at times t_1 and t_2, respectively ($t_1 \leq t \leq t_2$). Note that from t_0 to t_1 and from t_2 to t_3, the deformation is elastic, and $\dot{\mathbf{C}}^{-1}(\boldsymbol{\epsilon}^{\mathrm{p}})$ is zero. If $\delta t = t_2 - t_1$ and approaches zero, then

$$\mathbf{C}^{-1}(\boldsymbol{\epsilon}^{\mathrm{p}})_{t_2} \approx \mathbf{C}^{-1}(\boldsymbol{\epsilon}^{\mathrm{p}})_{t_1} + \dot{\mathbf{C}}^{-1}(\boldsymbol{\epsilon}^{\mathrm{p}})_{t_1}\,\delta t \tag{5.146}$$

$$\mathbf{C}^{-1}(\boldsymbol{\epsilon}^{\mathrm{p}})_t \approx \mathbf{C}^{-1}(\boldsymbol{\epsilon}^{\mathrm{p}})_{t_1} + \dot{\mathbf{C}}^{-1}(\boldsymbol{\epsilon}^{\mathrm{p}})_{t_1}\eta, \qquad t = t_1 + \eta,\ 0 \leq \eta \leq \delta t \tag{5.147}$$

$$\dot{\mathbf{C}}^{-1}(\boldsymbol{\epsilon}^{\mathrm{p}})_t \approx \dot{\mathbf{C}}^{-1}(\boldsymbol{\epsilon}^{\mathrm{p}})_{t_1} + \ddot{\mathbf{C}}^{-1}(\boldsymbol{\epsilon}^{\mathrm{p}})_{t_1}\eta, \qquad t = t_1 + \eta,\ 0 \leq \eta \leq \delta t \tag{5.148}$$

where $\eta = t - t_1$. After denoting $\mathbf{C}^{-1}(\boldsymbol{\epsilon}^{\mathrm{p}})t_1$, $\dot{\mathbf{C}}^{-1}(\boldsymbol{\epsilon}^{\mathrm{p}})t_1$, and $\dot{\mathbf{C}}^{-1}(\boldsymbol{\epsilon}^{\mathrm{p}})t_1$ by $\mathbf{C}_1^{-1}$, $\dot{\mathbf{C}}_1^{-1}$, and $\ddot{\mathbf{C}}_1^{-1}$, respectively, and substituting Eqs. (5.146) through (5.148) into Eq. (5.145), we have

$$\begin{aligned} W = &\int_{C_\sigma}(\boldsymbol{\sigma} - \boldsymbol{\sigma}^0):\mathbf{C}_1^{-1}:\dot{\boldsymbol{\sigma}}\,dt + \int_{t_1}^{t_2}(\boldsymbol{\sigma} - \boldsymbol{\sigma}^0):\left[\dot{\mathbf{C}}_1^{-1}:\boldsymbol{\sigma} + \dot{\boldsymbol{\epsilon}}^{\mathrm{p}}\right]dt \\ &+\eta\int_{t_1}^{t_2}(\boldsymbol{\sigma} - \boldsymbol{\sigma}^0):\left[\dot{\mathbf{C}}_1^{-1}:\boldsymbol{\sigma} + \ddot{\mathbf{C}}_1^{-1}:\dot{\boldsymbol{\sigma}}\right]dt \\ &+\delta t\int_{t_2}^{t_3}(\boldsymbol{\sigma} - \boldsymbol{\sigma}^0):\dot{\mathbf{C}}_1^{-1}:\dot{\boldsymbol{\sigma}}\,dt \geq 0 \end{aligned} \tag{5.149}$$

The final two terms in this equation can be neglected when $\delta t \to 0$, since they are higher-order quantities of δt. Furthermore the first term in Eq. (5.149) can be regarded as the elastic energy stored over a closed loading unloading path for a material whose elastic compliance $\mathbf{C}^{-1}$ is constant and equal to $\mathbf{C}_1^{-1}$. According to Eq. (5.124) this term is also zero. Equation (5.149) reduces to

$$W = \int_{t_1}^{t_2} (\boldsymbol{\sigma} - \boldsymbol{\sigma}^0) : \left[\dot{\mathbf{C}}_1^{-1} : \boldsymbol{\sigma} + \dot{\boldsymbol{\epsilon}}^{\mathrm{p}} \right] dt > 0 \tag{5.150}$$

Again, assuming that $\delta t \to 0$, the integrand can be approximated by

$$\begin{aligned}
(\boldsymbol{\sigma} - \boldsymbol{\sigma}^0) &: \left[\dot{\mathbf{C}}_1^{-1} : \boldsymbol{\sigma} + \dot{\boldsymbol{\epsilon}}^{\mathrm{p}} \right] \\
&= \left[(\boldsymbol{\sigma} - \boldsymbol{\sigma}^0) : \left(\dot{\mathbf{C}}_1^{-1} : \boldsymbol{\sigma} + \dot{\boldsymbol{\epsilon}}^{\mathrm{p}} \right) \right]_{t_1} \\
&\quad + \tfrac{1}{2} \left[\dot{\boldsymbol{\sigma}} : \left(\dot{\mathbf{C}}_1^{-1} : \boldsymbol{\sigma} + \dot{\boldsymbol{\epsilon}}^{\mathrm{p}} \right) + (\boldsymbol{\sigma} - \boldsymbol{\sigma}^0) : \left(\dot{\mathbf{C}}_1^{-1} : \dot{\boldsymbol{\sigma}} + \ddot{\boldsymbol{\epsilon}}^{\mathrm{p}} \right) \right]_{t_1} \delta t
\end{aligned} \tag{5.151}$$

and Eq. (5.151) becomes

$$\begin{aligned}
W &= \left[(\boldsymbol{\sigma} - \boldsymbol{\sigma}^0) : \left(\dot{\mathbf{C}}_1^{-1} : \boldsymbol{\sigma} + \dot{\boldsymbol{\epsilon}}^{\mathrm{p}} \right) \right]_{t_1} \delta t \\
&\quad + \tfrac{1}{2} \left[\dot{\boldsymbol{\sigma}} : \left(\dot{\mathbf{C}}_1^{-1} : \boldsymbol{\sigma} + \dot{\boldsymbol{\epsilon}}^{\mathrm{p}} \right) + (\boldsymbol{\sigma} - \boldsymbol{\sigma}^0) : \left(\dot{\mathbf{C}}_1^{-1} : \dot{\boldsymbol{\sigma}} + \ddot{\boldsymbol{\epsilon}}^{\mathrm{p}} \right) \right]_{t_1} \delta t^2
\end{aligned} \tag{5.152}$$

By using the same rationale as for Eq. (5.128), we can derive

$$(\boldsymbol{\sigma} - \boldsymbol{\sigma}^0) : \left(\dot{\mathbf{C}}^{-1} : \boldsymbol{\sigma} + \dot{\boldsymbol{\epsilon}}^{\mathrm{p}} \right) > 0 \tag{5.153}$$

if $\boldsymbol{\sigma}^0$ is not on the yield surface at $t = t_0$, and

$$\dot{\boldsymbol{\sigma}} : \left(\dot{\mathbf{C}}^{-1} : \boldsymbol{\sigma} + \dot{\boldsymbol{\epsilon}}^{\mathrm{p}} \right) > 0 \tag{5.154}$$

if $\boldsymbol{\sigma}^0$ is on the yield surface at $t_2 = t_0$; the subscript 1 and t_1 have been dropped for simplicity. Equation (5.141) is obtained directly from Eq. (5.154) and the two basic assumptions made by Drucker (1950, 1951) presented earlier. In addition, since $\dot{\boldsymbol{\epsilon}}^{\mathrm{p}}$ is no longer normal to the yield surface in the case of elastic-plastic coupling, the yield surface will no longer be convex. Note that if there is no elastic-plastic coupling, $\dot{\mathbf{C}}_1^{-1} = \mathbf{0}$, the normality of $\dot{\boldsymbol{\epsilon}}^{\mathrm{p}}$ and the convexity of the yield surface can be derived from Eqs. (5.153) and (5.154). Obviously it is also the case if the elastic-plastic coupling is negligible.

3. For perfectly plastic materials, Eq. (5.129) holds, but Eq. (5.130) should be modified to

$$d\boldsymbol{\sigma} : d\boldsymbol{\epsilon}^{\mathrm{p}} = 0 \tag{5.155}$$

since the yield surface is fixed in stress space and a change of the stress $d\boldsymbol{\sigma}$ during plastic deformation can occur only on the tangential plane of the yield surface. Convexity and normality conditions are valid for these materials because there is no elastic-plastic coupling.

5.8 ISOTROPIC HARDENING

Drucker's stability postulate can be used to determine the plastic potential for hardening or stable materials and perfectly plastic materials (sometimes called neutrally stable materials). Two general conclusions can be drawn from the Drucker's postulate:

1. The yield surface is convex.
2. The plastic strain rate or increment is normal to the yield surface.

The plastic strain rate can be written as

$$\dot{\boldsymbol{\epsilon}}^{\mathrm{p}} = \dot{\lambda}\frac{\partial F}{\partial \boldsymbol{\sigma}} \tag{5.156}$$

Thus

$$Q = F \tag{5.157}$$

Recall that these statements are true only for stable and neutrally stable materials with no elastic-plastic coupling. Unless explicitly stated otherwise, these conditions are assumed to be valid for materials under discussion.

For isotropic, pressure-insensitive materials, the yield function for isotropic hardening is

$$F = f(J_2', J_3') - \kappa(\alpha) = 0 \tag{5.158}$$

where J_2' and J_3' are the second and third invariants of the deviatoric stress tensor S. Only one hardening parameter α is necessary to characterize this isotropic hardening; κ is a monotonically increasing function of α. For the sake of simplicity and clarity, the function $f(J_2', J_3')$ is assumed to be a von Mises type. It then follows that

$$F = J_2' - \tfrac{1}{3}\sigma_{\mathrm{Y}}^2(\alpha) = 0 \tag{5.159}$$

where σ_{Y} is the current uniaxial yield strength that increases with α.

Usually there are two measures of hardening that can be used as α. The first one is defined by

$$\alpha = \epsilon_e^p = \int d\epsilon_e^p \tag{5.160}$$

where $d\epsilon_e^p$ is given by

$$\begin{aligned} d\epsilon_e^p &= \frac{2}{\sqrt{3}}\left(\frac{1}{2}d\boldsymbol{\epsilon}^p : d\boldsymbol{\epsilon}^p\right)^{1/2} \\ &= \frac{2}{\sqrt{3}}\left[(d\epsilon_1^p - d\epsilon_2^p)^2 + (d\epsilon_2^p - d\epsilon_3^p)^2 + (d\epsilon_3^p - d\epsilon_1^p)^2\right]^{1/2} \end{aligned} \tag{5.161}$$

ϵ_e^p is called the *equivalent* or *effective plastic strain*, but it should not be confused with $\sqrt{2}\,(\boldsymbol{\epsilon}^p : \boldsymbol{\epsilon}^p)^{1/2} / \sqrt{3}$.

Equation (5.161) represents the assumption that the amount of hardening depends only on the effective plastic strain or the length of the trajectory in the plastic strain space but *not* on the strain path. This called the *strain-hardening hypothesis*.

The evolution equation for α can be obtained as follows: From Eq. (5.161),

$$d\alpha = d\epsilon_e^p = \frac{2}{\sqrt{6}}\dot{\lambda}\left(\frac{\partial f}{\partial \boldsymbol{\sigma}} : \frac{\partial f}{\partial \boldsymbol{\sigma}}\right)^{1/2} \tag{5.162}$$

where Eqs. (5.159) and the normality condition of $d\dot{\boldsymbol{\epsilon}}^p$ have been used. It can be seen that once $\dot{\lambda}$ is determined, the evolution of α can be calculated from this equation. The calculation of $\dot{\lambda}$ will be discussed later.

To determine the function $\sigma_Y(\alpha)$, recall that for uniaxial tension Eq. (5.159) reduces to

$$\sigma = \sigma_Y(\epsilon^p) \tag{5.163}$$

so that $\sigma_Y(\epsilon^p)$ can be obtained by using the experimental uniaxial tension curve σ with respect to ϵ^p, as shown in Fig. 5.10. By simply replacing ϵ^p in Eq. (5.163) by ϵ_e^p defined by Eq. (5.160), the subsequent yield surface due to isotropic strain hardening is determined by

$$J_2' = \tfrac{1}{3}\sigma_Y^2(\epsilon_e^p) \tag{5.164}$$

The second measure used as the isotropic hardening parameter is the total plastic work defined by

$$\alpha = W^p = \int \boldsymbol{\sigma} : d\boldsymbol{\epsilon}^p \tag{5.165}$$

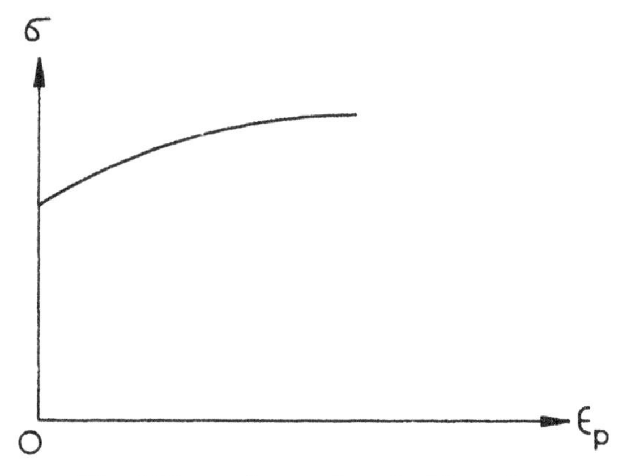

Fig. 5.10 σ-ϵ^{p} curve in uniaxial tension

Thus, as in Eq. (5.164), the yield condition can be written

$$J_2' = \tfrac{1}{3}\sigma_{\mathrm{Y}}^2(W^{\mathrm{p}}) \tag{5.166}$$

This is called the *work-hardening hypothesis*, which shows that the amount of hardening depends only on the total plastic work and is not dependent on the loading path. The evolution of W^{p} can be obtained by differentiating Eq. (5.165),

$$dW^{\mathrm{p}} = \boldsymbol{\sigma} : d\boldsymbol{\epsilon}^{\mathrm{p}} = \mathbf{S} : d\boldsymbol{\epsilon}^{\mathrm{p}} = \dot{\lambda}\frac{\partial f}{\partial \boldsymbol{\sigma}} : \mathbf{S} \tag{5.167}$$

Again, once $\dot{\lambda}$ is determined, dW^{p} and $W^{\mathrm{p}} = \int dW^{\mathrm{p}}$ can be calculated, and the subsequent yield surface at any stage of plastic deformation can be completely specified. It is easy to show that if the von Mises yield criterion or the Prandtl-Reuss equation is used, the work hardening and strain hardening are equivalent. Indeed from Eq. (5.119) it follows that

$$W^{\mathrm{p}} = \int \boldsymbol{\sigma} : d\boldsymbol{\epsilon}^{\mathrm{p}} = \int \sigma_{\mathrm{e}} d\epsilon_{\mathrm{e}}^{\mathrm{p}} \tag{5.168}$$

Then Eq. (5.166) is equivalent to

$$J_2' = \tfrac{1}{3}\sigma_{\mathrm{Y}}^2\,(W^{\mathrm{p}}) = \tfrac{1}{3}\sigma_{\mathrm{Y}}^2\left(\int \sigma_{\mathrm{e}} d\epsilon_{\mathrm{e}}^{\mathrm{p}}\right) = \tfrac{1}{3}\sigma_{\mathrm{Y}}^2(\epsilon_{\mathrm{e}}^{\mathrm{p}}) \tag{5.169}$$

The form of the function $\sigma_{\mathrm{Y}}(W^{\mathrm{p}})$ can also be obtained from the uniaxial tension test. For example, in the case of linear hardening, as shown in Fig. 5.11, where E^{p} is the slope of the linear hardening portion and σ_{Y}^0 is the

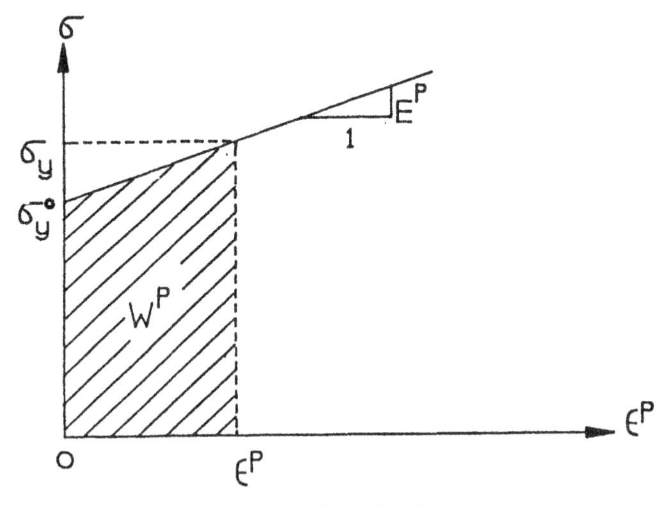

Fig. 5.11 Linear hardening

initial yield stress. Hence

$$\sigma_Y^2 = \sigma_Y^{0^2} + 2E^{\mathrm{p}}W^{\mathrm{p}} \tag{5.170}$$

Substituting Eq. (5.170) into Eq. (5.166) gives the subsequent yield surface for isotropic linear work-hardening materials

$$W^{\mathrm{p}} = \frac{1}{2E^{\mathrm{p}}}\left(\sigma_Y^2 - \sigma_Y^{0^2}\right) \tag{5.171}$$

$$J_2' = \tfrac{1}{3}\left(\sigma_Y^{0^2} + 2E^{\mathrm{p}}W^{\mathrm{p}}\right) \tag{5.172}$$

Compared to the work-hardening model, the strain hardening is simpler to use, but the work-hardening hypothesis is more general.

Note that the isotropic hardening implies the existence of a universal function between a certain measure of stress, called the *equivalent stress*, and a certain measure of strain, called the *equivalent strain* or *equivalent plastic strain*. Also the equivalent stress and strain do not have to be σ_e and ϵ_e or ϵ_e^{p} as defined earlier. Their definitions will depend on the yield criterion to be used. The function is called *universal* because it holds true for any stress and strain state and loading path. For example,

$$\sigma_e = \sigma_Y(\epsilon_e^{\mathrm{p}}) \tag{5.173}$$

based on the von Mises criterion, and

$$\tau_{max} = \tau_Y(\gamma_{max}^{\mathrm{p}}) \tag{5.174}$$

based on the maximum shear stress or Tresca criterion. The function form of σ_Y can be determined from the uniaxial tension test, and $\tau_Y(\gamma^p_{max})$ from the pure shear test. Finally, it should be mentioned that the existence of such a universal function for infinitesimal deformation has been established beyond any doubt. However, in finite deformation there is confusion about the experimental results. This problem will be discussed in Chapter 7.

5.9 KINEMATIC HARDENING

The term "kinematic hardening" was introduced by Prager (1955) in connection with his kinematic model. As stated in Section 4.5.3, the kinematic model assumes that during the process of plastic loading the yield surface translates in the stress space and its shape and size remain unchanged. This is motivated by the Bauschinger effect in the uniaxial tension-compression.

Suppose that the initial yield surface is described by

$$F = f(\boldsymbol{\sigma}) - \kappa = 0 \tag{5.175}$$

Then, due to the kinematic hardening in the process of plastic deformation, the subsequent yield surface takes the form

$$f(\boldsymbol{\sigma} - \boldsymbol{\alpha}) - \kappa = 0 \tag{5.176}$$

where $\boldsymbol{\alpha}$ is a tensorial hardening parameter, usually called *back stress*, that represents the center of the yield surface in the stress space. κ is a material constant representing the size of the yield surface. It can be seen that as $\boldsymbol{\alpha}$ changes due to plastic strain hardening, the yield surface translates in the stress space while maintaining its initial shape and size. Thus the formulation of a kinematic hardening model requires a specification for the evolution of $\boldsymbol{\alpha}$ in terms of $\boldsymbol{\epsilon}^p$, $\boldsymbol{\sigma}$, or $\boldsymbol{\alpha}$ itself, as discussed in Section 5.1.

In modeling the kinematic hardening behavior observed in experimental investigations, Prager proposed the following linear constitutive equation for the back stress $\boldsymbol{\alpha}$:

$$d\boldsymbol{\alpha} = c\,d\boldsymbol{\epsilon}^p \tag{5.177}$$

where c is a material constant. According to this model the yield surface keeps its original shape and size and moves in the direction of the plastic strain rate or strain increment, or in the direction of the normal to the yield surface at the loading point, due to the normality condition for the plastic strain rate or strain increment. Using Eq. (5.136), Eq. (5.177) can be rewritten as

$$d\boldsymbol{\alpha} = c\dot{\lambda}\frac{\partial F}{\partial \boldsymbol{\sigma}} \tag{5.178}$$

According to Ziegler, one deficiency of Prager's linear kinematic model is that it does not give consistent results for three-dimensional and two-dimensional cases. The reason is that the yield function F takes different forms for one-, two-, or three-dimensional cases. As a result, if the yield surface in three-dimensional space moves in the direction of the normal to the yield surface at the loading point, then the yield locus in the two-dimensional space does not move in the same direction (Shield and Ziegler 1958; Ziegler 1959).

Of interest to us here is the transverse softening effect introduced by Prager's model for uniaxial tension. Suppose that the axial load is applied in the 1-direction, causing the plastic flow. Obviously then

$$\begin{aligned} d\epsilon_1^{\mathrm{p}} &= d\epsilon^{\mathrm{p}} \\ d\epsilon_2^{\mathrm{p}} &= d\epsilon_3^{\mathrm{p}} = -\tfrac{1}{2} d\epsilon^{\mathrm{p}} \end{aligned} \tag{5.179}$$

By Eq. (5.177) it follows that

$$\begin{aligned} d\alpha_1 &= c\, d\epsilon^{\mathrm{p}} \\ d\alpha_2 &= d\alpha_3 = -\tfrac{1}{2} d\alpha_1 = -\tfrac{1}{2} c\, d\epsilon^{\mathrm{p}} \end{aligned} \tag{5.180}$$

Thus, as the yield surface moves toward the positive direction of loading causing kinematic hardening of the material, it moves toward the negative 2- and 3-directions causing the transverse softening, although no load is applied along these directions. This transverse softening or hardening effect is not desirable, since the experimental results do not support this point of view.

To overcome these shortcomings, Ziegler proposed the following modification of the Prager's model (1959):

$$d\boldsymbol{\alpha} = (\boldsymbol{\sigma} - \boldsymbol{\alpha})\, d\mu \tag{5.181}$$

where $d\mu$ is a proportionality scalar constant determined by the yield criterion. (The determination of $d\mu$ will be discussed later.) Instead of assuming that the yield surface moves along the normal direction, as in Eqs. (5.177) and (5.178), Ziegler's model assumes that the movement takes place in the radial direction determined by the vector $\boldsymbol{\sigma} - \boldsymbol{\alpha}$. The schematic representations of Prager's and Ziegler's models are shown in Fig. 5.12. Also Shield and Ziegler (1958) showed that using von Mises yield criterion, the yield surface *does* move along the radial direction $\boldsymbol{\sigma} - \boldsymbol{\alpha}$.

According to Ziegler, the advantages of his model are that it gives consistent results in one-, two-, or three-dimensional loading and does not show transverse softening or hardening effect. If several components of the stress tensor $\boldsymbol{\sigma}$ are zero, the corresponding components of $\boldsymbol{\alpha}$ are also zero. It should be pointed out that in three-dimensional case for the von Mises criterion, Prager's model is the same as Ziegler's model, since the radial direction is the same as the normal direction in this case, as can be seen from

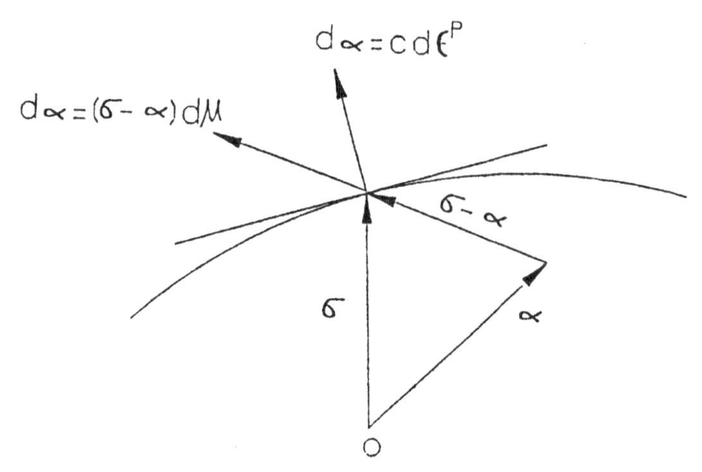

Fig. 5.12 Prager's and Ziegler's kinematic hardening model

Fig. 5.13. However, in the two-dimensional case the radial and the normal directions are obviously different, as shown in Fig. 5.14. Note that for the Tresca criterion the radial and normal directions are generally different, even in a three-dimensional case; see Fig. 5.15.

Compared with Prager's model, Ziegler's model is simpler to use, since the mathematical expression of $\partial F/\partial \boldsymbol{\sigma}$ is in general more complicated than that of $\boldsymbol{\sigma} - \boldsymbol{\alpha}$. In Prager's model, if c is a constant, it is called the *linear*

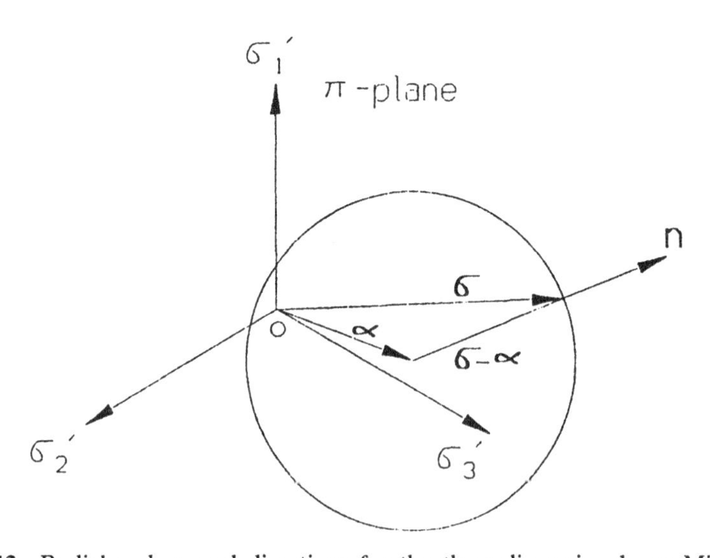

Fig. 5.13 Radial and normal directions for the three-dimensional von Mises yield surface

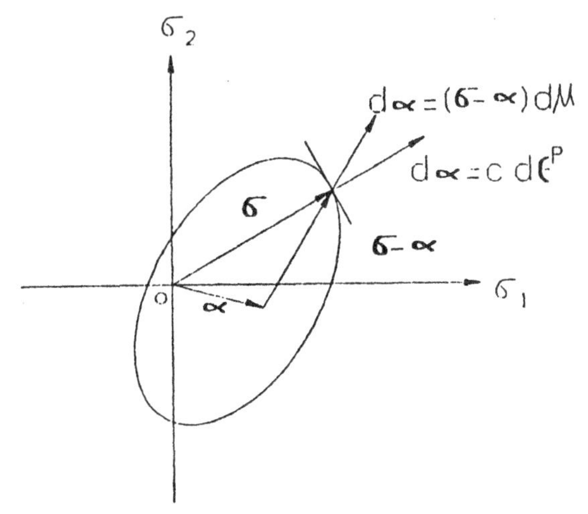

Fig. 5.14 Prager's and Ziegler's models for the two-dimensional von Mises criterion

hardening rule. However, c can be considered a function of the deformation history, thus representing a nonlinear hardening rule.

The constitutive equations for $\boldsymbol{\alpha}$ given in this section, Eqs. (5.178) and (5.181), are not in the general form presented by Eq. (5.36) of Section 5.1. However, as we pointed out earlier, there is an unspecified proportional factor λ or $d\mu$ in the constitutive equation for $\boldsymbol{\alpha}$, and its determination involves the yield criterion. It will be shown in the next section that after solving λ or $d\mu$ from the so-called consistency condition and substituting it into Eq. (5.178) or (5.181), the general form (5.36) for $\dot{\boldsymbol{\alpha}}$ can be derived.

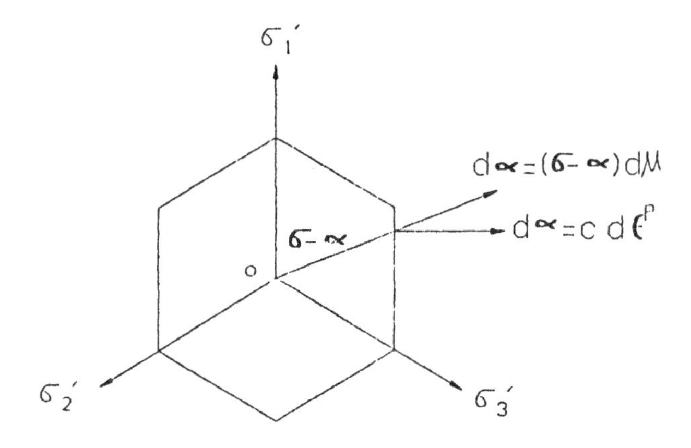

Fig. 5.15 Prager's and Ziegler's models for the Tresca criterion

5.10 GENERAL STRESS-STRAIN RELATIONS FOR PLASTIC DEFORMATION

In Section 5.5 to Section 5.9 we derived the results needed for establishing general stress-strain relations for plastic deformation. Now we use these results to obtain a general stress-strain relationship. As we noted in Section 5.1, according to the characteristics of plastic deformation, this relationship for plastic flow must be in the incremental form, or equivalently in the rate form. It can be written as

$$d\boldsymbol{\sigma} = \mathbf{C}^{ep} : d\boldsymbol{\epsilon} \quad \text{or} \quad \dot{\boldsymbol{\sigma}} = \mathbf{C}^{ep}\dot{\boldsymbol{\epsilon}} \tag{5.182}$$

where $\mathbf{C}^{ep}$ represents the elastic-plastic stiffness tensor which is of the fourth order and independent of time. The superscript "ep" indicates elastic-plastic deformation. By this convention the stress-strain relation (5.10) for elastic deformation can be rewritten in the following increment or rate form:

$$d\boldsymbol{\sigma} = \mathbf{C}^{e} : d\boldsymbol{\epsilon}^{e} \quad \text{or} \quad \boldsymbol{\sigma} = \mathbf{C}^{e} : \boldsymbol{\epsilon}^{e} \tag{5.183}$$

where $\mathbf{C}^{e}$ is the elastic stiffness tensor, which is assumed to be constant during the plastic deformation.

To establish the incremental equations for elastic-plastic hardening materials, several conditions are necessary to ensure an appropriate representation of the plastic flow. The following four assumptions were given by Prager (1949):

1. *Continuity*. It is assumed that the neutral loading does not cause plastic deformation.
2. *Uniqueness*. It ensures that for a given set of infinitesimal increment of surface traction applied to a body, the resulting stress and strain increments will be unique. This condition is described mathematically by

$$\dot{\boldsymbol{\sigma}} : \dot{\boldsymbol{\epsilon}}^{p} > 0 \quad \text{or} \quad d\boldsymbol{\sigma} : d\boldsymbol{\epsilon}^{p} > 0 \tag{5.184}$$

3. *Irreversibility*. It requires that the increment of the plastic work must be positive:

$$\boldsymbol{\sigma} : \dot{\boldsymbol{\epsilon}}^{p} > 0 \quad \text{or} \quad \boldsymbol{\sigma} : d\boldsymbol{\epsilon}^{p} > 0 \tag{5.185}$$

4. *Consistency*. During the plastic deformation the stress state or the stress point in the stress space must remain on the subsequent yield surface. In other words, loading from a plastically deformed state will result in another plastically deformed state.

In addition to these four assumptions there are two more that arise from the derivation of the normality condition in Drucker's postulate. They are the existence of the yield surface and the linearity of the relation between $d\boldsymbol{\sigma}$ and $d\boldsymbol{\epsilon}$, as presented by Eq. (5.182). Assumptions 1, 2, and 3 have been discussed previously. In fact assumptions 1 and 3 can be considered as consequences of Drucker's postulate. Assumption 4 will be discussed in detail now.

Suppose that the current plastic state is represented by the stress state $\boldsymbol{\sigma}$ and the hardening parameter $\boldsymbol{\alpha}$. For simplicity, only one tensorial hardening parameter is assumed. Then $\boldsymbol{\sigma}$ and $\boldsymbol{\alpha}$ must satisfy the yield condition

$$F(\boldsymbol{\sigma}, \boldsymbol{\alpha}) = 0 \tag{5.186}$$

Now consider the plastic deformation caused by an infinitesimal increment $d\boldsymbol{\sigma}$. The stress state becomes $\boldsymbol{\sigma} + d\boldsymbol{\sigma}$, and because of plastic deformation the hardening parameter will be $\boldsymbol{\alpha} + d\boldsymbol{\alpha}$. According to the consistency condition, $\boldsymbol{\sigma} + d\boldsymbol{\sigma}$ and $\boldsymbol{\alpha} + d\boldsymbol{\alpha}$ must be on the subsequent yield surface, namely they should satisfy

$$F(\boldsymbol{\sigma} + d\boldsymbol{\sigma}, \boldsymbol{\alpha} + d\boldsymbol{\alpha}) = 0 \tag{5.187}$$

Since $d\boldsymbol{\sigma}$, and hence $d\boldsymbol{\alpha}$, are infinitesimal, Eq. (5.187) can be expanded into

$$F(\boldsymbol{\sigma} + d\boldsymbol{\sigma}, \boldsymbol{\alpha} + d\boldsymbol{\alpha}) = F(\boldsymbol{\sigma}, \boldsymbol{\alpha}) + \frac{\partial F}{\partial \boldsymbol{\sigma}} : d\boldsymbol{\sigma} + \frac{\partial F}{\partial \boldsymbol{\alpha}} : d\boldsymbol{\alpha} = 0 \tag{5.188}$$

A comparison of Eq. (5.188) with Eqs. (5.186) and (5.187) gives the consistency condition

$$\dot{F} = \frac{\partial F}{\partial \boldsymbol{\sigma}} : d\boldsymbol{\sigma} + \frac{\partial F}{\partial \boldsymbol{\alpha}} : d\boldsymbol{\alpha} = 0 \tag{5.189}$$

If the hardening parameter is scalar, Eq. (5.189) is written as

$$\dot{F} = \frac{\partial F}{\partial \boldsymbol{\sigma}} : d\boldsymbol{\sigma} + \frac{\partial F}{\partial \alpha} : d\alpha = 0 \tag{5.190}$$

Furthermore, if there are n hardening parameters, m of these are tensorial and the remaining are scalar, then the consistency condition becomes

$$\dot{F} = \frac{\partial F}{\partial \boldsymbol{\sigma}} : d\boldsymbol{\sigma} + \sum_{i=1}^{m} \frac{\partial F}{\partial \boldsymbol{\alpha}_i} : d\boldsymbol{\alpha}_i + \sum_{j=1}^{n-m} \frac{\partial F}{\partial \alpha_j} d\alpha_j \tag{5.191}$$

Using the results in Sections 5.5 through 5.9 and the assumptions presented above, a general incremental stress-strain relation for plastic deformation can be derived.

5.10.1 Isotropic Hardening Materials

For isotropic hardening materials, it is already known that

$$d\boldsymbol{\epsilon}^{\mathrm{e}} = \mathbf{C}^{\mathrm{e}^{-1}} : d\boldsymbol{\sigma}, \quad d\boldsymbol{\epsilon}^{\mathrm{p}} = \dot{\lambda}\frac{\partial F}{\partial \boldsymbol{\sigma}} \tag{5.192}$$

$$F(\boldsymbol{\sigma}, \alpha) = f(\boldsymbol{\sigma}) - \kappa(\alpha) = 0 \tag{5.193}$$

where

$$\alpha = \epsilon_{\mathrm{e}}^{\mathrm{p}} \quad \text{or} \quad \alpha = W^{\mathrm{p}} \tag{5.194}$$

depending on whether the strain hardening or the work hardening is used. It is assumed that the function form $\kappa = \kappa(\alpha)$ already is known from the uniaxial tension test. According to the consistency condition and the strain-hardening assumption,

$$\begin{aligned} \dot{F} &= \frac{\partial f}{\partial \boldsymbol{\sigma}} : d\boldsymbol{\sigma} - \frac{d\kappa}{d\epsilon_{\mathrm{e}}^{\mathrm{p}}} d\epsilon_{\mathrm{e}}^{\mathrm{p}} \\ &= \frac{\partial f}{\partial \boldsymbol{\sigma}} : d\boldsymbol{\sigma} - \frac{d\kappa}{d\epsilon_{\mathrm{e}}^{\mathrm{p}}} \frac{2}{\sqrt{6}} \dot{\lambda} \left(\frac{\partial f}{\partial \boldsymbol{\sigma}} : \frac{\partial f}{\partial \boldsymbol{\sigma}} \right)^{1/2} = 0 \end{aligned} \tag{5.195}$$

where Eq. (5.192) has been used. The proportionality scalar factor $\dot{\lambda}$ can be determined from this equation to give

$$\dot{\lambda} = \frac{l}{(2/\sqrt{6})(d\kappa/d\epsilon_{\mathrm{e}}^{\mathrm{p}})(\partial f/\partial \sigma : \partial f/\partial \sigma)^{1/2}} = \frac{\sqrt{6}}{2d\kappa/d\epsilon_{\mathrm{e}}^{\mathrm{p}}} \mathbf{n} : d\sigma \tag{5.196}$$

Here

$$l = \frac{\partial f}{\partial \sigma} : d\sigma \quad \text{and} \quad \mathbf{n} = \frac{\partial f/\partial \sigma}{(\partial f/\partial \sigma : \partial f/\partial \sigma)^{1/2}} \tag{5.197}$$

where $\mathbf{n}$ is the unit normal to the yield surface. Then the plastic strain increment can be expressed as

$$d\boldsymbol{\epsilon}^{\mathrm{p}} = \frac{\sqrt{6}\,\mathbf{n}}{2d\kappa/d\epsilon_{\mathrm{e}}^{\mathrm{p}}} l = \frac{\sqrt{6}}{2d\kappa/d\epsilon_{\mathrm{e}}^{\mathrm{p}}} (\mathbf{n} : d\sigma) \frac{\partial f}{\partial \sigma} \tag{5.198}$$

and

$$d\alpha = d\epsilon_e^p = \frac{1}{d\kappa/d\epsilon_e^p} l \tag{5.199}$$

which conform with the general forms for $d\epsilon^p$ and $d\alpha$ given by Eqs. (5.36) and (5.37). From

$$d\boldsymbol{\epsilon} = d\boldsymbol{\epsilon}^e + d\boldsymbol{\epsilon}^p \tag{5.200}$$

and Eqs. (5.183) and (5.198), we get

$$d\boldsymbol{\epsilon} = \left[\mathbf{C}^{e^{-1}} + \frac{\sqrt{6}}{2\, d\kappa/d\epsilon_e^p} \frac{\partial f}{\partial \boldsymbol{\sigma}} \otimes \mathbf{n} \right] : d\boldsymbol{\sigma} \tag{5.201}$$

or

$$d\sigma = \left[\mathbf{C}^{e^{-1}} + \frac{\sqrt{6}}{2\, d\kappa/d\epsilon_e^p} \frac{\partial f}{\partial \boldsymbol{\sigma}} \otimes \mathbf{n} \right]^{-1} : d\boldsymbol{\epsilon} \tag{5.202}$$

Obviously

$$\mathbf{C}^{ep} = \left[\mathbf{C}^{e^{-1}} + \frac{\sqrt{6}}{2\, d\kappa/d\epsilon_e^p} \frac{\partial f}{\partial \sigma} \otimes \mathbf{n} \right]^{-1} \tag{5.203}$$

Since for hardening materials $d\kappa/d\epsilon_e^p \neq 0$, the above expressions are meaningful.

If the work hardening is assumed, Eq. (5.195) becomes

$$\dot{F} = l - \frac{d\kappa}{dW^p} \dot{\lambda} \frac{\partial f}{\partial \sigma} : \mathbf{S} = 0 \tag{5.204}$$

Thus

$$\dot{\lambda} = \frac{l}{(d\kappa/dW^p)(\partial f/\partial \sigma) : \mathbf{S}} \tag{5.205}$$

and

$$d\boldsymbol{\epsilon}^p = \frac{\partial f/\partial \boldsymbol{\sigma}}{(d\kappa/dW^p)(\partial f/\partial \sigma) : \mathbf{S}} l, \quad d\alpha = dW^p = \frac{1}{d\kappa/dW^p} \tag{5.206}$$

which are also congruent with Eqs. (5.36) and (5.37). Similarly we can derive

$$d\boldsymbol{\epsilon} = \left[\mathbf{C}^{e^{-1}} + \frac{(\partial f/\partial \boldsymbol{\sigma})(\partial f/\partial \boldsymbol{\sigma})}{(d\kappa/dW^{p})(\partial f/\partial \boldsymbol{\sigma}):\mathbf{S}}\right]:d\boldsymbol{\sigma} \quad \text{and} \quad d\boldsymbol{\sigma} = \mathbf{C}^{ep}:d\boldsymbol{\epsilon} \tag{5.207}$$

where

$$\mathbf{C}^{ep} = \left[\mathbf{C}^{e^{-1}} + \frac{(\partial f/\partial \boldsymbol{\sigma})(\partial f/\partial \boldsymbol{\sigma})}{(d\kappa/dW^{p})(\partial f/\partial \boldsymbol{\sigma}):\mathbf{S}}\right]^{-1} \tag{5.208}$$

For von Mises yield criterion

$$J_2' = \tfrac{1}{3}\sigma_Y^2(\epsilon_e^p) \tag{5.209}$$

$$\frac{\partial f}{\partial \boldsymbol{\sigma}} = \frac{\partial J_2'}{\partial \boldsymbol{\sigma}} = \mathbf{S}$$

$$\frac{d\kappa}{d\epsilon_e^p} = \frac{1}{3}\frac{d(\sigma_Y^2)}{d\epsilon_e^p} = \frac{2}{3}\frac{d\sigma_Y}{d\epsilon_e^p}\sigma_Y \tag{5.210}$$

Therefore

$$\mathbf{C}^{ep} = \left[\mathbf{C}^{e^{-1}} + \frac{9\mathbf{S}\mathbf{S}}{4\sigma_y^2(d\sigma_Y/d\epsilon_e^p)}\right]^{-1} \tag{5.211}$$

(for strain-hardening material). If $\sigma_Y = \sigma_Y(\alpha) = \sigma_Y(W^p)$ in Eq. (5.209), it can be shown that

$$\frac{d\kappa}{dW^p} = \frac{2}{3}\frac{d\sigma_Y}{dW^p}\sigma_Y \tag{5.212}$$

Then

$$\mathbf{C}^{ep} = \left[\mathbf{C}^{e-1} + \frac{9\mathbf{S}\mathbf{S}}{4(d\sigma_Y/dW^p)\sigma_Y^3}\right]^{-1} \tag{5.213}$$

(for work-hardening materials). In the case of linear hardening shown in Fig. 5.11, from Eq. (5.171) we can determine

$$\frac{d\sigma_Y}{dW^p} = \frac{E^p}{\sigma_Y} \tag{5.214}$$

where

$$E^p = \frac{d\sigma_Y}{d\epsilon_e^p} = \frac{d\sigma_Y}{d\epsilon^p}$$

for one-dimensional loading. Substituting $d\sigma_Y/dW^p$ and $d\sigma_Y/d\epsilon_e^p$ into Eqs. (5.211) and (5.213), we see that they are the same for the von Mises criterion.

5.10.2 Kinematic Hardening Materials

For kinematic hardening materials, the yield criterion is given by

$$F(\boldsymbol{\sigma}, \boldsymbol{\alpha}) = f(\boldsymbol{\sigma} - \boldsymbol{\alpha}) - \kappa_0 = 0 \tag{5.215}$$

where $\boldsymbol{\alpha}$ is the back stress and κ_0 is a constant. The constitutive or evolution equation for $\boldsymbol{\alpha}$ is

$$d\boldsymbol{\alpha} = c\,d\boldsymbol{\epsilon}^p \qquad (c > 0) \tag{5.216}$$

according to Prager, or

$$d\boldsymbol{\alpha} = (\boldsymbol{\sigma} - \boldsymbol{\alpha})\,d\mu \tag{5.217}$$

according to Ziegler.

Prager's kinematic hardening model is considered first. The consistency condition for Eq. (5.215) is

$$\frac{\partial f}{\partial \boldsymbol{\sigma}} : d\boldsymbol{\sigma} + \frac{\partial f}{\partial \boldsymbol{\alpha}} : d\boldsymbol{\alpha} = 0 \tag{5.218}$$

Since from Eq. (5.215)

$$\frac{\partial f}{\partial \boldsymbol{\sigma}} = -\frac{\partial f}{\partial \boldsymbol{\alpha}} \tag{5.219}$$

Eq. (5.218) can be rewritten as

$$\frac{\partial f}{\partial \boldsymbol{\sigma}} : d\boldsymbol{\sigma} = \frac{\partial f}{\partial \boldsymbol{\sigma}} : c\dot{\lambda}\frac{\partial f}{\partial \boldsymbol{\sigma}} \tag{5.220}$$

where Eq. (5.192) is used. Now the proportionality scalar factor $\dot{\lambda}$ is obtained by solving the equation above:

$$\dot{\lambda} = \frac{1}{c}\frac{\partial f/\partial \boldsymbol{\sigma} : d\boldsymbol{\sigma}}{\partial f/\partial \boldsymbol{\sigma} : \partial f/\partial \boldsymbol{\sigma}} = \frac{1}{c}\frac{l}{\partial f/\partial \boldsymbol{\sigma} : \partial f/\partial \boldsymbol{\sigma}} \tag{5.221}$$

Then the increments of the back stress and the plastic strain are determined by

$$d\boldsymbol{\alpha} = \frac{\partial f/\partial \boldsymbol{\sigma}}{\partial f/\partial \boldsymbol{\sigma} : \partial f/\partial \boldsymbol{\sigma}} l = (\mathbf{n} : d\boldsymbol{\sigma})\mathbf{n} \tag{5.222}$$

and

$$d\boldsymbol{\epsilon}^{\mathrm{p}} = \frac{1}{\mathrm{c}} \frac{\partial f/\partial \boldsymbol{\sigma}}{\partial f/\partial \boldsymbol{\sigma} : \partial f/\partial \boldsymbol{\sigma}} l = \frac{1}{\mathrm{c}} (\mathbf{n} : d\boldsymbol{\sigma})\mathbf{n} \tag{5.223}$$

Again, it can be seen that these constitutive equations conform to the general forms given by Eqs. (5.36) and (5.37).

The total increment of the strain is obtained by

$$\begin{aligned} d\boldsymbol{\epsilon} &= d\boldsymbol{\epsilon}^{\mathrm{e}} + d\boldsymbol{\epsilon}^{\mathrm{p}} \\ &= \left[\mathbf{C}^{\mathrm{e}^{-1}} + \frac{1}{\mathrm{c}}\mathbf{nn}\right] : d\boldsymbol{\sigma} \end{aligned} \tag{5.224}$$

and the elastic-plastic stiffness tensor $\mathbf{C}^{\mathrm{ep}}$ is

$$\mathbf{C}^{\mathrm{ep}} = \left[\mathbf{C}^{\mathrm{e}^{-1}} + \frac{1}{\mathrm{c}}\mathbf{nn}\right]^{-1} \tag{5.225}$$

For the von Mises criterion

$$\tfrac{1}{2}(\mathbf{S} - \boldsymbol{\alpha}) : (\mathbf{S} - \boldsymbol{\alpha}) = \tfrac{1}{3}\sigma_{\mathrm{Y}}^{0^2} \tag{5.226}$$

where σ_{Y}^{0} is the initial yield stress,

$$\mathbf{C}^{\mathrm{ep}} = \left[\mathbf{C}^{\mathrm{e}^{-1}} + \frac{1}{\mathrm{c}} \frac{(\mathbf{S} - \boldsymbol{\alpha})(\mathbf{S} - \boldsymbol{\alpha})}{(\mathbf{S} - \boldsymbol{\alpha}) : (\mathbf{S} - \boldsymbol{\alpha})}\right]^{-1} \tag{5.227}$$

When the elastic response is isotropic, $\mathbf{C}^{\mathrm{ep}}$ takes the following form in the principal stress space:

$$\left[C_{ij}^{\mathrm{ep}}\right] = \left\{ \begin{bmatrix} \frac{1}{E} & -\frac{\nu}{E} & -\frac{\nu}{E} \\ -\frac{\nu}{E} & \frac{1}{E} & -\frac{\nu}{E} \\ -\frac{\nu}{E} & -\frac{\nu}{E} & \frac{1}{E} \end{bmatrix} + \frac{1}{c(\mathbf{S} - \boldsymbol{\alpha}) : (\mathbf{S} - \boldsymbol{\alpha})} \times \begin{bmatrix} (S_{11} - \alpha_{11})^2 & (S_{11} - \alpha_{11})(S_{22} - \alpha_{22}) & (S_{11} - \alpha_{11})(S_{33} - \alpha_{33}) \\ (S_{22} - \alpha_{22})(S_{11} - \alpha_{11}) & (S_{22} - \alpha_{22})^2 & (S_{22} - \alpha_{22})(S_{33} - \alpha_{33}) \\ (S_{33} - \alpha_{33})(S_{11} - \alpha_{11}) & (S_{33} - \alpha_{33})(S_{22} - \alpha_{22}) & (S_{33} - \alpha_{33})^2 \end{bmatrix} \right\}^{-1} \tag{5.228}$$

where ν is the elastic Poisson ratio.

For uniaxial tension in one direction, $\alpha_{22} = \alpha_{33} = -\frac{1}{2}\alpha_{11}$ by Prager's model, and $(\mathbf{S} - \boldsymbol{\alpha}):(\mathbf{S} - \boldsymbol{\alpha}) = \frac{3}{2}(S_{11} - \alpha_{11})^2$. Therefore from Eq. (5.228),

$$d\epsilon_{11} = \left(\frac{1}{E} + \frac{2}{3c}\right) d\sigma_{11} \tag{5.229}$$

On the other hand,

$$d\epsilon_{11} = \frac{1}{E^{t}} d\sigma_{11} \tag{5.230}$$

where E_t is the tangential slope of the uniaxial stress-strain curve. Comparing Eq. (5.230) with Eq. (5.229), we obtain

$$\frac{2}{3c} = \frac{1}{E^{t}} - \frac{1}{E} = \frac{1}{E^{p}} \tag{5.231}$$

$$c = \frac{2}{3}E^{p} \tag{5.232}$$

where Eq. (5.14) is used and E^{p} is the slope of $\sigma - \epsilon^{p}$ in an uniaxial case. Thus the material parameter c used in Prager's model can be determined from the uniaxial tension or compression test in terms of the slope of the $\sigma - \epsilon^{p}$ curve.

Now for Ziegler's model given by Eq. (5.217), the consistency condition is

$$\frac{\partial f}{\partial \boldsymbol{\sigma}} : d\boldsymbol{\sigma} + \frac{\partial f}{\partial \boldsymbol{\alpha}} : (\boldsymbol{\sigma} - \boldsymbol{\alpha})\, d\mu = 0 \tag{5.233}$$

or

$$\frac{\partial f}{\partial \boldsymbol{\sigma}} : d\boldsymbol{\sigma} = \frac{\partial f}{\partial \boldsymbol{\sigma}} : (\boldsymbol{\sigma} - \boldsymbol{\alpha})\, d\mu \tag{5.234}$$

As a result the proportional scalar $d\mu$ is determined by

$$d\mu = \frac{l}{\partial f/\partial \boldsymbol{\sigma} : (\boldsymbol{\sigma} - \boldsymbol{\alpha})} \tag{5.235}$$

Substituting this into Eq. (5.217), we obtain

$$d\boldsymbol{\alpha} = \frac{(\boldsymbol{\sigma} - \boldsymbol{\alpha})}{\partial f/\partial \boldsymbol{\sigma} : (\boldsymbol{\sigma} - \boldsymbol{\alpha})} l \tag{5.236}$$

which is a special form of Eq. (5.235).

Note from Eq. (5.235) that the evolution of $\boldsymbol{\alpha}$ is not related to the plastic strain $\boldsymbol{\epsilon}^{p}$ nor its increment $d\boldsymbol{\epsilon}^{p}$, so $d\boldsymbol{\epsilon}^{p}$ does not enter the consistency condition and hence $\dot{\lambda}$ cannot be determined from the consistency condition. However, since the normality condition holds,

$$(\mathbf{n}:d\boldsymbol{\sigma})\,\mathbf{n} = K^{\mathrm{p}} d\boldsymbol{\epsilon}^{\mathrm{p}} \qquad (K^{\mathrm{p}} > 0) \tag{5.237}$$

where K^{p} is a proportionality factor and is called the *plastic modulus*. This means that the component of $d\boldsymbol{\sigma}$ in the normal direction to the yield surface is equal to K^{p} times $d\boldsymbol{\epsilon}^{\mathrm{p}}$, since $d\boldsymbol{\epsilon}^{\mathrm{p}}$ is also in the normal direction of the yield surface. Then

$$d\boldsymbol{\epsilon}^{\mathrm{p}} = \frac{1}{K^{\mathrm{p}}}(\mathbf{n}:d\boldsymbol{\sigma})\mathbf{n} = \frac{1}{K^{\mathrm{p}}} \frac{l}{\partial f/\partial\boldsymbol{\sigma} : \partial f/\partial\boldsymbol{\sigma}} \frac{\partial f}{\partial\boldsymbol{\sigma}} \tag{5.238}$$

After comparing this with the associated flow rule (5.192), we conclude that

$$\dot{\lambda} = \frac{1}{K^{\mathrm{p}}} \frac{l}{\partial f/\partial\boldsymbol{\sigma} : \partial f/\partial\boldsymbol{\sigma}} \tag{5.239}$$

The parameter K^{p} can be determined from the uniaxial tension or compression test, as discussed before. In fact it can be shown that for the von Mises yield criterion

$$K^{\mathrm{p}} = \tfrac{2}{3}E^{\mathrm{p}} \tag{5.240}$$

which is the same as c given by Eq. (5.232), since Eqs. (5.238) and (5.223) are similar. This is the reason why K^{p} is considered the plastic modulus. It is necessary, however, to point out that despite this similarity, the meanings of K^{p} and c are different in the sense that c is the parameter entering the evolution equation for $\boldsymbol{\alpha}$, while K^{p} represents hardening behavior of the material and is independent of $d\mu$, the parameter representing the movement of the yield surface. It is interesting to note that the parameter c in Prager's model and the $d\mu$ in Ziegler's model are determined differently. The former is determined from the consistency condition and the latter from the uniaxial test.

The relation between $d\boldsymbol{\epsilon}$ and $d\boldsymbol{\sigma}$ for Ziegler's model is the same as Eq. (5.224), with c replaced by K^{p}. Nevertheless, Eqs. (5.238) and (5.239) are very general and hold for any hardening material, according to Drucker's assumptions, and even can be extended to softening materials with a negative value of K^{p}. With an appropriate choice of K^{p}, all the equations derived previously for $d\boldsymbol{\epsilon}^{\mathrm{p}}$ can be reproduced.

5.10.3 Combined Isotropic and Kinematic Hardening Materials

To simplify the discussion, only a combination of the work hardening and Prager's kinematic hardening model is considered here. The yield function is now written as

$$f(\boldsymbol{\sigma} - \boldsymbol{\alpha}) - \kappa(W^{\mathrm{p}}) = 0 \tag{5.241}$$

During the plastic deformation the consistency condition is

$$\frac{\partial f}{\partial \boldsymbol{\sigma}} : d\boldsymbol{\sigma} + \frac{\partial f}{\partial \boldsymbol{\alpha}} : d\boldsymbol{\alpha} - \frac{d\kappa}{dW^{\mathrm{p}}} dW^{\mathrm{p}} = 0 \tag{5.242}$$

Using Eqs. (5.178), (5.192), and (5.168), this equation can be rewritten as

$$\frac{\partial f}{\partial \boldsymbol{\sigma}} : d\boldsymbol{\sigma} = c\dot{\lambda} \frac{\partial f}{\partial \boldsymbol{\sigma}} : \frac{\partial f}{\partial \boldsymbol{\sigma}} + \dot{\lambda} \frac{d\kappa}{dW^{\mathrm{p}}} \frac{\partial f}{\partial \boldsymbol{\sigma}} : \mathbf{S} \tag{5.243}$$

From Eq. (5.243) the proportionality scalar $\dot{\lambda}$ can be solved:

$$\dot{\lambda} = \frac{\partial f/\partial \boldsymbol{\sigma} : d\boldsymbol{\sigma}}{c(\partial f/\partial \boldsymbol{\sigma} : \partial f/\partial \boldsymbol{\sigma}) + (d\kappa/dW^{\mathrm{p}})\partial f/\partial \boldsymbol{\sigma} : \mathbf{S}} = \frac{1}{K^{\mathrm{p}}} \frac{1}{\partial f/\partial \boldsymbol{\sigma} : \partial f/\partial \boldsymbol{\sigma}} \tag{5.244}$$

where K^{p} is defined by

$$K^{\mathrm{p}} = c + \frac{(d\kappa/dW^{\mathrm{p}})\partial f/\partial \boldsymbol{\sigma} : \mathbf{S}}{\partial f/\partial \boldsymbol{\sigma} : \partial f/\partial \boldsymbol{\sigma}} \tag{5.245}$$

Then the plastic strain increment $d\boldsymbol{\epsilon}^{\mathrm{p}}$ is given by

$$d\boldsymbol{\epsilon}^{\mathrm{p}} = \dot{\lambda} \frac{\partial f}{\partial \boldsymbol{\sigma}} = \frac{1}{K^{\mathrm{p}}} (\mathbf{n} : d\boldsymbol{\sigma})\mathbf{n} \tag{5.246}$$

and the relation between $d\boldsymbol{\sigma}$ and $d\boldsymbol{\epsilon}$ is

$$d\boldsymbol{\epsilon} = \left[C^{\mathrm{e}^{-1}} + \frac{1}{K^{\mathrm{p}}} \mathbf{nn} \right] : d\boldsymbol{\sigma} \tag{5.247}$$

These are obtained by adding Eq. (5.246) and the inverse of Eq. (5.183). Also, for cases involving the von Mises criterion and uniaxial tension in 1-direction, the same procedures as in Eqs. (5.226) through (5.229) should be followed so that

$$K^{\mathrm{p}} = c + \frac{d\kappa}{dW^{\mathrm{p}}} = \frac{2}{3} E^{\mathrm{p}} \tag{5.248}$$

where $\partial f/\partial \boldsymbol{\sigma} = \mathbf{S}$ is used to simplify the expression (5.245) for K^{p}. It is seen that the kinematic hardening modulus c, the isotropic modulus $d\kappa/dW^{\mathrm{p}}$, and the plastic modulus K^{p} are related. If any two of them are determined experimentally, the third one can be obtained by using Eq. (5.248).

Note that only the sum of c and $d\kappa/dW^{\mathrm{p}}$ can be determined from the uniaxial tension or compression test. Unfortunately, only one test of this kind is not enough to separate the hardening effects due to isotropic hardening represented by $d\kappa/dW^{\mathrm{p}}$ from kinematic hardening denoted by c. In fact, if only monotonic forward loading in one direction is considered, it is impossible to obtain the isotropic and kinematic hardening contributions, and either effect—whether alone or combined—will give a satisfactory description of the hardening behavior for such loading. The difference between isotropic and kinematic hardening can be identified only if the loading path direction is reversed or altered. The separation of c and $d\kappa/dW^{\mathrm{p}}$ requires loading-unloading experiments.

5.10.4 Perfectly Plastic Materials

If the material is perfectly plastic, the previous derivation in this section is not valid and Eq. (5.207), where $d\kappa/dW^{\mathrm{p}} = 0$ for perfectly plastic materials, becomes meaningless. As stated previously, for these materials Drucker's postulate should be modified to

$$d\boldsymbol{\sigma} : d\boldsymbol{\epsilon}^{\mathrm{p}} = 0 \tag{5.249}$$

This means that the plastic deformation occurs when the stress increment is in the tangential plane to the yield surface. In one-dimensional loading the plastic flow will continue without any limit when the stress is equal to the yield stress. Mathematically there will be no one-to-one correspondence between the stress and the strain. The general relation between $d\boldsymbol{\sigma}$ and $d\boldsymbol{\epsilon}$ becomes singular (more precisely indeterminate since $d\boldsymbol{\sigma}$ is also zero) if the material is perfectly plastic. Physically for such material the strain increment $d\boldsymbol{\epsilon}$ will be infinite for a given increment of stress $d\boldsymbol{\sigma}$ at any plastic state of deformation. The mathematical difficulty is in accord with this physical feature. However, the relation between $d\boldsymbol{\sigma}$ and $d\boldsymbol{\epsilon}$ for perfectly plastic materials can be established by taking the elastic deformation into account. The details are given below.

For perfectly plastic materials, the yield function is given by

$$F(\boldsymbol{\sigma}) = f(\boldsymbol{\sigma}) - \kappa_0 = 0 \tag{5.250}$$

where κ_0 is a constant. The consistency condition in this case is

$$\frac{\partial f}{\partial \boldsymbol{\sigma}} : d\boldsymbol{\sigma} = 0 \tag{5.251}$$

Comparing this equation with Eq. (5.249) gives the plastic strain increment

$$d\boldsymbol{\epsilon}^{p} = \dot{\lambda}\frac{\partial f}{\partial \boldsymbol{\sigma}} \tag{5.252}$$

which is the associated flow rule. Using the basic assumption that the total increment of strain can be decomposed into the elastic and plastic components, we can write

$$d\boldsymbol{\epsilon} = \mathbf{C}^{e^{-1}} : d\boldsymbol{\sigma} + \dot{\lambda}\frac{\partial f}{\partial \boldsymbol{\sigma}} \tag{5.253}$$

Multiplying both sides of the equation by $\partial f/\partial \boldsymbol{\sigma} : \mathbf{C}^{e}$ results in

$$\frac{\partial f}{\partial \boldsymbol{\sigma}} : \mathbf{C}^{e} : d\boldsymbol{\epsilon} = \frac{\partial f}{\partial \boldsymbol{\sigma}} : d\boldsymbol{\sigma} + \dot{\lambda}\frac{\partial f}{\partial \boldsymbol{\sigma}} : \mathbf{C}^{e} : \frac{\partial f}{\partial \boldsymbol{\sigma}} \tag{5.254}$$

Using Eq. (5.254), the proportionality scalar $\dot{\lambda}$ is obtained:

$$\dot{\lambda} = \frac{\partial f/\partial \boldsymbol{\sigma} : \mathbf{C}^{e} : d\boldsymbol{\epsilon}}{\partial f/\partial \boldsymbol{\sigma} : \mathbf{C}^{e} : \partial f/\partial \boldsymbol{\sigma}} \tag{5.255}$$

Substituting $\dot{\lambda}$ into Eq. (5.253) gives

$$d\boldsymbol{\epsilon}^{p} = \frac{\partial f/\partial \boldsymbol{\sigma} : \mathbf{C}^{e} : d\boldsymbol{\epsilon}}{\partial f/\partial \boldsymbol{\sigma} : \mathbf{C}^{e} : \partial f/\partial \boldsymbol{\sigma}}\frac{\partial f}{\partial \boldsymbol{\sigma}} \tag{5.256}$$

It should be pointed out that $d\boldsymbol{\epsilon}^{p}$ is only theoretically determined by this equation, since for perfectly plastic materials, both $d\boldsymbol{\epsilon}^{p}$ and $d\boldsymbol{\epsilon}$ are indeterminate for a given $d\boldsymbol{\sigma}$. Now Eq. (5.253) becomes

$$d\boldsymbol{\epsilon} = \mathbf{C}^{e^{-1}} d\boldsymbol{\sigma} + \frac{\partial f}{\partial \boldsymbol{\sigma}}\frac{\partial f/\partial \boldsymbol{\sigma} : \mathbf{C}^{e} : d\boldsymbol{\epsilon}}{\partial f/\partial \boldsymbol{\sigma} : \mathbf{C}^{e} : \partial f/\partial \boldsymbol{\sigma}} \tag{5.257}$$

Multiplying both sides of the equation by $\mathbf{C}^{e}$ and rearranging results in

$$d\boldsymbol{\sigma} = \left[\mathbf{C}^{e} - \frac{(\mathbf{C}^{e} : \partial f/\partial \boldsymbol{\sigma})(\partial f/\partial \boldsymbol{\sigma} : \mathbf{C}^{e})}{\partial f/\partial \boldsymbol{\sigma} : \mathbf{C}^{e} : \partial f/\partial \boldsymbol{\sigma}}\right] d\boldsymbol{\epsilon} = \mathbf{C}^{ep} : d\boldsymbol{\epsilon} \tag{5.258}$$

and the definition of $\mathbf{C}^{ep}$ is obvious.

It can be shown that the tensor $\mathbf{C}^{ep}$ given in Eq. (5.258) is singular and that $(\mathbf{C}^{ep})^{-1}$ does not exist. This is in accord with the physical characteristics for the perfectly plastic material.

Equation (5.258) can be coded in a finite element program. The commonly used algorithm for the elastic-plastic finite element analysis is the displacement method where the expression for stiffness matrix of a material is needed. The key in determining $\dot{\lambda}$ for a perfectly plastic material is that the elastic response of the material must be included along with plastic deformation.

It should be noted that the foregoing procedure can be followed to derive $\mathbf{C}^{ep}$ for hardening materials. The $\mathbf{C}^{ep}$ is obtained directly without the need of inverse of a fourth-ordered tensor, as shown in the last section.

5.10.5 Nonassociated Flow Rule

Up to now, the general relation between $d\boldsymbol{\sigma}$ and $d\boldsymbol{\epsilon}$ has been derived by assuming that the flow rule is associated. The results can be generalized to the nonassociated flow rule

$$d\boldsymbol{\epsilon}^{\mathrm{p}} = \dot{\lambda}\frac{\partial Q}{\partial \boldsymbol{\sigma}} \tag{5.259}$$

In the following, the relation between $d\boldsymbol{\epsilon}$ and $d\boldsymbol{\sigma}$ is derived for combined isotropic and kinematic hardening materials. The basic equations used are

$$d\boldsymbol{\epsilon} = d\boldsymbol{\epsilon}^{\mathrm{e}} + d\boldsymbol{\epsilon}^{\mathrm{p}}$$

$$f(\boldsymbol{\sigma} - \boldsymbol{\alpha}) - \kappa(W^{\mathrm{p}}) = 0$$

$$d\boldsymbol{\epsilon}^{\mathrm{e}} = \mathbf{C}^{\mathrm{e}^{-1}} : d\boldsymbol{\sigma}$$

The consistency condition for nonassociated flow rule is

$$\frac{\partial f}{\partial \boldsymbol{\sigma}} : d\boldsymbol{\sigma} - \frac{\partial f}{\partial \boldsymbol{\sigma}} : \dot{\lambda} c \frac{\partial Q}{\partial \boldsymbol{\sigma}} - \frac{d\kappa}{dW^{\mathrm{p}}} \mathbf{S} : \dot{\lambda} \frac{\partial Q}{\partial \boldsymbol{\sigma}} = 0 \tag{5.260}$$

where the work hardening and Prager's model have been introduced. From this equation the proportionality scalar $\dot{\lambda}$ can be obtained:

$$\dot{\lambda} = \frac{\partial f/\partial \boldsymbol{\sigma} : d\boldsymbol{\sigma}}{c(\partial f/\partial \boldsymbol{\sigma} : \partial Q/\partial \boldsymbol{\sigma}) + (d\kappa/dW^{\mathrm{p}})\mathbf{S} : \partial Q/\partial \boldsymbol{\sigma}} \tag{5.261}$$

The plastic strain increment $d\boldsymbol{\epsilon}^{\mathrm{p}}$ can be determined by substituting $\dot{\lambda}$ into Eq. (5.259):

$$d\boldsymbol{\epsilon}^{\mathrm{p}} = \frac{\partial f/\partial \boldsymbol{\sigma}}{c(\partial f/\partial \boldsymbol{\sigma} : \partial Q/\partial \boldsymbol{\sigma}) + (d\kappa/dW^{\mathrm{p}})\mathbf{S} : \partial Q/\partial \boldsymbol{\sigma}} l \tag{5.262}$$

Therefore the relation between $d\boldsymbol{\epsilon}$ and $d\boldsymbol{\sigma}$ is easily derived

$$d\boldsymbol{\epsilon} = \left[\mathbf{C}^{e-1} + \frac{(\partial f/\partial \boldsymbol{\sigma})(\partial Q/\partial \boldsymbol{\sigma})}{c(\partial f/\partial \boldsymbol{\sigma} : \partial Q/\partial \boldsymbol{\sigma}) + (d\kappa/dW^{\mathrm{p}})\mathbf{S} : \partial Q/\partial \boldsymbol{\sigma}}\right] : d\boldsymbol{\sigma} \quad (5.263)$$

Once the plastic potential Q is determined experimentally, the incremental stress-strain relation or the constitutive equation for plastic deformation can be obtained using this equation.

5.11 DEFORMATION THEORY OF PLASTICITY

Thus far the stress-strain relations given in this chapter have been in incremental form and hence considered incremental theories of plasticity. The basis for these incremental theories, as discussed at the beginning of this chapter, is that plastic deformation is dependent on the loading path, so it is necessary to integrate the incremental stress-strain relation along the loading path to obtain the final state of deformation.

However, there are still several plasticity theories that provide relations between the total components of the stress and strain, or the final or current stress and strain. These theories are called the *deformation theories of plasticity*. The first deformation theory was proposed by Hencky (1924) and was further developed by Ilyushin (1943, 1947). It will be briefly presented below.

The Hencky-Ilyushin deformation theory proposes that the total plastic strain tensor $\boldsymbol{\epsilon}^{\mathrm{p}}$ is proportional to the total deviatoric stress tensor $\mathbf{S}$:

$$\boldsymbol{\epsilon}^{\mathrm{p}} = \phi \mathbf{S} \quad (5.264)$$

This equation is similar to the Prandtl-Ruess equation, except that instead of the incremental plastic strain $d\boldsymbol{\epsilon}^{\mathrm{p}}$ it uses the total plastic strain $\boldsymbol{\epsilon}^{\mathrm{p}}$. This means that $\boldsymbol{\epsilon}^{\mathrm{p}}$ is coaxial with $\mathbf{S}$, and hence $\boldsymbol{\sigma}$, since the principal axes of $\mathbf{S}$ and $\boldsymbol{\sigma}$ are the same. Note that $\boldsymbol{\epsilon}^{\mathrm{p}}$ given by this equation is a deviatoric tensor.

Multiplying each side of Eq. (5.264) by itself leads to

$$\boldsymbol{\epsilon}^{\mathrm{p}} : \boldsymbol{\epsilon}^{\mathrm{p}} = \phi^2 \mathbf{S} : \mathbf{S} \quad (5.265)$$

Making use of the definition for effective stress σ_{e} and $\epsilon_{\mathrm{e}}^{\mathrm{p}}$,

$$\sigma_{\mathrm{e}} = \sqrt{\tfrac{3}{2}}\,(\mathbf{S} : \mathbf{S})^{1/2} \quad (5.266)$$

$$\epsilon_{\mathrm{e}}^{\mathrm{p}} = \sqrt{\tfrac{2}{3}}\,(\boldsymbol{\epsilon}^{\mathrm{p}} : \boldsymbol{\epsilon}^{\mathrm{p}})^{1/2} \quad (5.267)$$

the proportional function ϕ can be solved in terms of σ_e and ϵ_e^p

$$\phi = \frac{3\epsilon_e^p}{2\sigma_e} \tag{5.268}$$

Equation (5.264), after substitution of ϕ into it, becomes

$$\boldsymbol{\epsilon}^p = \frac{3\epsilon_e^p}{2\sigma_e}\mathbf{S} \tag{5.269}$$

As part of the theory, it is assumed that there exists a universal function or relation between σ_e and ϵ_e^p:

$$\sigma_e = \sigma_e(\epsilon_e^p) \tag{5.270}$$

that is independent of the loading path and therefore can be determined from simple uniaxial tension or compression test experiments.

The elastic strain $\boldsymbol{\epsilon}^e$ is governed by Hooke's law, as mentioned in Section 5.2. The total strain $\boldsymbol{\epsilon}$ is assumed to be the sum of $\boldsymbol{\epsilon}^e$ and $\boldsymbol{\epsilon}^p$ and is given by

$$\boldsymbol{\epsilon} = \frac{\mathbf{S}}{2G} + \frac{3\epsilon_e^p}{2\sigma_e}\mathbf{S} + \frac{3\sigma_m}{K}\mathbf{I} \tag{5.271}$$

where $\sigma_m = J_1/3$ is the mean stress, K is the elastic bulk modulus, and $\mathbf{I}$ is the unit tensor. If $\boldsymbol{\epsilon}^e$ is small compared to $\boldsymbol{\epsilon}^p$ and can be neglected, the constitutive equation based on the deformation theory is

$$\boldsymbol{\epsilon} = \frac{3\epsilon_e}{2\sigma_e}\mathbf{S} \tag{5.272}$$

and

$$\sigma_e = \sigma_e(\epsilon_e) \tag{5.273}$$

These equations for plastic deformation are the generalization of Eq. (5.55) for elastic deformation. If the deformation is elastic, $\sigma_e = 3G\epsilon_e$, and Eq. (5.272) reduces to Hooke's law.

Since plastic deformation in general is path dependent, the application of the deformation theory is very limited. As discussed previously, plastic deformation has two characteristics: the nonlinear response to the load and the irreversibility of the plastic flow. The irreversibility is responsible for the path dependence and hence results in the need for incremental constitutive equations. However, for proportional loading paths a deformation theory, which is basically a nonlinear relation between stress and strain, should provide a satisfactory description of the plastic deformation. In fact, as shown below, for proportional or simple loading (i.e., all the stress components are increasing in the same ratio), the incremental theory reduces to the deformation theory. In proportional or radial loading the stress increases in ratio

$$\boldsymbol{\sigma} = k\boldsymbol{\sigma}^0 \tag{5.274}$$

where k is a monotonically increasing proportionality factor and $\boldsymbol{\sigma}^0$ is an arbitrary reference stress state. From this we can obtain

$$\sigma_e = k\sigma_e^0 \quad \text{and} \quad \mathbf{S} = k\mathbf{S}^0 \tag{5.275}$$

The incremental Prandtl-Reuss equation can be written for this case as

$$d\boldsymbol{\epsilon}^p = \frac{3d\epsilon_e^p}{2\sigma_e}\mathbf{S} = \frac{3d\epsilon_e^p}{2\sigma_e^0}\mathbf{S}^0 \tag{5.276}$$

which can be easily integrated to become

$$\boldsymbol{\epsilon}^p = \frac{3\epsilon_e^p}{2\sigma_e^0}\mathbf{S}^0 = \frac{3\epsilon_e^p}{2\sigma_e}\mathbf{S} \tag{5.277}$$

This is the Hencky-Ilyushin deformation constitutive model, according to which plastic strain is only a function of the final or current state of the stress and is independent of the loading path.

The application of the deformation theory is not limited to proportional loading. In fact the limitation of the theory is still an open question. It is emphasized here that because of the irreversibility of the plastic deformation, the deformation theory must be used cautiously. When a loading path differs substantially from the proportional loading path, serious errors may result.

REFERENCES

Bridgman, P. W. 1923. The compressibility of thirty metal as a function of pressure and temperature. *Proc. Am. Acad. Arts Sci.* **58**:165.

Bridgman, P. W. 1949a. The physics of high pressure. London: Bell and Sons.

Bridgman, P. W. 1949b. Linear compressions to 30000 kglcm^2, including relatively incompressible substances. *Proc. Am. Acad. Arts Sci.* **77**:189.

de Saint-Venant, B. 1870. Memoire sur l'établissement des equations différentielles mouvements intérieurs opérés dans les corps solids ductiles au delà des limites ou l'élasticité pourrait les ramener a leur premier état. *C. R. Acad. Sci., Paris* **70**:473.

Drucker, D. C. 1951. A more fundamental approach to plastic stress-strain relations. *Proc. First U.S. Nat. Congr. Appl. Mech.*:487.

Drucker, D. C. 1950. Some implications of work hardening and ideal plasticity. *Quart. Appl. Math.* **7**:411.

Hencky, H. 1924. Zur Theorie plastischer Deformationen und der hierdurch im Material hervorgerufenen Nachspannungen. *Z. Angew. Math. Mech.* **4**:323.

Hooke, R. 1678. Lectures De Potential Restitutiva, or of Spring, Explaining the Power of Springing Bodies. London: John Martyn.

Ilyushin, A. A. 1960. On the increment of plastic deformation and the yield surface. *PMM* **24**:663.

Ilyushin, A. A. 1947. Theory of plasticity at simple loading of the bodies exhibiting plastic hardening. *PMM* **11**:291.

Ilyushin, A. A. 1943. Some problems of plastic deformations. *PMM* **7**:245.

Levy, M. 1871. Extrait du memoire sur les equations generales des mouvements interieurs des corps solides ductiles au dela des limites ou l'élasticite pourrait les ramener a leur premier état. *J. Math. Pure Appl.* **16**:369.

Levy, M. 1870. Memoire sur des equations generales des mouvements interieurs des corps solides ductiles au dela limites ou l'élasticite pourrait les ramener a leur premier état. *C. R. Acad. Sci. Paris* **70**:1323.

Liams, G., and H. Ford. 1957. An experimental investigation of the yield criterion and the stress-strain law. *J. Mech. Phys. Solids* **5**:215.

Lode, W. 1926. Versuche über den Empfluss der mittleren Hauptspannung auf des Fliessen der Metalle—Eisen, Kupfer und Nickel. *Z. Tech. Phys.* **36**:11.

Prandtl. L. 1924. Spannungsverteilung in plastischen Koerpern. *Proc. 1st Int. Congr. on Appl. Mech.*, Delft, p. 43.

Prager, W. 1955. The theory of plasticity: A survey of recent achievements. *Proc. Inst. Mech. Eng.* **169**:41.

Prager, W. 1949. Recent developments in the mathematical theory of plasticity. *J. Appl. Phys.* **20**:235.

Prager, W. 1945. Strain hardening under combined stress. *J. Appl. Phys.* **16**:837.

Reuss, A. 1930. Beruecksichtigung der elastischen Formaenderungen in der Plastizitaetstheorie. *Z. Angen. Math. Mech.* **10**:266.

Shield, R., and H. Ziegler. 1958. On Prager's hardening rule. *Z. Angow. Math. Phys.* **9a**:260.

Taylor, G. I., and H. Quinney. 1931. The plastic distortion of metals. *Phil. Trans. R. Soc., London* **A230**:323.

von Mises, R. 1928. Mechnik der plastischen Formanderung von Kristallen. *Z. Angen. Math. Mech.* **8**:161.

von Mises, R. 1913. Mechanik der festen Lorper in Plastisch deformabllem Zustand. *Nachr. Gess. Wiss. Gottingen*:582.

Ziegler, H. 1959. A modification of Prager's hardening rule. *Quart. Appl. Math.* **17**:55.

6

RECENT DEVELOPMENTS IN PLASTICITY

In the last chapter we discussed the four elements for a rate-independent plasticity theory based on the concept of yield surface. These were the yield function or yield surface, the constitutive equation for $\dot{\epsilon}^p$ or the flow rule, the hardening behavior characterized by the evolution equations for hardening parameters (e.g., κ and α), and finally, the loading and unloading criterion. It has been well established that for metals the von Mises criterion gives a fairly reasonable description of yielding. It has been proved theoretically and verified experimentally that for stable materials, in Drucker's sense, and, even for materials with softening behavior but without elastic-plastic coupling, the plastic strain rate $\dot{\epsilon}^p$ is governed by the associated flow rule or normality flow rule. The loading and unloading criterion for stable or hardening materials is also established using the concept of the yield surface. For softening materials, confusion arises between the further plastic loading and the elastic unloading if the formulation is in the stress space, as discussed in the last chapter. This chapter examines this problem. Strain space formulation for overcoming this difficulty will be discussed in Chapter 8. Clearly the way to a good plasticity theory is by the modeling of hardening behavior of materials. This is the focus of most recent developments in the plasticity area.

Classical isotropic hardening and Prager's or Ziegler's kinematic hardening models (described in Chapter 5) offer a reasonable explanation of the hardening properties of materials for the case of proportional loading where the load is increasing monotonically and no unloading occurs. Recent experiments in cyclic loading have revealed that the material responses under this loading condition are much more complex than under monotonic loading and cannot be modeled by the isotropic hardening rule and Prager's or Ziegler's kinematic hardening rule. Indeed a realistic analysis of the response of engineering materials under symmetric, unsymmetric, or random cyclic loading reversals in the plastic range is a very difficult and important subject in plasticity theory, and it has attracted many researchers for years. Within the last half-century some models have been proposed to meet the challenge. In this chapter four of these developments are presented and discussed. The models are the multisurface model proposed by Mroz (1967), the two-surface model by Dafalias and Popov (1976), the nonlinear kinematic hardening model initiated by Armstrong and Frederick (1966), and recently developed further by Chaboche (1977), and the endochronic theory proposed by Valanis (1971) and developed further by Watanabe and Atluri (1986). These four models were chosen because they are representative of the most current plasticity theories and because they have different characteristics that reflect a wide spectrum of current constitutive models for plastic deformation. The advantages as well as the disadvantages of these models will be discussed in this chapter.

The fundamental experimental observations in cyclic loading are presented first, in order to understand the motivation for these models. The development of the four constitutive models is then discussed one by one, including the basic formulations and simplified expressions in the uniaxial tension and pure shear, the main features and capabilities of each model, and determination of the material constants. Finally, the predictions using each model are compared to experimental data. Although the discussion of each model is comprehensive, it is impossible to explore all their aspects.

All four theories consider the deformation in the infinitesimal range and accept the decomposition of the total strain $\boldsymbol{\epsilon}$ into the elastic and plastic parts, $\boldsymbol{\epsilon}^e$ and $\boldsymbol{\epsilon}^p$:

$$\boldsymbol{\epsilon} = \boldsymbol{\epsilon}^e + \boldsymbol{\epsilon}^p$$

For the first three models (i.e., all except the endochronic theory) another common feature is the use of the concept of the yield surface which, in the stress space, can be written

$$F(\sigma, \alpha_i) = 0 \qquad (i = 1, 2, \ldots, n)$$

where α_i are the hardening parameters. The endochronic model does not use the yield surface concept in its development, although the concept of the yield surface can be retrieved by the model. This will be shown later. Last, it should be mentioned that the cyclic loading considered here is restricted to the low number of cycles and fatigue behavior is thus excluded from consideration.

6.1 OBSERVATIONS DURING CYCLIC LOADING OF METALS

The cyclic loading tests for uniaxial cases can be classified into the following three different types:

1. Fixed strain loading where the ϵ_{max} and ϵ_{min} are prescribed and $\Delta\epsilon = \epsilon_{max} - \epsilon_{min}$ remains constant during the test.
2. Fixed stress loading where the σ_{max} and σ_{min} are prescribed and $\Delta\sigma = \sigma_{max} - \sigma_{min}$ remains constant during the test.
3. Random loading where the σ_{max}, σ_{min}, ϵ_{max}, and ϵ_{min} change randomly.

This classification is based on whether the strain or the stress is controlled during the test. Other classifications are possible. For example, symmetric or unsymmetric cyclic loading may be defined according to whether $\sigma_{mean} = \frac{1}{2}(\sigma_{max} + \sigma_{min})$ is zero or not zero.

Figure 6.1. shows typical experimental results for uniaxial random cyclic loading on grade 60 steel specimen (Dafalias 1975). Despite the obvious complexity several features of the material behaviors under cyclic loading can be observed. First, except for the initial yield point A, which shows a sharp change from the elastic state to the plastic state, followed by a plateau, all other elastic-plastic curves show a smooth elastic-plastic transition, such as the curves between points E_3 and F_3 and between C_3 and D_3. Second, for the stress cycles with nonzero σ_{mean}, the plastic strain accumulates with the repetition of such cycles. This can be seen from loops 1 and 2 of Fig. 6.1, where the stresses at points D_1, D_2, B_1, F_1, B_2, and F_2 are approximately the same. This phenomenon is called the *cyclic creep* or *ratcheting effect*.

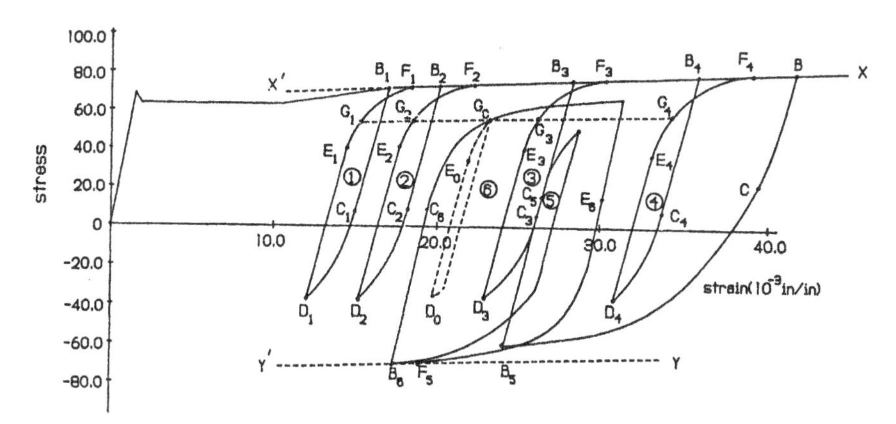

Fig. 6.1 Experimental results for uniaxial random cyclic loading on grade 60 steel specimen (Dafalias 1975)

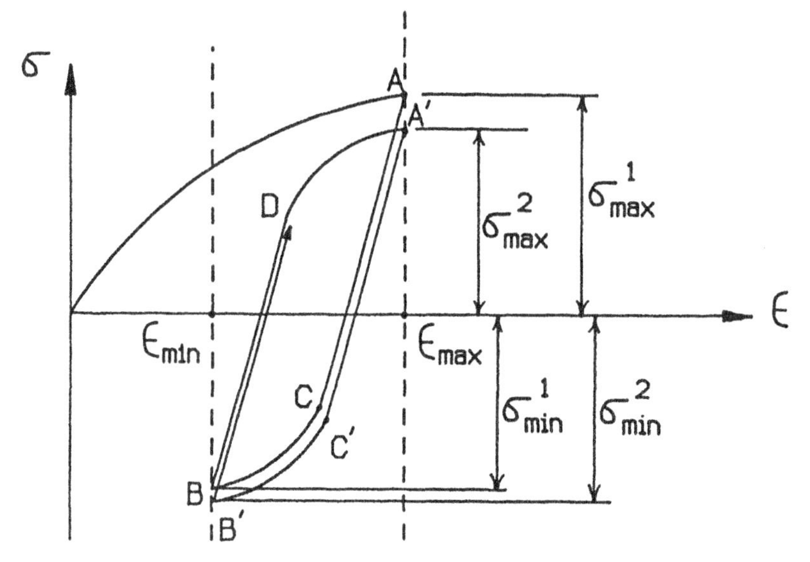

Fig. 6.2 Mean stress relaxation under nonsymmetric strain cycles

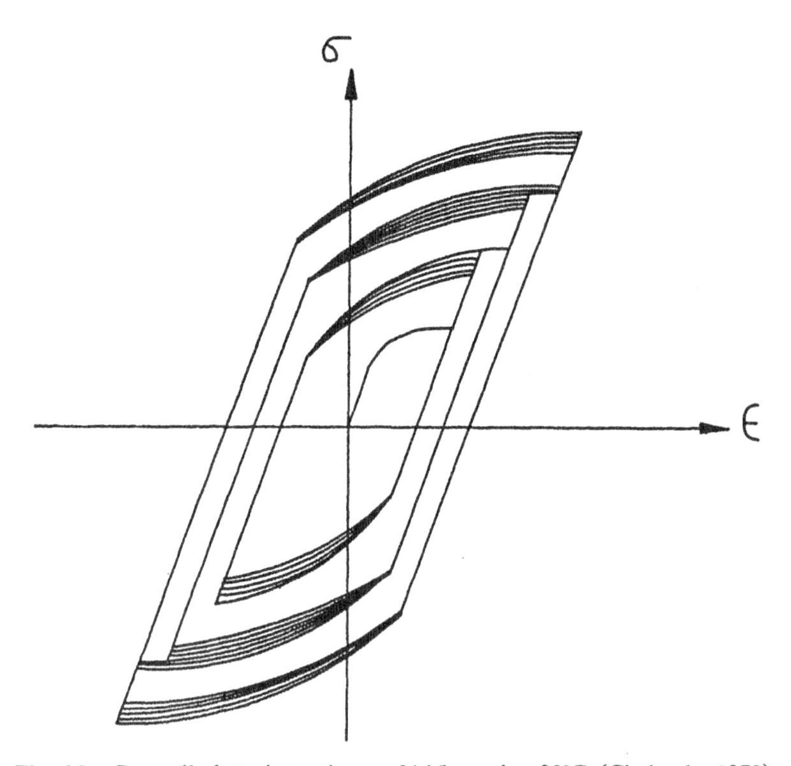

Fig. 6.3 Controlled strain testing on 316 L steel at 20°C, (Chaboche 1979)

A similar phenomena not included in Fig. 6.1 is the mean stress relaxation effect under nonsymmetric strain cycles. This is shown schematically in Fig. 6.2. Here the constant elastic region is assumed so that AC, $A'C'$, and BD have the same length. As a result of the strain cycling, the magnitude of σ_{max} will decrease (i.e., $\sigma_{max}^2 < \sigma_{max}^1$) and the magnitude of σ_{min} will increase ($|\sigma_{min}^2| > |\sigma_{min}^1|$) in such a way that the σ_{mean} stress will approach zero. This phenomenon is typical even when the elastic region is not constant. The third characteristic observed in Fig. 6.1 is that the loading and unloading curves tend to approach the two definite bounding lines $X'X$ and $Y'Y$, which in this case appear to be straight and parallel. Other features in Fig. 6.1 are the formation of the stress-strain loops under cyclic loading, the presence of the Bauschinger effect, the constancy of the Young's modulus E, and the approximate constancy of the elastic range.

Figure 6.3 shows a typical experimental result for uniaxial symmetric cyclic loading at increasing strain levels on a 316 L steel specimen. During the test the strain magnitude was controlled. At each strain level, the cyclic loading was repeated, and a definite increase in peak stress observed. This effect represents isotropic hardening. The cyclic loading is continued until the peak stress is stabilized (i.e., the stabilization of the isotropic hardening effect). Except for the smooth elastic-plastic transition and the Bauschinger effect, which are also observed from Fig. 6.1, the following two new conclusions can be drawn from Fig. 6.3:

1. The isotropic hardening effect will stabilize eventually and will not continue to increase without bound for a given strain level.
2. The stabilized value for the isotropic hardening depends on the strain level and will increase with it.

From the experimental observations Drucker and Palgen (1981) suggested four basic requirements for a cyclic plasticity model, which were further elaborated by Dafalias (1984) into the following five requirements related to the hardening behavior of metals under cyclic loading:

1. Unsymmetric stress cycles will cause cyclic creep or ratcheting effect in the direction of mean stress.
2. Unsymmetric strain cycles will cause progressive relaxation of the mean stress to zero.
3. The model must predict as accurately as possible the variation of the plastic modulus during random cyclic loading. That means the constitutive equation should be able to model the smooth elastic-plastic transition.
4. Under symmetric stress or strain cycles the material hardens or softens toward a properly defined stabilized state with only kinematic hardening.

5. Extensive plastic loading overwhelms and wipes out many, if not all, of the past effects. This can be seen in Fig. 6.1. No matter how many cycles the material has experienced, its stress-strain curve will approach one of the two bounding lines if the plastic deformation goes on extensively in one direction. After the convergence with the bounding line, the stress-strain curve will appear as if no cyclic loading had ever occurred. In other words, the material has forgotten most of its previous events because of the extensive plastic deformation in one direction.

These five requirements can be satisfied by any constitutive model describing complex material behavior under cyclic loading. The first three requirements concern kinematic hardening and the fourth, isotropic hardening; the last one is a general requirement.

Obviously an isotropic hardening model cannot describe the material behaviors mentioned above. For the uniaxial case, the corresponding stress-strain curve generated by an isotropic hardening model is shown in Fig. 6.4. It is seen that

$$\sigma_B = |\sigma_B'| > \sigma_A \tag{6.1}$$

according to the isotropic hardening assumption, where σ_A is the initial yield stress.

For any stress cycle the loading between B and B' corresponds to the elastic response, namely along the straight line BB'. On the other hand, for any strain cycles (e.g., the loading and unloading between ϵ_{max} and ϵ_{min} shown in the figure) the material will continue harden at points $1(B'), 2, 3, 4, 5, 6, 7, 8, \ldots,$ due to the accumulation of the equivalent plastic strain, so that eventually the elastic state will be reached asymptomatically after an infinite number of cycles. The maximum yield stress due to the hardening effect is bounded, of course, by the ultimate stress σ_{u}. Therefore the isotropic hardening model is primarily applied to monotonic loading, since it is not capable of even simulating the hysteresis loop observed under cyclic loading and fails to meet the five requirements listed previously.

Prager's linear kinematic hardening model for a uniaxial cyclic loading case is depicted in Fig. 6.5. According to the model the yield surface will translate in the stress space but will not change its size. As a result the length AA' is the same as lengths CC' and BB'. Also the stress-strain curve after the initial yielding is a straight line (AB) because of the constancy of the parameter c in the model:

$$\dot{\alpha} = c\dot{\epsilon}^{\text{p}} \tag{6.2}$$

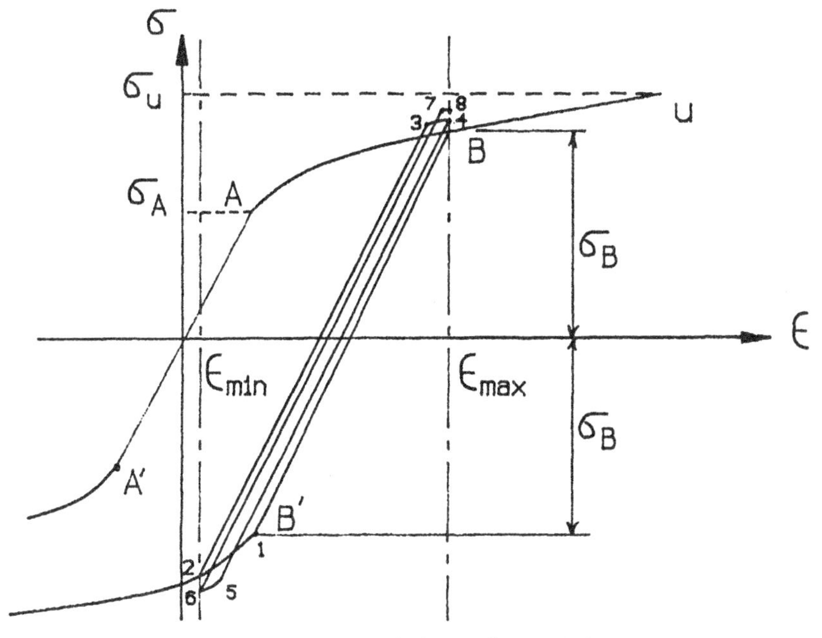

Fig. 6.4 Isotropic hardening model

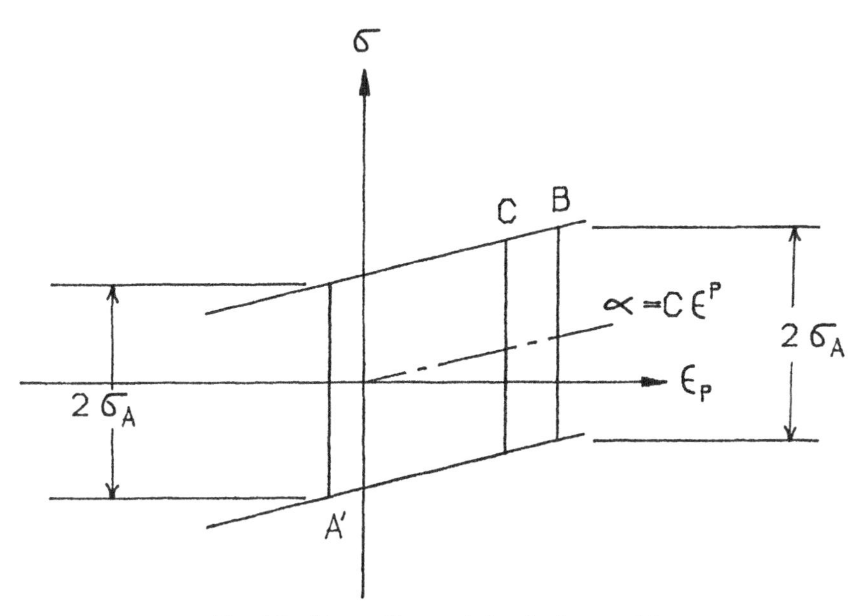

Fig. 6.5 Linear kinematic hardening model

Integrating the equation yields

$$\alpha = c\epsilon^{p} \tag{6.3}$$

This is shown in Fig. 6.5 for the uniaxial loading case as a straight line representing the locus of the origin O during the plastic deformation. Therefore the model fails to predict the smooth elastic-plastic transition. Moreover the stress cycles between σ_B and $\sigma_{c'}$ and the strain cycles between ϵ_B and $\epsilon_{C'}$ give the same closed parallelogram $CBB'C'$; therefore the model does not include the ratcheting effect or a cyclic creep and mean stress relaxation.

The model (6.3) can be generalized for by considering c as a nonlinear function of the (equivalent) plastic strain so that a varying tangential modulus is obtained. This modification does not provide a good solution to the linearity of the postyield behavior of the model. It causes inconsistency in the reversal of plastic flow because α depends uniquely on ϵ^{p}:

$$\boldsymbol{\alpha} = c(\boldsymbol{\epsilon}^{p})\boldsymbol{\epsilon}^{p} \tag{6.4}$$

The results of this generalization are shown in Fig. 6.6. Although the response to the loading has improved in the figure, the response during the reversal of loading is totally unacceptable.

Recall that the hardening parameters used in the first three models are the back stress tensor $\boldsymbol{\alpha}$ representing the center of the yield surface and the isotropic hardening parameter κ, sometimes called the *drag stress*, to characterize the size of the yield surface. With these parameters the general form of

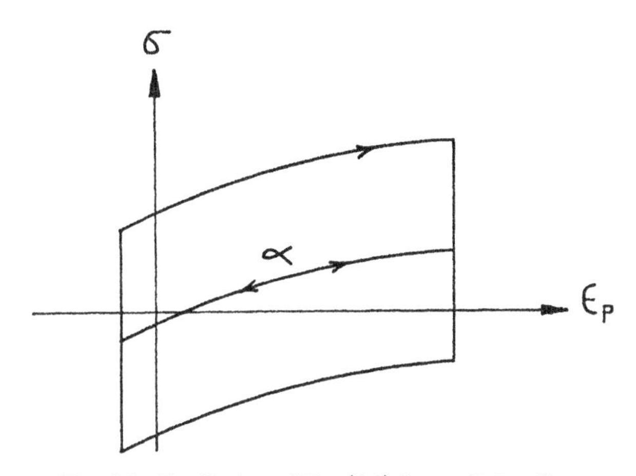

Fig. 6.6 Prediction of Eq. (6.4) for cyclic loading

the yield function (5.20) can be specified as

$$F(\boldsymbol{\sigma}, \boldsymbol{\alpha}, \kappa) = 0 \tag{6.5}$$

What differentiates the three models from each other are the evolution equations for $\boldsymbol{\alpha}$ and κ, which are discussed in the following sections.

6.2 MROZ'S MULTISURFACE MODEL

6.2.1 One-Dimensional Considerations

Any complex constitutive model is partly, if not completely, motivated by uniaxial tension-compression test results. This was also the case when Mroz proposed the multi-yield surface model by introducing the concept of a "field of work-hardening moduli" instead of the single modulus c used in Prager's model (6.2). Here the constitutive equation aims to model the nonlinear hardening behavior and the smooth transition from the elastic to plastic deformation. Fig. 6.7 shows a single loop of the stress-strain curve under a uniaxial tension–compression test. For the sake of simplicity and clarification

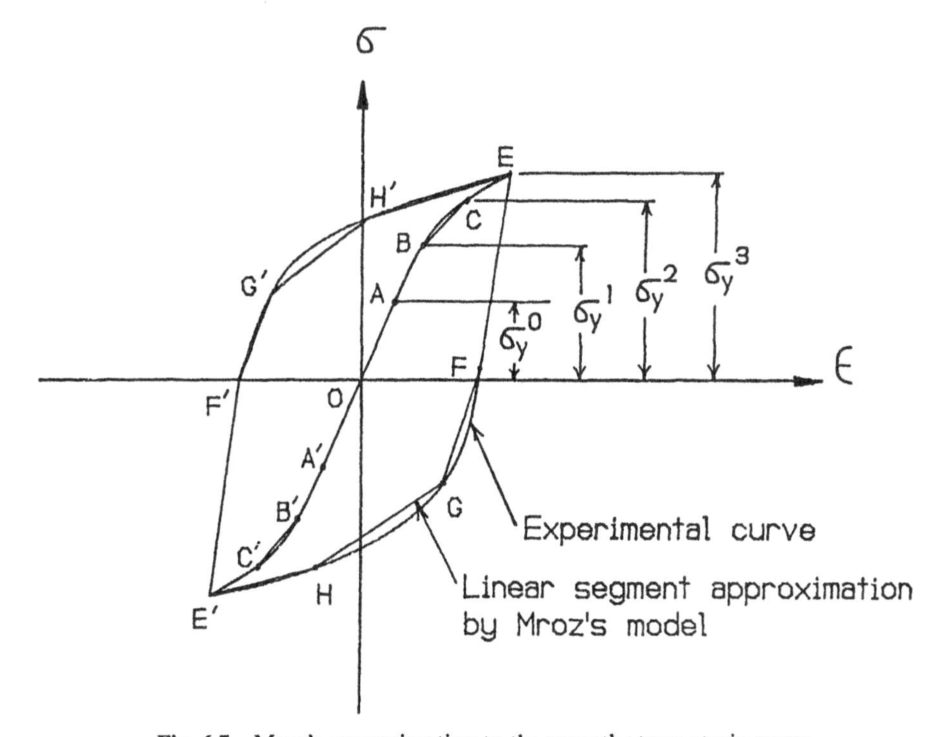

Fig. 6.7 Mroz's approximation to the smooth stress-strain curve

of the presentation, the pure kinematic response is assumed. Isotropic hardening will be described later with the general mathematical formularization of the model. The model discretizes the smooth elastic-plastic stress-strain curve in any of the loading directions in n linear segments ($n = 3$ is shown in Fig. 6.7) or pieces. Each segment has a constant tangential modulus E_i^{t} $(i = 1, 2, \ldots, n)$. The stress-strain relation for any one of the linear segments is given by

$$d\epsilon_i = d\epsilon_i^{\mathrm{e}} + d\epsilon_i^{\mathrm{p}} \qquad (i = 1, 2, \ldots, n) \tag{6.6}$$

and

$$\begin{aligned} d\epsilon_i^{\mathrm{e}} &= \frac{d\sigma}{E} \\ d\epsilon_i^{\mathrm{p}} &= \frac{d\sigma}{E_i^{\mathrm{p}}} \qquad (i = 1, 2, \ldots, n) \end{aligned} \tag{6.7}$$

where E and E^{p} are the elastic and plastic moduli, respectively, and the subscript i stands for the ith linear segment in the plastic region. Note the following definition:

$$d\epsilon = \frac{d\sigma}{E_i^{\mathrm{t}}} \tag{6.8}$$

This equation and Eq. (6.6) give the following relation between E_i^{t}, E_i^{p}, and E:

$$\frac{1}{E_i^{\mathrm{t}}} = \frac{1}{E_i^{\mathrm{p}}} + \frac{1}{E} \tag{6.9}$$

Since elastic modulus E remains constant during the deformation, this equation means that the association of the tangential modulus E_i^{t} to each of the linear segments is equivalent to assigning E_i^{p} to the ith segment.

The choice of different E_i^{t} or E_i^{p} used in the stress-strain relation (6.7) or (6.8) is governed by the stress level σ_{yi} $(i = 0, 1, 2, \ldots, n)$ measured at the discrete points on the stress-strain curve shown in Fig. 6.7. σ_{y0} is the initial yield stress. Since kinematic hardening is assumed, the difference between σ_E and σ_F is exactly twice σ_A. For the same reason the values of $|\sigma_G - \sigma_F|$, $|\sigma_H - \sigma_G|$, and $|\sigma_{E'} - \sigma_H|$ are also exactly twice that of $|\sigma_B - \sigma_A|$, $|\sigma_C - \sigma_B|$ and $|\sigma_E - \sigma_c|$, respectively. This will become clearer later when the model is generalized to the three-dimensional case using the concept of multi-yield surfaces. As a result of this observation, after completing the cycle at point E, the branch $E'F'G'H'E$ is twice the initial tension curve $OABCE$. This is also true for the branch $EFGHE'$ and the initial compression curve $OA'B'C'E'$.

Clearly then Mroz's model for a one-dimensional case is a simple one. It approximates the nonlinear stress-strain curve for the linear segments and determines E_i^{t} or E_i^{p} $(i = 1, 2, \ldots, n)$ for each segment and σ_{yi} $(i = 0, 1, \ldots, n - 1)$ for every adjacent point between any two linear segments. All of these can be easily obtained from the uniaxial cyclic loading experimental results. The model is put into the following complete form: For elastic loading or unloading

$$d\epsilon = \frac{d\sigma}{E} \tag{6.10}$$

For elastic-plastic loading

$$d\epsilon = \left(\frac{1}{E} + \frac{1}{E_i^{\mathrm{p}}}\right) d\sigma \tag{6.11}$$

For the forward-loading direction

$$\sigma_{yi-1} \leq \sigma < \sigma_{yi} \tag{6.12}$$

or for reverse loading direction

$$\sigma_{yi-1} \geq \sigma > \sigma_{yi} \tag{6.13}$$

6.2.2 Generalization to Multiaxial Case

To obtain a clear picture of the model, the geometrical illustration of the model will be presented first, followed by its mathematical representation. The multiaxial generalization of the model in stress space can be obtained by using several or multi-yield surfaces $F_0, F_1, F_2, \ldots, F_n$:

$$F_i = f_i(\boldsymbol{\sigma} - \boldsymbol{\alpha}_i) - \boldsymbol{\kappa}_{0i} = 0 \tag{6.14}$$

corresponding to the yield stress $\sigma_{y0}, \sigma_{y1}, \ldots, \sigma_{yn}$ in the uniaxial case. Here F_0 is the initial yield surface, $F_1, F_2, \ldots, F_n$ define regions of constant hardening moduli, σ_{yi} $(i = 0, 1, \ldots, n)$ separate different linear hardening segments in uniaxial loading, and α_i, κ_{0i} denote the centers and radii of yield surfaces F_i $(i = 0, 1, 2, \ldots, n)$. For initial isotropic materials, all the surfaces F_i $(i = 1, 2, \ldots, n)$ are assumed to be concentric.

These surfaces are represented by circles in the σ_1-σ_2-plane in Fig. 6.8; they are similar to the loci of von Mises yield surfaces on the π-plane. Now consider the process of uniaxial loading, unloading, and reverse loading of an initially isotropic material. Figure 6.8(a) shows the surfaces at the initial state and the loading path OE and unloading followed by the reverse loading path EK. The stress space (here it is the reduced stress space in the $\sigma_1 - \sigma_2$

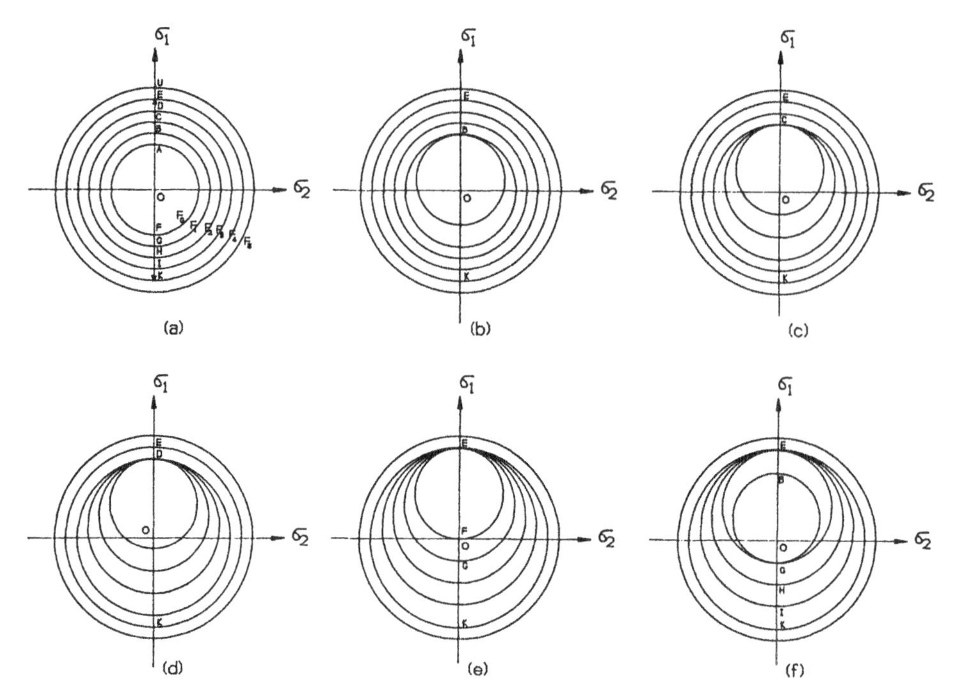

Fig. 6.8 Illustration of Multi-yield surfaces model

plane) is separated by these surfaces into $n + 1$ regions ($n = 5$ is shown in Fig. 6.8), and a different hardening modulus is attached to each region, thus forming a "field of work-hardening moduli" in the stress space. The innermost region, enclosed by F_0, is the elastic region, and the outermost surface, F_n, is the bounding surface.

Beginning from zero stress (point O) until the initiation of yielding (point A in Fig. 6.8), all the circles remain fixed and are centered about O. When the specimen is further loaded, the surface F_0 begins to move along σ_1 axis until it touches the surface F_1 at point B. During this process, other surfaces, except F_0, remain stationary. If the specimen is further loaded, F_0 and F_1 will move together along the σ_1 axis while others remain still. This process will continue until F_0 and F_1 contact F_2 at point C. Surfaces F_0, F_1, and F_2 will then move together along the σ_1 axis if the load continues to increase. The next two steps required to reach points D and E are shown in Fig. 6.8(d) and (e), respectively. If the unloading and reverse loading process EK is initiated, beginning from E, all circles initially remain stationary during the phase of elastic unloading, (i.e., until the stress point moves from point E to point F). Then reverse loading begins, and F_0 will move alone along the σ_1 axis until point F reaches point G, as shown in Fig. 6.8(f). The motion of the F circles for later steps of reverse loading up to point K, and subsequent

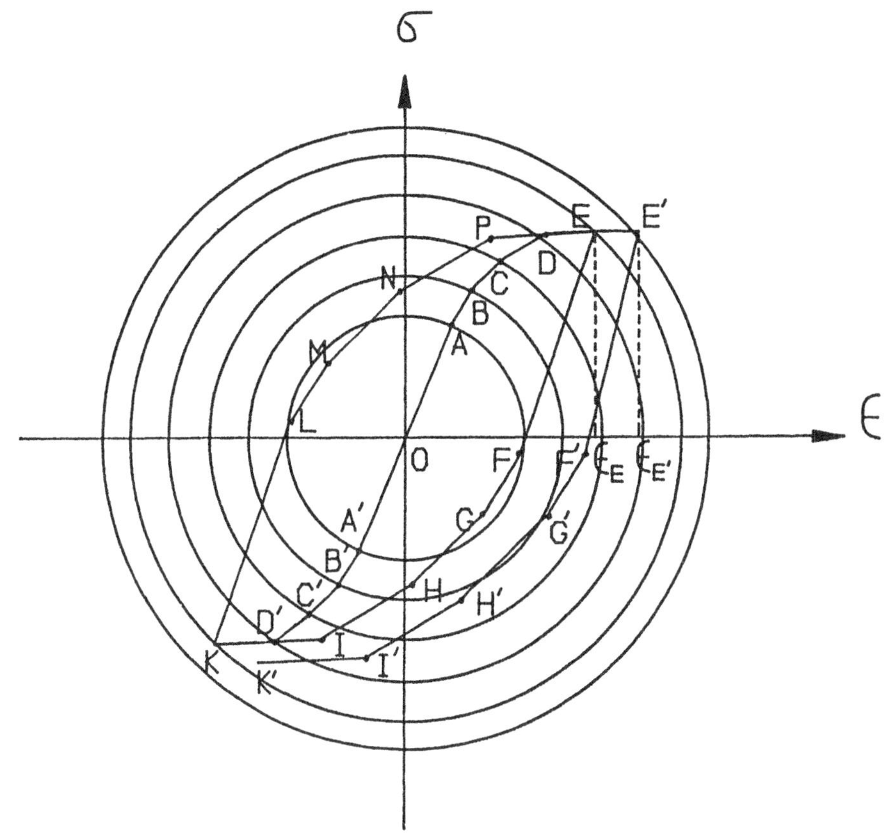

Fig. 6.9 Uniaxial stress-strain curve in multi-yield surfaces model

loading along *KLMN ... E*, as shown in Fig. 6.9, may be constructed in a similar manner. The stress-strain curve is obtained in this way in Fig. 6.9. From Figs. 6.8 and 6.9 it is easy to see that the distances between points A and O, B and A, C, and B, D and C, and E and D of Fig. 6.8 correspond to σ_A, $\sigma_B - \sigma_A$, $\sigma_C - \sigma_B$, $\sigma_D - \sigma_C$, and $\sigma_E - \sigma_D$ of Fig. 6.9, respectively. During reverse loading the elastic region is exactly twice the original one, and the distances between G and F, H and G, I and H, and K and I, as shown in Fig. 6.8(f), are also exactly twice the original ones shown in Fig. 6.8(a). This is why, as pointed out previously, the branch of the curves *EFGHIK* and *KLMNPE* can be deduced by a 200% magnification of branches *OA'B'C'D'K* and *OABCDE*, respectively.

Let us consider the mathematical representation of this phenomenon. During the plastic loading between surface F_l and F_{l+1}, $0 \leq l < n$, the current yield or active surface $F_l = 0$ is called the *loading surface*, and all the inner surfaces $F_0, F_1, \ldots, F_{l-1}$ are assumed to be tangent to F at the loading

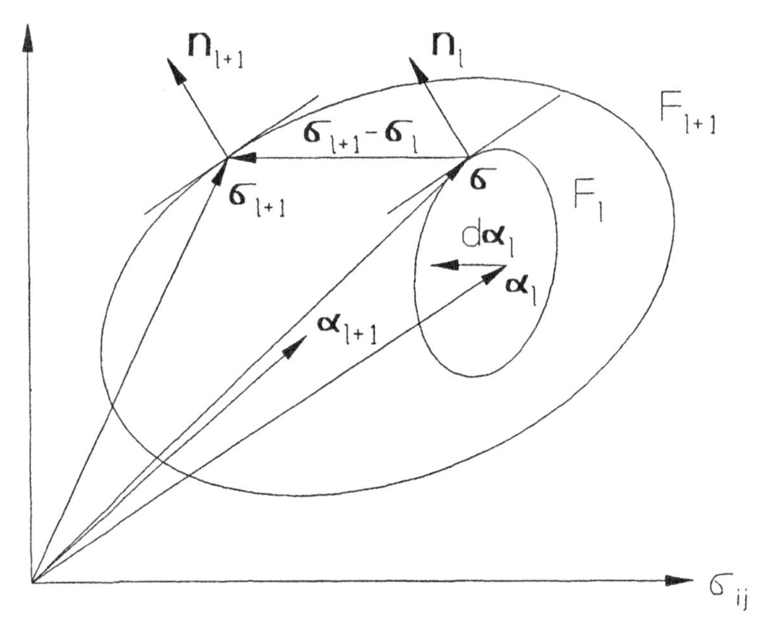

Fig. 6.10 Translation of the loading surface F_l

point so that

$$F_0 = F_1 = \cdots = F_{l-1} = F_l = 0 \tag{6.15}$$

The tangency requirement of $F_0, F_1, \ldots, F_{l-1}$ to F_l at the loading point is to ensure that these surfaces will not intersect each other. As a result they have the same normal at the contact point

$$\lambda_0 \frac{\partial f_0}{\partial \boldsymbol{\sigma}} = \lambda_1 \frac{\partial f_1}{\partial \boldsymbol{\sigma}} = \cdots = \lambda_1 \frac{\partial f_1}{\partial \boldsymbol{\sigma}} \tag{6.16}$$

where yield surfaces given by Eq. (6.14) have been used.

Figure 6.10 depicts two typical yield surfaces F_l, F_{l+1}, with centers α_l and α_{l+1}, respectively. These surfaces are expressed as

$$f_l(\boldsymbol{\sigma} - \boldsymbol{\sigma}_1) - \kappa_{0l} = 0 \tag{6.17}$$

$$f_{l+1}(\boldsymbol{\sigma} - \boldsymbol{\alpha}_{l+1}) - \kappa_{0l+1} = 0 \tag{6.18}$$

where κ_{0l} and κ_{0l+1} represent sizes of F_l and F_{l+1} and are assumed to be constant.

The current stress $\boldsymbol{\sigma}$ (subscript l is dropped to emphasize its current state), which represents the loading point, is on F_l. The outer unit normal to F_l at the loading point is $\mathbf{n}_l$. Also shown in Fig. 6.10 is the stress point $\boldsymbol{\sigma}_{l+1}$

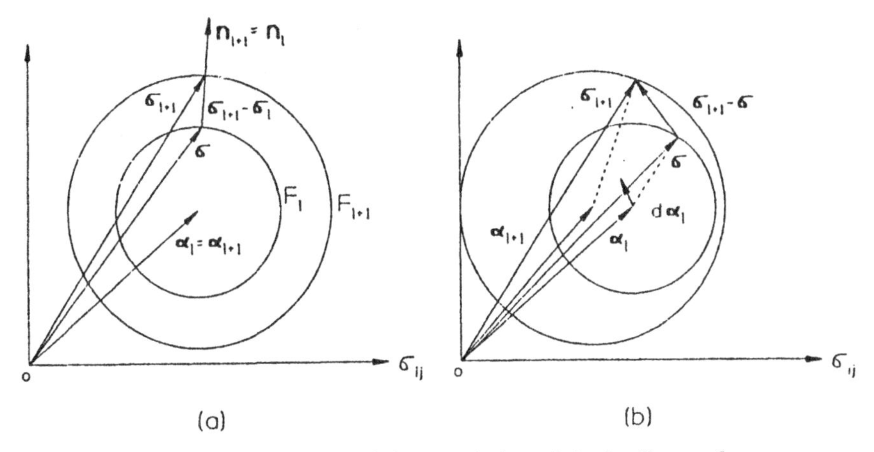

Fig. 6.11 Special cases of the translation of the loading surface

on the surface F_{l+1}, corresponding to the same direction of the outward normal (i.e., $\mathbf{n}_{l+1} = \mathbf{n}_l$). To ascertain the coincident outward normals, when the two surfaces contact each other, the evolution equations for $\boldsymbol{\alpha}$ are used

$$d\boldsymbol{\alpha}_l = d\mu(\boldsymbol{\sigma}_{l+1} - \boldsymbol{\sigma}_l) \tag{6.19}$$

This is shown by vector $d\alpha_l$ in Fig. 6.10, which is parallel to the vector $\boldsymbol{\sigma}_{l+1} - \boldsymbol{\sigma}$. Note that, in general, the moving direction of the yield surface F_l is neither in the normal $\mathbf{n}_l$ direction, as in Prager's model, nor in the radial direction, as in Ziegler's model. Only for cases where the centers of F_l and F_{l+1} coincide and the functions F_l and F_{l+1} are of the same type does the hardening rule (6.19) reduce to Ziegler's model (5.181). This is schematically shown in Fig. 6.11(a). Note that in the case of circular locus of the yield surface, Ziegler's model is the same as Prager's model, which is the case shown in Fig. 6.11(a). Assuming that F_l and F_{l+1} are of the same type and are homogeneous functions of order 2 of their arguments, we have

$$f(\boldsymbol{\sigma} - \boldsymbol{\alpha}_l) - (\sigma_{0l})^2 = 0 \tag{6.20}$$

$$F(\boldsymbol{\sigma} - \boldsymbol{\alpha}_{l+1}) - (\sigma_{0l+1})^2 = 0 \tag{6.21}$$

where σ_{0l} and σ_{0l+1} denote the size of F_l and F_{l+1} and can be identified as the yield stresses in uniaxial loading. Mroz (1967) showed that the stress point $\boldsymbol{\sigma}_{l+1}$ can be found by drawing a vector $\boldsymbol{\sigma}_{l+1} - \boldsymbol{\alpha}_{l+1}$, parallel to the vector $\boldsymbol{\sigma} - \boldsymbol{\alpha}_l$. This is expressed by

$$\boldsymbol{\sigma}_{l+1} - \boldsymbol{\alpha}_{l+1} = \frac{\sigma_{0l+1}}{\sigma_{0l}}(\boldsymbol{\sigma} - \boldsymbol{\alpha}_l) \tag{6.22}$$

Substituting this into Eq. (6.19) gives the translation of F_l:

$$d\boldsymbol{\alpha}_l = \frac{d\mu}{\sigma_{0l}}\left[(\sigma_{0l+1} - \sigma_{0l})\boldsymbol{\sigma} - (\boldsymbol{\alpha}_l\sigma_{0l+1} - \boldsymbol{\alpha}_{l+1}\sigma_{0l})\right] \tag{6.23}$$

which is shown in Fig. 6.11(*b*). If $\alpha_l = \alpha_{l+1}$, this becomes

$$d\boldsymbol{\alpha}_l = d\mu \frac{\sigma_{0l+1}}{\sigma_{0l}}(\boldsymbol{\sigma} - \boldsymbol{\alpha}_l) \tag{6.24}$$

which is a model of Ziegler's type. Now consider the general form (6.19). The proportionality factor $d\mu_l$ can be obtained from the consistency condition $dF_l = 0$:

$$\begin{aligned}\frac{\partial f_l}{\partial \boldsymbol{\sigma}} : d\boldsymbol{\sigma} + \frac{\partial f_l}{\partial \boldsymbol{\alpha}} : d\boldsymbol{\alpha} &= \frac{\partial f_l}{\partial \boldsymbol{\sigma}} : d\boldsymbol{\sigma} - \frac{\partial f_l}{\partial \boldsymbol{\sigma}} : d\boldsymbol{\alpha} \\ &= \frac{\partial f_l}{\partial \boldsymbol{\sigma}} : d\boldsymbol{\sigma} - \frac{\partial f}{\partial \boldsymbol{\sigma}} : (\boldsymbol{\sigma}_{l+1} - \boldsymbol{\sigma})\, d\mu_l = 0\end{aligned} \tag{6.25}$$

Then

$$d\mu_l = \frac{\partial f_l/\partial \boldsymbol{\sigma} : d\boldsymbol{\sigma}}{\partial f_l/\partial \boldsymbol{\sigma} : (\boldsymbol{\sigma}_{l+1} - \boldsymbol{\sigma})} = \frac{\mathbf{n}_l : d\boldsymbol{\sigma}}{\mathbf{n}_l : (\boldsymbol{\sigma}_{l+1} - \boldsymbol{\sigma})} \tag{6.26}$$

During the process of loading all the inner surfaces $F_0, F_l, \ldots, F_{l+1}$ are tangential to F_l at the loading point so that

$$\boldsymbol{\alpha}_m = \boldsymbol{\sigma} - \frac{\sigma_{0m}}{\sigma_{0l}}(\boldsymbol{\sigma} - \boldsymbol{\alpha}_l) \qquad (m = 0, 1, \ldots, l-1) \tag{6.27}$$

This is derived the same way as Eq. (6.22) for similar yield surfaces. On the other hand, if the loading path is reversed to cause plastic loading in the reverse direction, Eq. (6.27) is replaced by

$$d\boldsymbol{\alpha}_m = 0 \qquad (m \neq 0) \tag{6.28}$$

and $d\boldsymbol{\alpha}_0$ is determined by Eqs. (6.19) and (6.20).

The plastic strain increment $d\boldsymbol{\epsilon}^{\mathrm{p}}$ is assumed to be given by the normality or the associated flow rule. As in Ziegler's model, instead of a multiplier λ which is to be determined by the consistency condition, the plastic modulus K_i^{p}, $i = 1, 2, \ldots, n$, is introduced to define the plastic strain increment for the stress $\boldsymbol{\sigma}$ on F_l:

$$d\boldsymbol{\epsilon}^{\mathrm{p}} = \frac{1}{K_l^{\mathrm{p}}}(\mathbf{n}_l : d\boldsymbol{\sigma})\mathbf{n}_l \tag{6.29}$$

where $\mathbf{n}_l$ is the outward unit normal and can be expressed in terms of the yield surface and stress state as

$$\mathbf{n}_l = \frac{\partial f_l / \partial \boldsymbol{\sigma}}{[\partial f_l / \partial \boldsymbol{\sigma} : \partial f_l / \partial \boldsymbol{\sigma}]^{1/2}} \tag{6.30}$$

From Eqs. (6.26) and (6.30), the following relations can be derived:

$$d\mu_l = K_l^{\mathrm{p}} \frac{d\boldsymbol{\epsilon}^{\mathrm{p}} : \mathbf{n}_l}{\mathbf{n}_l : (\boldsymbol{\sigma}_{l+1} - \boldsymbol{\sigma})} \tag{6.31}$$

and

$$K_l^{\mathrm{p}} = \frac{\mathbf{n}_l : d\boldsymbol{\sigma}}{[d\boldsymbol{\epsilon}^{\mathrm{p}} : d\boldsymbol{\epsilon}^{\mathrm{p}}]^{1/2}} \tag{6.32}$$

Isotropic hardening is introduced by writing the κ_i in yield functions as the function of ϵ_e^{p}:

$$F_i(\boldsymbol{\sigma}, \boldsymbol{\alpha}, \boldsymbol{\epsilon}_l^{\mathrm{p}}) = f_i(\boldsymbol{\sigma} - \boldsymbol{\alpha}) - \kappa_i(\boldsymbol{\epsilon}_e^{\mathrm{p}}) = 0 \qquad (i = 0, 1, \ldots, n) \tag{6.33}$$

where ϵ_e^{p} is the effective plastic strain defined by Eq. (5.160). Then, instead of Eq. (6.25), the following consistency condition holds for the stress state $\boldsymbol{\sigma}$ on F_l $(0 \le l < n)$:

$$\frac{\partial f_l}{\partial \boldsymbol{\sigma}} : d\boldsymbol{\sigma} - \frac{\partial f}{\partial \boldsymbol{\sigma}} : (\boldsymbol{\sigma}_{l+1} - \boldsymbol{\sigma})\, d\mu_l - \frac{d\kappa_l}{d\epsilon_e^{\mathrm{p}}} d\epsilon_e^{\mathrm{p}} = 0 \tag{6.34}$$

Equation (6.26) becomes

$$\begin{aligned} d\mu_l &= \frac{\partial f_l / \partial \boldsymbol{\sigma} : d\boldsymbol{\sigma} - (d\kappa_l / d\epsilon_e^{\mathrm{p}})\, d\epsilon_e^{\mathrm{p}}}{\partial f_l / \partial \boldsymbol{\sigma} : (\boldsymbol{\sigma}_{l+1} - \boldsymbol{\sigma})} \\ &= \left[1 - \frac{\sqrt{2/3}\,(d\kappa_l / d\epsilon_e^{\mathrm{p}})}{K_l^{\mathrm{p}} [\partial f_l / \partial \boldsymbol{\sigma} : \partial f_l / \partial \boldsymbol{\sigma}]^{1/2}}\right] \frac{\mathbf{n}_l : d\boldsymbol{\sigma}}{\mathbf{n}_l : (\boldsymbol{\sigma}_{l+1} - \boldsymbol{\sigma})} \end{aligned} \tag{6.35}$$

where Eqs. (5.161) and (6.29) were used to arrive at the final expression. During the process of loading, surfaces that are in mutual contact at the loading point, $F_0, F_1, \ldots, F_l$, undergo both translation and expansion, while others F_j, $l + 1 \le j \le n$, which are not reached by the loading point, only expand uniformly.

The application of the model in the case of a von Mises material gives the following particular forms of equations for stress state on F_l:

$$\left[\tfrac{3}{2}(\mathbf{S}-\boldsymbol{\alpha}_i'):(\mathbf{S}-\boldsymbol{\alpha}_i')\right] = \sigma_{yi}^2(\epsilon_e^p) \qquad (i = 0, 1, \ldots, n) \tag{6.36}$$

$$\mathbf{n}_i = \sqrt{\frac{3}{2}}\,\frac{\mathbf{S}-\boldsymbol{\alpha}_i'}{\sigma_{yi}} \qquad (i = 0, 1, \ldots, l) \tag{6.37}$$

$$d\boldsymbol{\epsilon}^p = \frac{3/2}{K_l^p\sigma_{yl}^2}\left[(\mathbf{S}-\boldsymbol{\alpha}_l'):d\boldsymbol{\sigma}\right](\mathbf{S}-\boldsymbol{\alpha}_l') \tag{6.38}$$

$$d\boldsymbol{\alpha}_l = d\mu_l(\boldsymbol{\sigma}_{l+1}-\boldsymbol{\sigma}) \tag{6.39}$$

$$d\mu_l = \left(1 - \frac{2d\sigma_{yl}/d\epsilon_e^p}{3K_l^p}\right)\frac{(\mathbf{S}-\boldsymbol{\alpha}_l'):d\boldsymbol{\sigma}}{(\mathbf{S}-\boldsymbol{\alpha}_l'):(\boldsymbol{\sigma}_{l+1}-\boldsymbol{\sigma})} \tag{6.40}$$

For uniaxial loading in one direction, $S_{22} = S_{33} = -\frac{1}{2}S_{11}$, $\alpha_{22}' = \alpha_{33}' = -\frac{1}{2}\alpha_{11}'$ $(\alpha_{11} \neq 0)$, $\alpha_{22} = \alpha_{33} = 0$, and according to Eq. (6.39), the equations above become

$$\sigma_{11} - \alpha_{11} = \sigma_{yi}(\epsilon_{11}^p) \qquad (i = 0, 1, \ldots, n) \tag{6.41}$$

$$d\epsilon_{11}^p = \frac{2}{3}\frac{d\sigma_{11}}{K_l^p} = \frac{d\sigma_{11}}{E_l^p} \tag{6.42}$$

$$d\alpha_{11} = d\mu_l[\sigma_{yl+1} - \sigma_{11}] \tag{6.43}$$

$$d\mu_l = \left(1 - \frac{2}{3}\frac{d\sigma_{yl}/d\epsilon_{11}^p}{K_l^p}\right)\frac{d\sigma_{11}}{(\sigma_{yl+} - \sigma_{11})} \tag{6.44}$$

Equation (6.42) gives the relation between the plastic modulus K_l^p for a multiaxial case and its counterpart in uniaxial loading E_l^p:

$$K_l^p = \frac{2}{3}E_l^p \tag{6.45}$$

which is the same as that derived in the last chapter.

Now consider the determination of the material constants or functions that appear in the model. From the above discussion, it can be seen that these include the following:

1. Yield stresses, σ_{yi}, $i = 0, 1, \ldots, n$, including their evolution with the effective plastic strain ϵ_e^p. These will specify the contribution of the isotropic hardening of the material.
2. Plastic moduli K_i^p, $i = 1, 2, \ldots, n-1$. These constitute a "field of work-hardening moduli" and are used to determine the magnitude of the plastic strain increments.

Another set of material parameters $d\mu_i$, $i = 1, 2, \ldots, n$, which represents the translation of the center of the yield surfaces in the course of plastic loading, can be calculated using Eq. (6.35) in general, or Eq. (6.40) in particular, if σ_{yi} and K_i^{p} are known based on experimental results.

Determining the "field of work-hardening moduli" is quite simple and straightforward. After approximating the nonlinear uniaxial stress-strain curve, shown in Fig. 6.7, by n linear segments, the tangential moduli E_i^{t}, $i = 1, 2, \ldots, n$, can be measured easily from these n linear segments. The plastic moduli E_i^{p} in uniaxial loading can be calculated by Eq. (6.9), and hence the plastic moduli K_i^{p} for multiaxial cases can be determined using Eq. (6.45). But, instead of plotting σ versus ϵ for the uniaxial loading, a σ versus ϵ^{p} plot can be directly generated. Plastic moduli E_i^{p} are then directly measured from the curve and transformed into K_i^{p} using Eq. (6.45).

The determination of $\sigma_{yi}(\epsilon_{\mathrm{e}}^{i})$, $i = 0, 1, \ldots, n$, requires using their initial values and their variation with $\epsilon_{\mathrm{e}}^{\mathrm{p}}$. The initial values are easily obtained by measuring stress levels at the adjacent points, such as σ_A, σ_B, σ_C, σ_D, and σ_E, shown in Fig. 6.9. The variation of σ_{yi} as a function of the effective plastic strain can be determined, as pointed out by Mroz (1967), by comparing stress differences corresponding to segments OA and EF, AB and FG, BC and GH, and so on. In fact, for purely kinematic work hardening as described by the model in Fig. 6.9, these should be $\sigma_E - \sigma_F = 2\sigma_A$, $\sigma_F - \sigma_G = 2(\sigma_B - \sigma_A)$, and $\sigma_G - \sigma_H = 2(\sigma_C - \sigma_B)$, and so on, as discussed before. Therefore deviations from these equalities can be attributed to the isotropic hardening effect, which causes the variations in σ_{yi}. Variations determined in this way represent the isotropic hardening at a particular strain level, say, ϵ_E in Fig. 6.9. Then, applying a cyclically varying loading (i.e., a cyclical loading with unloading at different strain levels), different unloading and reverse loading curves can be obtained. Then these curves can be used to determine the variations of σ_{yi} as the strain level changes. A second unloading and reverse loading curve at a higher strain level $\epsilon_{E'}$ is shown in Fig. 6.7. After the variations of σ_{yi} as a function of strain are known, their variations with ϵ_{p} can be obtained easily. Young's modulus remains constant, and the elastic strain can be subtracted out from the total strain to give uniaxial plastic strain ϵ^{p}. Finally, a replacement of the effective plastic strain $\epsilon_{\mathrm{e}}^{\mathrm{p}}$ by ϵ^{p} gives the dependence of σ_{yi} on $\epsilon_{\mathrm{e}}^{\mathrm{p}}$. As pointed out before, the isotropic hardening will be stabilized, and these functions $\sigma_{yi}(\epsilon_{\mathrm{e}}^{\mathrm{p}})$ will approach constant values, which leaves kinematic hardening as the only hardening effect. The specific model to describe the stabilization of isotropic hardening will be discussed later in this chapter.

The comprehensive evaluation of any proposed model is a complicated task. The capability of the model to predict various aspects observed in the cyclic loading as well as complex nonproportional loading experiments must be considered carefully. The multisurface model presented in this section has been applied to metals and soils under cyclic loading condition, and satisfactory results have been obtained. The interested reader can refer to papers by Mroz (1969, 1972), Prevost (1977, 1978), and Mroz et al. (1979).

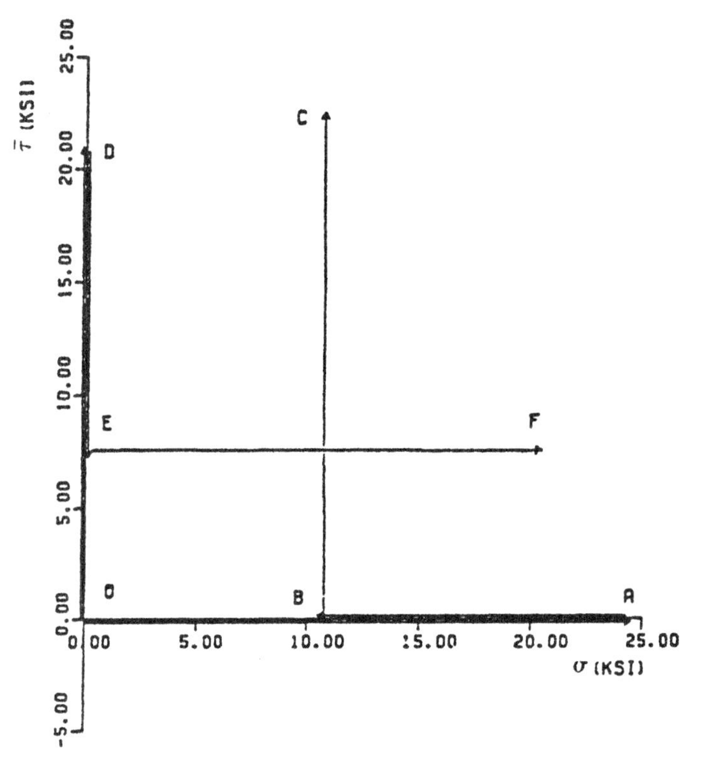

Fig. 6.12 Loading paths in the stress subspace

The following discussion presents work done by Khan and Wang (1988). To verify among other things, the ability of the multisurface model to predict correctly the direction of the plastic strain increment, especially after a sudden directional change in the loading path, an experimental investigation of the combined tension-torsion loading was conducted. The thin-walled tubes of annealed polycrystalline copper were subjected to loading in tension exceeding 10% or in shear to 15% of strain, followed by partial unloading and reloading in a different direction (shear or tension depending on whether the preloading was tension or shear). These two nonproportional loading paths are schematically shown in Fig. 6.12.

To compare the prediction of the direction of the plastic strain increment using this model with the experimental results, the yield surfaces of von Mises type were used. For combined tension-torsion,

$$F_i = (\sigma - \alpha_{\sigma i})^2 + (\bar{\tau} - \overline{\alpha_{\tau i}})^2 - \sigma_{yi}^2 = 0 \qquad (i = 0, 1, \ldots, n) \quad (6.46)$$

where σ and α_σ are axial stress and back stress associated with σ, respectively, and $\bar{\tau} = \sqrt{3}\,\tau$, $\overline{\alpha_\tau} = \sqrt{3}\,\alpha_\tau$, where τ and α_τ are the shear and the corresponding back stress, and where σ_{yi} represents the radius of the ith yield surface and can be identified as the stress at the turning point on the linear segment approximation to the nonlinear uniaxial stress strain curve. The ratio between $d\epsilon^p$ and $d\bar{\gamma}^p = d\gamma^p/\sqrt{3}$ can be obtained by using the associated flow rule:

$$\frac{d\bar{\gamma}^p}{d\epsilon^p} = \frac{\bar{\tau} - \bar{\alpha}_{\tau l}}{\sigma - \alpha_{\sigma l}} \tag{6.47}$$

for the stress state or loading point on F_l.

The growth laws for the center of loading surface $F_l = 0$ can be expressed as

$$d\sigma_{\sigma l} = d\mu_l(\sigma_{l+1} - \sigma) = d\mu_l\left[\frac{\sigma_{yl+1}}{\sigma_{yl}}(\sigma - \alpha_{\sigma l}) + \alpha_{\sigma l+1} - \sigma\right] \tag{6.48}$$

$$d\sigma_{\tau l} = d\mu_l(\bar{\tau}_{l+1} - \bar{\tau}) = d\mu_l\left[\frac{\sigma_{yl+1}}{\sigma_{yl}}(\bar{\tau} - \bar{\alpha}_{\tau l}) + \alpha_{\tau l+1} - \bar{\tau}\right] \tag{6.49}$$

$$d\mu_l = \frac{(\sigma - \alpha_{\sigma l})\,d\sigma + (\bar{\tau} - \bar{\alpha}_{\tau l})\,d\bar{\tau}}{\sigma_{yl+1}\sigma_{yl} + (\sigma - \alpha_{\sigma l})(\alpha\sigma_{\alpha l+1} - \sigma) + (\bar{\tau} - \bar{\alpha}_{\tau l})(\bar{\alpha}_{\tau l+1} - \bar{\tau})} \tag{6.50}$$

Note that in monotonic uniaxial loading, $\sigma - \alpha_{\sigma l} = \sigma_{yl}$ and $\alpha_{\sigma l+1} = 0$. Since stress σ_{yl+1} has not been reached yet, Eq. (6.48) is reduced to Eq. (6.43).

The prediction of the model and experimental results are shown in Fig. 6.13. The strains measured in the tests are based on the partially unloaded reference configurations, which are the configuration corresponding to points B and E, respectively.

In Fig. 6.13(a) the experimental data (as shown by the hexagonal symbols) are compared with the prediction of Mroz's multisurface kinematic hardening model in strain subspace ($\bar{\gamma}^p - \epsilon^p$) for loading path BC. Also shown are predictions obtained using Ziegler's models and the endochronic theory. Two different scales are used in the plots; in the one on the right, the same scales are used for both axes ϵ^p ($EPAX$) and $\bar{\gamma}^p$ ($GAMMAP$), while on the left-hand side, the axial strains are magnified about 13 times in relation to the shear strains, in order to see the changes more clearly in the initial reloading part. The origin refers to the partially unloaded reference configuration. Similar plots for loading path EF are shown in Fig. 6.13(b). It is seen from the figure that although Mroz's model does not fully agree with the experimental observation, the direction of plastic strain increment obtained by it is in good agreement with the experimental data as the strain level increases.

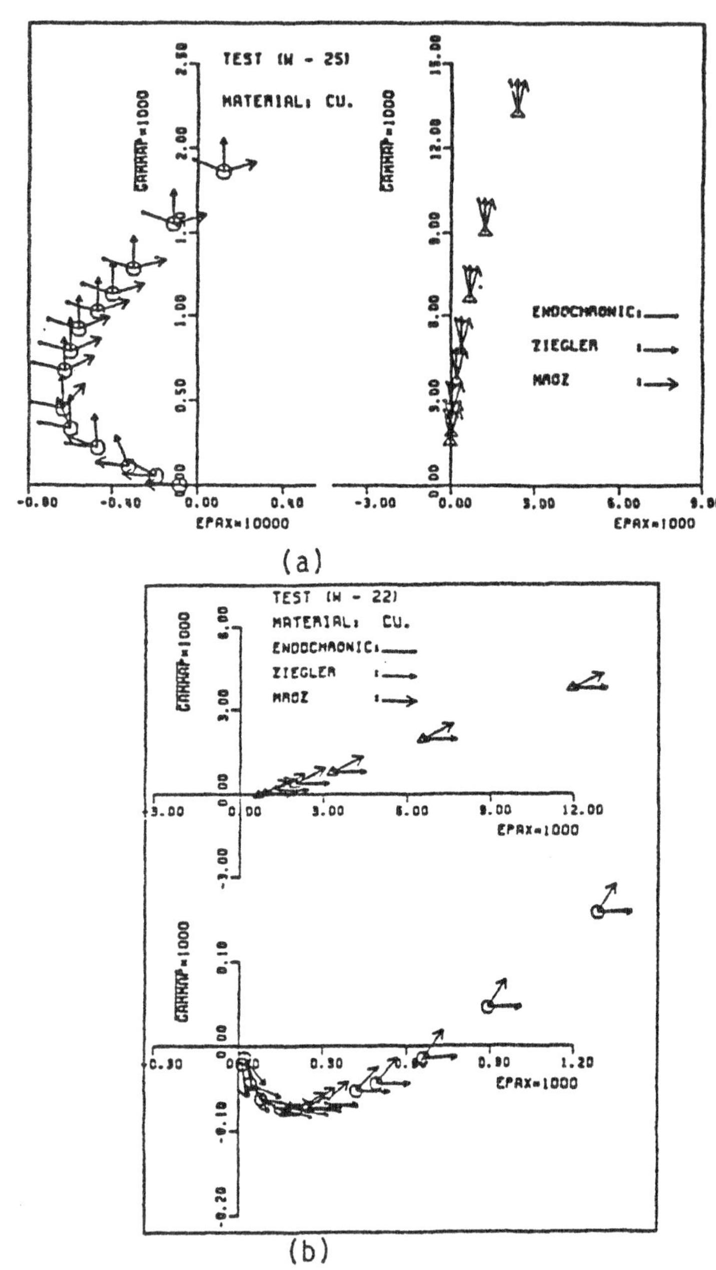

Fig. 6.13 Comparison between experimental results and predictions by using three models: (*a*) path *BC*; (*b*) path *EF*

As pointed out by Chaboche (1986), the Mroz multisurface model is able to describe

1. The correct nonlinearity of the stress-strain loops.
2. The Bauschinger effect.
3. Cyclic hardening or softening of the material.

The Mroz model (the version discussed in this chapter) gives no ratcheting at all and cannot describe the mean stress relaxation in the course of nonsymmetric strain cycles (neither does Prager's linear kinematic rule). The predicted elastic-plastic stress strain curve is piecewise linear because of the field of constant plastic moduli. To produce a smooth nonlinear curve, a large number of yield surfaces are necessary. For the requirements listed in Section 6.1, it can be seen that Mroz's model satisfies requirements 3, 4, and 5, but not requirements 1 and 2.

6.3 DAFALIAS AND POPOV'S TWO-SURFACE MODEL

Dafalias and Popov (1975) proposed a two-surface model using the plastic internal variables or (piv) concept. This model was motivated by the need for correctly describing the uniaxial cyclic experimental curves shown in Fig. 6.1. In the following presentation, the model used is based on the works of Dafalias and Popov (1976), Dafalias (1984), and Chaboche (1986). Compared with the multisurface Mroz model discussed above, this model uses only two surfaces and defines a continuous variation of the plastic modulus E^p between these two surfaces (rather than a piecewise constant plastic moduli field).

6.3.1 Modeling of Uniaxial Stress-Strain Curve

The fundamental point of departure from other models is the practical mathematical statement that any curve is completely described by its differential equation with appropriate initial or boundary conditions. Thus, basically, to describe any uniaxial stress-strain curve, we need to know its differential equation in the form

$$\frac{d\sigma}{d\epsilon} = E^t \tag{6.51}$$

where E^t is the slope of the curve, or the tangential modulus of the curve at any loading point. Obviously during plastic deformation E^t will vary; however, according to Eq. (6.9), E^t is a function of the plastic modulus E^p, since the elastic modulus E is constant. Therefore the key to describing a stress-strain curve correctly is to find out

$$\frac{d\sigma}{d\epsilon^{p}} = E^{p} \tag{6.52}$$

as a function of plastic strain. The determination of the evolution of E^{p} with the elastic-plastic straining becomes very important. E^{p} will change during the course of plastic deformation. In general, Dafalias and Popov (1975, 1976) assumed that it is a function of plastic internal variables (denoted collectively as piv) in the form

$$E^{p} = \hat{E}^{p}\,(piv) \tag{6.53}$$

It becomes essential to identify the piv as phenomenological quantities and to specify the particular form of the function $\hat{E}^{p}$ (piv). To this end, the experimental curves shown in Fig. 6.1 were examined carefully, and the following features were found to be of primary importance in modeling the real material responses under cyclic loading:

1. Irrespective of previous loading history, the different stress-strain curves tend to converge with two definite bounding lines XX' and YY', which, in this particular case, appear to be straight and parallel. This suggests the existence of an outer or bounding surface;
2. At points G_1 to G_4, which are equal distance from the bound line XX', the curves have almost the same slope as the stress-strain curve. This suggests that the distance from the stress-strain curve to the bounding line XX', or the distance between the yield or the loading surface (sometimes even called the *inner surface*) and the bounding surface if the multiaxial case is considered, can be identified as one of the piv.
3. At point G_0, which is the same distance from the line XX' as points G_1 to G_4, the slope is quite different from those at points G_1 to G_4. Yet for points G_1 to G_4 the most recent yielding initiated at E_1 to E_4 (which are also about the same distance from XX'), while for G_0 yielding initiated at C_6 is a much larger distance. This suggests that differences in recent loading history strongly affect the material response, and piv is required to reflect it.

These three observations have been consolidated into the following main hypothesis proposed by Dafalias and Popov:

$$E^{p} = \hat{E}^{p}(\delta_{\text{in}}, \delta, \epsilon_{e}^{p}) \tag{6.54}$$

with the limit value

$$\delta \to \delta_{in}, \quad E^p \to \infty \tag{6.55}$$

and

$$\delta \to 0, \quad E^p = \overline{E^p}(\epsilon_e^p) \tag{6.56}$$

where δ_{in} is the value of δ at the initiation of yielding and plays the role of a discrete memory parameter for the most recent plastic loading. It can be updated each time a new plastic flow is initiated during loading, unloading, and reloading. δ is the distance from the loading point to the line XX', as mentioned in the second observation above. $\bar{E}^p$ in Eq. (6.56) represents the plastic hardening modulus on the bounding line XX'. Equation (6.55) can be deduced from Eq. (6.9). It is seen that in elastic region, $E^t = E$, so that $E^p = \infty$ according to Eq. (6.9). At the initiation of the plastic flow, $\delta \to \delta_{in}$, the plastic modulus E^p should approach infinity, in order to produce a smooth transition from the elastic the elastic-plastic region. The quantities used in Eqs. (6.54) to (6.56) are schematically illustrated in Fig. 6.14 along with some other quantities that will be explained next.

In Fig. 6.14 the uniaxial cyclic stress-strain curve is plotted in the stress–plastic-strain space. The two bound lines are assumed to be straight and parallel so that $\bar{E}^p$ is a constant. This means the bounds will only undergo the kinematic hardening. In general, $\bar{E}^p$ can be a function of ϵ_e^p,

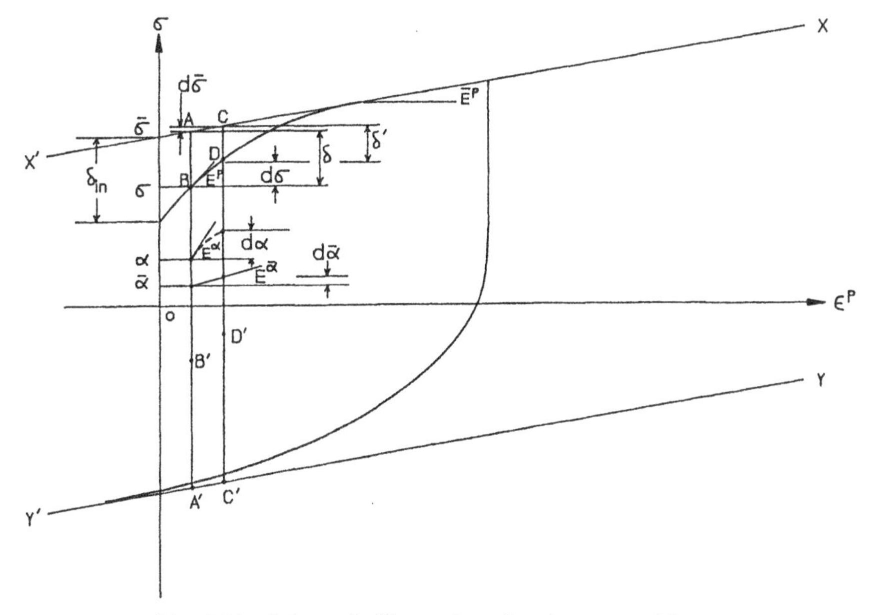

Fig. 6.14 Schematic illustration of various quantities

resulting in nonlinear bounds. In the figure the line segment BB' represents the elastic region which is enclosed in a larger segment AA' defined by two bounds A and A'. During the plastic deformation the elastic region, as well as the region enclosed by the bounds, can translate and expand, for example, into the regions DD' and CC', respectively. Mathematically these can be expressed by the loading or yield function F and bounding function $\bar{F}$ in the following forms for uniaxial loading:

$$F = (\sigma - \alpha)^2 - \sigma_y^2(\epsilon^p) = 0 \tag{6.57}$$

$$\bar{F} = (\bar{\sigma} - \bar{\alpha})^2 - \bar{\sigma}_y^2(\epsilon^p) = 0 \tag{6.58}$$

where σ and $\bar{\sigma}$, α and $\bar{\alpha}$ denote the stresses on F and $\bar{F}$ and their centers, respectively, as shown in Fig. 6.14, while their respective sizes are σ_y and $\bar{\sigma}_y$. From the figure, it can be derived that

$$d\epsilon^p = \frac{d\sigma}{E^p} \tag{6.59}$$

$$d\alpha = E^\alpha d\epsilon^p = \frac{E^\alpha}{E^p}\, d\sigma \tag{6.60}$$

$$d\bar{\sigma} = \bar{E}^p\, d\epsilon^p = \frac{\bar{E}^p}{E^p}\, d\sigma \tag{6.61}$$

$$d\bar{\alpha} = E^{\bar{\alpha}}\, d\epsilon^p = \frac{E^{\bar{\alpha}}}{E^p}\, d\sigma \tag{6.62}$$

where E^p, $\bar{E}^p$ are as defined earlier and E^α and $E^{\bar{\alpha}}$ are the corresponding hardening moduli associated with the translation of the centers α and $\bar{\alpha}$, as shown in Fig. 6.14. These two moduli are not independent and can be expressed in terms of E^p and $\sigma_y(\epsilon^p)$ [or $\bar{E}^p$ and $\bar{\sigma}(\epsilon^p)$] by using the consistency condition $dF = 0$ and $d\bar{F} = 0$. This gives

$$E^\alpha = E^p - \frac{d\sigma_y}{d\epsilon^p} \tag{6.63}$$

and

$$E^{\bar{\alpha}} = \bar{E}^p - \frac{d\bar{\sigma}_y}{d\epsilon^p} \tag{6.64}$$

It can be seen that if the elastic region does not change its size (i.e., only kinematic hardening occurs), $E^\alpha = E^p$. Similarly, if the distance between two bounds does not change, $E^{\bar{\alpha}} = \bar{E}^p$.

This model is novel because it relates E^p to δ and uses δ_{in} as a discrete memory of the most recent unloading-reloading event. δ has to be updated each time a drastic change in loading direction occurs. The parameter and the updating procedure are necessary to account for the drastic difference in the plastic modulus E^p when the loading direction is reversed. In addition δ_{in} is an indirect measure of the amount of the loading in the opposite direction, since a large δ_{in} implies that an extensive plastic loading before stress reversal. This point can be seen by comparing the different stress-strain loops in Fig. 6.1.

Also from Fig. 6.14 it is easy to show that

$$\begin{aligned} \delta &= \bar{\sigma} - \sigma = \overline{\sigma_y} + \bar{\alpha} - \sigma \\ &= \bar{\sigma}_y - \sigma_y + \bar{\alpha} - \alpha \end{aligned} \tag{6.65}$$

or

$$\begin{aligned} \delta &= d\sigma + \delta' - d\bar{\sigma} \\ &= d\sigma + \delta' - \bar{E}^p d\epsilon^p \\ &= \delta' + \left(1 - \frac{\bar{E}^p}{E^p}\right) d\sigma \end{aligned} \tag{6.66}$$

Thus

$$d\delta = \delta' - \delta = \left(\frac{\bar{E}^p}{E^p} - 1\right) d\sigma \tag{6.67}$$

These equations are useful if the differential equation (6.52) is to be integrated to obtain the closed form solution. They also justify the need to introduce the bounding surface in addition to a single loading surface: A bounding surface is necessary, among other things, to define the distance δ, as shown by Eqs. (6.65) to (6.67), which is a very important parameter or internal variable in describing the state of the plastic deformation.

The quantities $d\epsilon^p$, $d\alpha$, $d\bar{\alpha}$, and $d\delta$ can be calculated by using Eqs. (6.59) to (6.67) for a given increment of stress $d\sigma$, if the plastic modulus E^p, its limiting values on bounds $\bar{E}^p$, and the constitutive equations for σ_y and $\bar{\sigma}_y$ are known. These four material functions must be determined experimentally. σ_y and $\bar{\sigma}_y$ represent the isotropic hardening of the elastic region and the bounds; they can be determined using the uniaxial experimental data after subtracting the kinematic hardening effect. Once the bounding line or curve is determined experimentally, $\bar{E}^p$ can be measured directly from it. The function form of $\bar{E}^p$ proposed by Dafalias and Popov (1976) is

$$E^p = \bar{E}^p + h(\delta_{in}) \left(\frac{\delta}{\delta_{in} - \delta}\right) \tag{6.68}$$

where

$$h = \hat{h}(\delta_{\text{in}}) \tag{6.69}$$

is a model parameter function of δ_{in} controlling the "steepness" of the stress-strain curve. Obviously the larger it is, the steeper the curve. It is seen that the form (6.65) satisfies Eqs. (6.55) and (6.56).

The particular form of h used by Dafalias and Popov to model the complex random cyclic loading curve presented in Fig. 6.1 is

$$h = \frac{a}{1 + br^m} \tag{6.70}$$

where a, b, and m are three constants to be determined experimentally, $r = \delta_{\text{in}}/\sigma_{\text{r}}$ is a dimensionless variable, and σ_{r} is a reference stress nondimensionalizing δ_{in} that has the dimension of stress. The stress-strain curve predicted by this model—for $a = 164\bar{E}^{\text{p}}$, $b = 46$, $m = 3$, and σ_{r} is the distance between the experimentally determined bounds—is in agreement with the experimental result. The interested reader can refer to the original work by Dafalias and Popov (1976) for more details.

6.3.2 Multiaxial Case

Now the two-surface model based on the uniaxial consideration can be generalized into the multiaxial case. The loading surface F and bounding surface $\bar{F}$ are written in the general forms

$$F = f(\boldsymbol{\sigma} - \boldsymbol{\alpha}) - \boldsymbol{\kappa} = 0 \tag{6.71}$$

$$\bar{F} = \bar{f}(\bar{\boldsymbol{\sigma}} - \bar{\boldsymbol{\alpha}}) - \bar{\boldsymbol{\kappa}} = 0 \tag{6.72}$$

where $\boldsymbol{\sigma}$ and $\bar{\boldsymbol{\sigma}}$ are stresses on F and $\bar{F}$, respectively; $\boldsymbol{\alpha}$ and $\bar{\boldsymbol{\alpha}}$, and $\boldsymbol{\kappa}$ and $\bar{\boldsymbol{\kappa}}$ represent the centers and sizes of F and $\bar{F}$, respectively. The isotropic hardening of the loading and bounding surfaces can be accounted for by assuming that $\boldsymbol{\kappa}$ and $\bar{\boldsymbol{\kappa}}$ are functions of the effective plastic strain $\epsilon_{\text{e}}^{\text{p}}$:

$$\boldsymbol{\kappa} = \boldsymbol{\kappa}(\boldsymbol{\epsilon}_{\text{e}}^{\text{p}}) \tag{6.73}$$

$$\bar{\boldsymbol{\kappa}} = \bar{\boldsymbol{\kappa}}(\boldsymbol{\epsilon}_{\text{e}}^{\text{p}}) \tag{6.74}$$

where, as defined earlier,

$$\boldsymbol{\epsilon}_{\text{e}}^{\text{p}} = \int d\boldsymbol{\epsilon}_{\text{e}}^{\text{p}} = \int \left(\frac{2}{3} d\boldsymbol{\epsilon}^{\text{p}} : d\boldsymbol{\epsilon}^{\text{p}} \right)^{1/2} \tag{6.75}$$

The form of the functions $\kappa(\epsilon_{\text{e}}^{\text{p}})$ and $\bar{\kappa}(\epsilon_{\text{e}}^{\text{p}})$ need to be specified using the uniaxial experimental data after subtraction of the effect of the kinematic hardening. The particular observation of this will be discussed in the next section. Here it is assumed that κ and $\bar{\kappa}$ are already determined.

As discussed in the case of uniaxial loading, the field of plastic moduli K_i^p, is replaced with a generalized plastic modulus K^p, which is related to E^p in the uniaxial case:

$$K^p = K^p(\delta_{in}, \delta, \epsilon_e^p) \tag{6.76}$$

The parameter δ is a generalization of the uniaxial concept and is defined by

$$\delta = [(\bar{\sigma} - \sigma) : (\bar{\sigma} - \sigma)]^{1/2} \tag{6.77}$$

It represents the distance between the current stress state on the loading surface and the stress state $\bar{\sigma}$ on the bounding surface. A reasonable rule seems to be that the point σ on F and the point $\bar{\sigma}$ on $\bar{F}$ have the same outward normal so that

$$\lambda \frac{\partial f}{\partial \sigma} = \bar{\lambda} \frac{\partial \bar{f}}{\partial \bar{\sigma}} \tag{6.78}$$

The plastic modulus K^p varies continuously between

$$K^p(\delta_{in}, \delta_{in}, \epsilon_e^p) = \infty \quad \text{and} \quad K^p(\delta_{in}, 0, \epsilon_e^p) = \bar{K}^p(\epsilon_e^p) \tag{6.79}$$

where $\bar{K}^p$ is the value of K^p on the bounds. This is a direct generalization of Eqs. (6.55) and (6.56). Again, to describe the most recent plastic loading event, δ_{in} has to be updated each time the plastic loading initiates after a reversal of, or more generally a change in, the loading path.

The plastic strain is determined by the normality flow rule

$$d\boldsymbol{\epsilon}^p = \frac{1}{K^p}(\mathbf{n} : d\boldsymbol{\sigma})\mathbf{n} \tag{6.80}$$

This equation, which we derived in the last chapter, implies that the plastic flow is caused by the normal component of $d\boldsymbol{\sigma}$ and is in the direction of the normal to the loading or yielding surface $\mathbf{n}$, which denotes the unit normal to F at the loading point. Then

$$d\epsilon_e^p = \left(\frac{2}{3} d\epsilon^p : d\epsilon^p\right)^{1/2} = \sqrt{\frac{2}{3}} \frac{1}{K^p}(\mathbf{n} : d\boldsymbol{\sigma}) \tag{6.81}$$

As in the Mroz model, at the contact point, the two surfaces must have the same normal so that they will not intercept each other. To ensure this, we assume that

$$d\bar{\boldsymbol{\alpha}} - d\boldsymbol{\alpha} = d\mu(\bar{\boldsymbol{\sigma}} - \boldsymbol{\sigma}) \tag{6.82}$$

This means that the relative translation of the two surfaces follows the rule proposed in the Mroz model. The relative translation is used here because the translation of the bounding surface is allowed, while in the Mroz model the outer surface only can undergo isotropic expansion before the inner surfaces contact it.

There is no restriction on the translation direction of the loading or yield surface, and it can be selected by the user; the evolution law for $\boldsymbol{\alpha}$ is proposed as

$$d\boldsymbol{\alpha} = \frac{K_\alpha}{K^{\mathrm{p}}} \frac{(\mathbf{n} : d\boldsymbol{\sigma})}{\mathbf{n} : \boldsymbol{\nu}} \boldsymbol{\nu} \tag{6.83}$$

where K_α is a hardening modulus for $\boldsymbol{\alpha}$ and $\boldsymbol{\nu}$ is a unit vector representing the direction of the translation of $\boldsymbol{\alpha}$; the factor $1/K^{\mathrm{p}}(\mathbf{n} : d\boldsymbol{\sigma}/\mathbf{n} : \boldsymbol{\nu})$ is introduced to simplify the manipulation. K_α is determined from the consistency condition of the loading surface $dF = 0$,

$$\frac{\partial f}{\partial \boldsymbol{\sigma}} : d\boldsymbol{\sigma} - \frac{\partial f}{\partial \boldsymbol{\sigma}} : d\boldsymbol{\alpha} - \frac{d\kappa}{d\epsilon_{\mathrm{e}}^{\mathrm{p}}} d\epsilon_{\mathrm{e}}^{\mathrm{p}} = 0 \tag{6.84}$$

Substituting Eq. (6.83) into (6.84) yields

$$K^\alpha = K^{\mathrm{p}} - \sqrt{\frac{2}{3}} \frac{d\kappa/d\epsilon_{\mathrm{e}}^{\mathrm{p}}}{(\partial f/\partial \boldsymbol{\sigma} : \partial f/\partial \boldsymbol{\sigma})^{1/2}} \tag{6.85}$$

where Eq. (6.81) has been used to simplify the result. It can be seen that $K^\alpha = K^{\mathrm{p}}$ if the elastic region remains constant $(d\kappa/d\epsilon_{\mathrm{e}}^{\mathrm{p}} = 0)$.

Similarly the multiplier $d\mu$ in Eq. (6.82) can be obtained by the consistency condition of the bounding surface $d\bar{F} = 0$:

$$\frac{\partial \bar{f}}{\partial \bar{\boldsymbol{\sigma}}} : d\bar{\boldsymbol{\sigma}} - \frac{\partial \bar{f}}{\partial \bar{\boldsymbol{\sigma}}} : d\bar{\boldsymbol{\alpha}} - \frac{d\bar{\kappa}}{d\epsilon_{\mathrm{e}}^{\mathrm{p}}} d\epsilon_{\mathrm{e}}^{\mathrm{p}} = 0 \tag{6.86}$$

After substituting $d\bar{\boldsymbol{\alpha}}$ into this equation and rearranging the result, we can show that

$$d\mu = \frac{\left(\partial \bar{f}/\partial \boldsymbol{\sigma} : d\boldsymbol{\alpha}\right) + \left(d\bar{\kappa}/d\epsilon_{\mathrm{e}}^{\mathrm{p}}\right)\left(\sqrt{2/3}\right)\left(1/K^{\mathrm{p}}\right)\left(\mathbf{n} : d\boldsymbol{\sigma}\right) - \left(\partial \bar{f}/\partial \bar{\boldsymbol{\sigma}} : d\bar{\boldsymbol{\sigma}}\right)}{\partial \bar{f}/\partial \bar{\boldsymbol{\sigma}} : (\bar{\boldsymbol{\sigma}} - \boldsymbol{\sigma})} \tag{6.87}$$

Note that in Eq. (6.78), F and $\bar{F}$ have the same outward normal at $\boldsymbol{\sigma}$ and $\bar{\boldsymbol{\sigma}}$; Eq. (6.61) can be generalized into the multiaxial case as

$$\mathbf{n} : d\bar{\boldsymbol{\sigma}} = \bar{K}^{\mathrm{p}} \sqrt{\frac{3}{2}}\, d\epsilon_{\mathrm{e}}^{\mathrm{p}} = \frac{\bar{K}^{\mathrm{p}}}{K^{\mathrm{p}}} \mathbf{n} : d\boldsymbol{\sigma} \tag{6.88}$$

Eq. (6.87) can be simplified into

$$d\mu = \frac{K^{\alpha} - K^{\bar{\alpha}}}{K^{\mathrm{p}}} \frac{\mathbf{n} : d\boldsymbol{\sigma}}{\mathbf{n} : (\bar{\boldsymbol{\sigma}} - \boldsymbol{\sigma})} \tag{6.89}$$

where

$$K^{\bar{\alpha}} = \bar{K}^{\mathrm{p}} - \frac{\sqrt{2/3}\,(d\bar{\kappa}/d\epsilon_{\mathrm{e}}^{\mathrm{p}})}{\left(\partial \bar{f}/\partial \bar{\boldsymbol{\sigma}} : \partial \bar{f}/\partial \bar{\boldsymbol{\sigma}}\right)^{1/2}} \tag{6.90}$$

Then

$$d\bar{\boldsymbol{\alpha}} = d\boldsymbol{\alpha} - \frac{K^{\alpha} - K^{\bar{\alpha}}}{K^{\mathrm{p}}} \frac{(\mathbf{n} : d\boldsymbol{\sigma})}{\mathbf{n} : (\bar{\boldsymbol{\sigma}} - \boldsymbol{\sigma})} (\bar{\boldsymbol{\sigma}} - \boldsymbol{\sigma}) \tag{6.91}$$

Again, $K^{\bar{\alpha}} = \bar{K}^{\mathrm{p}}$ if the bounding surface remains constant in size.

As we mentioned earlier, the direction of the translation of the loading surface, represented by $\boldsymbol{\nu}$ in the expression for $d\boldsymbol{\alpha}$, can be specified freely. A number of choices can be made. For example, the direction of loading, or the direction of the plastic strain increment, is one possibility. Of course this choice must concur with the experimental observation. In the literature this direction is usually chosen as in the Mroz model, whereby Eqs. (6.83) and (6.91) would be reduced to

$$d\boldsymbol{\alpha} = \frac{K^{\alpha}}{K^{\mathrm{p}}} \frac{(\mathbf{n} : d\boldsymbol{\sigma})}{\mathbf{n} : (\bar{\boldsymbol{\sigma}} - \boldsymbol{\sigma})} (\bar{\boldsymbol{\sigma}} - \boldsymbol{\sigma}) \tag{6.92}$$

$$d\bar{\boldsymbol{\alpha}} = \frac{K^{\bar{\alpha}}}{K^{\mathrm{p}}} \frac{(\mathbf{n} : d\boldsymbol{\sigma})}{\mathbf{n} : (\bar{\boldsymbol{\sigma}} - \boldsymbol{\sigma})} (\bar{\boldsymbol{\sigma}} - \boldsymbol{\sigma}) \tag{6.93}$$

where K^{α}, $K^{\bar{\alpha}}$ are given by Eq. (6.85) and (6.90). If it is assumed that $\bar{F}$ is of the same type as F (or $\bar{F}$ is similar to F according to Dafalias and Popov), then

$$\bar{\boldsymbol{\sigma}} - \bar{\boldsymbol{\alpha}} = \eta(\boldsymbol{\sigma} - \boldsymbol{\alpha}) \tag{6.94}$$

where η is a similarity factor and represents the ratio of their sizes. This assumption of similarity is realistic, since in practice $\bar{F}$ and F are usually considered to be the same functional form. For example, both the loading and bounding surfaces can be assumed to be of von Mises type:

$$F = \frac{3}{2}(\mathbf{S} - \boldsymbol{\alpha}') : (\mathbf{S} - \boldsymbol{\alpha}') = \sigma_y^2 \tag{6.95}$$

$$\bar{F} = \frac{3}{2}(\bar{\mathbf{S}} - \bar{\boldsymbol{\alpha}}') : (\bar{\mathbf{S}} - \bar{\boldsymbol{\alpha}}') = \bar{\sigma}_y^2 \tag{6.96}$$

Hence Eq. (6.94) becomes

$$\bar{\boldsymbol{\sigma}} - \bar{\boldsymbol{\alpha}} = \frac{\bar{\sigma}_y}{\sigma_y}(\boldsymbol{\sigma} - \boldsymbol{\alpha}) \tag{6.97}$$

which is the same as Eq. (6.22). Thus $\bar{\boldsymbol{\sigma}}$ can be calculated in terms of $\boldsymbol{\alpha}$, $\bar{\boldsymbol{\alpha}}$, σ_y, and $\bar{\sigma}_y$ (which are given by the constitutive equations), and $\boldsymbol{\sigma}$ is the current stress state on the loading surface.

The loading and bounding surfaces F and $\bar{F}$, the flow rule (6.80), the constitutive equations for $\boldsymbol{\alpha}$ and $\bar{\boldsymbol{\alpha}}$, given either by Eqs. (6.83) and (6.91) or

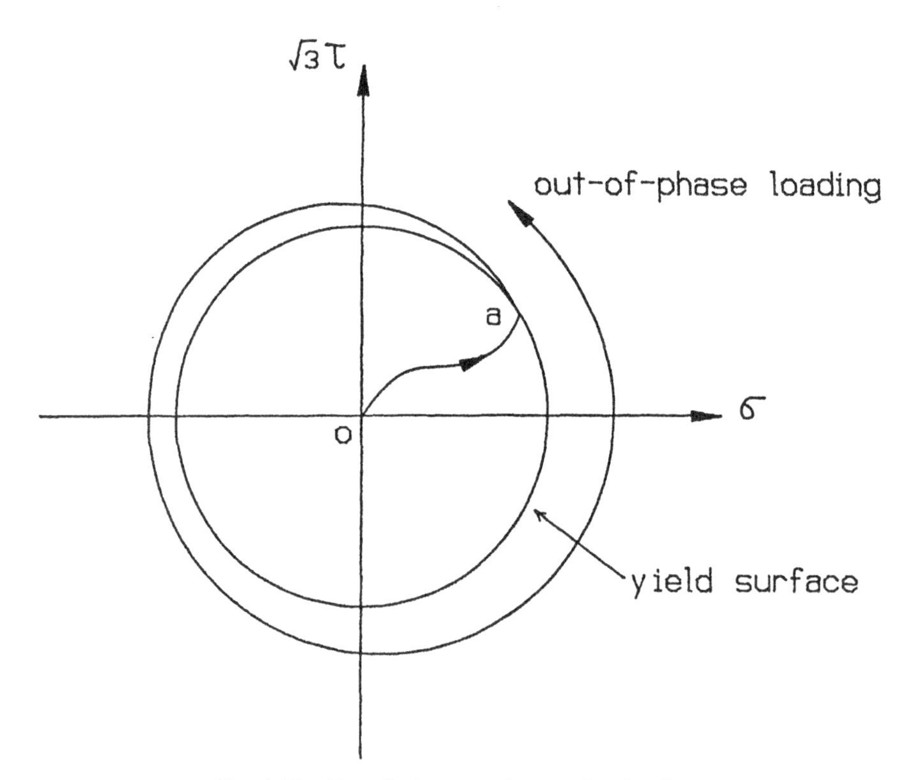

Fig. 6.15 Out-of-phase tension-torsion loading

by Eqs. (6.92) and (6.93), the constitutive equations for κ and $\bar{\kappa}$, and the particular form for K satisfying the conditions given by Eq. (6.79) complete the two surface model proposed by Dafalias and Popov. In summary, the model has the following material functions which have to be determined experimentally:

1. The isotropic hardening functions, $\kappa(\epsilon_e^p)$ and $\bar{\kappa}(\epsilon_e^p)$.
2. The plastic modulus K^p $(\delta_{in}, \delta, \epsilon_e^p)$, including its limiting value on bounds, $\bar{K}^p = K^p(\delta_{in}, 0, \epsilon_e^p)$. As derived in the last section, in particular, by Eq. (6.45), K^p is related to its uniaxial counterpart E^p by

$$K^p = \frac{2}{3}E^p \quad \text{and} \quad \bar{K}^p = \frac{2}{3}\bar{E}^p \tag{6.98}$$

K^p and $\bar{K}^p$ can thus be determined from the uniaxial cyclic loading test. The recommended form for K^p is

$$K^p = \bar{K}^p + h(\delta_{in})\frac{\delta}{\delta_{in} - \delta} \tag{6.99}$$

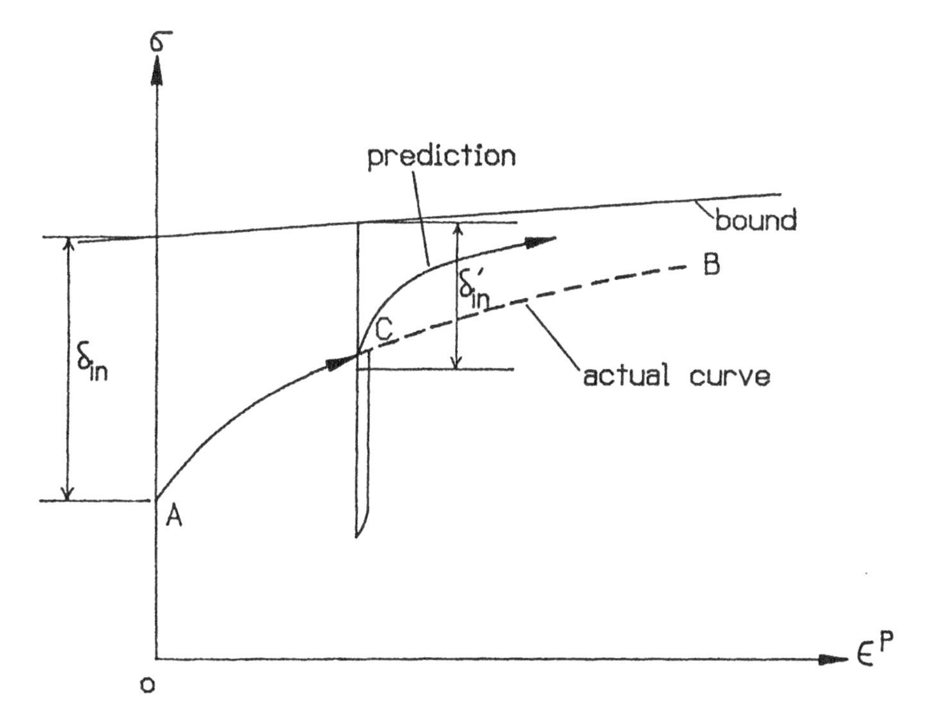

Fig. 6.16 Overshooting of the tensile curve after unloading

The two-surface models presented here can satisfy all five requirements listed in Section 6.1. One of the disadvantages of this model is that it needs significant memory capability for updating procedures in the computational code. Also, as Fig. 6.15 shows, the updating procedure can fail in the special case of out-of-phase multiaxial loading because no unloading takes place although cycling (the reverse loading) is evident.

Chaboche (1986) pointed out another inconsistency of this model that occurs in uniaxial load-unload-reload situations where plastic flow is very small during unloading. This is shown in Fig. 6.16. The reloading is computed with updated quantities. According to the two-surface models presented here, the distance δ_{in} is reinitialized, which gives an initial infinite plastic modulus. This leads to the overshooting of the new tensile curve indicated in the figure.

6.4 NONLINEAR KINEMATIC HARDENING MODEL

Modeling of the nonlinear stress-strain curve is the main objective of all recently developed plasticity theories. In the Mroz model the nonlinearity is introduced by defining a field of hardening moduli associated with several surfaces which are originally concentric that can translate rigidly and expand uniformly. In the Dafalias and Popov model the nonlinearity is described by modeling the continuous variation of the hardening modulus based on the concept of two surfaces, the loading or yield surface and the bounding surface. The introduction of this concept is necessary to define a distance δ in the stress space that represents the difference between the current loading state and the state represented by the bounds. This distance variable has been found to be very important in defining the variation of the hardening modulus.

There is, however, another approach to modeling the nonlinearity of the stress-strain curve that is a direct generalization of Prager and Ziegler's linear kinematic hardening model. It does not use the concept of the bounding surface, although this concept, as pointed out by Chaboche (1986), can be deduced from the nonlinear constitutive equation of the kinematic hardening variable, namely the back stress $\boldsymbol{\alpha}$.

6.4.1 Foundation of the Nonlinear Kinematic Hardening Model

The nonlinear kinematic hardening model was originally proposed by Armstrong and Frederick (1966); it has recently been developed by Chaboche (1977, 1986). This presentation is based on Chaboche's work. In the following derivation the von Mises yield criterion is assumed and can be written as

$$F = \frac{3}{2}(\mathbf{S} - \boldsymbol{\alpha}') : (\mathbf{S} - \boldsymbol{\alpha}') - \sigma_y^2(\epsilon_e^p) = 0 \qquad (6.100)$$

where, as usual, $\boldsymbol{\alpha}$ and σ_y are the kinematic and isotropic hardening variables, and S and $\boldsymbol{\alpha}'$ represent the deviatoric quantities of $\boldsymbol{\sigma}$ and $\boldsymbol{\alpha}$. The associated flow rule is used and takes the form

$$d\boldsymbol{\epsilon}^{\mathrm{p}} = \frac{1}{K^{\mathrm{p}}}(\mathbf{n} : d\boldsymbol{\sigma})\mathbf{n} \tag{6.101}$$

where $\mathbf{n}$ is the outward unit normal to the yield surface and is given by

$$\mathbf{n} = \frac{\partial f/\partial \boldsymbol{\sigma}}{|\partial f/\partial \boldsymbol{\sigma}|} = \frac{3(\mathbf{S} - \boldsymbol{\alpha}')}{[9(\mathbf{S} - \boldsymbol{\alpha}') : (\mathbf{S} - \boldsymbol{\alpha}')]^{1/2}} = \sqrt{\frac{3}{2}}\,\frac{\mathbf{S} - \boldsymbol{\alpha}'}{\sigma_y} \tag{6.102}$$

As we noted earlier, the classical linear kinematic hardening model by Prager can only give the linear hardening curve and hence is not capable of describing the nonlinear elastic-plastic behavior observed in cyclic loading tests. To modify this, Armstrong and Frederick (1966) and Chaboche (1977) introduced an additional nonlinear term to the model:

$$d\boldsymbol{\alpha} = \frac{2}{3}c\,d\boldsymbol{\epsilon}^{\mathrm{p}} - \gamma\boldsymbol{\alpha}\,d\epsilon_{\mathrm{e}}^{\mathrm{p}} \tag{6.103}$$

where c and γ are two material constants. In general, they can be assumed to be functions of the effective plastic strain, while $d\epsilon_{\mathrm{e}}^{\mathrm{p}}$ is the increment of the effective plastic strain defined by Eq. (6.81). The key in this model is the second term on the right-hand side, called the *recall term*, which will affect the plastic flow differently for tensile or compressive loading because it depends on $|d\boldsymbol{\epsilon}^{\mathrm{p}}|$ and is very important in predicting the nonlinear stress-strain loop under cyclic loading. This will soon become clear as the uniaxial case is considered.

The consistency condition is now written

$$3(\mathbf{S} - \boldsymbol{\alpha}') : d\boldsymbol{\sigma} - 3(\mathbf{S} - \boldsymbol{\alpha}') : d\boldsymbol{\alpha}' - 2\sigma_y \frac{d\sigma_y}{d\epsilon_{\mathrm{e}}^{\mathrm{p}}} d\epsilon_{\mathrm{e}}^{\mathrm{p}} = 0 \tag{6.104}$$

Making use of Eqs. (6.101)–(6.104) and (6.81) and noting that

$$(\mathbf{S} - \boldsymbol{\alpha}') : d\boldsymbol{\alpha}' = (\mathbf{S} - \boldsymbol{\alpha}') : d\boldsymbol{\alpha} \tag{6.105}$$

we obtain the expression

$$K^{\mathrm{p}} = \frac{2}{3}c - \gamma\frac{(\mathbf{S} - \boldsymbol{\alpha}') : \boldsymbol{\alpha}}{\sigma_y} + \frac{2}{3}\frac{d\sigma_y}{d\epsilon_{\mathrm{e}}^{\mathrm{p}}} \tag{6.106}$$

or

$$K^{\mathrm{p}} = \frac{2}{3}c - \gamma\frac{(\mathbf{S} - \boldsymbol{\alpha}') : \boldsymbol{\alpha}'}{\sigma_{\mathrm{y}}} + \frac{2}{3}\frac{d\sigma_{\mathrm{y}}}{d\epsilon_{\mathrm{e}}^{\mathrm{p}}} \tag{6.107}$$

since $(\mathbf{S} - \boldsymbol{\alpha}') : \boldsymbol{\alpha} = (\mathbf{S} - \boldsymbol{\alpha}') : \boldsymbol{\alpha}'$. In fact for cases where the hydrostatic part of the stress has no effect on the yielding, the back stress $\boldsymbol{\alpha}$ can be considered as a deviator, since in the stress space the yield surface is a cylinder that is orthogonal to the deviatoric plane and the translation in the hydrostatic direction does not change the surface. In this book $\boldsymbol{\alpha}$ is not taken as a deviator in order to make the development of the theory applicable to most general cases.

Now some features of the nonlinear kinematic hardening model can be explored. From Eq. (6.107), the model (6.103) relates the variation of the plastic hardening modulus to the relative orientation of the effective stress tensor represented by $\mathbf{S} - \boldsymbol{\alpha}'$ and the kinematic hardening variable $\boldsymbol{\alpha}$. Note that the direction of $\mathbf{S} - \boldsymbol{\alpha}'$ is of a radius connecting the center to the loading point on the yield surface. Thus the variation of the plastic modulus K^{p} with the change of the loading direction is taken into account to some extent. This is explained more clearly in Fig. 6.17. Consider the stress state A denoted by σ on the yield surface. It can be seen that the vector $\mathbf{S}_A - \boldsymbol{\alpha}'$ in the stress space forms an acute angle $0 \leq \varphi < \pi/2$ with the vector $\boldsymbol{\sigma}$ so that $(\mathbf{S}_A - \boldsymbol{\alpha}') : \boldsymbol{\alpha} > 0$.

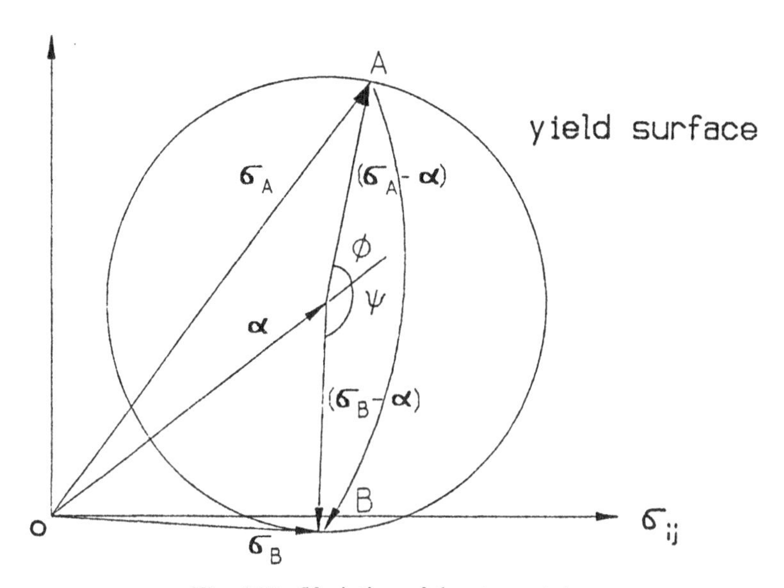

Fig. 6.17 Variation of the stress state

Now assume that the stress state changes from point A to point B following the path within the yield surface shown in the figure. Suppose that σ_B is such that $\mathbf{S}_B - \boldsymbol{\alpha}'$ forms an obtuse angle $\psi > \pi/2$ with σ. As a result $(\mathbf{S}_B - \boldsymbol{\alpha}') : \boldsymbol{\alpha} < 0$ at point B. Assuming that $\gamma > 0$, from Eq. (6.107), it follows that $K_A^p < K_B^p$, which is reasonable and qualitatively in accord with the experimental observations. In the special case where $\varphi = 0$ and $\psi = \pi$, the model gives a higher plastic modulus in the reversed direction than in the positive direction, which is the typical phenomenon observed in the cyclic loading tests.

In the case of uniaxial loading Eqs. (6.100), (6.101), (6.103), and (6.107) can be reduced to

$$F = |\sigma - \alpha| - \sigma_y = 0 \tag{6.108}$$

$$d\epsilon^p = \frac{2/3}{K^p} d\sigma = \frac{1}{E^p} d\sigma \tag{6.109}$$

$$d\alpha = c\, d\epsilon^p - \gamma \alpha |d\epsilon^p| \tag{6.110}$$

$$E^p = \frac{3}{2} K^p = c \mp \gamma \alpha + \frac{d\sigma_y}{d\epsilon^p} \tag{6.111}$$

where $-$ and $+$ correspond to loading and reverse loading, respectively. Note the essential difference resulting from the two plastic strain increment terms in Eq. (6.103), giving $d\epsilon^p$ and $|d\epsilon^p|$, respectively, in Eq. (6.110). During the tensile or positive loading, it follows that

$$d\sigma = (c - \gamma\alpha) d\epsilon^p \quad \text{and} \quad E_t^p = c - \gamma\alpha + \frac{d\sigma_y}{d\epsilon^p} \tag{6.112}$$

When the loading direction is reversed, the plastic flow is compressive:

$$d\sigma = (c + \gamma\alpha) d\epsilon^p \quad \text{and} \quad E_c^p = c + \gamma\alpha + \frac{d\sigma_y}{d\epsilon^p} \tag{6.113}$$

where the subscripts t and c stand for tensile and compressive loading, respectively. This is shown schematically in Fig. 6.18.

As is clear from the figure, the effects introduced by the recall term on tensile and compressive plastic flow are different. Assuming that $\gamma > 0$, the hardening modulus for reverse plastic flow is higher than for original loading at a given plastic strain ϵ^p. Therefore the relation between α and ϵ^p is nonunique, and the nonlinearity and the concavity of the stress-strain curve is correctly reproduced.

One of the merits of the nonlinear kinematic hardening model (6.103) is that for uniaxial loading, it can be integrated to obtain the analytical solution. Indeed, beginning with any ϵ_0^p and α_0 shown in Fig. 6.18 for c and γ

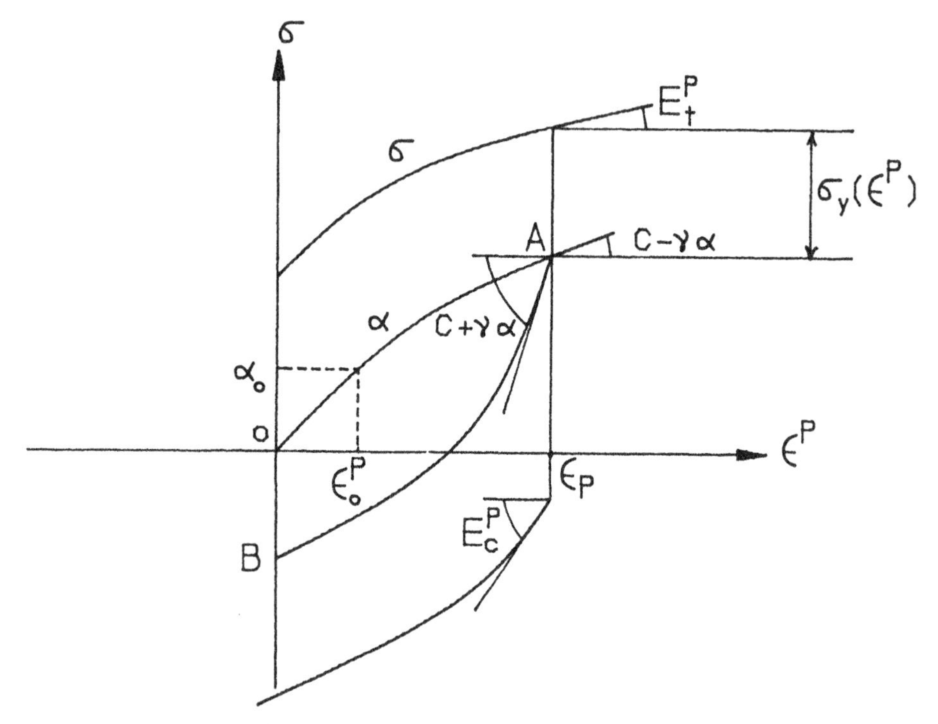

Fig. 6.18 Illustration of hardening modulus for loading and reverse loading

constant, we can integrate Eq. (6.112) to get

$$\alpha(\epsilon^{\mathrm{p}}) = \frac{c}{\gamma} + \left(\alpha_0 - \frac{c}{\gamma}\right) e^{-\gamma(\epsilon^{\mathrm{p}} - \epsilon_0^{\mathrm{p}})} \qquad (\epsilon^{\mathrm{p}} > \epsilon_0^{\mathrm{p}}) \tag{6.114}$$

where $\alpha(\epsilon^{\mathrm{p}})$ and ϵ^{p} represent any point at the half-cycle OA. Similarly, for another half-cycle AB, we obtain the following:

$$\alpha(\epsilon^{\mathrm{p}}) = -\frac{c}{\gamma} + \left(\alpha_0 + \frac{c}{\gamma}\right) e^{\gamma(\epsilon^{\mathrm{p}} - \epsilon_0^{\mathrm{p}})} \qquad (\epsilon^{\mathrm{p}} < \epsilon^{\mathrm{p}}) \tag{6.115}$$

We can combine these two equations as

$$\alpha(\epsilon^{\mathrm{p}}) = \pm\frac{c}{\gamma} + \left(\alpha_0 \mp \frac{c}{\gamma}\right) e^{\mp\gamma(\epsilon^{\mathrm{p}} - \epsilon_0^{\mathrm{p}})} \tag{6.116}$$

where the sign on the top corresponds to tensile plastic flow and that below to the compressive flow. By substituting Eq. (6.116) into Eq. (6.108), we can now write the stress-strain curve for uniaxial cases:

$$\sigma = \alpha(\epsilon^{\mathrm{p}}) \pm \sigma_{\mathrm{y}}(\epsilon^{\mathrm{p}})$$

$$= \pm\frac{c}{\gamma} + \left(\alpha_0 \mp \frac{c}{\gamma}\right)e^{\pm\gamma(\epsilon^{\mathrm{p}}-\epsilon_0^{\mathrm{p}})} \pm \sigma_{\mathrm{y}}(\epsilon^{\mathrm{p}}) \tag{6.117}$$

where the sign convention is the same as explained above. Equations (6.116) and (6.117) are useful in curve fitting the actual stress-strain curve to determine the material constants c and γ. Note that these equations are obtained under the assumption that c and γ remain constant in the course of plastic flow. For real materials, c and γ may be material functions and vary with the plastic deformation. In fact, as shown by the second part of Eqs. (6.112) and (6.113), they can be represented by the plastic modulus for tensile and compressive plastic flow at a given plastic strain. According to general requirement 4, the isotropic hardening will stabilize eventually during the cyclic loading (i.e., $d\sigma_{\mathrm{y}}/d\epsilon^{\mathrm{p}} = 0$). Suppose that $E_{\mathrm{c}}^{\mathrm{p}}$ and $E_{\mathrm{t}}^{\mathrm{p}}$ are taken from such a cyclic curve where the stabilization of the isotropic hardening has been reached; then it follows from Eqs. (6.112) and (6.113) that

$$c(\epsilon^{\mathrm{p}}) = \frac{E_{\mathrm{t}}^{\mathrm{p}} + E_{\mathrm{c}}^{\mathrm{p}}}{2} \tag{6.118}$$

$$\gamma(\epsilon^{\mathrm{p}}) = \frac{E_{\mathrm{c}}^{\mathrm{p}} - E_{\mathrm{t}}^{\mathrm{p}}}{2\boldsymbol{\alpha}_s} \tag{6.119}$$

where $\boldsymbol{\alpha}_{\mathrm{s}}$ is the value of the kinematic hardening variable at the load reversal point. It can be estimated that c and γ values determined this way will vary with the plastic strain, since $E_{\mathrm{t}}^{\mathrm{p}}$ and $E_{\mathrm{c}}^{\mathrm{p}}$ (as can be seen from Fig. 6.18) are not constant during plastic flow.

It is interesting to note that the present nonlinear rule can be considered a two-surface model. The yield surface is obviously $F = 0$, represented by Eq. (6.100), and the bounding surface can be deduced as follows: Using the nonlinear rule (6.103), we reach the ultimate value of $\boldsymbol{\alpha}$ when $d\boldsymbol{\alpha} = 0$, which is expressed

$$\boldsymbol{\alpha}^{\mathrm{u}} = \frac{2}{3}\frac{c}{\gamma}\frac{d\boldsymbol{\epsilon}^{\mathrm{p}}}{d\epsilon_{\mathrm{e}}^{\mathrm{p}}} \tag{6.120}$$

where the superscript u represents the ultimate value. We then multiply both sides of this equation by themselves to get

$$\left(\frac{3}{2}\boldsymbol{\alpha}^{u} : \boldsymbol{\alpha}^{u}\right)^{1/2} = \frac{c}{\gamma} \tag{6.121}$$

Note that $\boldsymbol{\alpha}^{\mathrm{u}}$ is the ultimate value of $\boldsymbol{\alpha}$; therefore, in general,

$$\left(\frac{3}{2}\boldsymbol{\alpha}:\boldsymbol{\alpha}\right)^{1/2} \le \frac{c}{\gamma} \tag{6.122}$$

where the equality holds if $\boldsymbol{\alpha}$ reaches its ultimate value. On the other hand, note that $\boldsymbol{\alpha}$ is a deviator. When von Mises yield criterion is used, the inequality

$$\left[\frac{3}{2}(\mathbf{S}-\boldsymbol{\alpha}):(\mathbf{S}-\boldsymbol{\alpha})\right]^{1/2} \le \sigma_{\mathrm{y}} \tag{6.123}$$

holds and represents an elastic domain within yield surface. Recall that $[(\mathbf{S}-\boldsymbol{\alpha}):(\mathbf{S}-\boldsymbol{\alpha})]^{1/2}$ is the usual euclidean distance in the stress space, and according to fundamental euclidean geometry,

$$\begin{aligned}\left[\frac{3}{2}\mathbf{S}:\mathbf{S}\right]^{1/2} &= \left[\frac{3}{2}(\mathbf{S}-\boldsymbol{\alpha}+\boldsymbol{\alpha}):(\mathbf{S}-\boldsymbol{\alpha}+\boldsymbol{\alpha})\right]^{1/2} \\ &\le \left[\frac{3}{2}(\mathbf{S}-\boldsymbol{\alpha}):(\mathbf{S}-\boldsymbol{\alpha})\right]^{1/2} + \left(\frac{3}{2}\boldsymbol{\alpha}:\boldsymbol{\alpha}\right)^{1/2}\end{aligned} \tag{6.124}$$

Substituting Eqs. (6.122) and (6.123) into the equation above results in the inequality

$$\left(\frac{3}{2}\mathbf{S}:\mathbf{S}\right)^{1/2} \le \sigma_{\mathrm{y}} + \frac{c}{\gamma} \tag{6.125}$$

In essence, this expression means that the stress state obtained using the nonlinear kinematic hardening rule (6.103) is bounded by the surface

$$\left(\frac{3}{2}\mathbf{S}:\mathbf{S}\right)^{1/2} = \sigma_{\mathrm{y}} + \frac{c}{\gamma} \tag{6.126}$$

which is the equation for the bounding surface. This idea is depicted in Fig. 6.19. Note that the bounding surface can expand uniformly but cannot translate in the stress space.

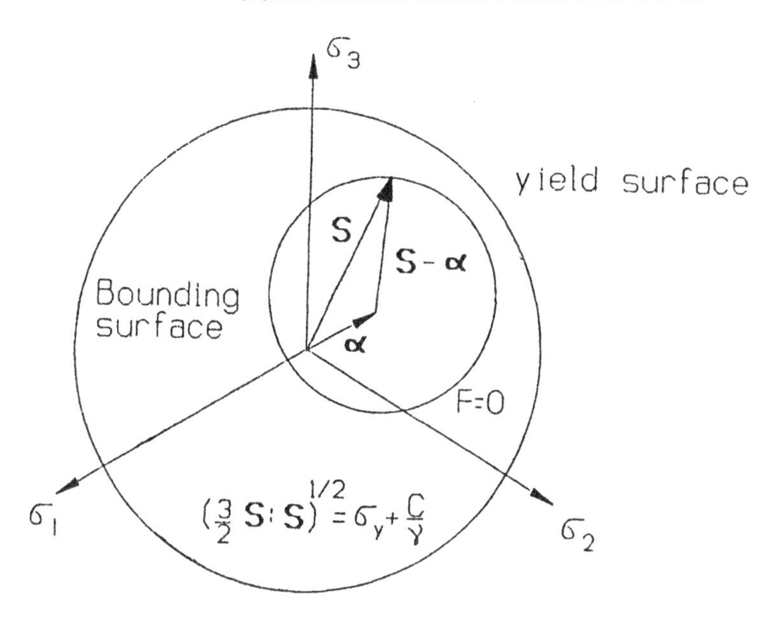

Fig. 6.19 Yield and bounding surfaces

6.4.2 Formulation of Isotropic Hardening

In previous sections we used different models to describe the evolution of the kinematic variable $\boldsymbol{\alpha}$. Although we included the isotropic hardening variable $\kappa(\epsilon_e^p)$ or $\sigma_y(\epsilon_e^p)$ in the formulas we developed, we did not consider any specific form of the evolution equation for it. We will now address this topic. The model we present here to describe the isotropic hardening can be used along with any kinematic hardening models in this chapter. κ or σ_y is normally assumed to be a function of the effective plastic strain, i.e., strain hardening is assumed. The generalization to work hardening or others is simple and straightforward.

Recall that isotropic hardening will eventually stabilize at a certain value. A simple constitutive equation was used by Zavenl and Lee (1978) and Chaboche (1977):

$$d\kappa = b(\kappa_s - \kappa)\,d\epsilon_e^p \tag{6.127}$$

where b and κ_s are two material constants. We can easily integrate this equation to get

$$\kappa(\epsilon_e^p) = \kappa_s\left(1 - e^{-b\epsilon_e^p}\right) \tag{6.128}$$

As can be seen from this equation, the isotropic hardening variable κ, sometimes called the *drag stress*, varies with the accumulation of effective plastic strain ϵ_e^p. After a certain number of cycles during which ϵ_e^p increases monotonically, it stabilizes at the value κ_s. So κ_s represents the stabilized value of κ. It has been shown experimentally that the number of cycles needed to reach κ_s depends on the magnitude of strain; the larger the strain magnitude, the fewer the needed cycles.

Earlier we showed in Fig. 6.3 (and discussed in Section 6.1) that the stabilized value κ_s is not a constant. This value also depends on the magnitude of the strain. For example, after the stabilization in a strain cycle at a lower strain level, the isotropic hardening will commence again if the strain level of the cyclic loading is increased and will approach a new stabilized value, usually higher than the previous one. This observation was considered by Chaboche (1979) for the concept of a plastic strain memory surface to record the maximum plastic strain range experienced by the material. Selected evolution equations were proposed for the center and radius of the memory surface, and these in turn appeared in an evolution equation for the stabilized value κ_s. Thus κ_s is related to the history of the maximum plastic strain range and varies as the maximum plastic strain range changes. This model will not be discussed here. The interested reader can refer to the original paper for a full description.

6.5 ENDOCHRONIC THEORY OF PLASTICITY

The plasticity theories we have discussed so far have one thing in common: They all were based on the concept of the yield surface. The existence of the yield surface was the first fundamental assumption. As a result the stress space (or strain space) was divided into two domains: the elastic domain within which the deformation is purely elastic and the plastic domain where elastic-plastic deformation occurs. These two domains are separated by the yield surface, whose accurate determination is fairly important for these theories.

However, as we pointed out in Chapter 1, for some materials such as aluminum or stainless steel, the stress-strain curve in uniaxial tension or compression is nonlinear almost from the start. Therefore it is very hard, if not impossible, to define accurately the yield point on the curve representing the onset of the plastic deformation. In fact, for various engineering applications, the conventional definitions of the yield point or yield stress may substantially differ from one another. For example, the following criteria may be used to define the yield point, as proposed by Haythornthwaite (1968).

1. Departure from linearity.
2. Measurable or offset plastic strain.
3. Slope of the curve equal to a given fraction of the initial slope (e.g., 90–95%).
4. Intersection of postyield slope with the stress axis.
5. Intersection of postyield and elastic slopes.
6. Offset elastic slope.

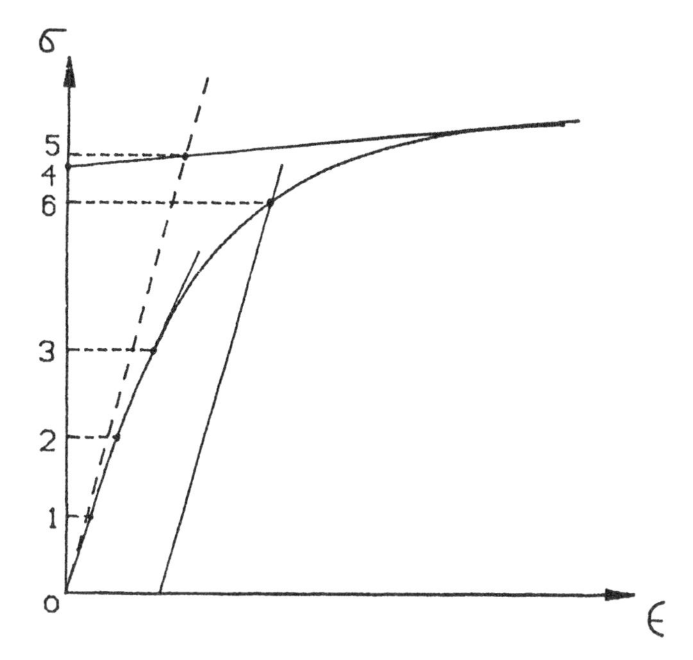

Fig. 6.20 Various definitions for yield point

These definitions are schematically shown in Fig. 6.20. As can be seen from the diagram, the differences between these definitions are large and may increase as the curvature of the nonlinear curve increases. Furthermore different definitions may render different conclusions for the kinematic or isotropic hardening behavior and the cross-effect of yielding, and so on, as pointed out by Valanis (1971).

Motivated by the foregoing critical considerations and by a development in irreversible thermodynamics based on the concept of internal variables, Valanis (1971) proposed a plasticity theory without yield surface which he called the *endochronic theory*. The word "endochronic" has a Greek root, *endo*, meaning inner or intrinsic, and *chronic*, meaning time. The constitutive equations relating $\boldsymbol{\sigma}$ and $\boldsymbol{\epsilon}$ are rigorously derived from the irreversible thermodynamics using the notation of internal variables and are analogous to those for viscoelastic materials. The difference is that the real time variable t that appeared in the constitutive equations for viscoelastic materials is replaced by the intrinsic time or inner time z, which will be discussed later in this text.

The endochronic theory has undergone development and modification in recent years. The original version by Valanis (1971) uses an intrinsic time measure that overestimates the unloading slope, although it satisfactorily predicts plastic responses of materials provided there are no reversals in the stress rate. Valanis (1980) modified his original version by introducing a new measure of intrinsic time which has made the theory more versatile and powerful in the analysis of plastic deformation under monotonic, as well as cyclic, loading conditions. The yield surface, which is a prior assumed element in classical plasticity theory, is a derived result of the endochronic theory.

The following presentation of the theory is based on the work by Valanis (1980, 1984) and Watanabe and Atluri (1986).

6.5.1 General Formulation of the Theory

In this section we assume that the constitutive equation governing the volume change is elastic. We will focus on the shape change or the relationship between the deviators $\mathbf{S}$ and $\boldsymbol{\epsilon}^{\mathrm{p}}$.

The endochronic theory derived by Valanis (1971, 1980) based on the irreversible thermodynamics consideration for initially isotropic materials at infinitesimal strain is

$$\mathbf{S} = 2G\int_0^z \rho(z - z')\frac{d\boldsymbol{\epsilon}^p}{dz'}dz' \tag{6.129}$$

where $\mathbf{S}$ and $\boldsymbol{\epsilon}^{\mathrm{p}}$ are the deviators of the stress tensor $\boldsymbol{\sigma}$ and the plastic strain tensor, respectively, and G is the elastic shear modulus. The kernel $\rho(z)$ is a material function and z appearing in the equation is the intrinsic time whose differential increment is given by

$$dz = \frac{d\zeta}{f(\zeta)} \tag{6.130}$$

where $f(\zeta)$ is a nonnegative function called the intrinsic time scale with $f(0) = 1$. ζ is the intrinsic time measure defined by

$$d\zeta = (d\boldsymbol{\epsilon}^{\mathrm{p}} : d\boldsymbol{\epsilon}^{\mathrm{p}})^{1/2} \tag{6.131}$$

It should be noted that $d\boldsymbol{\epsilon}^{\mathrm{p}}$ is deviatoric in nature, since the volume change is assumed to be elastic. It has been proved by Valanis (1971, 1980) that the kernel can be written in the form

$$\rho(z) = \rho_0\delta(z) + \rho_1(z) \tag{6.132}$$

where $\delta(z)$ is a Dirac delta function, ρ_0 is a material constant, and $\rho_1(z)$ is a nonsingular function. Substituting Eq. (6.132) into Eq. (6.129) yields

$$\mathbf{S} = S_{\mathrm{y}}^0\frac{d\boldsymbol{\epsilon}^{\mathrm{p}}}{dz} + 2G\int_0^z \rho_z(z - z')\frac{d\boldsymbol{\epsilon}^{\mathrm{p}}}{dz'}dz' \tag{6.133}$$

where

$$S_y^0 = 2G\rho_0 \tag{6.134}$$

whose physical meaning will be explained later. Let

$$\boldsymbol{\alpha}(z) = 2G\int_0^z \rho_1(z - z')\frac{d\boldsymbol{\epsilon}^{\mathrm{p}}}{dz'}dz' \tag{6.135}$$

Equation (6.133) can be rewritten as

$$\mathbf{S} = S_y^0\frac{d\boldsymbol{\epsilon}^{\mathrm{p}}}{dz} + \alpha(z) \tag{6.136}$$

or

$$\mathbf{S} - \boldsymbol{\alpha} = S_y^0\frac{d\boldsymbol{\epsilon}^{\mathrm{p}}}{d\zeta}f(\zeta) \tag{6.137}$$

From this expression two cases can be derived:

1. $f(\zeta) = 1$. According to Eqs. (6.131) and (6.137),

$$(\mathbf{S} - \boldsymbol{\alpha}):(\mathbf{S} - \boldsymbol{\alpha}) = \left(S_y^0\right)^2 \tag{6.138}$$

$$d\boldsymbol{\epsilon}^{\mathrm{p}} = \frac{1}{S_y^0}(\mathbf{S} - \boldsymbol{\alpha})d\zeta \tag{6.139}$$

   Equation (6.138) corresponds to the von Mises yield function, and Eq. (6.139) shows that the plastic strain increment $d\epsilon^{\mathrm{p}}$ is normal to the yield surface. Since the radius of the yield surface, S_y^0 is a constant (which can be related to the yield stress in the uniaxial loading, as will be explained later), this is a plasticity model with purely kinematic hardening. The constitutive equation for the evolution of $\boldsymbol{\alpha}$, which corresponds to the back stress or kinematic hardening variable in the classical plasticity theories and represents the center of the yield surface, is given by Eq. (6.135) and, as can be seen from it, is determined by the material function $\rho_1(z)$.

2. $f(\zeta) \neq 1$ but increases monotonically. In this case Eqs. (6.138) and (6.139) are replaced by

$$(\mathbf{S} - \boldsymbol{\alpha}):(\mathbf{S} - \boldsymbol{\alpha}) = \left(S_y^0\right)^2 f^2(\zeta) \tag{6.140}$$

$$d\boldsymbol{\epsilon}^{\mathrm{p}} = \frac{1}{S_y^0 f(\zeta)}(\mathbf{S} - \boldsymbol{\alpha})d\zeta \tag{6.141}$$

Thus the yield surface, defined by Eq. (6.140), can translate and expand at the same time. The kinematic hardening variable $\boldsymbol{\alpha}$ is still given by Eq. (6.135), depending on the particular form of the kernel $\rho_1(z)$, and the isotropic hardening is characterized by the intrinsic time scale function $f(\zeta)$, as can be seen from the right side of Eq. (6.140). Again, the plastic strain increment $d\boldsymbol{\epsilon}^{\mathrm{p}}$, defined by Eq. (6.141), is normal to the yield surface. When $f(\zeta) \neq 1$, the endochronic theory corresponds to combined isotropic and kinematic hardening. Various hardening rules can be deduced from the general equations of the theory, Eqs. (6.129)–(6.135), by choosing different forms of the intrinsic time scale function $f(\zeta)$ and the kernel $\rho_1(z)$.

It should be pointed out that the purely isotropic hardening model can be retrieved by choosing the kernel $\rho_1(z)$ in Eq. (6.135) so that $\boldsymbol{\alpha} = 0$ at all times. From Eqs. (6.140) and (6.141), we obtain

$$\mathbf{S}:\mathbf{S} = \left(S_{\mathrm{y}}^{0}\right)^{2} f^{2}(\zeta) \tag{6.142}$$

$$d\boldsymbol{\epsilon}^{\mathrm{p}} = \frac{1}{S_{\mathrm{y}}^{0} f(\zeta)} \mathbf{S}\, d\zeta \tag{6.143}$$

These are the equations for isotropic hardening materials. The strain or work softening response can be easily modeled by assuming that $f(\zeta)$ is a monotonic decreasing function of ζ. Note that here the yield surface is a derived concept, not a priori assumed one, and the regions of the material behavior can be characterized as

$$\begin{aligned} (\mathbf{S}-\boldsymbol{\alpha}):(\mathbf{S}-\boldsymbol{\alpha}) &< \left(S_{\mathrm{y}}^{0}\right)^{2} f^{2}(\zeta) \qquad \text{(Elastic response)} \\ (\mathbf{S}-\boldsymbol{\alpha}):(\mathbf{S}-\boldsymbol{\alpha}) &= \left(S_{\mathrm{y}}^{0}\right)^{2} f^{2}(\zeta) \qquad \text{(Elastic response)} \end{aligned} \tag{6.144}$$

but

$$(\mathbf{S}-\boldsymbol{\alpha}):d\boldsymbol{\epsilon}^{\mathrm{p}} \leq 0 \tag{6.145}$$

$$(\mathbf{S}-\boldsymbol{\alpha}):(\mathbf{S}-\boldsymbol{\alpha}) = \left(S_{\mathrm{y}}^{0}\right)^{2} f^{2}(\zeta) \qquad \text{(Plastic response)}$$

and

$$(\mathbf{S}-\boldsymbol{\alpha}):d\boldsymbol{\epsilon}^{\mathrm{p}} > 0 \tag{6.146}$$

6.5.2 Incremental Form of the Theory

The general equations given above are in the integral form. In some applications, especially in implementing the equations into computational codes, the incremental or rate form is more convenient to use. The incremental form of the endochronic theory as derived from the work of Valanis (1980) is discussed below.

According to the fundamental assumption that

$$d\boldsymbol{\epsilon}' = d\boldsymbol{\epsilon}^{e'} + d\boldsymbol{\epsilon}^{p} \tag{6.147}$$

and the constancy of the elastic modulus, it follows that

$$d\mathbf{S} = 2\mu_0 d\boldsymbol{\epsilon}^{e'} = 2\mu_0(d\boldsymbol{\epsilon}' - d\boldsymbol{\epsilon}^{p}) = 2\mu_0 d\boldsymbol{\epsilon}' - 2\mu_0 d\boldsymbol{\epsilon}^{p} \tag{6.148}$$

where $d\boldsymbol{\epsilon}'$ and $d\boldsymbol{\epsilon}^{e'}$ are the deviatoric parts of $d\boldsymbol{\epsilon}$ and $d\boldsymbol{\epsilon}^{e}$, respectively, and $d\boldsymbol{\epsilon}^{p}$ is deviatoric in nature. Recall that $d\boldsymbol{\epsilon}^{p}$ is given by Eq. (6.141) in general. The equation above is rewritten as

$$d\mathbf{S} = 2\mu_0 d\boldsymbol{\epsilon}' - \frac{2\mu_0}{S_y^0 f(\zeta)}(\mathbf{S} - \boldsymbol{\alpha})\,d\zeta \tag{6.149}$$

We want to find a relation between the increment of stress $\boldsymbol{\sigma}$ or $\mathbf{S}$ and that of $\boldsymbol{\epsilon}$. But first we must eliminate $d\zeta$ in the equation. We do this by the following procedure: We differentiate both sides of Eq. (6.140) to get the consistency condition

$$(\mathbf{S} - \boldsymbol{\alpha}) : d\mathbf{S} - (\mathbf{S} - \boldsymbol{\alpha}) : d\boldsymbol{\alpha} = \left(S_y^0\right)^2 f(\zeta) f'(\zeta)\,d\zeta \tag{6.150}$$

where $f'(\zeta) = df'(\zeta)/d\zeta$. Then we obtain the differential equation for the evolution of $\boldsymbol{\alpha}$ by differentiating both sides of Eq. (6.135) with respect to the intrinsic time z:

$$\frac{d\boldsymbol{\alpha}}{dz} = 2\mu_0 \rho_1(0)\frac{d\boldsymbol{\epsilon}^{p}}{dz} + 2\mu_0 \int_0^z \frac{d\rho_1(z - z')}{dz}\frac{d\boldsymbol{\epsilon}^{p}}{dz'}dz' \tag{6.151}$$

Using Eq. (6.130), this can be modified to

$$d\boldsymbol{\alpha} = 2\mu_0 \rho_1(0)\,d\boldsymbol{\epsilon}^{p} + \frac{2\mu_0 \mathbf{h}}{f(\zeta)}d\zeta \tag{6.152}$$

where

$$\mathbf{h} = \int_0^z \frac{d\rho_1(z - z')}{dz'}\frac{d\boldsymbol{\epsilon}}{dz'}dz' \tag{6.153}$$

Substituting this into Eq. (6.150) results in

$$(\mathbf{S} - \boldsymbol{\alpha}) : d\mathbf{S} - 2\mu_0 \rho_1(0)(\mathbf{S} - \boldsymbol{\alpha}) : d\boldsymbol{\epsilon}' + \rho_1(0)(\mathbf{S} - \boldsymbol{\alpha}) : d\mathbf{S}$$
$$- \frac{2\mu_0(\mathbf{S} - \boldsymbol{\alpha}) : \mathbf{h}}{f(\zeta)}d\zeta = \left(S_y^0\right)^2 f(\zeta) f'(\zeta)\,d\zeta \tag{6.154}$$

Now we can derive $d\zeta$ from this equation and substitute it into Eq. (6.149) to obtain the relation between $d\mathbf{S}$ and $d\boldsymbol{\epsilon}$. It is easier and more straightforward, however, to introduce Eq. (6.149) into Eq. (6.154) and eliminate $d\zeta$ on both sides and rearrange the resulting equation:

$$d\mathbf{S} = 2\mu_0 d\boldsymbol{\epsilon}' - \frac{2\mu_0[(\mathbf{S}-\boldsymbol{\alpha}):d\boldsymbol{\epsilon}']}{C\left(S_y^0\right)^2 f^2(\zeta)}(\mathbf{S}-\boldsymbol{\alpha}) \tag{6.155}$$

where C is defined by

$$C = 1 + \rho_1(0) + \frac{S_y^0 f'(\zeta)}{2\mu_0} + \frac{(\mathbf{S}-\boldsymbol{\alpha}):\mathbf{h}}{S_y^0 f^2(\zeta)} \tag{6.156}$$

Equations (6.155) and (6.152) are the incremental forms of the endochronic theory. Equation (6.155) can be put into a more compact form that is entirely analogous in structure to that of the classical theory of plasticity. To this end, we compare Eq. (6.155) with Eq. (6.149) to obtain

$$d\zeta = \frac{(\mathbf{S}-\boldsymbol{\alpha}):d\boldsymbol{\epsilon}'}{CS_y^0 f(\zeta)} = \frac{1}{C}\mathbf{n}:d\boldsymbol{\epsilon}' = \frac{1}{C}\mathbf{n}:d\boldsymbol{\epsilon} \tag{6.157}$$

since $\mathbf{n}$ is deviatoric and defined by

$$\mathbf{n} = \frac{\mathbf{S}-\boldsymbol{\alpha}}{S_y^0 f(\zeta)} \tag{6.158}$$

Obviously this is the outward unit normal to the yield surface. Using this notation, we give the flow rule of Eq. (6.141) by endochronic theory as

$$d\boldsymbol{\epsilon}^{\mathrm{p}} = \frac{1}{C}(\mathbf{n}:d\boldsymbol{\epsilon})\mathbf{n} \tag{6.159}$$

and recast Eq. (6.155) in the form

$$d\mathbf{S} = 2\mu_0\left[d\boldsymbol{\epsilon} - \frac{1}{C}(\mathbf{n}:d\boldsymbol{\epsilon})\mathbf{n}\right] \tag{6.160}$$

For easy reference the constitutive equation for α is rewritten below in a slightly different fashion:

$$d\boldsymbol{\alpha} = 2\mu_0\rho_1(0)d\boldsymbol{\epsilon}^{\mathrm{p}} + \frac{2\mu_0\mathbf{h}}{f(\zeta)}(d\boldsymbol{\epsilon}^{\mathrm{p}}:d\boldsymbol{\epsilon}^{\mathrm{p}})^{1/2} \tag{6.161}$$

Equations (6.159)–(6.161) are the incremental forms of the endochronic theory that are very useful in applications to the numerical algorithm, such as the finite element method. They apply in the process of plastic flow in that condition (6.146) holds. In the process of the elastic deformation, the linear Hookes law holds, and the volume change is governed by the linear elastic rule.

6.5.3 Kinematic and Isotropic Hardening

As we mentioned previously, the endochronic theory is very general and versatile, and it contains various isotropic and kinematic hardening rules for special cases, depending on the choice of the time scale function $f(\zeta)$ and the kernel $\rho_1(z)$. These features of the theory are explored below.

We consider the kinematic hardening rule first. Valanis (1971, 1980) proved that the general form for the kernel is given by the series

$$\boldsymbol{\rho}_1(z) = \sum_{i=1}^{\infty} \boldsymbol{\rho}_{1i} e^{-\xi_i z} \tag{6.162}$$

where $\rho_{1i}\xi_i$, $i = 1, 2, \ldots,$ are material constants that have to be determined experimentally. Theoretically the number of terms chosen should be infinite in order to obtain a perfect description of the real material response in the plastic flow. Obviously this is impossible in real practice, and only a finite number of terms can be used in the practical modeling. In most applications, even those including cyclic loading conditions, as few as two terms may be sufficient to give fairly satisfactory results. For example,

$$\boldsymbol{\rho}_1(z) = C_0 + C_1 e^{-\xi z} \tag{6.163}$$

involves only three material constants, C_0, C_1, and ξ. This greatly simplifies the work to determine the unknown material constants. This and the physical meaning of C_0, C_1, and α will be discussed later. To retain the generality of the derived equations, Eq. (6.162) is used and substituted into Eqs. (6.135) and (6.153) to get

$$\boldsymbol{\alpha} = \sum_{i=1}^{\infty} 2\mu_0 \int_0^z \rho_{1i} e^{-\xi_i(z-z')} \frac{d\boldsymbol{\epsilon}^{\mathrm{p}}}{dz'} dz' = \sum_{i=1}^{\infty} \boldsymbol{\alpha}^{(i)} \tag{6.164}$$

$$\mathbf{h} = \sum_{i=1}^{\infty} \left(-\frac{\xi_i}{2\mu_0} \right) \boldsymbol{\alpha}^{(i)} = \sum_{i=1}^{\infty} \mathbf{h}^{(i)} \tag{6.165}$$

where

$$\boldsymbol{\alpha}^{(i)} = 2\mu_0 \int_0^z \rho_{1i} e^{-\xi_i(z-z')} \frac{d\boldsymbol{\epsilon}^{\mathrm{p}}}{dz'} dz' \tag{6.166}$$

$$\mathbf{h}^{(i)} = -\frac{\xi_i}{2\mu_0} \boldsymbol{\alpha}^{(i)} \qquad \text{(No sum over } i\text{)} \tag{6.167}$$

As pointed out by Watanabe et al. (1986), $\alpha^{(i)}$ can be considered as a secondary internal variable or kinematic hardening variable. Note that

$$\rho_1(0) = \sum_{i=1}^{\infty} \rho_{1i} \tag{6.168}$$

By substituting Eq. (6.162) into Eq. (6.152) and making use of Eqs. (6.164)–(6.168), we obtain the increment $d\boldsymbol{\alpha}$ as

$$\begin{aligned} d\boldsymbol{\alpha} &= \sum_{i=1}^{\infty} d\boldsymbol{\alpha}^{(i)} \\ &= \sum_{i=1}^{\infty} \left(2\mu_0 \rho_{1i} d\boldsymbol{\epsilon}^{\mathrm{p}} - \frac{\xi_i}{f(\zeta)} \boldsymbol{\alpha}^{(i)} d\zeta \right) \\ &= \sum_{i=1}^{\infty} \left[2\mu_0 \rho_{1i} d\boldsymbol{\epsilon}^{\mathrm{p}} - \frac{\xi_i}{f(\zeta)} \boldsymbol{\alpha}^{(i)} \left(d\boldsymbol{\epsilon}^{\mathrm{p}} : d\boldsymbol{\epsilon}^{\mathrm{p}} \right)^{1/2} \right] \end{aligned} \tag{6.169}$$

where the definition of $d\boldsymbol{\alpha}^{(i)}$ is obvious. The kinematic hardening rule defined by this equation is very general and can retrieve various existing rules, provided that $\boldsymbol{\rho}_{1i}$, $\boldsymbol{\alpha}_i$, and $f(\zeta)$ are chosen correctly. For example, it is easy to verify that Prager's linear kinematic rule is obtained if $\rho_1(\zeta) = \rho$ is a constant. Indeed it follows directly from Eq. (6.135) that

$$\boldsymbol{\alpha} = 2\mu_0 \rho \boldsymbol{\epsilon}^{\mathrm{p}} = c \boldsymbol{\epsilon}^{\mathrm{p}} \tag{6.170}$$

On the other hand, if only one exponential term is chosen for $\rho_1(\zeta)$, according to Eq. (6.169), $d\boldsymbol{\alpha}$ will become

$$d\boldsymbol{\alpha} = 2\mu_0 \rho_{1i} d\boldsymbol{\epsilon}^{\mathrm{p}} - \frac{\xi_1 \boldsymbol{\alpha}}{f(\zeta)} \sqrt{\frac{3}{2}}\, d\epsilon_{\mathrm{e}}^{\mathrm{p}} \tag{6.171}$$

where Eq. (6.81) is used as the definition for $d\epsilon_{\mathrm{e}}^{\mathrm{p}}$. Comparing this equation with Eq. (6.104), we see immediately that the equations are equivalent, although the coefficient $\xi_1/f(\zeta)(\sqrt{3/2})$ of the recall term here is a function of the intrinsic time measure ζ, while the coefficient γ in Eq. (6.104) is a constant. More complicated kinematic hardening rules can be derived from Eq. (6.169). The interested reader may refer to the work by Valanis (1980, 1984) and Watanabe et al. (1986) for full discussions.

We examine next the resulting isotropic hardening rule of the endochronic theory. As we pointed out earlier, in the endochronic theory the size of the derived yield surface is represented by

$$S_{\mathrm{y}} = S_{\mathrm{y}}^0 f(\zeta) \tag{6.172}$$

so the isotropic hardening is signified by the intrinsic time scale function $f(\zeta)$. Two commonly used forms of $f(\zeta)$ with $f(0) = 1$ are

$$f(\zeta) = 1 + \beta\zeta \tag{6.173}$$

which is called the linear form, and

$$f(\zeta) = a + (1 - a)e^{-b\zeta} \tag{6.174}$$

which is called the exponential or saturated form, where β, a, and b are material constants.

If Eq. (6.173) is used, the constitutive equation for the isotropic hardening variable $\mathbf{S}_y$ is given by

$$\begin{aligned} dS_y &= S_y^0 \beta d\zeta \\ &= \frac{\sqrt{3}}{2} S_y^0 \beta d\epsilon_e^p \end{aligned} \tag{6.175}$$

which signifies a linear variation for S_y, since both S_y^0 and β are constants.

On the other hand, if the saturated form (6.174) is used, the following hardening rule for S_y is derived:

$$\begin{aligned} dS_y &= b\left(S_y^0 a - S_y\right) d\zeta \\ &= \sqrt{\frac{3}{2}}\, b\left(S_{ys} - S_y\right) d\epsilon_e^p \end{aligned} \tag{6.176}$$

where $S_{ys} = S_y^0 a$ is the stabilized value, or the saturated value, of S_y. This equation is equivalent to Eq. (6.126) in that the stabilization of the isotropic hardening, which is commonly observed under cyclic loading conditions, can be described by endochronic theory if Eq. (6.174) is used.

6.5.4 Uniaxial Case, Determination of Material Constants

To determine material constants, the general constitutive equations developed previously are reduced to the uniaxial case, and then the experimental results for the uniaxial loading are used and compared with the derived formula. The following presentation is based on work by Watanabe and Atluri (1986).

We re-list the following equations for easy reference:

$$(\mathbf{S} - \boldsymbol{\alpha}):(\mathbf{S} - \boldsymbol{\alpha}) = \left(S_y^0\right)^2 f^2(\zeta) \tag{6.177}$$

$$\mathbf{S} = S_y^0 \frac{d\boldsymbol{\epsilon}^p}{dz} + \boldsymbol{\alpha} \tag{6.178}$$

$$\boldsymbol{\alpha} = 2G\int_0^z \rho_1(z - z')\frac{d\boldsymbol{\epsilon}^p}{dz'}dz' \tag{6.179}$$

$$dz = \frac{d\zeta}{f(\zeta)}, \quad d\zeta = \left(d\boldsymbol{\epsilon}^p : d\boldsymbol{\epsilon}^p\right)^{1/2} = \sqrt{\frac{3}{2}}\, d\epsilon_e^p \tag{6.180}$$

These equations are used along with the following particular forms of $f(\zeta)$ and $\rho_1(z)$:

$$f(\zeta) = 1 + \beta\zeta \tag{6.181}$$

$$\rho_1(z) = \frac{E_1}{E}e^{-\xi_1 z} + \frac{E_2}{E} \tag{6.182}$$

where β, α_1, E_1, and E_2 are material constants, and E is the Young's modulus of the material introduced to nondimensionalize E_1 and E_2. Equations (6.177)–(6.182) are a complete set of equations for the endochronic theory. As they indicate, the material response predicted by this model is governed by five material constants: S_1^0, β, ξ_1, E_1, and E_2. G and E appearing in the model are shear, and Young's moduli are known a priori.

For uniaxial loading in one direction, the components of $\boldsymbol{\sigma}$ and $d\boldsymbol{\epsilon}^p$ satisfy the following relations:

$$\sigma_{11} \neq 0, \quad \sigma_{ij} = 0 \tag{6.183}$$

Otherwise,

$$d\epsilon_{22}^p = d\epsilon_{33}^p = -\tfrac{1}{2}d\epsilon_{11}^p \quad d\epsilon_{ij}^p = 0 \qquad (\text{for } i \neq j) \tag{6.184}$$

Then it is easy to calculate that

$$d\zeta = \sqrt{\frac{3}{2}}\,|d\epsilon_{11}^p| \tag{6.185}$$

$$\zeta = \sqrt{\frac{3}{2}}\int |d\epsilon_{11}^p| \tag{6.186}$$

The yield function (6.177) reduces to

$$|\sigma_{11} - \alpha_{11}| = \sqrt{\frac{3}{2}}\, S_y^0 f(\zeta) \tag{6.187}$$

From this equation and $f(\zeta) = 1$, parameter $\sqrt{3/2}\, S_y^0$ can be identified as the initial yield stress σ_y^0 in uniaxial tension:

$$S_y^0 = \sqrt{\frac{2}{3}}\, \sigma_y^0 \tag{6.188}$$

which is the initial yield stress in pure shear.

Equation (6.178) is reduced for the uniaxial tension case to

$$\sigma_{11} = \frac{3}{2} S_y^0 \frac{d\epsilon_{11}^p}{dz} + 3G \int_0^z \rho_1(z - z') \frac{d\epsilon_{11}^p}{dz'} dz' \tag{6.189}$$

Note from Eqs. (6.180), (6.181), and (6.185) that for uniaxial tension

$$\frac{d\epsilon_{11}^p}{dz} = f(\zeta) \frac{d\epsilon_{11}^p}{|d\epsilon_{11}^p|} \sqrt{\frac{2}{3}} = \sqrt{\frac{2}{3}}\, f(\zeta) \tag{6.190}$$

Substituting Eqs. (6.181) and (6.182) and the result above into Eq. (6.189), we can calculate the expression for stress:

$$\sigma_{11} = \sigma_y^0 \left(1 + \beta \sqrt{\frac{3}{2}}\, \epsilon_{11}^p\right) + \sqrt{\frac{2}{3}} \left(\frac{3G}{E}\right) \frac{E_1}{\beta n_1} \left(1 + \beta \sqrt{\frac{3}{2}}\, \epsilon_{11}^p\right) \left\{1 - \left(1 + \beta \sqrt{\frac{3}{2}}\, \epsilon_{11}^p\right)^{-n_1}\right\} + \left(3 \frac{G}{E}\right) E^2 \epsilon_{11}^p \tag{6.191}$$

where

$$n_1 = 1 + \frac{\xi_1}{\beta} \tag{6.192}$$

This equation is valid if $\epsilon_{11}^p > 0$; otherwise, the elastic stress-strain relation should be used. This equation of the stress-plastic strain curve is highly nonlinear and has four more material constants besides the already determined initial yield stress σ_y^0: E_1, E_2, β, and n_1 (or ξ_1), which make the model fairly flexible and in agreement with experimental results.

Equation (6.191) shows that as ϵ_{11}^p increases, the term with power $-n_1$ may be omitted, and the stress-strain curve described by the asymptotic value of stress, which is denoted σ^{∞}, as

$$\sigma_{11}^{\infty} = \sigma_y^0\left(1 + \beta\sqrt{\frac{3}{2}}\,\epsilon_{11}^p\right) + \sqrt{\frac{2}{3}}\left(\frac{E_1}{\beta n_1}\right)$$

$$\left(1 + \beta\sqrt{\frac{3}{2}}\,\epsilon_{11}^p\right)\left(\frac{3G}{E}\right) + \left(\frac{3G}{E}\right)E_2\epsilon_{11}^p \tag{6.193}$$

This is an equation of a straight line.

To determine β, n_1, E_1, and E_2, it is helpful to derive the following two equations:

$$\frac{d\sigma_{11}}{d\epsilon_{11}^p} = \sqrt{\frac{3}{2}}\,\sigma_y^0\beta + \left(\frac{3G}{E}\right)\frac{E_1}{n_1}$$

$$\left\{1 + (n_1 - 1)\left(1 + \beta\sqrt{\frac{3}{2}}\,\epsilon_{11}^p\right)^{-n_1}\right\} + \left(\frac{3G}{E}\right)E_2 \tag{6.194}$$

and

$$\frac{d\sigma_{11}^{\infty}}{d\epsilon_{11}^p} = \sqrt{\frac{3}{2}}\,\sigma_y^0\beta + \left(\frac{3G}{E}\right)\frac{E_1}{n_1} + \left(\frac{3G}{E}\right)E_2 \tag{6.195}$$

From the experimental stress-plastic strain curve of uniaxial tension, it is possible to define the following quantities:

1. E_0^p is the initial tangential modulus of the stress-plastic strain curve.
2. E_{∞}^p is the tangential modulus of the asymptotic stress.
3. σ_0^{∞} is the stress measured at the interception of σ^{∞} with the stress axis.

These quantities are shown in Fig. 6.21.

According to the definitions of these quantities and Eqs. (6.193), (6.194), and (6.195), the following equations are derived and used to determine the material constants:

$$E_0^p = \left.\frac{d\sigma_{11}}{d\epsilon_{11}^p}\right|_{\epsilon_{11}^p=0} = \sqrt{\frac{3}{2}}\,\sigma_y^0\beta + \left(\frac{3G}{E}\right)E_1 + \left(\frac{3G}{E}\right)E_2 \tag{6.196}$$

$$E_{\infty}^p = \left.\frac{d\sigma_{11}^{\infty}}{d\epsilon_{11}^p}\right|_{\epsilon_{11}^p=0} = \sqrt{\frac{3}{2}}\,\sigma_y^0\beta + \left(\frac{3G}{E}\right)\frac{E_1}{n_1} + \left(\frac{3G}{E}\right)E_2 \tag{6.197}$$

$$\sigma_0^{\infty} = \sigma_{11}^{\infty}|_{\epsilon_{11}^p=0} = \sigma_y^0 + \sqrt{\frac{2}{3}}\left(\frac{3G}{E}\right)\frac{E_1}{\beta n_1} \tag{6.198}$$

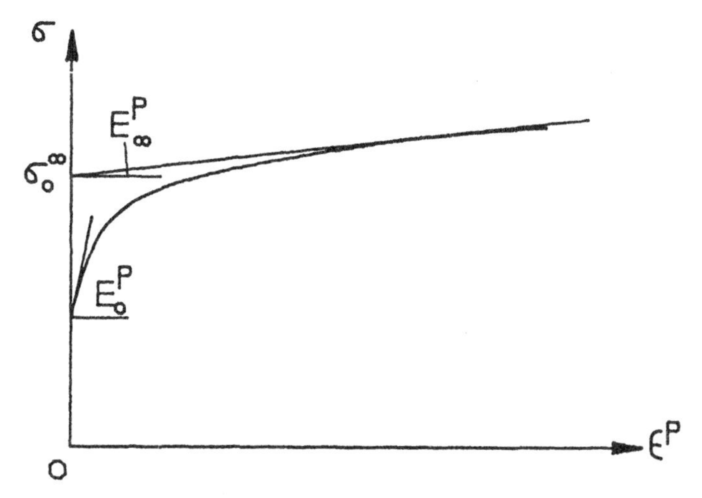

Fig. 6.21 Definitions of E_0^p and σ_0^∞

Note that there are four constants, β, n_1, E_1, and E_2, to be determined, and it is obvious that the three equations above are insufficient for this purpose. One more equation is needed. Recall that the function $f(\zeta)$ describes the isotropic expansion (or contraction) of the yield surface. It is natural to relate the constant β appearing in $f(\zeta) = 1 + \beta\zeta$ to this effect. As discussed before, isotropic hardening can be separated from kinematic hardening only when the load reversal is considered. Now look at the loading and unloading process shown in Fig. 6.22. The load is increased until point A is reached, with stress σ_{11}^A and plastic strain ϵ_A^p. The load is then reversed, and the initial plastic flow is detected to occur at point B, with stress σ_{11}^B and plastic strain ϵ_A^p. σ_{11}^A and σ_{11}^B are calculated using Eq. (6.189) as

$$\sigma_{11}^A = \sqrt{\frac{3}{2}}\,\sigma_y^0\left(\frac{d\epsilon_{11}^p}{dz}\right)_A + 3G\int_0^{z_A}\rho_1(z-z')\frac{d\epsilon_{11}^p}{dz'}dz' \qquad (6.199)$$

$$\sigma_{11}^B = \sqrt{\frac{3}{2}}\,\sigma_y^0\left(\frac{d\epsilon_{11}^p}{dz}\right)_B + 3G\int_0^{z_B}\rho_1(z-z')\frac{d\epsilon_{11}^p}{dz'}dz' \qquad (6.200)$$

Note that $z_A = z_B$, since they correspond to the same plastic strain level ϵ_A^p, and that

$$\left(\frac{d\epsilon_{11}^p}{dz}\right)_A = \sqrt{\frac{2}{3}}\,\frac{d\epsilon_{11}^p}{|d\epsilon_{11}^p|}f(\zeta_A) = \sqrt{\frac{2}{3}}\left(1 + \beta\sqrt{\frac{3}{2}}\,\epsilon_A^p\right) \qquad (6.201)$$

$$\left(\frac{d\epsilon_{11}^p}{dz}\right)_B = \sqrt{\frac{2}{3}}\,\frac{d\epsilon_{11}^p}{|d\epsilon_{11}^p|}f(\zeta_B) = -\sqrt{\frac{2}{3}}\left(1 + \beta\sqrt{\frac{3}{2}}\,\epsilon_A^p\right) \qquad (6.202)$$

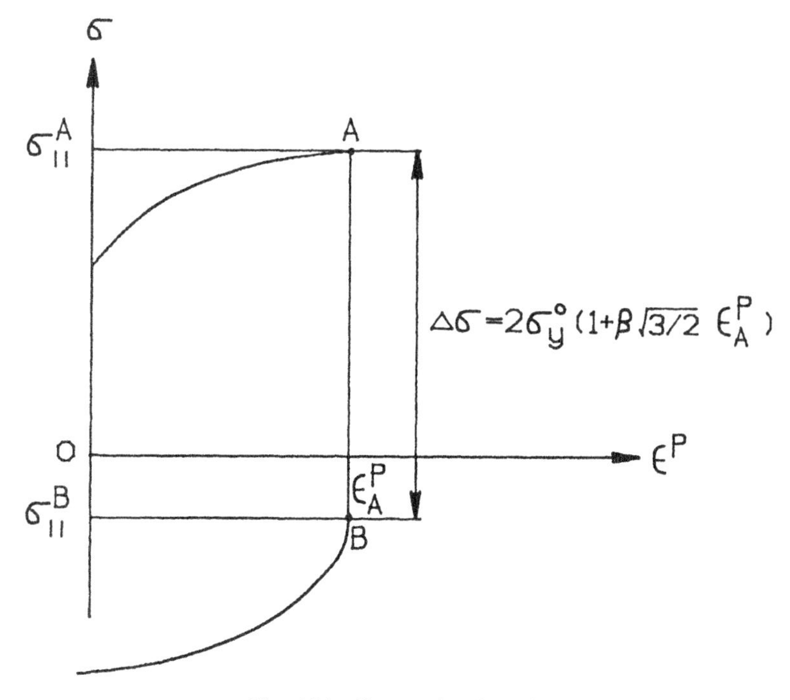

Fig. 6.22 Determination of β

Using these results in Eqs. (6.199) and (6.200), we can calculate the stress drop $\Delta\sigma$ as

$$\Delta\sigma = 2\sigma_y^0\left(1 + \beta\sqrt{\frac{3}{2}}\,\epsilon_A^p\right) \tag{6.203}$$

and this serves to determine β. Once β is determined from this equation and substituted into Eqs. (6.196), (6.197), and (6.198), the other three constants n_1, E_1, and E_2 can be calculated easily. The results are as follows:

$$\beta = \frac{\Delta\sigma - 2\sigma_y^0}{\sqrt{6}\,\sigma_y^0\epsilon_A^p} \tag{6.204}$$

$$n_1 = 1 + \sqrt{\frac{2}{3}}\left(\frac{\epsilon_0^p - \epsilon_\infty^p}{\sigma_0^\infty - \sigma_y^0}\right)\frac{1}{\beta} \tag{6.205}$$

$$E_1 = \left(\frac{E}{3G}\right)\left[(E_0^p - E_\infty^p) + \sqrt{\frac{3}{2}}\left(\sigma_0^\infty - \sigma_y^0\right)\beta\right] \tag{6.206}$$

$$E_2 = \left(\frac{E}{3G}\right)\left(E_\infty^p - \sqrt{\frac{3}{2}}\,\sigma_0^\infty\beta\right) \tag{6.207}$$

If Eq. (6.174) is used instead of Eq. (6.173), the constants to be determined will be the following six: S_y^0, ξ_1, E_1, E_2, b, and a. They can be calculated in a way similar to this procedure. A detailed discussion may be found in the work by Wu and Yip (1981).

It can be seen from Fig. 6.12, taken from Khan and Wang (1988), that the endochronic theory gives a poorer prediction for this particular case as compared with the Mroz model. Note that this study assumes the pure kinematic hardening model, the functions $f(\zeta)$ and $\rho_1(z)$ are Eqs. (6.174) and (6.163), respectively. If more terms for $\rho_1(z)$ are used, then better results are anticipated.

REFERENCES

Armstrong, P. J., and C. O. Frederick. 1966. A mathematical representation of the multiaxial Bauschinger effect. *G.E.G.B. Report RD/B/N* 731.

Chaboche, J. L. 1986. Time independent constitutive theories for cyclic plasticity. *Int. J. Plast.* **2**(2):149.

Chaboche, J. L., K. Dang-Van, and G. Cordier. 1979. Modelization of the strain memory effect on the cyclic hardening of 316 stainless steel. SMIRT-5, Division L., Berlin.

Chaboche, J. L. 1977. Viscoplastic constitutive equations for the description of cyclic and anisotropic behavior of metals. *Bull. de l' Acad. Polonaaise des Sciences, Sévie Sc. et Techn.* **25**(1):33.

Dafalias, Y. F. 1984. Modelling cyclic plasticity: Simplicity versus sophistication. *Mech. Engineering Materials*, ed. C. S. Desai and R. H. Gallagher. New York: Wiley, p. 153.

Dafalias, Y. F., and E. P. Popov. 1976. Plastic internal variables formalism of cyclic plasticity. *J. Appl. Mech.* **98**:645.

Dafalias, Y. F. 1975a. On cyclic and anisotropic plasticity. I. A general model including material behavior under stress reversals. II. Anisotropic hardening for initial orthotropic materials. Ph.D. dissertation. Department of Civil Engineering, University of California, Berkeley.

Dafalias, Y. F., and E. P. Popov. 1975b. A model of nonlinearly hardening materials for complex loading. *Acta Mech.* **21**:173.

Drucker, D. C., and L. Palgen. 1981. On stress-strain relations suitable for cyclic and other loadings. *J. Appl. Mech.* **48**:479.

Haythornthwaite, R. M. 1968. A more rational approach to strain-hardening data. *Engineering Plasticity*. Cambridge: Cambridge University Press.

Khan, A. S., and X. Wang. 1988. On non-proportional infinitermimal plastic deformation after finite plastic prestraining and partial unloading. *J. Mech. Phys. Solids* **36**:519.

Lee, D., and F. Zavenl, Jr. 1978. A generalized strain rate dependent constitutive equation for anisotropic metals. *Acta Metall.* **29**:1771.

Mroz, Z., V. A. Norris, and O. C. Zienkiewicz. 1979. Application of an anisotropic hardening model in the analyzers of elastoplastic deformation of souls. *Geotechnique* **29**:1.

Mroz, Z. 1972. A description of work-hardening of metals with application to variable loading. *Proc. Symp. Foundations of Plasticity*, ed. S. Sawczuk. Alphen aan den Rijn, The Netherlands: Noordhoff, p. 551.

Mroz, Z. 1969. An attempt to describe the behavior of metals under cyclic loads using a more general work hardening model. *Acta Mech.* **7**:199.

Mroz, Z. 1967. On the description of anisotropic work-hardening. *J. Mech. Phys. Solids* **15**:163.

Prevost, J. H. 1978. Plasticity theory for soil stress-strain behavior. *ASCE J. Eng. Mech.* **104**:1177.

Valanis, K. C. 1984. Continuum foundations of endochronic plasticity. *J. Eng. Mater. Tech.* **106**:367.

Valanis, K. C. 1980. Fundamental consequences of a new intrinsic time measure plasticity as a limit of the endochronic theory. *Arch. Mech.* **32**:171.

Valanis, K. E. 1971. A theory of viscoplasty without a yield surface. *Arch. Mech.* **23**:517.

Watanake, O., and S. N. Atluri. 1986. Internal time, general internal variable, and multiyield-surface theories of plasticity and creep: A unification of concepts. *Int. J. Plast.* **2**(1):37.

Wu, H. C., and M. C. Yip. 1981. Endochronic description of cyclic hardening behavior for metallic materials. *J. Eng. Mater. Tech.* **103**:212.

7

FINITE PLASTIC DEFORMATION

A detailed and consistent analysis of finite deformation is necessary for many engineering applications, such as ductile fracture, metal forming, or problems involving strain localization. Advancements in the computational capabilities of computers has made it possible to implement more complex and realistic constitutive models. As a result there has been considerable attention given to constitutive modeling of engineering materials for finite elastic-plastic deformation. In the mechanics community several somewhat controversial plasticity theories have been proposed, and the macroscopic constitutive formulation for finite strain has been vigorously debated.

In this chapter we consider the basic principles involved in formulating a plasticity theory for finite deformation. Our intention is to point out the differences in constitutive modeling between finite and infinitesimal deformations and to examine inherent difficulties. We discuss briefly a number of selected models for finite plasticity. At the end of this chapter we explore some additional interesting features of constitutive modeling in analyzing the problem of simple shear, and also illustrate one of the main controversies in this field: the oscillation of the shear stress with the increase in shear strain, which is a physically unacceptable phenomenon.

7.1 BASIC CONSIDERATIONS

The differences in constitutive modeling of finite elastic-plastic and infinitesimal elastic-plastic deformation can be expressed in the general form of plasticity theory as

$$F = f(\boldsymbol{\sigma}, \boldsymbol{\alpha}) + \kappa \qquad \text{(Yield function)} \tag{7.1}$$

$$d\dot{\boldsymbol{\epsilon}} = \dot{\lambda}\frac{\partial Q}{\partial \boldsymbol{\sigma}} \qquad \text{(Flow rule)} \tag{7.2}$$

$$\dot{\kappa} = \dot{\kappa}(\epsilon_e^p) \qquad \text{(Constitutive equations for hardening)} \tag{7.3}$$

$$\dot{\boldsymbol{\alpha}} = \boldsymbol{\varphi}(\boldsymbol{\sigma}, \boldsymbol{\alpha}) \cdot \dot{\boldsymbol{\sigma}} \tag{7.4}$$

The symbols have the same meaning as discussed in previous chapters. $\boldsymbol{\varphi}$ in Eq. (7.4) is a fourth-order tensor-valued function of $\boldsymbol{\alpha}$ and $\boldsymbol{\sigma}$.

Similarly, for finite deformation, any plasticity theory based on the concept of the yield surface must define a yield surface, a constitutive equation for plastic strain rate or increment, (i.e., the flow rule), and constitutive equations for the hardening parameters (e.g., the back stress $\boldsymbol{\alpha}$ and the isotropic hardening parameter κ). Equations (7.1)–(7.4) are established and justified for the case of infinitesimal elastic-plastic deformation. Unfortunately, the direct generalization of these for finite elastic-plastic deformation can cause several serious problems. The main points to be considered when making such generalizations are discussed below.

7.1.1 Stress and Strain Measures

As we explained in Chapter 2, several strain measures can be used to describe the deformation state of a continuum (although theoretically an infinite number of strain measures are possible). A couple of examples we have given earlier are the Almansi-Hemel strain tensor $\mathbf{E}$, which is the Lagrange description, and $\mathbf{e}$, which is the Euler description. The situation is similar for stress measures that describe the stress state of a continuum. The Cauchy stress tensor $\boldsymbol{\sigma}$ or the Kirchhoff stress tensor $\boldsymbol{\tau}$, which belong to the Euler description, and the Piola-Kirchhoff stress tensors $\boldsymbol{\Sigma}^{\mathrm{I}}$ and $\boldsymbol{\Sigma}^{\mathrm{II}}$, which belong to the Lagrange description, are the possible choices for formulating the constitutive model for finite elastic-plastic deformation. The reason different stress and strain measures are needed is to determine if the initial configuration or the current configuration should be used as the reference configuration. For infinitesimal deformation, this difference is negligible and therefore $\boldsymbol{\Sigma}^{\mathrm{I}}$, $\boldsymbol{\Sigma}^{\mathrm{II}}$, $\boldsymbol{\tau}$, and $\boldsymbol{\sigma}$ coincide; $\mathbf{E}$ and $\mathbf{e}$ become the same tensor and can be accurately approximated by the infinitesimal strain tensor $\boldsymbol{\epsilon}$. As a result, when formulating the plasticity theory for infinitesimal deformation, as given by Eqs. (7.1)–(7.4), it is not necessary to state what kind of stress and strain measures are used in the formulation.

Yet the specification of the stress and strain measure is necessary in the case of finite elastic-plastic deformation, since the difference between different stress and strain measures can no longer be ignored. For example, in the case of uniaxial tension or compression, the stress components in the loading direction for the Cauchy stress $\boldsymbol{\sigma}$, the Piola-Kirchhoff stresses $\boldsymbol{\Sigma}^{\mathrm{I}}$ and $\boldsymbol{\Sigma}^{\mathrm{II}}$, derived in Eqs. (2.143)–(2.145), are given by

$$\sigma_{11} = \frac{P}{\lambda_2^2 A_0} = \frac{P}{A} \tag{7.5}$$

$$\Sigma_{11}^{\mathrm{I}} = \frac{P}{A_0} \tag{7.6}$$

$$\Sigma_{11}^{\mathrm{II}} = \frac{P}{\lambda_1 A_0} \tag{7.7}$$

where A_0 and A are the original and current areas of the cross section, respectively, and λ_1 and λ_2 are the stretch ratios in the loading and transverse directions, respectively. From these equations it follows that

$$\frac{\sigma_{11}}{\Sigma_{11}^{\mathrm{I}}} = \frac{1}{\lambda_2^2}$$

$$\frac{\sigma_{11}}{\Sigma_{11}^{\mathrm{II}}} = \frac{\lambda_1}{\lambda_2^2} \tag{7.8}$$

We see from these expressions that when the deformation is infinitesimal, $\lambda_1 \approx 1$ and $\lambda_2 \approx 1$, so that $\sigma_{11} \approx \Sigma_{11}^{\mathrm{I}} \approx \Sigma_{11}^{\mathrm{II}}$. As λ_1, λ_2 increase, the difference between them substantially increases. Similar analysis can apply to the strain measures. In Section 1.1, we showed that

$$\mathbf{E}_{11}^{n} = \mathbf{e}_{11}^{l} + \frac{1}{2!}\left(\mathbf{e}_{11}^{l}\right)^2 + \frac{1}{3!}\left(\mathbf{e}_{11}^{l}\right)^3 + \cdots \tag{7.9}$$

where $\mathbf{e}^{l}$ and $\mathbf{E}^{n}$ are the logarithmic (natural, true) strain and nominal strain tensors, respectively. Again, $\mathbf{E}^{n}$and $\mathbf{e}^{l}$ will coincide only in the case of infinitesimal deformation.

One of the consequences of the difference between various stress or strain measures is that a certain behavior may hold only for certain measures. For example, it is well known that if the Cauchy stress $\boldsymbol{\sigma}$ is used for most of the properly annealed metals, the stress-strain curves for uniaxial tension and uniaxial compression will coincide, suggesting the existence of a universal or equivalent curve between the properly defined effective stress and effective strain, as discussed in Section 5.3. But, if the nominal or the first Piola-Kirchhoff stress $\boldsymbol{\Sigma}^{\mathrm{I}}$ is used, the stress-strain curves for uniaxial tension and compression will coincide only in the infinitesimal strain region and separate from one another when the strain becomes finite, as is schematically shown in

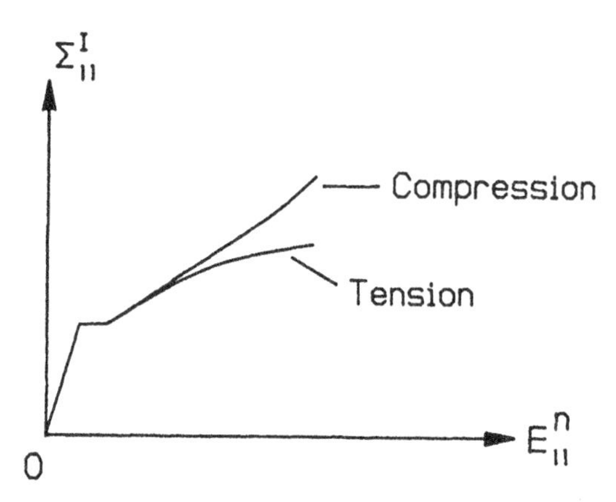

Fig. 7.1 Nominal stress-strain curves for uniaxial tension and compression

Fig. 7.1. Note that this conclusion holds true regardless of the choice of the strain measure.

Selecting the stress and strain measures to use in formulating a plasticity theory at finite deformation, or deciding whether the Lagrange or Euler description should be used, is more or less a matter of taste except for the following two considerations: First, the proper choice should make it relatively more convenient and simple to formulate a model and obtain solutions later. Second, and more important, the choice should make it easier to determine the actual physical phenomenon being scrutinized. For example, in an isotropic hardening plasticity theory where a universal function between the effective stress and strain is important, the Cauchy stress $\boldsymbol{\sigma}$ is preferred to the first Piola-Kirchhoff stress $\boldsymbol{\Sigma}^I$.

7.1.2 Objective Rates

In formulating a plasticity model, the stress rate is required, as can be seen from the consistency condition. The stress rate will appear in the expression for the proportionality factor $\boldsymbol{\lambda}$ and, in turn, in the expression of the constitutive equations for the hardening parameters $\boldsymbol{\kappa}$ and $\boldsymbol{\alpha}$. Also, as can be seen from Eq. (7.4), the rate of back stress $\boldsymbol{\alpha}$ is needed in formulating its constitutive equation. However, as discussed in Section 3.6, the material time derivative or rate of the Cauchy stress $\boldsymbol{\sigma}$ is not objective and hence is inappropriate for use in the formulation of constitutive equations for plastic deformation; these equations must be objective. It is common practice to assume that the back stress $\boldsymbol{\alpha}$ has the same characteristics as the Cauchy stress $\boldsymbol{\sigma}$, defined in reference to the current configuration, thus making the rate of $\boldsymbol{\alpha}$ also inappropriate for use in the constitutive equations. This means that the plastic deformation models for infinitesimal deformation, Eqs. (7.1)–(7.4), cannot be directly applied to finite deformation if the Cauchy stress $\boldsymbol{\sigma}$ is used. The material rate $\dot{\boldsymbol{\sigma}}$ and $\dot{\boldsymbol{\alpha}}$, among other things, must be replaced by a certain kind of objective rate. It was shown in Section 3.6 that there are several stress rates that are objective. Additional objective stress rates are listed in the literature. Here we need additional discussion on how to determine the optimum objective rate for use in constitutive equations of the plasticity theory for finite deformation.

For small deformation theories all the objective rates can be approximated by the material rate. For finite deformations the process is not so simple, since solutions obtained by different objectives rates are substantially different if the constitutive equations are not properly adjusted. Because all of these rates are objective, more than just the principle of objectivity or frame invariance must be considered. The underlying physics of the plastic flow at finite deformation is a primary factor. It should be pointed out that currently there is no general agreement as to which objective rate is best, and research on this problem is still in progress. A brief review is given below.

Truesdell (1955) was first to note that when a linear hypoelastic material is subjected to the finite simple shear deformation, and the Jaumann rate is used in the linear hypoelastic constitutive equation, an oscillatory shear stress response is obtained. Of course this oscillation is physically unacceptable, since the shear stress will increase monotonically as the shear strain increases. This problem was left unexplored until twenty-four years later when Dienes (1979) attributed this oscillation to the improper use of the Jaumann rate in the linear hypoelastic constitutive equation. He used a different objective rate, proposed by Green and Naghdi (1965) and sometimes known as the *Green-McInnis rate* (1961), and obtained the nonoscillatory solution for the shear stress. Prager (1961) also obtained a nonoscillatory solution to the simple shear problem by using the Jaumann rate and adding some extra nonlinear terms to the linear hypoelastic constitutive equation.

The problem of the oscillatory solution reappeared three years after Dienes's work, but this time in the field of finite plasticity theory with a linear kinematic hardening rule. Nagtegaal and de Jong (1982) presented a numerical solution to the simple shear problem for von Mises materials using the Prager-Ziegler linear kinematic hardening rule and the Jaumann rates for the stress and back stress. Similar to the case of hypoelasticity, the oscillatory solution for stress was obtained in response to a monotonically increasing shear strain. They concluded that this occurred because of the disadvantage of the Prager-Ziegler linear kinematic hardening rule. This anomaly has prompted a series of investigations by the mechanics community. It was pointed out by Lee et al. (1983) that the anomaly is caused by the improper use of the Jaumann rate rather than by the constitutive equation. They suggested a new corotational rate defined by the angular velocity or spin of a unit vector attached to the material and instantaneously oriented along the principal direction of the back stress. Indeed they showed that the oscillation of stress is eliminated by the use of this rate. Later Dafalias (1983) obtained the nonoscillatory solution to the linear kinematic hardening problem under finite simple shear using the corotational rates proposed by Lee et al. and used by Dienes for hypoelasticity. At the same time Dafalias proposed a new corotational rate based on the concept of the plastic spin and showed that the oscillation of stress can be removed by the use of this new rate. The nonoscillatory solution obtained by using the Green-McInnis rate was also retrieved by Johnson and Bammann (1984) who treated as analogous the anomalies that occurred in hypoelasticity and kinematic hardening plasticity. Recently Zbib and Aifantis (1988) introduced a new concept of relative spin based on another new corotational rate that they suggested. It reduces the rate derived from plastic spin, in the case of large plastic deformation, and it can also be used in the constitutive equation of hypoelasticity. They were able to show that the new corotational rate is more general than others and can recover the same results as other rates with an adjustment in the constant contained in he expression for relative spin.

Although the Zbib-Aifantis work was aimed at finding an appropriate corotational rate, it focused on modifying the linear constitutive equation. Atluri (1984) was able to show that in the case of hypoelasticity any one of the corotational rates (including the Jaumann rate) can be used to obtain a nonoscillatory solution to a large simple shear problem if the usual linear hypoelastic constitutive equation is generalized to contain certain terms of nonlinearity. This idea, similar to the one practiced by Prager (1961), was generalized to the finite plasticity theory to modify the linear kinematic hardening rule and suppress the oscillations in stress for a simple shear problem. Unfortunately, as pointed out by Zbib and Aifantis (1988), this method is more or less purely mathematical and does not provide any insight or systematic way to generalize the constitutive equation.

In this chapter we use an arbitrary objective rate in our presentation of the constitutive equation for finite plastic flow. Our intention is not to emphasize one rate over another, except in the discussion of simple shear where several corotational rates are compared. But rather our choice of rate was governed by the experimental evidence, and not by the frame invariant consideration or the oscillatory feature.

7.1.3 Decomposition of Strain Measure

In classical infinitesimal deformation elastic-plastic theory, the infinitesimal strain tensor $\boldsymbol{\epsilon}$ can be unambiguously divided into the elastic and plastic parts

$$\boldsymbol{\epsilon} = \boldsymbol{\epsilon}^{e} + \boldsymbol{\epsilon}^{p} \tag{7.10}$$

and

$$d\boldsymbol{\epsilon} = d\boldsymbol{\epsilon}^{e} + d\boldsymbol{\epsilon}^{p} \tag{7.11}$$

The plastic strain or its rate is defined on the physical ground of elastic unloading as the permanent strain or strain rate remains when the load is removed.

However, in finite elastic-plastic deformation this additive decomposition of the strain measure or its rate is not so unambiguously defined. In fact, this ambiguity has led to some controversity in recent years. Again, at the heart of the issue is the fact that there exits a coupling between elastic and plastic deformation in the case of finite elastic-plastic deformation. Further, it is possible to use different strain measures and corresponding rates for finite deformation, and this choice, as pointed out by Nimat-Nasser (1979), is more a matter of taste and convenience than anything else. This topic will be explored in more detail in the next section.

There are additional obstacles in generalizing the plasticity theory for infinitesimal deformation to the finite elastic-plastic deformation, including the existence and choice of the unstressed or intermediate configuration and the large rotation resulting from finite deformation. Clearly generalization of the plasticity theory to finite deformation is by no means trivial and involves much more than simply changing the notations.

Finally, it should be noted that there is no commonly accepted practical definition for finite deformation. The word "finite" is used relative to the word "infinitesimal." Infinitesimal refers to a difference between the original and current configurations that is so small that it can be ignored. Stress and strain are all therefore defined in reference to the original configuration. It thus seems reasonable to regard deformation as finite whenever the differences between the stress and strain defined in accordance to the original configuration and those to the current configuration are not negligible.

7.2 KINEMATICS OF FINITE PLASTIC DEFORMATION

This section presents the multiplicative decomposition of the deformation gradient $\mathbf{F}$ proposed by Lee et al. (1967) and the basis of it, the additive decomposition of the deformation rate.

When a work-hardening ductile material is stressed to the finite elastic-plastic deformation region, both elastic and plastic deformations are generated to give the total deformation. When the load is removed, the material is assumed to return to the zero stress state, and the elastic strain will be recovered. The remaining strain at the zero stress state is termed the *plastic strain*. The statements above can be translated into the following mathematical descriptions: As illustrated in Fig. 7.2, $\mathbf{X}$ represents the original or Lagrangian coordinate of any material point in the continuum under consideration, and $\mathbf{x}$ is the current or Eulerian coordinate of the same material points in the deformed body. Then, any infinitesimal line segment $d\mathbf{X}$ deforms into $d\mathbf{x}$ according to

$$d\mathbf{x} = \mathbf{F} \cdot d\mathbf{X} \tag{7.12}$$

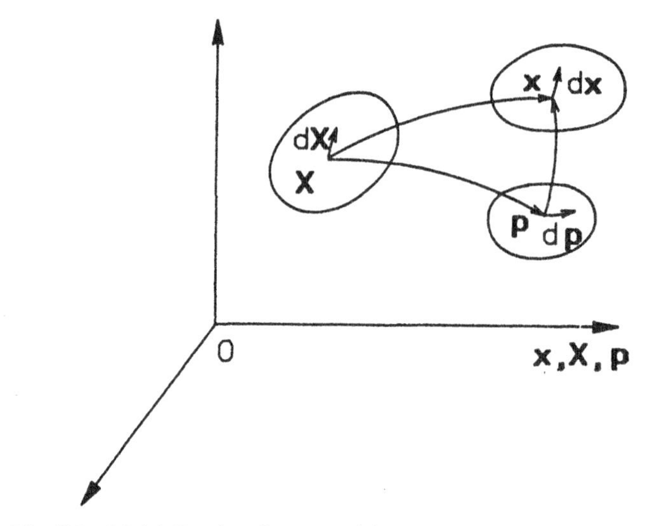

Fig. 7.2 Multiplicative decomposition of the deformation gradient

where $\mathbf{F}$ is the deformation gradient defined in Chapter 2. Now imagine an intermediate configuration described by the coordinate $\mathbf{p}$ which is obtained from the current configuration by unloading the continuum to a zero stress state, or from the original configuration by pure plastic deformation. The deformation of an infinitesimal line segment $d\mathbf{X}$, described by Eq. (7.12), can be accomplished with the following two steps:

1. $d\mathbf{X}$ is deformed into $d\mathbf{p}$ in the intermediate configuration by plastic flow

$$d\mathbf{p} = \mathbf{F}^{\mathrm{p}} \cdot d\mathbf{X} \tag{7.13}$$

where $\mathbf{F}^{\mathrm{p}}$ is the plastic deformation gradient from $\mathbf{X}$ to $\mathbf{p}$,

$$\mathbf{F}^{\mathrm{p}} = \frac{\partial \mathbf{p}}{\partial \mathbf{X}} \tag{7.14}$$

2. $d\mathbf{p}$ is further deformed into $d\mathbf{x}$ in the current configuration by elastic deformation

$$d\mathbf{x} = \mathbf{F}^{\mathrm{e}} \cdot d\mathbf{p} \tag{7.15}$$

where

$$\mathbf{F}^{\mathrm{e}} = \frac{\partial \mathbf{x}}{\partial \mathbf{p}} \tag{7.16}$$

is the elastic deformation gradient. Comparing Eqs. (7.13) and (7.15) with Eq. (7.12) gives the following well-known multiplicative decomposition of the deformation gradient:

$$\mathbf{F} = \mathbf{F}^e \cdot \mathbf{F}^p \tag{7.17}$$

As noted by Lee (1981), after a nonhomogeneous plastic deformation, unloading a body will not generally lead to a uniform zero stress state; instead, a residual stress state configuration will result. In this case the body can be imagined to be divided into an infinite number of infinitesimal elements, each of which can reach the zero stress state after releasing the residual stress. The unstressed configuration thus obtained is obviously noncontinuous, and $\mathbf{F}^p$ and $\mathbf{F}^e$ can then be defined as point-functions that relate the deformations in infinitesimal neighborhoods of a material particle, and Eq. (7.17) will still hold. $\mathbf{F}^e$ and $\mathbf{F}^p$ are continuous functions if the plastic flow is homogeneous.

It should be pointed out that the intermediate configuration described by $\mathbf{p}$ is, in general, not uniquely determined, since an arbitrary rigid rotation can be superimposed on it and leave it unstressed. This can be verified as follows: Let $\mathbf{Q}$ be an arbitrary orthogonal tensor and represent an arbitrary rigid rotation, then

$$\mathbf{F} = \mathbf{F}^e \cdot \mathbf{F}^p = \mathbf{F}^e \cdot \mathbf{Q}^T \cdot \mathbf{Q} \cdot \mathbf{F}^p = \hat{\mathbf{F}}^e \cdot \hat{\mathbf{F}}^p \tag{7.18}$$

where

$$\begin{aligned} \hat{\mathbf{F}}^e &= \mathbf{F}^e \cdot \mathbf{Q}^T \\ \hat{\mathbf{F}}^p &= \mathbf{Q} \cdot \mathbf{F}^p \end{aligned} \tag{7.19}$$

Obviously the configuration determined by $\hat{\mathbf{F}}^p$ remains unstressed because it is formed by a rigid rotation of an unstressed configuration. To solve this nonuniqueness, it is assumed that the elastic de-stressing will involve pure deformation only, so $\mathbf{F}^e$ is chosen to be

$$\mathbf{F}^e = \mathbf{V}^e \tag{7.20}$$

and thus is a symmetric tensor. With this additional requirement for $\mathbf{F}^p$, the above nonuniqueness problem is solved. Equation (7.20) will be used in the subsequent discussion.

From Eq. (7.17) and the definitions for the Almansi-Hemel strain tensors, Eqs. (2.55) and (2.56), it follows that

$$\mathbf{E} = \tfrac{1}{2}\left(\mathbf{F}^T \cdot \mathbf{F} - \mathbf{I}\right) = \mathbf{F}^{p^T} \cdot \mathbf{E}^e \cdot \mathbf{F}^p + \mathbf{E}^p \tag{7.21}$$

$$\mathbf{e} = \tfrac{1}{2}\left[\mathbf{I} - \left(\mathbf{F}^{-1}\right)^T \cdot \left(\mathbf{F}^{-1}\right)\right] = \mathbf{e}^e + \left(\mathbf{V}^{e^{-1}}\right)^T \cdot \mathbf{e}^p \cdot \left(\mathbf{V}^{e^{-1}}\right) \tag{7.22}$$

where

$$\mathbf{E}^{e} = \tfrac{1}{2}\left(\mathbf{V}^{e^2} - I\right)$$

$$\mathbf{E}^{p} = \tfrac{1}{2}\left(\mathbf{F}^{p^T} \cdot \mathbf{F}^{p} - \mathbf{I}\right) \tag{7.23}$$

$$\mathbf{e}^{e} = \tfrac{1}{2}\left[\mathbf{I} - \left(\mathbf{V}^{e^{-1}}\right)^T \cdot \left(\mathbf{V}^{e^{-1}}\right)\right]$$

$$\mathbf{e}^{p} = \tfrac{1}{2}\left[\mathbf{I} - \left(\mathbf{F}^{p^{-1}}\right)^T \cdot \left(\mathbf{F}^{p^{-1}}\right)\right] \tag{7.24}$$

Note that $\mathbf{E}^p$ is the strain tensor measured relative to the initial reference configuration and is based on the deformation from the original reference configuration to the intermediate or unstressed configuration (described by $\mathbf{F}^p$). Thus it can be considered pure plastic strain. On the other hand, $\mathbf{E}^e$ is the strain tensor measured relative to the intermediate configuration, for the deformation from the unstressed configuration to the current configuration (described by $\mathbf{V}^e$), and hence it represents the pure elastic strain. Similary $\mathbf{e}^e$ and $\mathbf{e}^p$ represent the pure elastic and plastic strain tensors measured in the current and unstressed configurations, respectively. It is seen from Eqs. (7.18) and (7.19) that in finite deformation the purely elastic and plastic strains are coupled geometrically. This is due to the fact that the elastic deformation occurs in a body that has been already plastically strained, and vice versa. If the total deformation gradient is infinitesimal, the following approximations hold:

$$\mathbf{F} = \mathbf{I} + \boldsymbol{\epsilon}$$

$$\mathbf{V}^e = \mathbf{I} + \boldsymbol{\epsilon}^e$$

$$\mathbf{F}^p = \mathbf{I} + \boldsymbol{\epsilon}^p \tag{7.25}$$

where $\boldsymbol{\epsilon}$, $\boldsymbol{\epsilon}^e$ and $\boldsymbol{\epsilon}^p$ are the total, elastic, and plastic infinitesimal strain tensors, respectively. Then it is easy to verify that if higher-order terms of $\boldsymbol{\epsilon}$, $\boldsymbol{\epsilon}^e$, and $\boldsymbol{\epsilon}^p$ are ignored, we have

$$\mathbf{E} = \mathbf{e} = \boldsymbol{\epsilon} = \boldsymbol{\epsilon}^e + \boldsymbol{\epsilon}^p \tag{7.26}$$

which is the well-known additive decomposition for strain at infinitesimal deformation.

The velocity gradient $\mathbf{L}$ can be calculated from Eq. (7.17) as

$$\mathbf{L} = \dot{\mathbf{F}} \cdot \mathbf{F}^{-1} = \dot{\mathbf{V}}^e \cdot \mathbf{V}^{e^{-1}} + \mathbf{V}^e \cdot \dot{\mathbf{F}}^p \cdot \mathbf{F}^{p^{-1}} \cdot \mathbf{V}^{e^{-1}} = \mathbf{L}^e + \mathbf{V}^e \cdot \mathbf{L}^p \cdot \mathbf{V}^{e^{-1}} \tag{7.27}$$

where the elastic and plastic velocity gradients $\mathbf{L}^e$ and $\mathbf{L}^p$ are given by

$$\mathbf{L}^e = \dot{\mathbf{V}}^e \cdot \mathbf{V}^{e^{-1}} = \mathbf{D}^e + \mathbf{W}^e \tag{7.28}$$

$$\mathbf{L}^p = \dot{\mathbf{F}}^p \cdot \mathbf{F}^{p^{-1}} = \mathbf{D}^p + \mathbf{W}^p \tag{7.29}$$

Where $\mathbf{D}^e, \mathbf{D}^p$ and $\mathbf{W}^e, \mathbf{W}^p$ are, respectively, the symmetric and antisymmetric parts of $\mathbf{L}^e$ and $\mathbf{L}^p$.

From $\mathbf{L} = \mathbf{D} + \mathbf{W}$ and the preceding results it is easy to show that

$$\mathbf{D} = \mathbf{D}^e + \left(\mathbf{V}^e \cdot \mathbf{D}^p \cdot \mathbf{V}^{e^{-1}}\right)_s + \left(\mathbf{V}^e \cdot \mathbf{W}^p \cdot \mathbf{V}^{e^{-1}}\right)_s \tag{7.30}$$

$$\mathbf{W} = \mathbf{W}^e + \left(\mathbf{V}^e \cdot \mathbf{D}^p \cdot \mathbf{V}^{e^{-1}}\right)_a + \left(\mathbf{V}^e \cdot \mathbf{W}^p \cdot \mathbf{V}^{e^{-1}}\right)_a \tag{7.31}$$

where the subscripts "s" and "a" denote the symmetric and antisymmetric part of a tensor. From these two equations it can be seen that in finite elastic-plastic deformation the additive decomposition for the deformation rate does not hold in general:

$$\mathbf{D} \neq \mathbf{D}^e + \mathbf{D}^p \tag{7.32}$$

here $\mathbf{D}^e$ and $\mathbf{D}^p$ are defined by Eqs. (7.28) and (7.29). However, for most of metals the elastic strain is negligible compared to the plastic strain in the case of large deformation, so it is reasonable to assume that the elastic stretch tensor $\mathbf{V}^e$ is not different from the unit tensor (i.e., $\mathbf{V}^e \approx \mathbf{I} + \boldsymbol{\epsilon}^e$), where $\boldsymbol{\epsilon}^e$ is the infinitesimal strain tensor. By ignoring the higher-order infinitesimal quantities, Eqs. (7.30) and (7.31) can be approximated by

$$\begin{aligned} \mathbf{D} &= \mathbf{D}^e + \mathbf{D}^p \\ \mathbf{W} &= \mathbf{W}^e + \mathbf{W}^p \end{aligned} \tag{7.33}$$

which is a commonly used assumption in the plasticity theories for finite deformation. It should be noted that according to Eq. (7.21), the identity

$$\mathbf{E} = \mathbf{E}^e + \mathbf{E}^p \tag{7.34}$$

does not hold, even when the elastic deformation is infinitesimal. As discussed earlier, this identity is valid only if both elastic and plastic deformations are infinitesimal, Eqs. (7.25) and (7.26), or if the plastic deformation is infinitesimal compared with the elastic deformation, which has no physical meaning in the case of finite deformation.

It should be emphasized that this conclusion on the additive decomposition for strain $\mathbf{E}$ is based on the definitions for $\mathbf{E}^e$ and $\mathbf{E}^p$ given by Eq. (7.23). However, one could define the plastic and "elastic" strain—so that Eq. (7.34) holds—by sacrificing of the physical meaning for elastic strain in finite deformation. In the general plasticity theory for finite deformation developed by Green and Naghdi (1965), which is based on continuum thermodynamics theory, Eq. (7.34) would hold in general. What is most important in this formulation is that among the three variables $\mathbf{E}$, $\mathbf{E}^e$, and $\mathbf{E}^p$, only $\mathbf{E}$ is kinematically expressible in terms of the deformation gradient by

$$\mathbf{E} = \tfrac{1}{2}(\mathbf{F}^T \cdot \mathbf{F} - \mathbf{I}) = \tfrac{1}{2}(\mathbf{C} - \mathbf{I}) \tag{7.35}$$

and $\mathbf{E}^e$ and $\mathbf{E}^p$ are not defined kinematically by displacements. Instead, the plastic strain $\mathbf{E}^p$ is defined by a constitutive equation, and then the tensor $\mathbf{E}^e$ is defined by

$$\mathbf{E}^e = \mathbf{E} - \mathbf{E}^p \tag{7.36}$$

Note the difference between the definitions of Eqs. (7.36) and (7.23) which, in general, can be written as

$$\mathbf{E}^e = \tfrac{1}{2}(\mathbf{F}^{e^T} \cdot \mathbf{F}^e - \mathbf{I}) = \tfrac{1}{2}(\mathbf{C}^e - \mathbf{I}) \tag{7.37}$$

As emphasized by Green and Naghdi, the variable introduced by Eq. (7.36) is not an ordinary elastic strain tensor. In other words, it is introduced merely in order to represent the difference between the total strain $\mathbf{E}$ given by Eq. (7.35) and the plastic strain $\mathbf{E}^p$ given by the constitutive equation. It is labeled elastic strain because of the analogy of Eq. (7.37) to the corresponding one in infinitesimal deformation. Green and Naghdi (1966, 1971) pointed out that the introduction of a tensor $\mathbf{E}^e$ is unnecessary and that the use of two kinematic variables $\mathbf{E}$ and $\mathbf{E}^p$ is sufficient in a general theory. This theory is discussed in detail in the above-cited references and will not be repeated here.

We finally note that a theory using $\mathbf{D}$, $\mathbf{D}^P$, and $\mathbf{D}^e$ is the spatial or Eulerian description and that a theory using $\mathbf{E}$, $\mathbf{E}^p$, and $\mathbf{E}^e$ is the material or Lagrangian description. Most parts of this chapter use the spatial or Eulerian description.

7.3 HYPOELASTICITY AND SIMPLE SHEAR

As discussed in chapter 5 for infinitesimal deformation, the total strain rate is divided into the elastic and plastic parts and different constitutive rules are proposed for each. The rule governing $\dot{\epsilon}^p$ is the flow rule and the rule governing $\dot{\boldsymbol{\epsilon}}^e$ is the generalized Hooke's law.

For finite elastic-plastic deformation, as discussed in the last section, it is generally assumed that

$$\mathbf{D} = \mathbf{D}^e + \mathbf{D}^p \tag{7.38}$$

Recall that in finite deformation the elastic deformation is generally much smaller than the plastic deformation. This decomposition of the deformation rate $\mathbf{D}$ calls for constitutive laws for both $\mathbf{D}^e$ and $\mathbf{D}^p$. The constitutive law for $\mathbf{D}^e$, called *hypoelasticity theory*, is discussed in this section and the law for $\mathbf{D}^p$ (i.e., the flow rule at finite deformation) will be considered in the next section.

As shown by Rivlin (1955) and Truesdell (1955), the general form of the objective stress-rate and strain-rate relation for hypoelasticity can be postulated as

$$\mathring{\boldsymbol{\sigma}} = \mathbf{f}(\boldsymbol{\sigma}, \mathbf{D}) \tag{7.39}$$

where $\boldsymbol{\sigma}$ is the Cauchy stress tensor, $\mathbf{D}$ is the deformation rate, which is equal to the elastic rate $\mathbf{D}^e$ if the elastic-plastic deformation is considered. $\mathring{\boldsymbol{\sigma}}$ represents any one of the objective stress rates (not necessarily the Jaumann rate), as listed in Section 3.6, or in the literature. As we noted at the beginning of this chapter, the reason for the use of $\mathring{\boldsymbol{\sigma}}$ is that the material rate of $\boldsymbol{\sigma}$ is not objective and cannot be used in a constitutive equation, which must be frame indifferent.

The simplest constitutive equation conforming to the general form (7.39) for elastic isotropic materials is the following linear relation between $\mathring{\boldsymbol{\sigma}}$ and $\mathbf{D}$:

$$\mathring{\boldsymbol{\sigma}} = \alpha_1 \operatorname{tr} \mathbf{D}\mathbf{I} + \alpha_2 \mathbf{D} \tag{7.40}$$

where α_1 and α_2 are functions of the invariants of tensors $\boldsymbol{\sigma}$, $\mathbf{D}^e$ and the joint invariants of them. The variation of the elastic constants due to finite deformation can be incorporated by these two material functions. If it is further assumed that the elastic property remains constant during finite elastic-plastic deformation, then Eq. (7.40) becomes

$$\mathring{\boldsymbol{\sigma}} = \lambda \operatorname{tr}(\mathbf{D})\ \mathbf{I} + 2G\mathbf{D} \tag{7.41}$$

where λ and G are the conventional Lame elastic constants. Equation (7.41) can be viewed as a generalization of the infinitesimal elasticity theory (Eq. 7.42 below) to finite deformation by replacing $\boldsymbol{\sigma}$ by an objective stress rate $\mathring{\boldsymbol{\sigma}}$ and $\boldsymbol{\epsilon}^e$ by the deformation rate $\mathbf{D}^e$.

$$\boldsymbol{\sigma} = \lambda \operatorname{tr} \boldsymbol{\epsilon}\mathbf{I} + 2G\boldsymbol{\epsilon} \tag{7.42}$$

However, Eq. (7.41) is not derived from Eq. (7.42) since, in general, $\mathbf{D} \neq \mathring{\boldsymbol{\epsilon}}$: As we noted earlier, which objective rate one should choose is still an one problem. In fact an infinite number of objective stress rates can be defined, and different results can be obtained, by using different rates in the constitutive equation. To illustrate this point, Eq. (7.41) is used to obtain the solution to the simple shear problem, whose kinematics were been discussed in

Section 7.1. But various objective rates have been used in obtaining the solution:

1. Jaumann rate

$$\overset{\circ}{\boldsymbol{\sigma}} = \dot{\boldsymbol{\sigma}} - \mathbf{W} \cdot \boldsymbol{\sigma} + \boldsymbol{\sigma} \cdot \mathbf{W} \tag{7.43}$$

2. Truesdell rate

$$\overset{\diamond}{\boldsymbol{\sigma}} = \overset{\circ}{\boldsymbol{\sigma}} - \mathbf{L} \cdot \boldsymbol{\sigma} - \boldsymbol{\sigma} \cdot \mathbf{L}^{\mathrm{T}} + \boldsymbol{\sigma} \operatorname{tr} \mathbf{D} \tag{7.44}$$

3. Green-McInnis rate

$$\overset{\triangledown}{\boldsymbol{\sigma}} = \dot{\boldsymbol{\sigma}} + (\dot{\mathbf{R}} \cdot \mathbf{R}^{\mathrm{T}}) \cdot \boldsymbol{\sigma} - \boldsymbol{\sigma} \cdot (\dot{\mathbf{R}} \cdot \mathbf{R}^{\mathrm{T}})^{\mathrm{T}} \tag{7.45}$$

where $\mathbf{L} = \mathbf{D} + \mathbf{W}$ and $\mathbf{R}$ is the rotation tensor as defined in Chapter 2.

As we showed in Section 2.11, for

$$x_1 = X_1 + \gamma X_2, \quad x_2 = X_2, \quad x_3 = X_3 \tag{7.46}$$

$$\boldsymbol{\sigma} = \begin{bmatrix} \sigma_{11} & \sigma_{12} & 0 \\ \sigma_{12} & \sigma_{22} & 0 \\ 0 & 0 & 0 \end{bmatrix}, \quad \mathbf{L} = \begin{bmatrix} 0 & \dot{\gamma} & 0 \\ 0 & 0 & 0 \\ 0 & 0 & 0 \end{bmatrix} \tag{7.47}$$

$$\mathbf{D} = \frac{1}{2} \begin{bmatrix} 0 & \dot{\gamma} & 0 \\ \dot{\gamma} & 0 & 0 \\ 0 & 0 & 0 \end{bmatrix}, \quad \mathbf{W} = \frac{1}{2} \begin{bmatrix} 0 & \dot{\gamma} & 0 \\ -\dot{\gamma} & 0 & 0 \\ 0 & 0 & 0 \end{bmatrix} \tag{7.48}$$

$$\mathbf{R} = \begin{bmatrix} \cos\phi & \sin\phi & 0 \\ -\sin\phi & \cos\phi & 0 \\ 0 & 0 & 1 \end{bmatrix} \tag{7.49}$$

where

$$\tan\phi = \frac{\gamma}{2} \tag{7.50}$$

Solutions for this problem were obtained by Dienes (1979), Johnson and Bammann (1984), and Atluri (1984). The evaluation of the stress response for these three rates consists in using Eqs. (7.47)–(7.49) in the appropriate stress rate equations given by Eqs. (7.43)–(7.45), substituting the rate into Eq. (7.41), and then solving the resulting differential equations. The details of the procedure are omitted, but the results are reproduced below. The reader can refer to the above-cited papers for details. The initial conditions of zero strain and stress are assumed for the solutions.

1. For the Jaumann rate it is easy to verify that the resulting differential equations are

$$\dot{\sigma}_{11} - \dot{\gamma}\sigma_{12} = 0 \tag{7.51}$$

$$\dot{\sigma}_{12} + \tfrac{1}{2}\dot{\gamma}(\sigma_{11} - \sigma_{22}) = \dot{\gamma}G \tag{7.52}$$

$$\dot{\sigma}_{22} + \dot{\gamma}\sigma_{12} = 0 \tag{7.53}$$

These equations are solved as

$$\sigma_{11} = -\sigma_{22} = G(1 - \cos\gamma) \quad \sigma_{12} = G\sin\gamma \tag{7.54}$$

Obviously in this case the stress response to the monotonically increasing shear strain γ is sinusoidly oscillatory, which is physically unacceptable. Indeed, this is the reason why the Jaumann rate is regarded as inappropriate for use in the constitutive equations.

2. For the Truesdell rate the resulting differential equations are

$$\dot{\sigma}_{11} - 2\dot{\gamma}\sigma_{12} = 0 \tag{7.55}$$

$$\dot{\sigma}_{12} - \dot{\gamma}\sigma_{22} = \dot{\gamma}G \tag{7.56}$$

$$\dot{\sigma}_{22} = 0 \tag{7.57}$$

which can be integrated easily to give

$$\sigma_{11} = G\gamma^2, \quad \sigma_{12} = G\gamma, \quad \sigma_{22} = 0 \tag{7.58}$$

This solution does not exhibit oscillatory behavior, but the linear increase of σ_{12} with γ is unsuitable to use in modeling the real material response. There are also other disadvantages of the Truesdell rate. For example, the vanishing of $\overset{\diamond}{\boldsymbol{\sigma}}$ does not ensure that the invariants of $\boldsymbol{\sigma}$ are constants, as noted by Johnson and Bammann (1984), while the Jaumann rate $\overset{\circ}{\boldsymbol{\sigma}}$ does, as shown by Prager (1961). However, the yield function is formulated by using invariants of $\boldsymbol{\sigma}$. If these invariants vary, this suggests a change in the yield surface, which is usually caused by plastic deformation. It was noted by Prager (1961) that in formulating a finite plasticity theory, it is preferable to use a stress rate that will ensure a one-to-one correspondence between the vanishing of this rate and the constancy of the invariants of $\boldsymbol{\sigma}$, and hence the vanishing of further plastic flow. Therefore the use of $\overset{\diamond}{\boldsymbol{\sigma}}$ in an elastic-plastic theory has been limited.

3. The solution to the simple shear problem using the Green-McInnis rate was extensively examined by Dienes (1979). The results are given below:

$$\sigma_{11} = -\sigma_{22} = 4G\left[\cos 2\phi \ln(\cos \phi) + \phi \sin 2\phi - \sin^2 \phi\right] \quad (7.59)$$

$$\sigma_{12} = 2G \cos 2\phi\left[2\phi - 2\tan 2\phi(\cos \phi) - \tan \phi\right] \quad (7.60)$$

where ϕ is defined by Eq. (7.50). Dienes showed that the stress components given by these equations are monotonically increasing functions of γ. Therefore the rate $\overset{\nabla}{\boldsymbol{\sigma}}$ is considered to be appropriate for use in the constitutive equation. Later we will show that a nonoscillatory response can be obtained in plasticity theory with linear kinematic hardening when σ is used in the constitutive equation for the back stress $\boldsymbol{\alpha}$ and in the consistency condition for Cauchy stress $\boldsymbol{\sigma}$.

7.4 PLASTICITY THEORY FOR FINITE DEFORMATION

In this section we introduce plasticity theory involving kinematic and stress measures in the current configuration, which is the Euler description. Both isotropic and kinematic hardening are considered.

7.4.1 Prandtl-Reuss Equation at Finite Strain

It is often assumed that the Prandtl-Reuss equation for infinitesimal deformation can be generalized for the finite strain case. That is, the plastic deformation rate $\mathbf{D}^{\mathrm{p}}$ is assumed to be coaxial with the deviator of the Cauchy stress $\boldsymbol{\sigma}$ (and thus $\mathbf{S}$):

$$\mathbf{D}^{\mathrm{p}} = \lambda \mathbf{S} \quad (7.61)$$

where λ is a proportionality factor and $\mathbf{S} = \boldsymbol{\sigma} - \frac{1}{3}\operatorname{tr}\boldsymbol{\sigma}\mathbf{I}$. As usual, the incompressibility of the plastic flow is implied by Eq. (7.61), since

$$\operatorname{tr}\mathbf{D}^{\mathrm{p}} = 0 \quad \text{or} \quad D^{\mathrm{p}}{}_{ii} = 0 \quad (7.62)$$

This is the mathematical expression for incompressibility of the plastic flow at finite deformation, and it reduces to Eq. (5.84) if the deformation is infinitesimal. As in infinitesimal deformation the Prandtl-Reuss Eq. (7.61) implies that yielding is governed by the von Mises criterion in terms of the Cauchy stress

$$\tfrac{3}{2}\mathbf{S} : \mathbf{S} = \sigma_{\mathrm{Y}}^2 \quad (7.63)$$

where σ_Y is yield strength measured from the uniaxial tension or compression experiment. Here it is assumed that

$$\sigma_Y = \sigma_Y\left(\int D_e^p \, dt\right) \tag{7.64}$$

where D_e^p is the effective plastic deformation rate defined by

$$D_e^p = \left[\tfrac{2}{3}\mathbf{D}^p : \mathbf{D}^p\right]^{1/2} \tag{7.65}$$

Using Eqs. (7.63) and (7.65), the proportional factor λ can be obtained as

$$\lambda = \frac{3D_e^p}{2\sigma_Y} \tag{7.66}$$

Then the Prandtl-Reuss Eq. (7.61) becomes

$$\mathbf{D}^p = \frac{3D_e^p}{2\sigma_Y}\mathbf{S} \tag{7.67}$$

which is a direct generalization of Eqs. (5.87) and (7.94) for finite deformation. When Eq. (7.67) is used, it is usually assumed that the elastic strain is small compared with the plastic one and hence can be neglected. It follows that

$$\mathbf{D} = \frac{3D_e}{2\sigma_Y}\mathbf{S} \tag{7.68}$$

If, on the other hand, elastic strain must be considered, the hypoelasticity theory discussed in the previous section can be used to calculate the elastic deformation rate. From Eq. (7.41) it is easy to verify that

$$\mathbf{D}^{e'} = \frac{\mathring{\boldsymbol{\sigma}}}{2G} \tag{7.69}$$

and that

$$\operatorname{tr}\mathring{\boldsymbol{\sigma}} = 3\lambda \operatorname{tr}\mathbf{D}^e + 2G\operatorname{tr}\mathbf{D}^e = (3\lambda + 2G)\operatorname{tr}\mathbf{D} \tag{7.70}$$

So the volume change is elastic and is given by

$$\operatorname{tr}\mathbf{D} = \frac{\operatorname{tr}\mathring{\boldsymbol{\sigma}}}{3K} \tag{7.71}$$

where K is the elastic bulk modulus defined by

$$K = \frac{3\lambda + 2G}{3} \tag{7.72}$$

Using Eqs. (7.69) and (7.71), we can express the total deformation rate $\mathbf{D}$ by

$$\mathbf{D} = \mathbf{D}^{e} + \mathbf{D}^{p} = \frac{\operatorname{tr} \mathring{\boldsymbol{\sigma}}}{9K}\mathbf{I} + \frac{\mathring{\boldsymbol{\sigma}}}{2G} + \frac{3D_{e}^{p}}{2\sigma_{Y}}\mathbf{S} \tag{7.73}$$

Note $\mathring{\boldsymbol{\sigma}}$ can be any one of the objective stress rates, as we noted earlier. The choice of stress rate should be determined by experimental evidence. Equation (7.73) is not convenient to use since $\mathbf{D}_{e}^{p}$ appears on the right-hand side, which is unknown at present. It is ideal to obtain a constitutive relation with $\mathbf{D}$ on one side and the objective stress rate $\mathring{\boldsymbol{\sigma}}$ on the other side. This can be achieved by using the plastic potential theory.

7.4.2 Plastic Potential Theory for Finite Strain

We will formulate here a plasticity theory for finite deformation involving kinematic and isotropic hardening within the most general framework of the plastic potential theory. Let the yield surface in the current configuration be

$$F(\boldsymbol{\sigma}, \boldsymbol{\alpha}, \kappa) = f(\boldsymbol{\sigma}, \boldsymbol{\alpha}, \kappa) - \kappa = 0 \tag{7.74}$$

where $\boldsymbol{\alpha}$ is the back stress and κ is the isotropic hardening parameter. The plastic potential theory can be generalized to relate the plastic deformation rate $\mathbf{D}^{p}$ to the Cauchy stress $\boldsymbol{\sigma}$ as

$$\mathbf{D}^{p} = \lambda \frac{\partial Q}{\partial \boldsymbol{\sigma}} \tag{7.75}$$

where Q is the plastic potential function and λ is the proportionality factor.

To simplify the discussion, we use simple constitutive equations for α and κ:

$$\mathring{\boldsymbol{\alpha}} = cD^{p} \tag{7.76}$$

$$\kappa = \kappa\left(\textstyle\int D_{e}^{p}\, dt\right) \tag{7.77}$$

More general forms can be assumed where the effective plastic deformation rate D_{e}^{p} is defined by Eq. (7.65), and $\mathring{\boldsymbol{\alpha}}$ is any one of the objective rates for $\boldsymbol{\alpha}$ defined similarly to $\boldsymbol{\sigma}$, since $\boldsymbol{\alpha}$ and $\boldsymbol{\sigma}$ are assumed to transform between the observer frame by the same law. For example, under the change of the frame of reference given by Eq. (3.74), the transformation law for $\boldsymbol{\alpha}$ is

$$\boldsymbol{\alpha}^* = \mathbf{Q}(t) \cdot \boldsymbol{\alpha} \cdot \mathbf{Q}^{\mathrm{T}}(t) \tag{7.78}$$

where $\boldsymbol{\alpha}^*$ and $\boldsymbol{\alpha}$ are the back stress in the frame $\mathbf{x}^*$ and the frame $\mathbf{x}$, respectively. The Jaumann rate for $\boldsymbol{\alpha}$ is defined by

$$\mathring{\boldsymbol{\alpha}} = \dot{\boldsymbol{\alpha}} - \mathbf{w} \cdot \boldsymbol{\alpha} + \boldsymbol{\alpha} \cdot \mathbf{w} \tag{7.79}$$

which is the same as the Jaumann rate for $\boldsymbol{\sigma}$, with $\boldsymbol{\sigma}$ replaced by $\boldsymbol{\alpha}$.

Using the yield function (7.74) and the constitutive equations (7.75), (7.76), and (7.77), we derive the stress rate (the deformation rate relation can be derived similar to the infinitesimal theory).

For a scalar function G, on the other hand,

$$\mathring{G} = \dot{G} \tag{7.80}$$

where $\mathring{G}$ is any one of the objective rates. Hence the consistency condition at finite deformation is written

$$\dot{\mathrm{F}} = \mathring{\mathrm{F}} = \frac{\partial f}{\partial \boldsymbol{\sigma}} : \mathring{\boldsymbol{\sigma}} + \frac{\partial f}{\partial \boldsymbol{\sigma}} : \mathring{\boldsymbol{\alpha}} - \dot{\kappa}\, dD_{\mathrm{e}}^{\mathrm{p}} = 0 \tag{7.81}$$

Note that in this expression the objective rates $\mathring{\boldsymbol{\sigma}}$ and $\mathring{\boldsymbol{\alpha}}$ are used instead of the usual material rate. According to Eqs. (7.65) and (7.75), $dD_{\mathrm{e}}^{\mathrm{p}}$ can be expressed as

$$dD_{\mathrm{e}}^{\mathrm{p}} = \sqrt{\frac{2}{3}}\, \lambda \left(\frac{\partial Q}{\partial \boldsymbol{\sigma}} : \frac{\partial Q}{\partial \boldsymbol{\sigma}} \right)^{1/2} \tag{7.82}$$

Therefore, after making use of Eqs. (7.75), (7.76), and (7.82), we rewrite Eq. (7.81) as

$$\frac{\partial f}{\partial \boldsymbol{\sigma}} : \mathring{\boldsymbol{\sigma}} + c\lambda \frac{\partial f}{\partial \boldsymbol{\alpha}} : \frac{\partial Q}{\partial \boldsymbol{\sigma}} - \sqrt{\frac{2}{3}}\, \lambda \dot{\kappa} \left(\frac{\partial Q}{\partial \boldsymbol{\sigma}} : \frac{\partial Q}{\partial \boldsymbol{\sigma}} \right)^{1/2} = 0 \tag{7.83}$$

From this we can derive the proportionality factor λ:

$$\lambda = \frac{\partial f / \partial \boldsymbol{\sigma} : \mathring{\boldsymbol{\sigma}}}{-c(\partial f / \partial \boldsymbol{\alpha}) : \partial Q / \partial \boldsymbol{\sigma} + \sqrt{2/3} \left[\dot{\kappa} (\partial Q / \partial \boldsymbol{\sigma} : \partial Q / \partial \boldsymbol{\sigma})^{1/2} \right]} \tag{7.84}$$

Substituting Eq. (7.84) into Eq. (7.75), we have

$$\mathbf{D}^{\mathrm{p}} = \frac{\partial Q/\partial \boldsymbol{\sigma} \cdot \partial f/\partial \boldsymbol{\sigma} : \overset{\circ}{\boldsymbol{\sigma}}}{-c(\partial f/\partial \boldsymbol{\alpha} : \partial Q/\partial \boldsymbol{\sigma}) + \sqrt{2/3}\left[\dot{\kappa}(\partial Q/\partial \boldsymbol{\sigma} : \partial Q/\partial \boldsymbol{\sigma})^{1/2}\right]} \tag{7.85}$$

This, along with the hypoelastic relation,

$$\mathbf{D}^{\mathrm{e}} = \frac{\operatorname{tr} \overset{\circ}{\boldsymbol{\sigma}}}{9\mathrm{K}} \mathbf{I} + \frac{\overset{\circ}{\boldsymbol{\sigma}}}{2G} \tag{7.86}$$

and $\mathbf{D} = \mathbf{D}^{\mathrm{e}} + \mathbf{D}^{\mathrm{p}}$ give the relationship between $\mathbf{D}$ and $\overset{\circ}{\boldsymbol{\sigma}}$ in the component form

$$D_{ij} = \left[\frac{\delta_{ij}\delta_{mn}}{9K} + \frac{\delta_{im}\delta_{jn}}{2G} + \frac{\dfrac{\partial Q}{\partial \sigma_{ij}} \dfrac{\partial f}{\partial \sigma_{mn}}}{-c\dfrac{\partial f}{\partial \alpha_{pq}} \dfrac{\partial Q}{\partial \sigma_{pq}} + \sqrt{\dfrac{2}{3}}\,\dot{\kappa}\left(\dfrac{\partial Q}{\partial \sigma_{pq}} \dfrac{\partial Q}{\partial \sigma_{pq}} \right)^{1/2}} \right] \overset{\circ}{\sigma}_{mn} \tag{7.87}$$

In the equation above K and G are the elastic constants. They can be determined experimentally in the elastic deformation range, and they assumed to be unchanged during the plastic deformation. The constant c and function $\dot{\boldsymbol{\kappa}}$ governing the kinematic and isotropic hardening, respectively, can be determined in a way similar to that used in the infinitesimal theory. It should be noted that in obtaining c and $\boldsymbol{\kappa}$ by the uniaxial tension or compression test, the experimental curve, in terms of the Cauchy stress and logarithmic strain, must be used.

As in infinitesimal deformation the specification of the plastic potential function Q is difficult to accomplish. Therefore the general approach is to use the associated flow rule $Q \equiv f$ and

$$\mathbf{D}^{\mathrm{p}} = \lambda \frac{\partial f}{\partial \boldsymbol{\sigma}} \tag{7.88}$$

For von Mises materials the yield function is written

$$\frac{1}{2}(\mathbf{s} - \boldsymbol{\alpha}') : (\mathbf{s} - \boldsymbol{\alpha}') = \frac{1}{3}\sigma_{\mathrm{Y}}^{2}\int D_{\mathrm{e}}^{\mathrm{p}}\, dt \tag{7.89}$$

Where $\boldsymbol{\alpha}'$ is the deviatoric part of $\boldsymbol{\alpha}$. From this equation it is easy to calculate

$$\frac{\partial f}{\partial \boldsymbol{\sigma}} = (\mathbf{s} - \boldsymbol{\alpha}') \tag{7.90}$$

$$\frac{\partial f}{\partial \boldsymbol{\alpha}} = -\frac{\partial f}{\partial \boldsymbol{\sigma}} \tag{7.91}$$

Equation (7.85) then reduces to

$$\mathbf{D}^{\mathrm{p}} = \frac{1}{K^{p}}(\mathbf{n} : \mathring{\boldsymbol{\sigma}})\mathbf{n} \tag{7.92}$$

where

$$\mathbf{n} = \frac{\partial f/\partial \boldsymbol{\sigma}}{(\partial f/\partial \boldsymbol{\sigma} : \partial f/\partial \boldsymbol{\sigma})^{1/2}} = \sqrt{\frac{3}{2}}\,\frac{\mathbf{s} - \boldsymbol{\alpha}'}{\sigma_{\mathrm{Y}}} \tag{7.93}$$

is the unit outward normal to the yield surface and K^{p} is the plastic modulus defined by

$$K^{\mathrm{p}} = c + \frac{2}{3}\frac{d\sigma_{\mathrm{Y}}}{D_{\mathrm{e}}^{\mathrm{p}}} \tag{7.94}$$

The expressions above have the same form as those derived in the plasticity theory at infinitesimal deformation except that the objective rates $\mathring{\boldsymbol{\sigma}}$ and $\mathring{\boldsymbol{\alpha}}$ are used, and the Cauchy stress $\boldsymbol{\sigma}$ and the deformation rate $\mathbf{D}$ are defined in reference to the current configuration.

For von Mises materials Eq. (7.87) becomes

$$D_{ij} = \left[\frac{\delta_{ij}\delta_{mn}}{9K} + \frac{\delta_{im}\delta_{jn}}{2G} + \frac{1}{K^{\mathrm{p}}}n_{ij}n_{mn}\right]\mathring{\sigma}_{mn} \tag{7.95}$$

where n_{ij} are the components of $\mathbf{n}$ defined by Eq. (7.93).

7.5 PLASTIC SPIN

As we mentioned earlier, the issue of which objective rate to use in the constitutive relations for finite deformation plasticity remains under discussion. There have been extensive studies by the mechanics community on the undesirable oscillatory responses of the stresses predicted by the improper use of the rate-type constitutive equations. Various objective rates have been proposed. Among these is one based on the concept of plastic spin by Dafalias (1983, 1985) which we will now discuss.

To describe the anisotropic hardening produced by plastic flow in the deformation process, the back stress tensor $\boldsymbol{\alpha}$ and its evolution with the plastic deformations are introduced into the rate-type stress-strain relations. The back stress $\boldsymbol{\alpha}$ (and the isotropic hardening parameter κ) represents macroscopically the effects of microstructural arrangements. However, it is defined directly at the macro level and determined in macroscopic experiments so that it measures an average effect of the change of the microstructure. Because of this microscopic origin of the back stress $\boldsymbol{\alpha}$, it is called the *internal structure variable*. Defalias (1983, 1985) recognized that it should be attached to the substructure of the medium and not to the continuum itself. Similarly the Cauchy stress is also supported by the substructure of the material. Therefore Dafalias proposed that the objective rate used in the constitutive relations should be corotational with the material substructure. If the spin of the substructue is denoted by $\boldsymbol{\omega}$, then the objective rates $\mathring{\boldsymbol{\sigma}}^{\omega}$ and $\mathring{\boldsymbol{\alpha}}^{\omega}$, which are corotational with the substructure, are given by

$$\mathring{\boldsymbol{\sigma}}^{\omega} = \dot{\boldsymbol{\sigma}} - \boldsymbol{\omega} \cdot \boldsymbol{\sigma} + \boldsymbol{\sigma} \cdot \boldsymbol{\omega} \tag{7.96}$$

and

$$\mathring{\boldsymbol{\alpha}}^{\omega} = \dot{\boldsymbol{\alpha}} - \boldsymbol{\omega} \cdot \boldsymbol{\alpha} + \boldsymbol{\alpha} \cdot \boldsymbol{\omega} \tag{7.97}$$

Dafalias (1983, 1985) showed that the substructure spin is the difference of the plastic spin $\mathbf{W}^{\mathrm{p}}$ from the total spin of the continuum $\mathbf{W}$, expressed by

$$\boldsymbol{\omega} = \mathbf{W} - \mathbf{W}^{\mathrm{p}} \tag{7.98}$$

which is analogous to the following expression obtained from the additive decomposition of the deformation rate:

$$\mathbf{D}^{\mathrm{e}} = \mathbf{D} - \mathbf{D}^{\mathrm{p}} \tag{7.99}$$

Using Eq. (7.98) in Eqs. (7.96) and (7.97), and noting the definition of the Jaumann rate, we obtain

$$\mathring{\boldsymbol{\sigma}}^{\omega} = \mathring{\boldsymbol{\sigma}} + \mathbf{W}^{\mathrm{p}} \cdot \boldsymbol{\sigma} - \boldsymbol{\sigma} \cdot \mathbf{W}^{\mathrm{p}} \tag{7.100}$$

$$\mathring{\boldsymbol{\alpha}}^{\omega} = \mathring{\boldsymbol{\alpha}} + \mathbf{W}^{\mathrm{p}} \cdot \boldsymbol{\sigma} - \boldsymbol{\sigma} \cdot \mathbf{W}^{\mathrm{p}} \tag{7.101}$$

here $\mathring{\boldsymbol{\sigma}}$ and $\mathring{\boldsymbol{\alpha}}$ are the Jaumann rates of $\boldsymbol{\sigma}$ and $\boldsymbol{\alpha}$, as defined in Chapter 3.

It can be seen from the preceding equations that a direct consequence of using the concept of the substructural spin $\boldsymbol{\omega}$ is the requirement for the constitutive equation of the plastic spin $\mathbf{W}^{\mathrm{p}}$, which appears in the objective rates and hence in the rate-type stress-strain relations. As a result, according to Dafalias, in the case of finite plastic deformation, it is not only necessary to propose a constitutive equation, such as Eq. (7.75), for the plastic deformation rate $\mathbf{D}^p$ but also necessary to define an evolution equation for the plastic spin $\mathbf{W}^p$. The general form of the constitutive equation for $\mathbf{W}^p$ can be written

$$\mathbf{W}^p = \lambda \boldsymbol{\Omega}^p(\boldsymbol{\sigma}, \boldsymbol{\alpha}) \tag{7.102}$$

where λ is the proportionality factor appearing in Eq. (7.75).

Dafalias showed that after applying the requirement of the frame indifference to the constitutive function $\mathbf{W}^p$, it turns out that $\boldsymbol{\Omega}^p$ should be an isotropic function of $\boldsymbol{\sigma}$ and $\boldsymbol{\alpha}$. Note that $\boldsymbol{\Omega}^p$ is antisymmetric. The general form of it can be obtained by using the representative theorem for an antisymmetric, isotropic tensorial function:

$$\boldsymbol{\Omega}^p = \eta_1(\boldsymbol{\alpha} \cdot \boldsymbol{\sigma} - \boldsymbol{\sigma} \cdot \boldsymbol{\alpha}) + \eta_2(\boldsymbol{\alpha}^2 \cdot \boldsymbol{\sigma} - \boldsymbol{\sigma} \cdot \alpha^2) + \eta_3(\alpha \cdot \sigma^2 - \boldsymbol{\sigma}^2 \cdot \boldsymbol{\alpha}) \tag{7.103}$$

$$+ \eta_4(\boldsymbol{\alpha} \cdot \boldsymbol{\sigma} \cdot \boldsymbol{\alpha}^2 - \boldsymbol{\alpha}^2 \cdot \boldsymbol{\sigma} \cdot \boldsymbol{\alpha}) + \eta_5(\boldsymbol{\sigma} \cdot \boldsymbol{\alpha} \cdot \boldsymbol{\sigma}^2 - \boldsymbol{\sigma}^2 \cdot \boldsymbol{\alpha} \cdot \boldsymbol{\sigma})$$

where $\boldsymbol{\eta}_i$, $i = 1, 2, 3, 4, 5$, are the scalar functions of the invariants of $\boldsymbol{\sigma}$ and $\boldsymbol{\alpha}$ and any other scaler variables, such as the effective plastic strain.

Equation (7.103) is a very general expression, but it is still difficult to evaluate the terms on its right-hand side, so the question of which way to formulate the objective rate remains an open one. The following simplified form of Eq. (7.103) has been used by Dafalias (1983, 1985) and Loret (1983) for the simple shear problem of rigid-plastic materials with kinematic hardening

$$\mathbf{W}^p = \lambda \eta_1(\boldsymbol{\alpha}' \cdot \mathbf{s} - \mathbf{s} \cdot \boldsymbol{\alpha}') \tag{7.104}$$

Using von Mises criterion, the equation above can be rewritten

$$\mathbf{W}^p = \lambda \eta_1[\boldsymbol{\alpha}' \cdot (\mathbf{s} - \boldsymbol{\alpha}') - (\mathbf{s} - \boldsymbol{\alpha}') \cdot \boldsymbol{\alpha}'] = \tfrac{1}{2}\eta(\boldsymbol{\alpha}' \cdot \mathbf{D}^p - \mathbf{D}^p \cdot \boldsymbol{\alpha}') \tag{7.105}$$

where $\boldsymbol{\eta}$ is an isotropic scaler-valued function of $\mathbf{s}$, $\boldsymbol{\alpha}'$, and other scaler variables and must be determined experimentally. Equation (7.105), along with the objective rates (7.100) and (7.101) will be used later to solve the simple shear problem of rigid-plastic materials with kinematic hardening.

7.6 COROTATIONAL INTEGRATION OF DEFORMATION RATE

In the plasticity theory based on the Eulerian configuration, the measure defined for deformation is the deformation rate $\mathbf{D}$. However, as we explained in Chapter 2, $\mathbf{D}$ is a measure of the changing rate of the line, area, and volume elements in the current configuration; it gives no information about deformation relative to the original configuration. One may desire to integrate **D** to obtain some kind of strain. However, due to the large rotation during finite deformation, it is generally meaningless to carry out the integration $\int \mathbf{D}\, dt$. This integration equals the logarithmic strain only when the principle axes of the stretches are fixed in space.

The problem of how to integrate **D** will also be encountered in certain constitutive equations. For example, if the endocrinic theory is to be generalized to the application of finite deformation, an integration involving **D** has to be defined. This process will be discussed in the next section. Now we turn our attention to the way to integrate the deformation rate **D**, the corotational integration.

The difficulty of defining the integration of **D** lies in the fact that it is a spatial tensor, not a material tensor. This means that it is not attached to specific material fibers that undergo large deformation and rotation. Tensor **D** is usually defined in reference to the fixed space coordinate system, while the principle axes of stretching of the material rotate relative to the system. To solve this probelm requires transforming the "spatial" deformation rate **D** into the "material" deformation rate $\mathbf{D}^{\mathrm{m}}$ which is defined in the coordinate frame as having the principal material stretch axes as the coordinate axes. The rate $\mathbf{D}^{\mathrm{m}}$ can then be integrated directly in this new coordinate system, and the final result can be obtained by transforming the integrated result back to the spatial coordinate system. The analytical procedures are given below.

Let the fixed spatial coordinate system be denoted by **x**, and the system fixed with principal material stretch directions by $\mathbf{x}^{\mathrm{m}}$. According to Chapter 2,

$$\mathbf{x}^{\mathrm{m}} = \mathbf{R}^{\mathrm{T}} \cdot \mathbf{x} \tag{7.106}$$

where **R** is the rotation tensor in the multiplicative decomposition of the deformation gradient **F**. Hence the "material" deformation rate $\mathbf{D}^{\mathrm{m}}$ can be expressed in terms of **R** and **D** as

$$\mathbf{D}^{\mathrm{m}} = \mathbf{R}^{\mathrm{T}} \cdot \mathbf{D} \cdot \mathbf{R} \tag{7.107}$$

Since $\mathbf{D}^{\mathrm{m}}$ is a material that is attached to the material frame of the principal stretches $\mathbf{x}^{\mathrm{m}}$, it can be integrated in this frame. Denote the integration by $\boldsymbol{\epsilon}_{\mathrm{R}}^{\mathrm{m}}$, then

$$\boldsymbol{\epsilon}_{\mathrm{R}}^{\mathrm{m}} = \int_0^t \mathbf{D}^{\mathrm{m}}(t')\, dt' = \int_0^t \mathbf{R}^{\mathrm{T}}(t') \cdot \mathbf{R}(t')\, dt' \tag{7.108}$$

This was called the *rotationless* deformation tensor by Storen and Rice (1975). In formulating a plasticity theory using an Eulerian description, all the quantities should be defined in the current configuration. The tensor $\boldsymbol{\epsilon}_R^m$ in the frame $\mathbf{x}^m$ should be transformed in the frame $\mathbf{x}$ in the same fashion as Eq. (7.107):

$$\boldsymbol{\epsilon}_R(t) = \mathbf{R}(t) \cdot \left[\int_0^t \mathbf{R}^T(t') \cdot \mathbf{D}(t') \cdot \mathbf{R}(t')\, dt' \right] \cdot \mathbf{R}^T(t) \qquad (7.109)$$

which is called the *corotational* integration of $\mathbf{D}$. Note that $\mathbf{R}(t)$ is the current, while $\mathbf{R}(t')$ is the rotation occurring at the moment $0 < t' < t$. In general, for any spatial tensor $\mathbf{a}$, its corotational integration can be written

$$\int_{\text{corotational } \mathbf{R}} \mathbf{a}(t')\, dt' = \mathbf{R}(t) \cdot \left[\int_0^t \mathbf{R}^T(t') \cdot \mathbf{a}(t') \cdot \mathbf{R}(t')\, dt' \right] \cdot \mathbf{R}^T(t) \qquad (7.110)$$

which is a corotational integral associated with the rigid rotation of the continuum. More generally, the corotational integral associated with any spin $\boldsymbol{\omega}$ given by

$$\boldsymbol{\omega} = \dot{\boldsymbol{\beta}} \cdot \boldsymbol{\beta}^T \qquad (7.111)$$

where $\boldsymbol{\beta}$, an arbitrary orthogonal tensor, is defined by

$$\int_{\text{corotational } \mathbf{R}} \mathbf{a}(t')\, dt' = \boldsymbol{\beta}(t) \cdot \left[\int_0^t \boldsymbol{\beta}^T(t') \cdot \mathbf{a}(t') \cdot \boldsymbol{\beta}(t')\, dt' \right] \cdot \boldsymbol{\beta}^T(t) \qquad (7.112)$$

If the spin tensor is given, for example, $\boldsymbol{\omega} = \mathbf{W} - \mathbf{W}^p$, the orthogonal tensor $\boldsymbol{\beta}$ can be obtained by integrating the following initial value problem:

$$\dot{\boldsymbol{\beta}} = \boldsymbol{\omega} \cdot \boldsymbol{\beta} \qquad (7.113)$$

The concept of the corotational integral will be used in the next section to generalize the endochronic theory for finite elastic-plastic deformation.

7.7 ENDOCHRONIC THEORY FOR FINITE DEFORMATION

In this section we take the endochronic theory originally developed by Valanis (1971) for infinitesimal deformation and generalize it for finite deformation using the concept of the corotational integral discussed above. We assume that the elastic deformation is infinitesimal and can be neglected, so the model we use is actually a constitutive description of the rigid plastic material.

According to the kinematics of finite elastic-plastic deformation presented in Section 7.1 and the polar decomposition of a nonsingular tensor, the deformation gradient can be written

$$\mathbf{F} = \mathbf{F}^{e} \cdot \mathbf{F}^{p} = \mathbf{V}^{e} \cdot \mathbf{R}^{e} \cdot \mathbf{R}^{p} \cdot \mathbf{U}^{p} \tag{7.114}$$

Since the elastic deformation is assumed to be infinitesimal,

$$\mathbf{V}^{e} \approx \mathbf{I} + \boldsymbol{\epsilon}^{e} \tag{7.115}$$

Hence Eq. (7.113) can be approximated by

$$\mathbf{F} = \mathbf{R} \cdot \mathbf{U}^{p} \tag{7.116}$$

where

$$\mathbf{R} = \mathbf{R}^{e} \cdot \mathbf{R}^{p} \tag{7.117}$$

is the total rigid rotation undergone by the material, and $\mathbf{U}^{p}$ is the plastic stretch tensor. From Eq. (7.117) it follows that

$$\mathbf{L} = \dot{\mathbf{F}} \cdot \mathbf{F}^{-1} = \left(\dot{\mathbf{R}} \cdot \mathbf{U}^{p} + \mathbf{R} \cdot \dot{\mathbf{U}}^{p}\right) \cdot \mathbf{U}^{p^{-1}} \cdot \mathbf{R}^{T} = \dot{\mathbf{R}} \cdot \mathbf{R}^{T} + \mathbf{R} \cdot \dot{\mathbf{U}}^{p} \cdot \mathbf{U}^{p^{-1}} \cdot \mathbf{R}^{T} \tag{7.118}$$

Then

$$\mathbf{D}^{p} = \mathbf{D} = \mathbf{R} \cdot \frac{1}{2}\left(\dot{\mathbf{U}}^{p} \cdot \mathbf{U}^{p^{-1}} + \mathbf{U}^{p^{-1}} \cdot \dot{\mathbf{U}}^{p}\right) \cdot \mathbf{R}^{T} \tag{7.119}$$

$$\mathbf{W} = \dot{\mathbf{R}} \cdot \mathbf{R}^{T} + \mathbf{R} \cdot \tfrac{1}{2}\left(\dot{\mathbf{U}}^{p} \cdot \mathbf{U}^{p^{-1}} - \mathbf{U}^{p^{-1}} \cdot \dot{\mathbf{U}}^{p}\right) \cdot \mathbf{R}^{T} = \dot{\mathbf{R}} \cdot \mathbf{R}^{T} + \mathbf{W}^{p} \tag{7.120}$$

where the definition for plastic spin is obvious. Comparing Eq. (7.119) with Eq. (7.107), we see that

$$\mathbf{D}^{*} = \frac{1}{2}\left(\dot{\mathbf{U}}^{p} \cdot \mathbf{U}^{p^{-1}} + \mathbf{U}^{p^{-1}} \cdot \dot{\mathbf{U}}^{p}\right) \tag{7.121}$$

is the "rotationless" deformation rate defined with respect to the rotating material frame

$$\mathbf{x}^{*} = \mathbf{R}^{T} \cdot \mathbf{x} \tag{7.122}$$

where $\mathbf{x}$ is the fixed space coordinate frame. The Cauchy stress $\boldsymbol{\sigma}$ in frame $\mathbf{x}$ is related to the corresponding one $\boldsymbol{\sigma}^{*}$ in $\mathbf{x}^{*}$ by

$$\boldsymbol{\sigma} = \mathbf{R} \cdot \boldsymbol{\sigma}^{*} \cdot \mathbf{R}^{T} \tag{7.123}$$

Since $\mathbf{D}^{*}$ is a material tensor and rotationless, it is reasonable to propose the following analogous constitutive equation for finite elastic-plastic deformation from the one for infinitesimal deformation discussed in Chapter 6:

$$\mathbf{S}^* = 2\mu \int_0^Z \rho(Z - Z') \frac{\mathbf{D}^*(Z')}{\dot{Z}'} \, dZ' \tag{7.124}$$

where the internal variables ζ and Z are defined by

$$\dot{\zeta} = \frac{d\zeta}{dt} = [\mathbf{D}^* : \mathbf{D}^*]^{1/2} \tag{7.125}$$

$$\dot{Z} = \frac{\dot{\zeta}}{f(\zeta)} \tag{7.126}$$

with $f(0) = 0$. Various function forms of $f(\zeta)$ used for infinitesimal deformation can be used here. For example,

$$f(\zeta) = 1 + \beta\zeta \tag{7.127}$$

or

$$f(\zeta) = a + (1 - a)e^{-\gamma\zeta} \tag{7.128}$$

where β or a and γ are material constants that must be determined experimentally. Introducing

$$\rho(Z) = \rho_0 \delta(Z) + \rho_1(Z) \tag{7.129}$$

where $\delta(Z)$ is the Dirac delta function, we rewrite Eq. (7.124) as

$$\mathbf{S}^* = 2\mu\rho_0 \frac{\mathbf{D}^*}{\dot{\mathbf{Z}}} + 2\mu \int_0^z \rho_1(Z - Z') \frac{\mathbf{D}^*(Z')}{\dot{Z}'} \, dZ' \tag{7.130}$$

which can be rearranged in the following form:

$$\mathbf{D}^* = \frac{\mathbf{S}^* - \boldsymbol{\alpha}^*}{\tau_Y f(\zeta)} \zeta \tag{7.131}$$

where

$$\boldsymbol{\alpha}^* = 2\mu \int_0^z \rho_1(Z - Z') \frac{\mathbf{D}^*}{\dot{Z}'} \, dZ' \tag{7.132}$$

$$\tau_Y = 2\mu\rho_0 \tag{7.133}$$

and $\tau_{Y'}$ as in Chapter 6, represents the initial yield stress in shear. Transforming Eq. (7.131) into the current configuration using Eqs. (7.119) and (7.123) yields

$$\mathbf{D} = \frac{\mathbf{S} - \boldsymbol{\alpha}}{\tau_Y f(\zeta)}\zeta \tag{7.134}$$

where

$$\alpha = 2\mu R(Z) \cdot \left[\int_0^z \rho_1(Z - Z')\frac{D^*(Z')}{\dot{Z}'}\, dZ'\right] \cdot R^T(Z) = R \cdot \alpha^* \cdot R^T \tag{7.135}$$

which represents the back stress tensor in the current configuration. It should be noted that the expression above can not be written in the form

$$\boldsymbol{\alpha} = 2\mu \int_0^z \rho_1(Z - Z')\frac{\mathbf{D}(Z')}{\dot{Z}'}\, dZ' \tag{7.136}$$

since $\mathbf{R}(Z)$ is the current rigid rotation and $\mathbf{D}^*(Z')$ in Eq. (7.135) is the plastic deformation rate at Z', for $0 \leq Z' \leq Z$.

Now the evaluation equation for $\boldsymbol{\alpha}$ can be obtained by taking the material derivative of Eq. (7.135):

$$\dot{\boldsymbol{\alpha}} = \dot{\mathbf{R}} \cdot \boldsymbol{\alpha}^* \cdot \mathbf{R}^T + \mathbf{R} \cdot \left[2\mu\rho_1(0)\mathbf{D}^* + 2\mu\left(\int_0^z \frac{d\rho_1(Z - Z')}{dZ}\frac{\mathbf{D}^*(Z')}{\dot{Z}'}\, dZ'\right)\frac{\zeta}{f(\zeta)}\right] \cdot \mathbf{R}^T + \mathbf{R} \cdot \boldsymbol{\alpha}^* \cdot \dot{\mathbf{R}}^T \tag{7.137}$$

This equation can be rearranged into

$$\dot{\boldsymbol{\alpha}} - \dot{\mathbf{R}} \cdot \mathbf{R}^T \cdot \mathbf{R} \cdot \boldsymbol{\alpha}^* \cdot \mathbf{R}^T - \mathbf{R} \cdot \boldsymbol{\alpha}^* \cdot \mathbf{R}^T \cdot \mathbf{R} \cdot \dot{\mathbf{R}}^T = 2\mu\left[\rho_1(0)\mathbf{D} + \frac{\mathbf{h}}{f(\zeta)}\zeta\right] \tag{7.138}$$

where

$$\mathbf{h} = \mathbf{R} \cdot \left[\int_0^z \frac{d\rho_1(Z - Z')}{dz}\frac{\mathbf{D}^*(Z')}{\dot{Z}'}\, dZ'\right] \cdot \mathbf{R}^T = R(Z) \cdot \left[\int_0^z \frac{d\rho_1(Z - Z')}{dZ}\frac{\mathbf{R}^T(Z') \cdot \mathbf{D}(Z') \cdot \mathbf{R}(Z')}{\dot{Z}'}\, dZ'\right] \cdot \mathbf{R}^T(Z) \tag{7.139}$$

Note that the left-hand side of Eq. (7.138) is the Green-McInnis corotational rate of $\boldsymbol{\alpha}$; hence it can be rewritten as

$$\overset{\nabla}{\boldsymbol{\alpha}} = 2\mu\left[\rho_1(0)\mathbf{D} + \frac{\mathbf{h}}{f(\zeta)}\zeta\right] \tag{7.140}$$

Equations (7.134) and (7.140) are the evolution equations for the plastic deformation rate $\mathbf{D}$ (assuming that $\mathbf{D}^e$ is negligible) and the back stress $\boldsymbol{\alpha}$. It should be emphasized that, in general, $\mathbf{h}$ in this equation must be calculated using Eq. (7.138); it cannot be written

$$\mathbf{h} = \int_0^z \frac{d\rho_1(Z - Z')}{dZ}\frac{\mathbf{D}(Z')}{\dot{Z}'}\,dZ' \tag{7.141}$$

Again, the reason is that in general $\mathbf{R}(Z')$ is not equal to $\mathbf{R}(Z)$ and $\mathbf{R}(Z) \cdot \mathbf{R}^{\mathrm{T}}(Z') \neq \mathbf{I}$.

However, if the kernel function $\rho_1(Z)$ is of the form

$$\rho_1(Z) = \sum_i \rho_{1i}\exp(-\xi_i Z) \tag{7.142}$$

the evolution equation for $\boldsymbol{\alpha}$ can then be further simplified. In fact, using Eq. (7.142), we can calculate

$$\frac{d\rho_1(Z)}{dZ} = -\sum_i \xi_i\rho_{1i}\exp(-\xi_i Z) \tag{7.143}$$

and

$$\alpha = R(Z) \cdot \left[\sum_i 2\mu\rho_{1i}\int_0^z e^{-\xi_i(Z-Z')}\frac{D^*(Z')}{\dot{Z}'}\,dZ'\right] \cdot R^T(Z) = \sum_i \alpha_i \tag{7.144}$$

where α_i is considered to be secondary back stress given by

$$\alpha_i = R(Z) \cdot \left[2\mu\rho_{1i}\int_0^z e^{-\xi_i(Z-Z')}\frac{D^*(Z')}{\dot{Z}'}\,dZ'\right] \cdot R^T(Z) = R \cdot \alpha_i^* \cdot R^T \tag{7.145}$$

Substituting Eq. (7.143) into Eq. (7.139), we have

$$h = R(Z) \cdot \left[\sum_i -\zeta_i\rho_{1i}\int_0^z e^{-\xi_i(Z-Z')}\frac{D^*(Z')}{\dot{Z}'}\,dZ'\right] \cdot R^T(Z) = -\frac{1}{2\mu}\sum_i \xi_i\alpha_i \tag{7.146}$$

where Eq. (7.145) has been used. Substituting this result into Eq. (7.140) gives the following evolution equation for α when the kernel function $\rho_1(Z)$ is given by Eq. (7.142):

$$\overset{\triangledown}{\boldsymbol{\alpha}} = \sum_i \overset{\triangledown}{\boldsymbol{\alpha}}_i = 2\mu \sum_i \left[\rho_{1i}\mathbf{D} - \frac{\zeta \xi_i}{2\mu f(\zeta)} \boldsymbol{\alpha}_i \right] \tag{7.147}$$

It can be seen by comparing Eq. (7.147) with Eq. (7.140) that the former is greatly simplified because of the use of the exponential form (7.144) of $\rho_1(Z)$. This is quite clear if only one term is chosen for $\rho_1(Z)$:

$$\rho_1(Z) = \rho_{10} e^{-\xi_0 Z} \tag{7.148}$$

Then Eq. (7.147) becomes

$$\overset{\triangledown}{\boldsymbol{\alpha}} = \sum_i \overset{\triangledown}{\boldsymbol{\alpha}}_i = 2\mu \left[\rho_{10}\mathbf{D} - \frac{\zeta \xi_0}{2\mu f(\zeta)} \boldsymbol{\alpha} \right] = 2\mu \left[\rho_{10}\mathbf{D} - \frac{\xi_0 \alpha}{2\mu f(\zeta)} (\mathbf{D}:\mathbf{D})^{1/2} \right] \tag{7.149}$$

where

$$\zeta = (\mathbf{D}^* : \mathbf{D}^*)^{1/2} = (\mathbf{D}:\mathbf{D})^{1/2} \tag{7.150}$$

is used to obtain the final expression. Equation (7.149) can be viewed as the generalization of Eq. (6.104) or Eq. (6.171) to the case of finite deformation under the assumption of rigid plasticity.

It should be pointed out here that, in general, tensor $\mathbf{h}$ must be calculated using Eq. (7.139) which requires the explicit expression for the rotation tensor $\mathbf{R}(Z)$ and a complicated integration. The tensor $\mathbf{h}$ so determined has to be substituted into Eq. (7.140) in order to find the evolution of the back stress $\boldsymbol{\alpha}$. In particular, if the kernel is assumed to be of the exponential form (7.142), the evolution of $\boldsymbol{\alpha}$ is determined by Eq. (7.147) which does not require an explicit expression for $\mathbf{R}$ and the corresponding integration.

Under the assumption of rigid plasticity, the corotational rate with the spin of the substructure can be introduced to replace the Green-McInnis rate. Comparing Eq. (7.98) with Eq. (7.120), we have

$$\boldsymbol{\omega} = \mathbf{W} - \mathbf{W}^{\mathrm{p}} = \dot{\mathbf{R}} \cdot \mathbf{R}^{\mathrm{T}} \tag{7.151}$$

Here $\dot{\boldsymbol{\omega}}$ represents the spin of the substructure. Introducing this term into Eq. (7.140) results in

$$\overset{\circ}{\boldsymbol{\alpha}} = \sum_i \overset{\circ}{\boldsymbol{\alpha}}_i = 2\mu \left[\rho_1(0)\mathbf{D} + \frac{\mathbf{h}}{f(\zeta)} \zeta \right] + \mathbf{W}^{\mathrm{p}} \cdot \boldsymbol{\alpha} - \boldsymbol{\alpha} \cdot \mathbf{W}^{\mathrm{p}} \tag{7.152}$$

Where $\mathring{\sigma}$ is the Jaumann rate of $\boldsymbol{\alpha}$, with respect to the continuum rotation $\mathbf{W}$.

Similarly, when the exponential for of $\rho_1(Z)$ is used, the equation reduces to

$$\mathring{\boldsymbol{\alpha}} = \sum_i \mathring{\boldsymbol{\alpha}}_i = 2\mu \sum_i \left[\rho_{1i}\mathbf{D} - \frac{\zeta \xi_i}{2\mu f(\zeta)} \boldsymbol{\alpha}_i \right] + \mathbf{W}^{\mathrm{p}} \cdot \boldsymbol{\alpha} - \boldsymbol{\alpha} \cdot \mathbf{W}^{\mathrm{p}} \quad (7.153)$$

where $\mathbf{W}^{\mathrm{p}}$ is the plastic spin and can be expressed in the form of Eq. (7.103), in general, or in the form of Eq. (7.105), in particular. Equation (7.153) was derived by Im and Atluri (1987) according to the concept of isoclinic configuration originated by Mandel (1973). Im and Atluri then used Eq. (7.153) to solve the problem of simple shear at finite deformation. Different material constants were tested, and satisfactory results were obtained by this model. The interested reader can refer to their work for more details. Only primary features of this theory are listed below:

1. The theory corresponds to the von Mises yield criterion in terms of the Cauchy stress. This can be easily verified by multiplying each side of Eq. (7.134) by itself to obtain

$$(\mathbf{S} - \boldsymbol{\alpha}) : (\mathbf{s} - \boldsymbol{\alpha}) = \tau_Y^2 f^2(\zeta) \quad (7.154)$$

   which is of the form of the von Mises yield function.
2. The isotropic hardening is described in the theory by choosing the proper form of the function $f(\zeta)$. If $f(\zeta)$ is a monotonically increasing function, it describes the hardening effect, since the subsequent yield surface (7.154) will expand uniformly in the course of the plastic deformation. On the other hand, the softening effect can be predicted by using a monotonically decreasing function $f(\zeta)$.
3. The anisotropy of the material induced by the plastic deformation is represented by the variable $\boldsymbol{\alpha}$. As in the case of infinitesimal plasticity theory, various kinds of rules for the translation of the center of the yield surface can be retrieved by the proper form of the kernel function $\rho_1(Z)$. This process has been explored in detail by Watanabe and Atluri (1986) and also in Chapter 6 of this book.

7.8. SIMPLE SHEAR

This section examines solutions for finite simple shear in the plastic region. We will use several corotational rates to illustrate the different results as we did with the hypoelasticity theory. Our discussion is analogous to that in Section 7.3, with one exception: Instead of the Cauchy stress $\boldsymbol{\sigma}$, we use the back stress $\boldsymbol{\alpha}$. The material here is assumed to be rigid plastic, and hardening is purely kinematic. Therefore we give the yield surface by

$$F = \tfrac{3}{2}(\mathbf{s} - \boldsymbol{\alpha}) : (\mathbf{s} - \boldsymbol{\alpha}) - \sigma_Y^{0^2} = 0 \tag{7.155}$$

where $\sigma_Y^{0^2}$ is a constant representing the size of the yield surface. Without loss of generality, the back stress $\boldsymbol{\alpha}$ is assumed to be deviatoric. The plastic deformation rate $\mathbf{D}^p$ is given by Eq. (7.92), rewritten as

$$\mathbf{D}^p = \frac{1}{K^p}(\mathbf{n} : \mathring{\boldsymbol{\sigma}})\mathbf{n} \tag{7.156}$$

where $\mathring{\boldsymbol{\sigma}}$ is any one of the corotational rates of $\boldsymbol{\sigma}$,

$$\mathrm{n} = \sqrt{\frac{3}{2}} \frac{\mathrm{s} - \alpha}{\sigma_Y^{0^2}} \tag{7.157}$$

and K^p is the plastic modulus. For rigid plastic materials,

$$\mathbf{D} = \mathbf{D}^p \tag{7.158}$$

For the evolution of the back stress $\boldsymbol{\alpha}$, we assume that

$$\mathring{\boldsymbol{\alpha}} = c\mathbf{D}^p = c\mathbf{D} \tag{7.159}$$

where c is a constant and, according to Eq. (7.94), $K^p = c$.

Let $\bar{\boldsymbol{\omega}}$ represent an arbitrary spin tensor. The objective corotational rate of $\boldsymbol{\sigma}$ and $\boldsymbol{\alpha}$ with $\bar{\boldsymbol{\omega}}$ can be written in the following general form:

$$\mathring{\boldsymbol{\sigma}} = \dot{\boldsymbol{\sigma}} - \bar{\boldsymbol{\omega}} \cdot \boldsymbol{\sigma} + \boldsymbol{\sigma} \cdot \bar{\boldsymbol{\omega}} \tag{7.160}$$

$$\mathring{\boldsymbol{\alpha}} = \dot{\boldsymbol{\alpha}} - \bar{\boldsymbol{\omega}} \cdot \boldsymbol{\alpha} + \boldsymbol{\alpha} \cdot \bar{\boldsymbol{\omega}} \tag{7.161}$$

The corotational rates used in this section include

1. Jaumann rate, $\bar{\boldsymbol{\omega}} = \mathbf{W}$
2. Green-McInnis rate, $\bar{\boldsymbol{\omega}} = \dot{\mathbf{R}} \cdot \mathbf{R}^{\mathrm{T}}$
3. Rate proposed by Dafalias, $\bar{\boldsymbol{\omega}} = \boldsymbol{\omega} = \mathbf{W} - \mathbf{W}^p$
4. Rate proposed by Lee et al. (1982), $\bar{\boldsymbol{\omega}} = \mathbf{W}^*$
   where

$$\mathbf{W}^* = \mathbf{W} + \mathbf{D} \cdot \mathbf{n}\mathbf{n}^{\mathrm{T}} - \mathbf{n}\mathbf{n}^{\mathrm{T}} \cdot \mathbf{D} \tag{7.162}$$

is the spin of the unit vector $\mathbf{n}$ attached to the material fiber which is instantaneously oriented along the principal direction of the back stress $\boldsymbol{\alpha}$. $\mathbf{W}^*$ was derived by Lee et al. (1983) using the standard method of continuum mechanics.

For simple shear it has been shown that

$$\mathbf{D} = \begin{bmatrix} 0 & \frac{\dot{\gamma}}{2} & 0 \\ \frac{\dot{\gamma}}{2} & 0 & 0 \\ 0 & 0 & 0 \end{bmatrix}, \quad \mathbf{W} = \begin{bmatrix} 0 & \frac{\dot{\gamma}}{2} & 0 \\ -\frac{\dot{\gamma}}{2} & 0 & 0 \\ 0 & 0 & 0 \end{bmatrix} \tag{7.163}$$

$$\mathbf{R} = \begin{bmatrix} \cos\phi & \sin\phi & 0 \\ -\sin\phi & \cos\phi & 0 \\ 0 & 0 & 0 \end{bmatrix} \tag{7.164}$$

where $\phi = \tan^{-1}(\gamma/2)$ and γ is shear strain at finite deformation.

Using Eq. (7.163), we can easily calculate from Eq. (7.156) and $\sigma_{33} = \sigma_{23} = \sigma_{13} = 0$ the following:

$$S_{11} = \alpha_{11} \tag{7.165}$$

$$S_{22} = \alpha_{22} \tag{7.166}$$

$$S_{33} = \alpha_{33} \tag{7.167}$$

$$\alpha_{13} = \alpha_{23} = 0 \tag{7.168}$$

The yield criterion in this case becomes

$$\sigma_{12} = \frac{\sigma_{Y0}}{\sqrt{3}} + \alpha_{12} \tag{7.169}$$

Therefore the stress state can be determined if the back stress $\boldsymbol{\alpha}$ is known. The key to this problem is to calculate the back stress $\boldsymbol{\alpha}$ using Eq. (7.159). Comparing Eq. (7.159) with the hypoelastic constitutive equation (7.41), and noting that $\operatorname{tr}\mathbf{D} = 0$ in the case of simple shear, it is obvious that Eq. (7.159) can be obtained from Eq. (7.41) by simply replacing $\boldsymbol{\sigma}$ and $\boldsymbol{\alpha}$ and $2G$ with c. The results for $\boldsymbol{\sigma}$ in Section 7.3 can be used to provide the solutions of $\boldsymbol{\alpha}$ to Eq. (7.159) with the above-mentioned change of variables if the corotational rates used in both cases are the same. Thus the solutions using the Jaumann rate and the Green-McInnis rate can be obtained immediately and are given below.

7.8.1 Solution for Jaumann Rate

Assume the material is initially isotropic, $\boldsymbol{\alpha}(0) = \mathbf{0}$. The only nonzero components of the back stress $\boldsymbol{\alpha}$ are α_{11}, α_{22}, and α_{12} and are given by

$$\alpha_{11} = -\alpha_{22} = \frac{c}{2}(1 - \cos\gamma), \quad \alpha_{12} = \frac{c}{2}\sin\gamma \tag{7.170}$$

From Eq. (7.167) and $\alpha_{33} = \sigma_{33} = 0$, it follows that

$$S_{33} = 0, \quad \sigma_{11} + \sigma_{22} = 0 \tag{7.171}$$

Therefore the nonzero components of the Cauchy stress $\boldsymbol{\sigma}$ are given by

$$S_{11} = \sigma_{11} = \alpha_{11} = \frac{c}{2}(1 - \cos\gamma) \tag{7.172}$$

$$S_{22} = \sigma_{22} = \alpha_{22} = \frac{c}{2}(1 - \cos\gamma) \tag{7.173}$$

$$\sigma_{12} = \frac{\sigma_{Y0}}{\sqrt{3}} + \frac{c}{2}\sin\gamma \tag{7.174}$$

These results are in accordance with those obtained by Nagtegaal et al. (1982) and indicate that the stress solution calculated using the Jaumann rate oscillates with a period of about six as the shear strain increases monotonically. The Jaumann rate therefore is considered inappropriate for use in the constitutive equations for finite elastic-plastic deformation.

7.8.2 Solution for Green-McInnis Rate

The nonzero components α_{11}, α_{22}, and α_{12} of the back stress $\boldsymbol{\alpha}$ can be expressed by

$$\alpha_{11} - \alpha_{22} = 2c\left[\cos 2\phi \ln(\cos\phi) + \phi\sin 2\phi - \sin^2 2\phi\right] \tag{7.175}$$

$$\alpha_{12} = c\cos 2\phi\left[2\phi - 2\tan 2\phi\ln(\cos\phi) - \tan\phi\right] \tag{7.176}$$

and the nonzero components of the Cauchy stress σ for σ_{11}, σ_{22} given by

$$\sigma_{11} = -\sigma_{22} = 2c\left[\cos 2\phi\ln(\cos\phi) + \phi\sin 2\phi - \sin^2 2\phi\right] \tag{7.177}$$

$$\sigma_{12} = c\cos 2\phi\left[2\phi - 2\tan 2\phi\ln(\cos\phi) - \tan\phi\right] \tag{7.178}$$

These solutions are similar to those obtained for hypoelastic materials under simple shear by using the Green-McInnis rate. It has been shown by

Dienes(1979) and Dafalias (1983) that σ_{11}, σ_{22}, and σ_{12} given above monotonically increase with the shear strain γ. Therefore the rate $\boldsymbol{\sigma}$ is considered an appropriate rate to use in the constitutive equations for finite elastic-plastic deformation.

7.8.3 Solution for $\bar{\boldsymbol{\omega}} = \mathbf{W} - \mathbf{W}^{\mathrm{p}}$

As we explained earlier, the plastic spin $\mathbf{W}^{\mathrm{p}}$ used by Dafalias (1983) is given by

$$\mathbf{W}^{\mathrm{p}} = \tfrac{1}{2}\eta(\boldsymbol{\alpha} \cdot \mathbf{D}^{\mathrm{p}} - \mathbf{D}^{\mathrm{p}} \cdot \boldsymbol{\alpha}) = \tfrac{1}{2}\eta(\boldsymbol{\alpha} \cdot \mathbf{D} - \mathbf{D} \cdot \boldsymbol{\alpha}) \tag{7.179}$$

Where η is a material constant. Note that the nonzero components of $\boldsymbol{\alpha}$ are $\alpha_{11} = -\alpha_{12}$ and α_{12}, and $\mathbf{D}^{\mathrm{p}} = \mathbf{D}$ is given by Eq. (7.163). It is easy to calculate that

$$W_{12}^{\mathrm{p}} = -W_{21}^{\mathrm{p}} = \tfrac{1}{2}\eta\alpha_{11}\dot{\gamma} \tag{7.180}$$

and all other components of $\mathbf{W}^{\mathrm{p}}$ are zero. Similarly the only nonzero components of $\bar{\boldsymbol{\omega}}$ are $\bar{\omega}_{12} = -\bar{\omega}_{21}$ and are given by

$$\bar{\omega}_{12} = -\bar{\omega}_{21} = \frac{\dot{\gamma}}{2}(1 - \eta\alpha_{11}) \tag{7.181}$$

Substituting these results into Eq. (7.159) leads to the following differential equations.

$$\frac{d\alpha_{11}}{d\gamma} = (1 - \eta\alpha_{11})\alpha_{12} \tag{7.182}$$

$$\frac{d\alpha_{12}}{d\gamma} = \frac{C}{2} - (1 - \eta\alpha_{11})\alpha_{11} \tag{7.183}$$

These were solved numerically by Defalias (1983) and the solutions are shown schematically in Fig. 7.3(a)–(c).

As these figures show, nonoscillatory results can be obtained by the proper choice of the constant ρ. For example, the solutions for $\sigma_{11} = -\sigma_{22}$ and σ_{12} using $\rho > 0.5$ are monotonically increasing with the shear strain γ and hence are physically plausible.

One advantage of using the corotational rate with the spin of the substructure is that it has some flexibility to represent the real behavior of the material under finite plastic deformation, since the material constant n can be chosen in such a way as to achieve the best fit between the theoretical results and the experimental curves. On the other hand, if the Jaumann rate

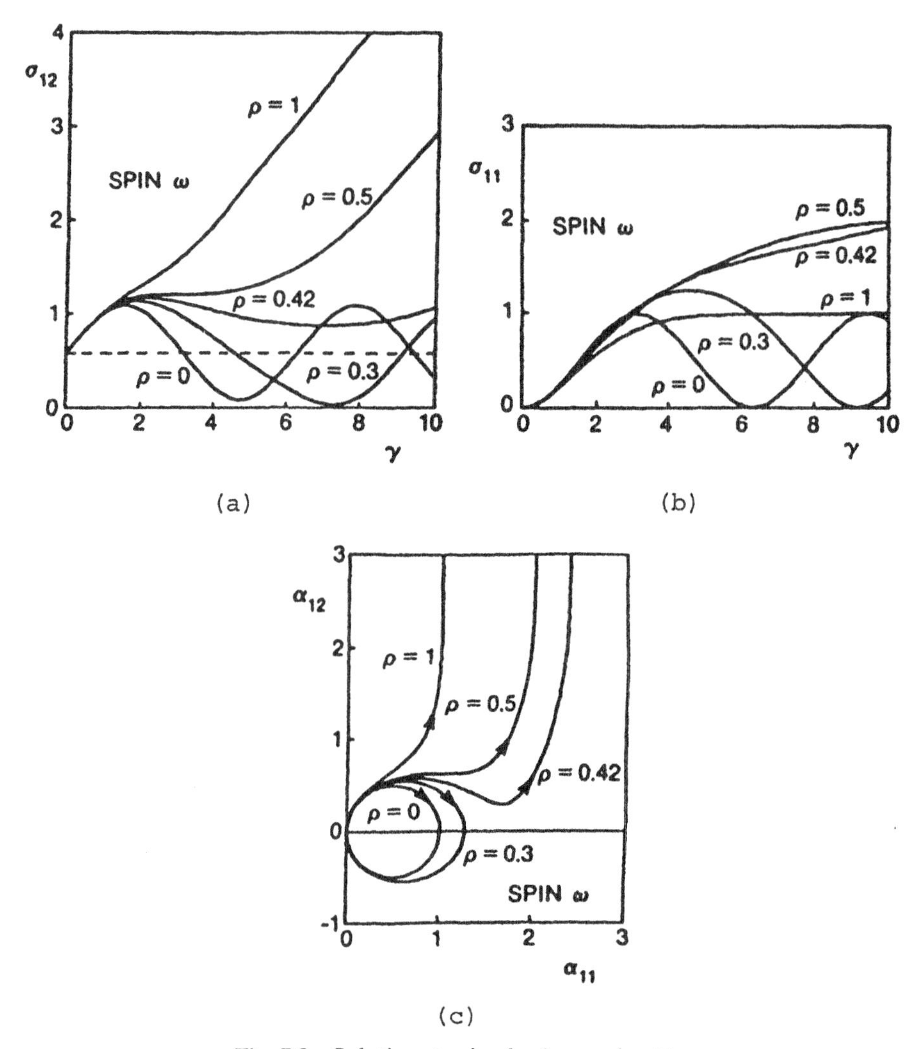

Fig. 7.3 Solutions to simple shear using $\overline{\omega}$

or the Green-Innis rate is used, no unidentified material constant is introduced, and the responses are less flexible.

7.8.4 Solution for $\overline{\omega} = \mathbf{W} + \mathbf{D} \cdot \mathbf{n}\mathbf{n}^{\mathrm{T}} - \mathbf{n}\mathbf{N}^{\mathrm{T}} \cdot \mathbf{D}$

It has been shown by Dafalias (1983) that the nonzero components of $\overline{\omega}$ for simple shear are $\overline{\omega}_{12} = \overline{\omega}_{21}$ and are given in the form

$$\overline{\omega}_{12} = -\overline{\omega}_{21} = \frac{\dot{\gamma}}{2}\left[1 - \frac{\alpha_{11}}{\left(\alpha_{11}^2 - \alpha_{12}^2\right)^{1/2}}\right] \tag{7.184}$$

Equation (7.159) leads to the following differential equations:

$$\frac{d\alpha_{11}}{d\gamma} = \alpha_{12}\left[1 - \frac{\alpha_{11}}{\left(\alpha_{11}^2 - \alpha_{12}^2\right)^{1/2}}\right] \tag{7.185}$$

$$\frac{d\alpha_{12}}{d\gamma} = \frac{C}{2} - \alpha_{11}\left[1 - \frac{\alpha_{11}}{\left(\alpha_{11}^2 - \alpha_{12}^2\right)^{1/2}}\right] \tag{7.186}$$

These were solved numerically by Lee et al. (1983), and the results are shown in Fig. 7.4(*a*)–(*c*).

In the figure the stresses obtained are clearly nonoscillatory. In fact they are increasing monotonically with shear strain γ. Thus the corotational rate

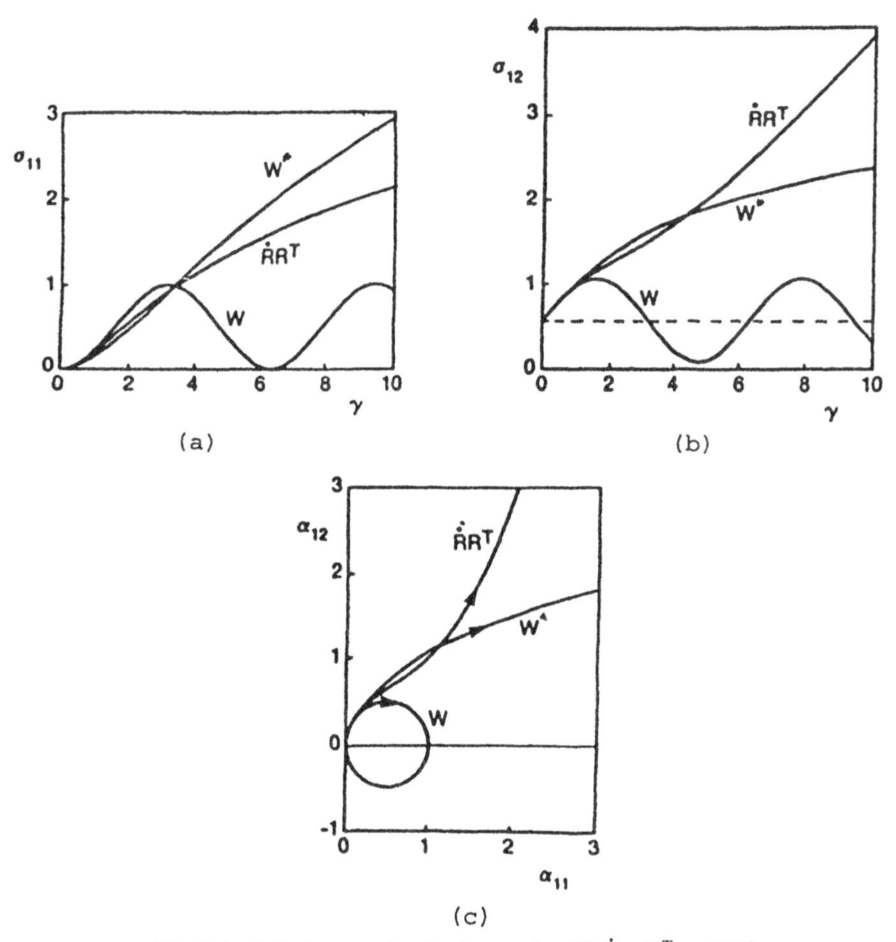

Fig. 7.4 Solutions to simple shear using **W**, $\dot{\mathbf{R}} \cdot \mathbf{R}^T$ and **W***

associated with **W*** effectively removes the oscillation effect and hence can be used in the constitutive equations for finite elastic-plastic deformation.

In concluding this chapter, we again emphasize that the particular choice of a corotational rate should not be governed solely by the requirements of objectivity and nonoscillation but also by experimental evidence. The main objective of a constitutive model is to provide the best possible description of the real behaviors of materials.

REFERENCES

Atluri, S. N. 1984. On constitutive relations at finite strain: Hypo-elasticity and elasto-plasticity with isotropic kinematic hardening. *Comput. Methods Appl. Mech. Eng.* **43**: 137.

Dafalias, Y. F. (1985) The plastic spin. *ASME J. Appl. Mech.* **52**: 865.

Defalias, Y. F. 1983. A missing link in the macroscopic constitutive formulation of large plastic deformations. *Plasticity Today*, ed. A. Sawczuk, and G. Bianchi. p. 135. Elsevier Applied Science Publishers, London.

Dienes, J. K. 1979. On the analysis of rotation and stress rate in deforming bodies. *Acta Mech.* **32**: 217.

Green, A. E., and B. C. McInnis. 1967. Generalized hypoelasticity. *Proc. R. Soc., Edinburgh* **A57**, pt. 3: 220.

Green, A. E., and P. M. Naghdi. 1971. Some remarks on elastic-plastic deformation at finite strain. *Int. J. Eng. Sc.*: 1219.

Green, A. E., and P. M. Naghdi. 1966. A thermodynamic development of elastic-plastic continua. *Proceedings, IUTRAM Symposium on Irreversible Aspects of Continuum Mechanics and Transfer of Physical Characteristics in Moving Fluids.* New York: Springer-Verlag, p. 117.

Green, A. E., and P. M. Naghdi. 1965. A general theory of an elastic-plastic continuum. *Arch. Rat. Mech. Anal.* **18**: 251.

Im, S., and S. N. Atluri, 1987. A study of two finite strain plasticity models: An internal time theory using Mandel's director concept and a general isotropic/kinematic hardening theory. *Int. J. Plast.* **3**: 163.

Johnson, G. C., and D. J. Bammann. 1984. A discussion of stress rate in finite deformation problems, *Int. J. Solids Struct.* **20**: 725.

Lee, E. H. 1981. Some comments on elastic-plastic analysis. *Int. J. Solids Struct.* **17**: 859.

Lee, E. H., and D. T. Lin. 1967. Finite-strain elastic-plastic theory particularly for plane wave analysis. *J. Appl. Phys.* **38**: 19.

Lee, E. H., R. L. Mallet, and T. B. Wertheimer. 1983. Stress analysis of anisotropic hardening in finite-deformation plasticity. *ASME, J. Appl. Mech.*, **50**: 554.

Mandel, J. 1973. Thermodynamics and plasticity. *Foundations of Continuum Thermodynamics*, ed. J. J. Delgado Domingos, M. N. R. Nina, and J. H. Whitelaw. New York: Wiley, p. 283.

Nagtegaal, J. C., and J. E. de Jong. 1982. Some aspects of nonisotropic work hardening in finite strain plasticity. *Plasticity of Metals at Finite Strain: Theory, Experiment and Computation*, ed. E. H. Lee, and R. L. Mallet. p. 65. Stanford University Press.

Nemat-Nasser, S. 1979. Decomposition of strain measures and their rates in finite deformation elastic-plasticity. *Int. J. Solids Struct.* **15**: 155.

Prager, W. 1961. Introduction to Mechanics of Continua. Boston: Ginn.

R. S. Rivlin, 1955. Further remarks on the stress-deformation relations for isotropic materials. *J. Rat. Mech. Anal.* **4**: 681.

Storen, S., and J. R. Rice, 1975. Localized necking in thin sheets. *J. Mech. Phys. Solids* **23**: 421.

C. Truesdell, 1955. The simplest rate theory of pure elasticity. *Comm. Pure Appl. Math.* **8**: 123.

Valansi, C. 1971. A theory of viscoplasticity without a yield surface. Part I. General theory. *Arch. Mech.* **23**: 517.

Zbib, H. M., and E. C. Aifantis. 1988. On the concept of relative and plastic spins and its implications to large deformation theories. Part I: Hypoelasticity and vertex-type plasticity, *Acta Mech.* **75**: 15.

8

STRAIN SPACE FORMULATIONS FOR PLASTIC DEFORMATION

Classical and recently developed plasticity theories that are formulated in the stress space have been discussed in previous chapters. In these theories the stress and its increment are treated as the independent variables, and the strain and its increment are assumed to be the dependent variables and are calculated from the stress and its increment. In the first mathematical representation of plastic flow by the Prandtl-Reuss Eq. (5.87), the plastic strain rate or increment is expressed in terms of the devitoric part of the stress and the increment of stress through the scaler factor λ. In addition the constitutive equations for the work hardening parameters α and κ, the yield surface, and the loading-unloading criteria are formulated with reference to the stress space.

While this stress space formulation represents the traditional approach in plasticity theory and has been commonly accepted in engineering applications, it has certain inherent disadvantages. Naghdi and Trapp (1975a, 1975b, 1975c) pointed out two significant shortcomings of the stress space formulation that restrict the theory's usefulness:

1. A plasticity theory based on stress space can not be directly reduced to represent the case of an elastic–perfectly plastic material, so a separate treatment is required.

As we noted in Section 5.10 for the behavior of perfect plastic deformation of materials, the Drucker's postulate concerning stable or hardening materials must be modified, and the elastic deformation has to be considered to derive the stress-strain increment relation. By Eq. (5.201) we could see that as $dk/d\epsilon_e^p$ approaches zero (i.e., no work hardening), the relation between $d\boldsymbol{\epsilon}$ and $d\boldsymbol{\sigma}$ becomes singular. It has been observed experimentally that for

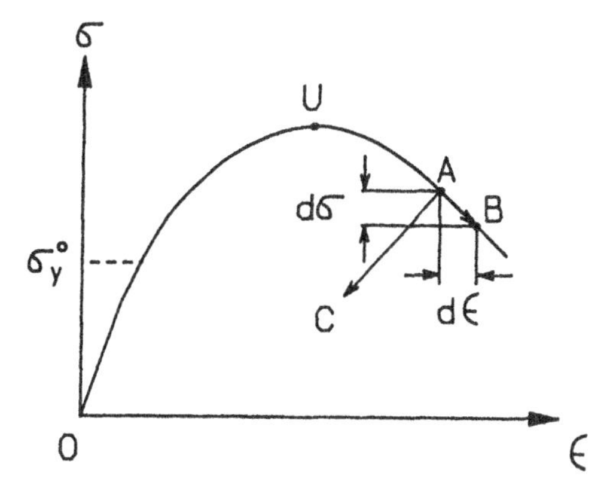

Fig. 8.1 Stress-strain curve of work-softening materials

many structural materials such as mild steel, the slope of the uniaxial stress-strain curve becomes zero at its ultimate strength and is so small that it nearly vanishes in the vicinity of its ultimate strength. Within that vicinity stress space formulations may not offer reliable results for the analysis of problems such as necking and instability.

2. There is some confusion in the formulation of the loading-unloading criteria in stress space for unstable or work-softening materials.

Figure 8.1 gives a schematic of the work-softening behavior in uniaxial loading. Let us consider a point A in the plastic deformation region that is in the descending part of the curve. Starting from point A, additional plastic deformation will occur along AB, or an elastic unloading will occur along AC. It is interesting to note that both the plastic deformation and the elastic unloading in this work-softening region are characterized by $d\sigma < 0$. However, according to the stress space formulation, the plastic loading occurs if $\partial f/\partial \boldsymbol{\sigma} : d\boldsymbol{\sigma} > 0$ and the elastic unloading happens whenever $\partial f/\partial \boldsymbol{\sigma} : d\boldsymbol{\sigma} < 0$ ($d\sigma > 0$ and $d\sigma < 0$ in the case of uniaxial loading). When this is compared with the uniaxial stress-strain curve during work-softening behavior, the contradiction is evident. This consideration can be readily generalized to the multiaxial case where the introduction of the yield surface is necessary.

Fig. 8.2(a) shows the expansion of the yield surface $F = 0$ into the subsequent one ($F' = 0$) due to the plastic loading caused by $d\boldsymbol{\sigma}$ for stable or work-hardening materials. For these materials further plastic loading from point A, which is on the yield surface $F = 0$, causes the yield surface to move outward, resulting in the subsequent yield surface $F' = 0$. According to the consistency condition, the current stress state B remains on $F' = 0$. On the other hand, unloading from a plastic state A will be caused by a stress increment $d\boldsymbol{\sigma}$ pointing inward from the yield surface $F = 0$, giving an elastic state C inside $F = 0$, as shown in Fig. 8.2(a). Obviously the condition or

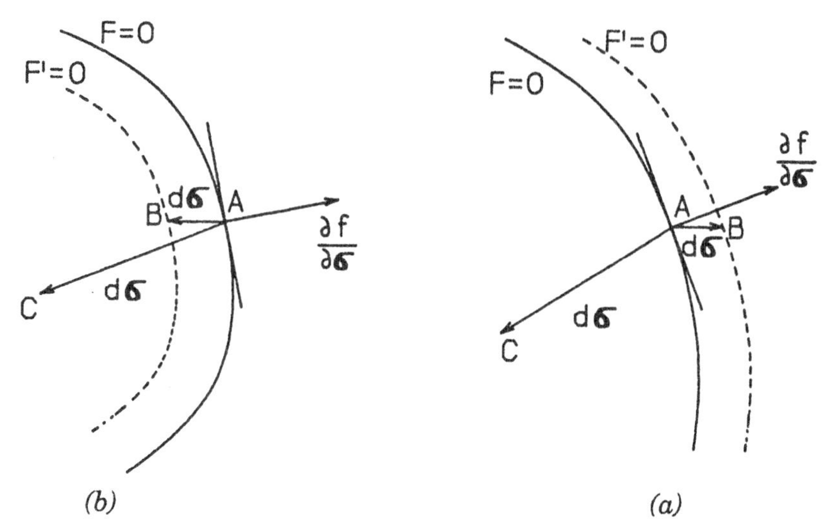

Fig. 8.2 Plastic loading and elastic unloading in the multiaxial case of (*a*) a work-hardening material and (*b*) work-softening material

criterion satisfied by plastic loading for work-hardening materials is $\partial f/\partial \boldsymbol{\sigma} : d\boldsymbol{\sigma} > 0$ and that for elastic unloading is $\partial f/\partial \boldsymbol{\sigma} : d\boldsymbol{\sigma} < 0$.

The plastic loading and elastic unloading for work-softening materials in a multiaxial case are shown in Fig. 8.2(*b*). In contrast to the work-hardening materials for which the yield surface moves outward during plastic deformation, the yield surface for work-softening materials will move inward due to the plastic deformation. This corresponds to the descending part of the uniaxial stress-strain curve, on which the stress at point B is smaller than that at point A. In the multiaxial case, as can be seen from Fig. 8.2(*b*), the subsequent yield surface $F' = 0$ is inside the previous yield surface $F = 0$; hence the stress increment $d\boldsymbol{\sigma}$ corresponding to this plastic deformation must point inward from the surface $F = 0$ to satisfy the consistency condition. Therefore $\partial f/\partial \boldsymbol{\sigma} : d\boldsymbol{\sigma} < 0$ holds for plastic loading. On the other hand, the elastic unloading occurs due to $d\boldsymbol{\sigma}$ along AC which points inward from the surface $F = 0$. Also $\partial f/\partial \boldsymbol{\sigma} : d\boldsymbol{\sigma} < 0$ holds in this case. Thus the plastic loading and elastic unloading cannot be distinguished from each other in the stress space for work-softening materials. It should be noted that the yield surface remains unchanged in the case of unloading, while the yield surface moves inward in the case of plastic loading.

The disadvantages of the stress space formulation can be eliminated if the plasticity theory is formulated in the strain space. Detailed explanations of this will be given in this chapter. The simplest case of uniaxial loading will be considered first because it reveals the basic ideas of the strain space formulation. The plasticity postulate proposed by Il'yushin (1961) will be discussed next because this postulate represents the initiation of the strain space formulation. The constitutive equations for plastic deformation will then be formulated in strain space. It is shown that the strain space formulation does provide a unified treatment for elastic-perfectly plastic materials and other work-hardening or softening materials. It also eliminates confusion in the definition of the loading-unloading criterion. Infinitesimal deformation is assumed in this chapter, except in the last section where we discuss finite deformation.

8.1 ONE-DIMENSIONAL LOADING CASE

To appreciate the basic idea of the strain space formulation and its difference from the stress space formulation, it is helpful to consider plastic deformation for a one-dimensional loading case. For the reader's convenience we summarize our discussion of stress space formulation from Chapter 5 below. A schematic representation is given in Fig. 8.3(*a*).

1. For any given loading history there exits a stress range on the σ axis within which the deformation is elastic; the stress-strain relation is given by Hooke's law, assuming linear elastic behavior. The stress level at the boundary of this range is the current yield stress σ_Y. Fig. 8.3(*a*) show the initial value of σ_Y^0 and the initial elastic range below σ_Y^0.
2. For work-hardening materials σ_Y increases with plastic deformation. The hardening behavior can be described by

$$\sigma_Y = \sigma(\epsilon^p) \tag{8.1}$$

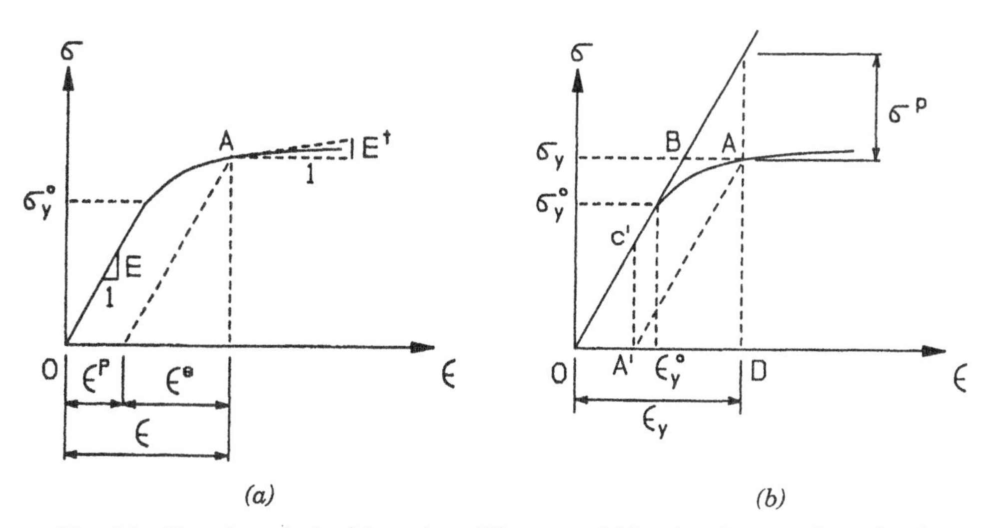

Fig. 8.3 Theories of plasticity using different variables for the one-dimensional loading case

3. The plastic flow occurs only if the stress is on the boundary points of the plastic region, and the stress increment is pointing outward from the region:

$$\sigma = \pm\sigma_Y \quad \text{and} \quad \sigma\, d\sigma > 0 \tag{8.2}$$

The first part of Eq. (8.2) is the yield condition, and the second is the loading criterion.

4. For any point in the plastic region, such as point A shown in Fig. 8.3(a), the strain consists of the elastic and plastic parts

$$\epsilon = \epsilon^{e} + \epsilon^{p} \tag{8.3}$$

The constitutive equations for both parts are formulated in terms of stress $\boldsymbol{\sigma}$:

$$\epsilon^{e} = \frac{\sigma}{E} \quad \text{and} \quad \epsilon^{p} = \int d\epsilon^{p} = \int \left(\frac{1}{E^{t}} - \frac{1}{E} \right) d\sigma \tag{8.4}$$

where E is the Young's modulus and E^{t} is the slope of the stress-strain curve, which can be considered as a function of stress. These constitutive statements are schematically shown in Fig. 8.3(a) and can be combined to give

$$\epsilon = \frac{\sigma}{E} + \epsilon^{p}(\sigma) \tag{8.5}$$

This formulation indicates that the total strain can be calculated as the sum of the elastic plastic strains caused by σ. It is important to see that stress σ is treated as the independent variable. The yield point, loading conditions, and strains ϵ^{e}, ϵ^{p}, and ϵ are all given in terms of σ.

However, the foregoing traditional treatment can be reversed. The strain ϵ can be used as the independent variable with the stress σ is defined in terms of ϵ. A parallel formulation to the four items above for the stress space formulation can be given for strain space formulation:

1. For any given loading history, there exits a strain range on the ϵ axis within which the deformation is elastic, and Hooke's law suffices to characterize the material behavior. The boundary of this range is the current yield strain ϵ_Y. The initial yield strain ϵ_Y^0 is shown in Fig. 8.3(b).
2. The value of ϵ_Y changes with plastic deformation. The relation between σ_Y and ϵ_Y, as can be seen from Fig. 8.3(b), is

$$\epsilon_Y = \frac{\sigma_Y}{E} + \epsilon^{p} = \epsilon_Y(\epsilon^{p}) \tag{8.6}$$

3. The plastic deformation is possible only if the strain is on the boundary and its increment is pointing outward from it:

$$\epsilon = \pm\epsilon_Y \quad \text{and} \quad \epsilon\, d\epsilon > 0 \tag{8.7}$$

Again, the first part of the equation represents the yield function and the second gives the loading criterion. The loading criterion $\epsilon d\epsilon > 0$ is obvious from both Fig. 8.1 and Fig. 8.3(*b*). For either hardening or softening materials, loading takes place if $\epsilon\, d\epsilon > 0$; unloading takes place if $\epsilon\, d\epsilon < 0$, and for neutral loading $\epsilon\, d\epsilon = 0$. This removes the deficiency of the loading criteria in the stress space formulation.

4. To determine stress at any point A in the plastic region, it is helpful to refer to Fig. 8.3(*b*). It can be seen that stress at point A is given by the height AD which, in turn, is obtained by subtracting the distance CA from the height CD. However, CD, denoted by σ^e, is the fictitious stress obtained by assuming a totally elastic deformation represented by the strain ϵ:

$$\sigma^e = E\epsilon \tag{8.8}$$

Denote the distance CA by σ^p, then the stress at point A is given by

$$\sigma = \sigma^e - \sigma^p = E\epsilon - \sigma^p \tag{8.9}$$

It is easy to see from Fig. 8.3(*b*) that $OA' = BA = \epsilon^p$ and hence that

$$\sigma^p = E\epsilon^p \tag{8.10}$$

This relation can also be derived from Eq. (8.5) by multiplying each term by Young's modulus E to give

$$E\epsilon = \sigma + E\epsilon^p \tag{8.11}$$

The superscript "p" in σ^p represents the correspondence between σ^p and ϵ^p, as given by Eq. (8.10).

Note that the yield condition, the loading criterion (8.7), and the elastic portion of the stress given by σ^e (8.8) are all formulated in terms of strain. To complete the theory, it is necessary to establish the constitutive equation for ϵ^p, or equivalently for σ^p, since these two are related by Eq. (8.10) in terms of strain increment. In the uniaxial loading case shown in Fig. 8.3(*b*),

$$d\epsilon^p = d\epsilon - d\epsilon^e = d\epsilon - \frac{d\sigma}{E} = d\epsilon - \frac{E^t d\epsilon}{E} = \left(1 - \frac{E^t}{E}\right) d\epsilon \tag{8.12}$$

This can be integrated to give the total plastic strain

$$\epsilon^{\mathrm{p}} = \int \left(1 - \frac{E^{\mathrm{t}}}{E}\right) d\epsilon \tag{8.13}$$

It should be noted that in these equations, E^{t} is a function of strain ϵ, $E^{\mathrm{t}} = E^{\mathrm{t}}(\epsilon)$. Equation (8.12) or Eq. (8.13) is the flow rule for ϵ^{p} in the strain space and a counterpart of Eq. (8.4), which is the flow rule for ϵ^{p} in the stress space. Using Eq. (8.10), Eqs. (8.12) and (8.13) can be changed to the following equivalent forms:

$$d\sigma^{\mathrm{p}} = \left(E - E^{\mathrm{t}}\right) d\epsilon \tag{8.14}$$

$$\sigma^{\mathrm{p}} = \int \left(E - E^{\mathrm{t}}\right) d\epsilon \tag{8.15}$$

Since the plastic strain is determined by strain history, the hardening rule for ϵ_Y, as a function of plastic strain given by Eq. (8.6), is also determined by strain history. A strain space formulation of the plasticity theory in the case of uniaxial loading can thus be completely established by Eqs. (8.9), (8.6), (8.7), and (8.14).

The above results can be extended to the multiaxial case (Yoder and Iwan 1981; Yoder 1981). At first, it is assumed that for any loading history there exits a domain in the strain space within which the deformation is elastic. The boundary surface of this elastic domain is called the *yield surface in strain space* and is expressed in the following general form:

$$\varphi(\epsilon, \sigma^{\mathrm{p}}, \alpha, \kappa) = 0 \tag{8.16}$$

where $\boldsymbol{\alpha}$ is the back stress tensor. The scaler function κ may be defined by

$$\kappa = \kappa(l), \quad l = \int \left(\frac{2}{3} \dot{\sigma}^{\mathrm{p}} : \dot{\sigma}^{\mathrm{p}}\right) dt \tag{8.17}$$

Analogous to the stress space formulation, the condition for further plastic loading strain space is

$$\varphi = 0 \quad \text{and} \quad \frac{\partial \varphi}{\partial \boldsymbol{\epsilon}} : d\boldsymbol{\epsilon} > 0 \tag{8.18}$$

This will be discussed in more detail later in the book.

The stress $\boldsymbol{\sigma}$ for any plastically deformed state is expressed in terms of strain in the form

$$\boldsymbol{\sigma} = \mathbf{C} : (\boldsymbol{\epsilon} - \boldsymbol{\epsilon}^{\mathrm{p}}) = \mathbf{C} : \boldsymbol{\epsilon} - \boldsymbol{\sigma}^{\mathrm{p}} \tag{8.19}$$

where $\mathbf{C}$ is the fourth-ordered elastic stiffness tensor and $\boldsymbol{\sigma}^{\mathrm{p}}$ is related to $\boldsymbol{\epsilon}^{\mathrm{p}}$ by

$$\boldsymbol{\sigma}^{\mathrm{p}} = \mathbf{C} : \boldsymbol{\epsilon}^{\mathrm{p}} \tag{8.20}$$

The flow rule for $\boldsymbol{\epsilon}^{\mathrm{p}}$, or equivalently σ^{p}, is needed in terms of strain as the independent variable. Equations (8.12)–(8.15) cannot be directly generalized to three-dimensional cases without considering other factors. In the next section the plasticity postulate proposed by Il'yushin (1961) will be detailed. Certain very important conclusions concerning the general form of the flow rule for σ^{p} and the shape of the yield surface in strain space be obtained from this postulate.

8.2 PLASTICITY POSTULATE AND ITS IMPLICATIONS

Il'yushin was the first person to explore the work postulate in strain space (1961). In discussing the postulates of plasticity (1960, 1961), he pointed out the limitation of Drucker's stability postulate and the inappropriateness of defining plastic deformation in stress space. He proposed an alternative plasticity postulate in strain space to provide a new definition for plastic strain. However, he did not formulate a complete plasticity theory in strain space, which was proposed later by Naghdi and coworkers. This postulate and its consequences will be explored in this chapter. Since a closed cycle of deformation or strain is used by the postulate, its definition needs to be introduced first. A closed cycle of strain is defined by a closed trajectory in strain space experienced by the material during the process of deformation. It starts from an equilibrium and compatible state ϵ^0 and σ^0, experiences a certain change in $\boldsymbol{\epsilon}$ and $\boldsymbol{\sigma}$ because of changes in the external loading, and then reverts to the original strain ϵ^0. But the stress $\boldsymbol{\sigma}$, however, may differ from $\boldsymbol{\sigma}^0$ at the end of the cycle. A closed cycle of strain $ABCD$ for a uniaxial loading case is shown in Fig. 8.4. Note that the path $ABCE$ is a closed cycle of stress used as used in Drucker's stability postulate. In addition it should be noted that

$$\epsilon_A = \epsilon_D \quad \text{and} \quad \sigma_A \neq \sigma_D \tag{8.21}$$

$$\epsilon_A \neq \epsilon_E \quad \text{and} \quad \sigma_A = \sigma_E \tag{8.22}$$

These equations reveal the difference between a stress cycle and a strain cycle. The two cycles are the same only when no plastic deformation occurs. For example, cycle ABA can be considered both a strain cycle and a stress cycle. However, under certain circumstances of deformation, a stress cycle cannot be achieved and only a closed cycle of strain is possible. This is shown in Fig. 8.5. Consider the plastic deformation process of a work-softening material starting from a point A on the descending part of the

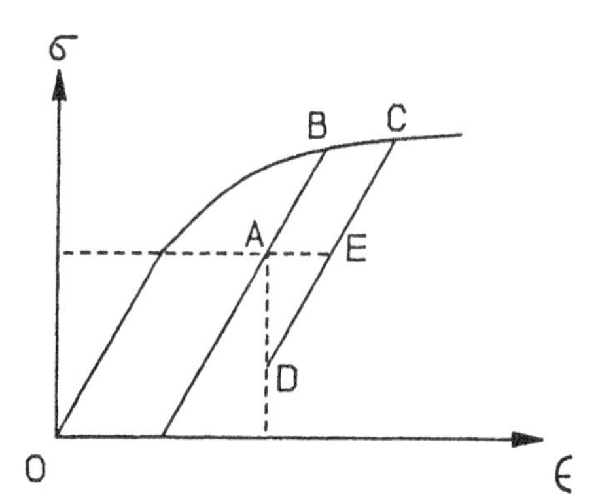

Fig. 8.4 Closed cycles of strain and stress

stress-strain curve. This plastic deformation ends at point B. Obviously $\sigma_B < \sigma_A$, and it is impossible for the stress to return to the same level as σ_A regardless of how the strain is changed. This means it is impossible to achieve a closed stress cycle for a deformation process starting from a plastic state of a work-softening material. On the other hand, a closed cycle of strain can be achieved with no difficulty since at point C, which is obtained by unloading from point B, the strain is the same as that at point A. Therefore Il'yushin's plasticity postulate using a closed cycle of deformation or strain is more general than the Drucker's stability postulate which is based on a closed cycle of loading or stress.

With the above discussion and the definition of a closed cycle of strain, Il'yushin's plasticity postulate can be stated as follows (see Il'yushin 1961): The deformation during a closed cycle of strain is accompanied by plastic deformation if the work of external forces on the closed cycle is positive, and it is purely elastic if the work is zero.

The preceding statement can be rephrased as the following commonly accepted postulate: The work done by the external forces on a material over a closed cycle of strain must be nonnegative. It is zero if only elastic deformation occurs during this cycle, and it is positive if plastic deformation

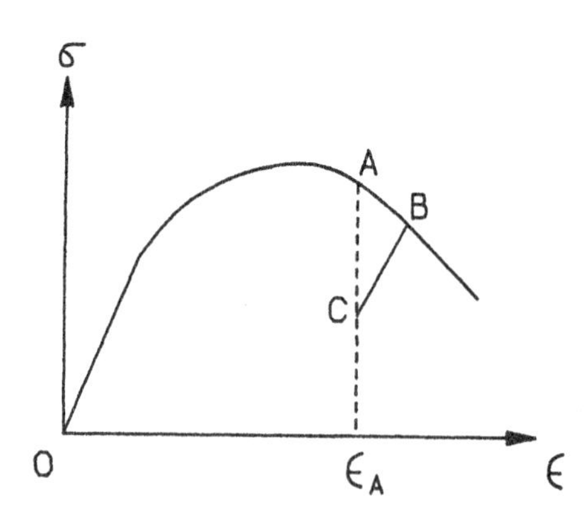

Fig. 8.5 Closed strain cycle for work-softening materials

takes place during this cycle. If a closed cycle of strain is denoted by C_ϵ, then this plasticity postulate can be formulated as

$$W = \int_{C_\epsilon} \boldsymbol{\sigma} : d\boldsymbol{\epsilon} \geq 0 \tag{8.23}$$

where the integral is over the cycle C_ϵ.

Now the consequences of the postulate will be examined. The details of the derivation are omitted here and only the results obtained by Il'yushin (1961) are explored. Interested readers can find the detailed derivation in Il'yushin's original work.

1. If the work of external forces over a stress cycle is denoted by W_D, as in Drucker's postulate, Il'yushin proved that

   $$W - W_D > 0 \quad \text{or} \quad W > W_D \tag{8.24}$$

   where W is the work expressed by Eq. (8.23). This result is quite clear if the uniaxial case is considered. As can be seen from Fig. 8.4, W is the area enclosed by the path $ABCED$, while W_D is the region enclosed by $ABCE$. Obviously $W - W_D$ is the area enclosed by AED which is positive. Equation (8.24) indicates that Il'yushin's postulate of plasticity in strain space is more general and less restrictive than Drucker's postulate in stress space and that the latter is a sufficient (but not necessary) condition in the framework of the former.
2. Using the relation

   $$d\boldsymbol{\sigma}^p = \mathbf{C} : d\boldsymbol{\epsilon}^p \tag{8.25}$$

   the following important result can be derived from the postulate of plasticity

   $$d\boldsymbol{\sigma}^p - d\mathbf{C} : \boldsymbol{\epsilon}^p = \eta \frac{\partial \varphi}{\partial \boldsymbol{\epsilon}} \tag{8.26}$$

   where $\boldsymbol{\epsilon}^p$ is the elastic strain, η is a proportionality scalar factor, φ is the yield function in the strain space as given by Eq. (8.16), and $d\mathbf{C}$ is the increment (or change) of the elastic stiffness tensor caused by the plastic deformation. If the elastic stiffness tensor remains constant during the course of plastic deformation (i.e., no elastic-plastic coupling), Eq. (8.26) becomes

   $$d\boldsymbol{\sigma}^p = \eta \frac{\partial \varphi}{\partial \boldsymbol{\epsilon}} \tag{8.27}$$

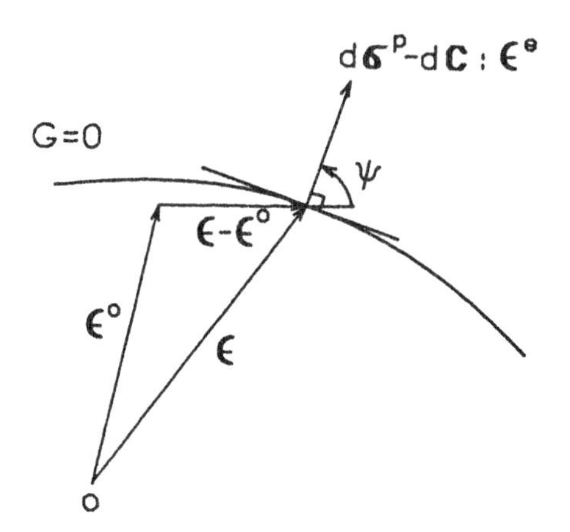

Fig. 8.6 Convexity of the yield surface in strain space

and is called the normality condition in the strain space. Equation (8.26) or (8.27) is very significant because it provides a general form of the constitutive equation for $\boldsymbol{\sigma}^{\mathrm{p}}$ [in turn, $\boldsymbol{\epsilon}^{\mathrm{p}}$ since $\boldsymbol{\sigma}^{\mathrm{p}}$ and $\boldsymbol{\epsilon}^{\mathrm{p}}$ are related by Eq. (8.25)]. In strain space formulation, to determine the constitutive law for $\boldsymbol{\sigma}^{\mathrm{p}}$, it is necessary to find the function form of the yield surface in the strain space, and the variation of the elastic stiffness tensor $\mathbf{C}$ with the plastic deformation. Equations (8.26) and (8.27) are the counterparts of Equations (5.141) and (5.156), respectively. The relationship between them will be discussed in a later section.

3. As illustrated in Fig. 8.6, the original strain is $\boldsymbol{\epsilon}^0$ and the current strain on the yield surface $\varphi = 0$ is $\boldsymbol{\epsilon}$. Another vector in the strain space, $d\boldsymbol{\sigma}^{\mathrm{p}} - d\mathbf{C} : \boldsymbol{\epsilon}^{\mathrm{e}}$, is normal to $\varphi = 0$ according to the discussion above. Il'yushin derived from the plasticity postulate that

$$\left(\boldsymbol{\epsilon} - \boldsymbol{\epsilon}^0\right) : \left(d\boldsymbol{\sigma}^{\mathrm{p}} - d\mathbf{C} : \boldsymbol{\epsilon}^{\mathrm{e}}\right) + \left(\boldsymbol{\epsilon} - \boldsymbol{\epsilon}^0\right) : d\mathbf{C} : \left(\boldsymbol{\epsilon} - \boldsymbol{\epsilon}^0\right) \geq 0 \quad (8.28)$$

and pointed out it does not imply the convexity of the surface $\varphi = 0$ in general, although a convex surface satisfies this condition. In most cases where the elastic stiffness tensor remains constant so that $d\mathbf{C} = 0$, Eq. (8.28) reduces to

$$\left(\boldsymbol{\epsilon} - \boldsymbol{\epsilon}^0\right) : d\boldsymbol{\sigma}^{\mathrm{p}} \geq 0 \quad (8.29)$$

It can be further expressed as

$$|\boldsymbol{\epsilon} - \boldsymbol{\epsilon}^0|\, |d\boldsymbol{\sigma}^{\mathrm{p}}| \cos\psi \geq 0 \quad (8.30)$$

where ψ is shown in Fig. 8.6. Since $|\boldsymbol{\epsilon} - \boldsymbol{\epsilon}^0| \geq 0$ and $|d\boldsymbol{\sigma}^{\mathrm{p}}| \geq 0$, this equation implies that

$$\cos\psi \geq 0 \quad (8.31)$$

and it follows that

$$-\frac{\pi}{2} \leq \psi \leq \frac{\pi}{2} \tag{8.32}$$

As discussed for Eq. (5.138), this means that the yield surface is convex in strain space.

The main consequences of Il'yushin's postulate of plasticity are now summarized. In strain space, for materials without elastic-plastic coupling ($\mathbf{C}$ does not change due to plastic deformation), the vector $d\boldsymbol{\sigma}^{\mathrm{p}}$ is normal to the yield surface (the so-called normality rule), and the yield surface is convex. For materials with elastic-plastic coupling, the vector that is normal to the yield surface in strain space is $d\boldsymbol{\sigma}^{\mathrm{p}} - d\mathbf{C}:\boldsymbol{\epsilon}^{\mathrm{e}}$, and the yield surface is not necessarily convex. The elastic-plastic coupling is always associated with the nonnormality condition for $d\boldsymbol{\sigma}^{\mathrm{p}}$, which is similar to the plastic strain increment $d\boldsymbol{\epsilon}^{\mathrm{p}}$ case in stress space formulation.

Before formulating the plasticity theory in strain space from the results obtained in this section, we will briefly discuss the yield surface in strain space.

8.3 YIELD SURFACE IN STRAIN SPACE

Usually experimental results concerning the yield behavior of materials are presented in the stress space, and the initial and subsequent yield points are defined by the stress states at which the phenomenon of yielding is observed. These studies provide knowledge about the shape, orientation, movement, and deformation of the yield surface in the stress space. In contrast, little has been done on the direct determination of the yield points (hence the yield surface) in terms of strain. However, the yield surface in the stress space can be transformed into the strain space (Naghdi et al. 1975a, 1975b, 1975c, 1981, 1983a, 1983b, 1984; Yoder and Iwan 1981; and Yoder 1981). In fact this can be accomplished easily in the case of infinitesimal deformation.

The yield surface in the stress space can be expressed as

$$F(\boldsymbol{\sigma}, \boldsymbol{\alpha}, \kappa) = 0 \tag{8.33}$$

where $\boldsymbol{\sigma}$, $\boldsymbol{\alpha}$, and κ are the stress tensor, the back stress tensor, and the isotropic hardening parameter, respectively.

For infinitesimal deformation, $\boldsymbol{\sigma}$ is related to $\boldsymbol{\epsilon}$ by Eq. (8.19). The yield surface in the strain space can be obtained by substituting Eq. (8.19) into Eq. (8.33):

$$F(\boldsymbol{\sigma}, \boldsymbol{\alpha}, \kappa) = F(\mathbf{C}:\boldsymbol{\epsilon} - \boldsymbol{\sigma}^{\mathrm{p}}, \boldsymbol{\alpha}, \kappa) = \varphi(\boldsymbol{\epsilon}, \boldsymbol{\sigma}^{\mathrm{p}}, \boldsymbol{\alpha}, \kappa) = 0 \tag{8.34}$$

It is clear that for any yield surface in stress space of the form of Eq. (8.33), there corresponds a unique yield surface in strain space of the form of Eq. (8.34). Introducing the variable

$$\boldsymbol{\alpha}_{\epsilon} = \mathbf{C}^{-1} : \boldsymbol{\alpha} \tag{8.35}$$

and using Eq. (8.20), the yield surface in strain space can be rewritten as

$$\varphi(\boldsymbol{\epsilon}, \boldsymbol{\epsilon}^{\mathrm{p}}, \boldsymbol{\alpha}_{\epsilon}, \kappa) = 0 \tag{8.36}$$

Since $\mathbf{C}$ is an elastic constant tensor, all the arguments in the above equation are strainlike variables.

Consider an isotropic, von Mises material for which the yield surface is given in stress space by

$$\frac{3}{2}\mathbf{S} : \mathbf{S} - \sigma_{\mathrm{Y}}^{2} = 0 \tag{8.37}$$

where $\mathbf{S}$ is the deviator of stress tensor $\boldsymbol{\sigma}$, and σ_{Y} is the uniaxial tensile yield stress. For infinitesimal deformation

$$S = 2G\epsilon'^{e} = 2G(\boldsymbol{\epsilon}' - \boldsymbol{\epsilon}^{\mathrm{p}}) = 2G\boldsymbol{\epsilon}' - \boldsymbol{\sigma}^{\mathrm{p}} \tag{8.38}$$

where $\boldsymbol{\epsilon}'$ is the deviator of strain tensor $\boldsymbol{\epsilon}$, and the plastic incompressibility $\boldsymbol{\epsilon}^{\mathrm{p}} = \boldsymbol{\epsilon}'^{\mathrm{p}}$ is implied. Substituting Eq. (8.38) into Eq. (8.37) leads to

$$4G^{2}(\boldsymbol{\epsilon}' - \boldsymbol{\epsilon}^{\mathrm{p}}) : (\boldsymbol{\epsilon}' - \boldsymbol{\epsilon}^{\mathrm{p}}) - \frac{2}{3}\sigma_{\mathrm{Y}}{}^{2} = 0 \tag{8.39}$$

which is the corresponding yield surface in strain space. Note that $\sigma_{\mathrm{Y}} = \sigma_{\mathrm{Y}}(\epsilon_{\mathrm{e}}^{\mathrm{p}})$ so that all variables in this equation are determined by strain.

According to the discussions here and in the previous section, the following can be summarized about the yield surface in strain space (Naghdi and Trapp 1976):

1. For materials whose elastic properties are not affected by plastic deformation, the yield surface φ is convex.
2. The shape and orientation of the initial yield surface in strain space are the same as those of the initial yield surface in stress space. This can be clearly seen in the preceding example of von Mises yield surface. The initial yield surface can be obtained from Eq. (8.39) by setting $\boldsymbol{\epsilon}^{\mathrm{p}} = 0$ and $\sigma_{\mathrm{Y}} = \sigma_{\mathrm{y}}^{0}$:

$$\frac{3}{2}\boldsymbol{\epsilon}' : \boldsymbol{\epsilon}' - \frac{1}{4G^{2}}\left(\sigma_{\mathrm{Y}}^{0}\right)^{2} = 0 \tag{8.40}$$

This has the same shape and orientation as the corresponding yield surface in stress space Eq. (8.37), differing by only a reduction factor of $1/(2G)$.

3. Comparing Eq. (8.39) with Eq. (8.37) shows that even the subsequent yield surface in strain space has exactly the same shape and orientation as those in stress space, except that the origin is shifted by an amount $\boldsymbol{\epsilon}^{\mathrm{p}}$. Although the subsequent yield surface in stress space only expands uniformly in this case of isotropic hardening, the corresponding surface in strain space will expand as well as translate due to the evolution of the plastic strain $\boldsymbol{\epsilon}^{\mathrm{p}}$ in the deformation process.
4. In the principal stress and principal strain space, the shape of the corresponding subsequent yield surfaces will differ from each other. This is because the principal directions of stress and strain do not coincide due to the presence of $\boldsymbol{\epsilon}^{\mathrm{p}}$ in Eq. (8.38). However, the shape and orientation of the initial yield surfaces, both in the principal strain space and in the principal stress space, are the same because prior to the initiation of plastic deformation, the plastic strain $\boldsymbol{\epsilon}^{\mathrm{p}} = 0$ and the principal axes of stress and strain coincide for initially isotropic materials.

8.4 CONSTITUTIVE LAWS IN STRAIN SPACE

In this section the constitutive equations of a plasticity theory in strain space will be formulated based on information presented in previous sections. The established results are listed first.

1. Yield surface:

$$\varphi(\boldsymbol{\epsilon}, \boldsymbol{\sigma}^{\mathrm{p}}, \boldsymbol{\alpha}, \kappa) = 0 \tag{8.41}$$

Here the variable σ^{p} is preferred to ϵ^{p} because the general flow rule in strain space derived from Il'yushin's postulate of plasticity is expressed in terms of σ^{p}, as seen below.

2. Flow rule:

$$d\boldsymbol{\sigma}^{\mathrm{p}} = \eta \frac{\partial \varphi}{\partial \epsilon} \tag{8.42}$$

where η is a scalar factor to be determined by the consistency condition. Note that in writing this equation, we assume that there is no elastic-plastic coupling.

3. Hardening rules: Recall that α and κ are introduced to represent the hardening effect caused by the plastic deformation. Therefore $d\alpha$ and

$d\kappa$ are zero if $d\sigma^p$ is zero. Thus it is general enough to assume that

$$d\boldsymbol{\alpha} = L(\boldsymbol{\epsilon}, \boldsymbol{\sigma}^p, \boldsymbol{\alpha}, \kappa) : d\boldsymbol{\sigma}^p \tag{8.43}$$

$$d\kappa = \mathbf{N}(\boldsymbol{\epsilon}, \boldsymbol{\sigma}^p, \boldsymbol{\alpha}, \kappa) : d\boldsymbol{\sigma}^p \tag{8.44}$$

where **L** and **N** are fourth-order and second-order tensors, respectively, representing the kinematic and isotropic hardening properties of materials. Their specific form should be determined experimentally. For the sake of simplicity, the linear kinematic hardening rule is used in the following formulation:

$$d\boldsymbol{\alpha} = m d\boldsymbol{\epsilon}^p = m\mathbf{C}^{-1} : d\boldsymbol{\sigma}^p \tag{8.45}$$

where m is a material constant and **C** is an elastic stiffness tensor. In addition Eq. (8.17) is used to characterize the evolution of κ, leading to

$$d\kappa = \kappa'\left(\frac{2}{3} d\boldsymbol{\sigma}^p : d\boldsymbol{\sigma}^p\right)^{1/2} \tag{8.46}$$

where

$$\kappa' = \frac{d\kappa}{dl} \tag{8.47}$$

and l is defined by Eq. (8.17).

4. Stress-strain relation:

$$d\boldsymbol{\sigma} = \mathbf{C} : d\boldsymbol{\epsilon} - d\boldsymbol{\sigma}^p \tag{8.48}$$

From this equation it is seen that the key to formulating a constitutive equation in strain space is to determine $d\sigma^p$ in terms of strain. From Eq. (8.42) it is clear that this can be done if the scalar factor η is determined in terms of ϵ and its increment $d\epsilon$. As in the stress space formulation discussed in Chapter 5, η has to be determined by using the following consistency condition:

$$\frac{\partial \varphi}{\partial \boldsymbol{\epsilon}} : d\boldsymbol{\epsilon} + \frac{\partial \varphi}{\partial \boldsymbol{\sigma}^p} : d\boldsymbol{\sigma}^p + \frac{\partial \varphi}{\partial \boldsymbol{\alpha}} : d\boldsymbol{\alpha} + \frac{\partial \varphi}{\partial \kappa} d\kappa = 0 \tag{8.49}$$

Introducing Eqs. (8.42), (8.45), and (8.46) into the above equation leads to

$$\frac{\partial \varphi}{\partial \boldsymbol{\epsilon}} : d\boldsymbol{\epsilon} + \eta \frac{\partial \varphi}{\partial \boldsymbol{\sigma}^p} : \frac{\partial \varphi}{\partial \boldsymbol{\epsilon}} + m\eta \frac{\partial \varphi}{\partial \boldsymbol{\alpha}} : C^{-1} : \frac{\partial \varphi}{\partial \boldsymbol{\epsilon}} + \eta \frac{\partial \varphi}{\partial \kappa} \kappa' \left(\frac{2}{3} \frac{\partial \varphi}{\partial \boldsymbol{\epsilon}} : \frac{\partial \varphi}{\partial \boldsymbol{\epsilon}}\right)^{1/2} = 0 \tag{8.50}$$

Thus the scalar factor η can be obtained as

$$\eta = \frac{1}{D}\frac{\partial\varphi}{\partial\boldsymbol{\epsilon}} : d\boldsymbol{\epsilon} \tag{8.51}$$

where

$$D = -\frac{\partial\varphi}{\partial\boldsymbol{\sigma}^{\mathrm{p}}} : \frac{\partial\varphi}{\partial\boldsymbol{\epsilon}} - m\frac{\partial\varphi}{\partial\boldsymbol{\alpha}} : \mathbf{C}^{-1}\frac{\partial\varphi}{\partial\boldsymbol{\epsilon}} - \frac{\partial\varphi}{\partial\kappa}\kappa'\left(\frac{2}{3}\frac{\partial\varphi}{\partial\boldsymbol{\epsilon}} : \frac{\partial\varphi}{\partial\boldsymbol{\epsilon}}\right)^{1/2} \tag{8.52}$$

Now the flow rule, Eq. (8.42), can be rewritten as

$$d\boldsymbol{\sigma}^{\mathrm{p}} = \frac{1}{D}\frac{\partial\varphi}{\partial\boldsymbol{\epsilon}}\frac{\partial\varphi}{\partial\boldsymbol{\epsilon}} : d\boldsymbol{\epsilon} \tag{8.53}$$

Substituting this result into Eq. (8.48) provides the following form of stress-strain relation in strain space:

$$d\boldsymbol{\sigma} = \left(\mathbf{C} - \frac{1}{D}\frac{\partial\varphi}{\partial\boldsymbol{\epsilon}}\frac{\partial\varphi}{\partial\boldsymbol{\epsilon}}\right) : d\boldsymbol{\epsilon} \tag{8.54}$$

Specific forms of the constitutive equation can be obtained from this equation by specifying the particular form of yield surface G. For example, in the case of von Mises materials, the yield surface in strain space is given by Eq. (8.37). Or, in terms of $\boldsymbol{\sigma}^{\mathrm{p}}$,

$$4G^2\left(\boldsymbol{\epsilon}' - \frac{1}{2G}\boldsymbol{\sigma}^{\mathrm{p}}\right) : \left(\boldsymbol{\epsilon}' - \frac{1}{2G}\boldsymbol{\sigma}^{\mathrm{p}}\right) - \frac{2}{3}\sigma_{\mathrm{Y}}^2 = 0 \tag{8.55}$$

From Eq. (8.55),

$$\frac{\partial\varphi}{\partial\boldsymbol{\epsilon}} = 8G^2\left(\boldsymbol{\epsilon}' - \frac{1}{2G}\boldsymbol{\sigma}^{\mathrm{p}}\right) \tag{8.56}$$

$$\frac{\partial\varphi}{\partial\boldsymbol{\sigma}^{\mathrm{p}}} = -4G\left(\boldsymbol{\epsilon}' - \frac{1}{2G}\boldsymbol{\sigma}^{\mathrm{p}}\right) \tag{8.57}$$

$$\frac{\partial\varphi}{\partial\boldsymbol{\alpha}} = 0 \tag{8.58}$$

$$\frac{\partial\varphi}{\partial\kappa} = \frac{\partial\varphi}{\partial\sigma_{\mathrm{Y}}} = -\frac{4}{3}\sigma_{\mathrm{Y}}, \quad \kappa' = \sigma_{\mathrm{Y}}' = \frac{d\sigma_{\mathrm{Y}}}{dl} \tag{8.59}$$

Using these derivatives, the following results can be obtained:

$$D = \frac{16}{3} G \sigma_Y^2 \left(1 + \frac{2}{3} \sigma'_Y\right) \tag{8.60}$$

$$d\boldsymbol{\sigma}^p = \frac{12G^3(\boldsymbol{\epsilon}' - \boldsymbol{\sigma}^p/2G) : d\boldsymbol{\epsilon}}{\sigma_Y{}^2(1 + 2\sigma'_Y/3)} \left(\boldsymbol{\epsilon}' - \frac{1}{2G} \boldsymbol{\sigma}^p\right) \tag{8.61}$$

or

$$d\boldsymbol{\sigma}^p = \frac{3G\mathbf{SS} : d\boldsymbol{\epsilon}}{\sigma_Y^2(1 + 2\sigma'_Y/3)} \tag{8.62}$$

where Eq. (8.36) has been used, and

$$d\boldsymbol{\sigma} = \left[\mathbf{C} - \frac{3G\mathbf{SS}}{\sigma_Y^2(1 + 2\sigma'_Y/3)}\right] : d\boldsymbol{\epsilon} \tag{8.63}$$

8.5 PERFECT PLASTICITY AND LOADING CRITERION IN STRAIN SPACE

It has been shown previously that a plasticity theory formulated in stress space is deficient for the case of an elastic-perfectly plastic material and for describing the loading-unloading criterion for work-softening materials. In this section it will be shown that a plasticity theory formulated in strain space can be directly reduced to an elastic–perfectly plastic theory without the need of a separate formulation. The loading-unloading criterion is also established which is valid for work-hardening, perfectly plastic, as well as work-softening materials.

The constitutive equations for perfectly plastic materials can be directly derived from the general theory, Eqs. (8.52)–(8.54), (Naghdi and Trapp 1975a, 1975b, 1975c). Because there is no hardening effect in this case,

$$m = \kappa' = 0 \tag{8.64}$$

Thus expression for D, as given in Eq. (8.52), reduces to

$$D = -\frac{\partial \varphi}{\partial \boldsymbol{\sigma}^p} : \frac{\partial \varphi}{\partial \boldsymbol{\epsilon}} \tag{8.65}$$

which is always nonzero. Therefore the following equations still hold for elastic–perfectly plastic materials:

$$d\boldsymbol{\sigma}^{\mathrm{p}} = \frac{1}{D}\left(\frac{\partial \varphi}{\partial \boldsymbol{\epsilon}} : d\boldsymbol{\epsilon}\right)\frac{\partial \varphi}{\partial \boldsymbol{\epsilon}} \tag{8.66}$$

$$d\boldsymbol{\sigma} = \left(\mathbf{C} - \frac{1}{D}\frac{\partial \varphi}{\partial \boldsymbol{\epsilon}}\frac{\partial \varphi}{\partial \boldsymbol{\epsilon}}\right) : d\boldsymbol{\epsilon} \tag{8.67}$$

It should be noted that, unlike the plasticity theory formulated in stress space, Eqs. (5.245), (5.246), and (5.247), the theory in strain space has no singular behavior in the region of zero rate of work hardening. This will become clearer if the general expression for D, Eq. (8.52), is compared with that for K^{p}, Eq. (5.245), which is conveniently rewritten as

$$K^{\mathrm{p}} = c + \frac{(d\kappa/dW^{\mathrm{p}})(\partial f/\partial \boldsymbol{\sigma}) : \mathbf{S}}{\partial f/\partial \boldsymbol{\sigma} : \partial f/\partial \boldsymbol{\sigma}} \tag{8.68}$$

Obviously $K^{\mathrm{p}} = 0$ in the case of elastic–perfectly plastic materials since $c = d\kappa/dW^{\mathrm{p}} = 0$. As a result Eqs. (5.246) and (5.247) for $d\boldsymbol{\epsilon}^{\mathrm{p}}$ and $d\boldsymbol{\epsilon}$, respectively, become singular in the case where the work hardening is zero. On the other hand, as shown by Eq. (8.65), D is nonzero even if the work-hardening rate of the material is zero. This removes the deficiency of singularity and enables the theory in strain space to be reduced to the elastic–perfectly plastic materials without any difficulty.

This can be further explored by examining von Mises materials. It has been derived that for these materials

$$d\boldsymbol{\sigma}^{\mathrm{p}} = \frac{3G\mathbf{S}\mathbf{S} : d\boldsymbol{\epsilon}}{\sigma_{\mathrm{Y}}{}^{2}(1 + (2/3)\sigma'_{\mathrm{Y}})} \tag{8.69}$$

According to the definition for $\boldsymbol{\sigma}^{\mathrm{p}}$, as given by Eq. (8.38),

$$d\boldsymbol{\sigma}^{\mathrm{p}} = 2G\,d\boldsymbol{\epsilon}^{\mathrm{p}} \tag{8.70}$$

Therefore Eq. (8.69) becomes

$$d\epsilon^{\mathrm{p}} = \frac{3\mathbf{S}\mathbf{S} : d\boldsymbol{\epsilon}}{2\sigma_{\mathrm{Y}}{}^{2}(1 + (2/3)\sigma'_{\mathrm{Y}})} \tag{8.71}$$

If it is further assumed that the material is elastic–perfectly plastic,

$$\sigma_{\mathrm{Y}} = \sigma_{\mathrm{Y}}{}^{0}, \quad \sigma'_{\mathrm{Y}} = 0 \tag{8.72}$$

$$d\boldsymbol{\epsilon}^{\mathrm{p}} = \frac{3\mathbf{S} : d\boldsymbol{\epsilon}}{2\sigma_{\mathrm{Y}}^{02}}\mathbf{S} \tag{8.73}$$

which is exactly the Prandtl-Reuss equation.

Now the loading-unloading criterion will be examined. As pointed out previously in the case of uniaxial loading, plastic loading is always characterized by $d\epsilon > 0$. As can be seen from Figs. 8.1 and 8.3, further plastic loading from point A in the plastic region results in $d\epsilon > 0$, although this process corresponds to a $d\sigma < 0$ for work-softening behavior shown in Fig. 8.1 and to a $d\sigma > 0$ for work-hardening behavior shown in Fig. 8.3. This observation can be generalized to the three-dimensional case to give the following loading-unloading criteria as established by Naghdi and coworkers (1975a, 1975b, 1975c, 1981, 1983a, 1983b, 1984):

1. Elastic unloading

$$\varphi = 0, \quad \frac{\partial \varphi}{\partial \boldsymbol{\epsilon}} : d\boldsymbol{\epsilon} < 0 \tag{8.74}$$

2. Neutral loading

$$\varphi = 0, \quad \frac{\partial \varphi}{\partial \boldsymbol{\epsilon}} : d\boldsymbol{\epsilon} = 0 \tag{8.75}$$

3. Plastic loading

$$\varphi = 0, \quad \frac{\partial \varphi}{\partial \boldsymbol{\epsilon}} : d\boldsymbol{\epsilon} > 0 \tag{8.76}$$

Note that $d\boldsymbol{\sigma}^{\mathrm{p}}$ will be nonzero if and only if the plastic loading takes place. Therefore it can be written

$$d\boldsymbol{\sigma}^{\mathrm{p}} = 0 \qquad \text{when } \varphi < 0 \tag{8.77}$$

$$d\boldsymbol{\sigma}^{\mathrm{p}} = 0 \qquad \text{when } \varphi = 0, \ \frac{\partial \varphi}{\partial \boldsymbol{\epsilon}} : d\boldsymbol{\epsilon} < 0 \tag{8.78}$$

$$d\boldsymbol{\sigma}^{\mathrm{p}} = 0 \qquad \text{when } \varphi = 0, \ \frac{\partial \varphi}{\partial \boldsymbol{\epsilon}} : d\boldsymbol{\epsilon} = 0 \tag{8.79}$$

$$d\boldsymbol{\sigma}^{\mathrm{p}} = \frac{1}{D}\left(\frac{\partial \varphi}{\partial \boldsymbol{\epsilon}} : d\boldsymbol{\epsilon}\right)\frac{\partial \varphi}{\partial \boldsymbol{\epsilon}} \qquad \text{when } \varphi = 0, \ \frac{\partial \varphi}{\partial \boldsymbol{\epsilon}} : d\boldsymbol{\epsilon} > 0 \tag{8.80}$$

The relation between the loading-unloading criteria in stress space and in strain space is as follows: Since there is a one-to-one correspondence between the yield surfaces in stress space and in strain space,

$$f(\boldsymbol{\sigma}, \boldsymbol{\alpha}, \kappa) = f(\boldsymbol{\sigma}(\boldsymbol{\epsilon}, \boldsymbol{\sigma}^{\mathrm{p}}), \boldsymbol{\alpha}, \kappa) = \varphi(\boldsymbol{\epsilon}, \boldsymbol{\sigma}^{\mathrm{p}}, \boldsymbol{\alpha}, \kappa) \tag{8.81}$$

Recall that $d\sigma^{\mathrm{p}}$, $d\alpha$, and $d\kappa$ are zero whenever there is no plastic loading.

Differentiating the equation above and applying it to this case leads to

$$\frac{\partial f}{\partial \boldsymbol{\sigma}} : d\boldsymbol{\sigma} = \frac{\partial \varphi}{\partial \boldsymbol{\epsilon}} : d\boldsymbol{\epsilon} \tag{8.82}$$

whenever

$$d\boldsymbol{\epsilon}^{\mathrm{p}} = d\boldsymbol{\sigma}^{\mathrm{p}} = d\boldsymbol{\alpha} = 0, \quad d\kappa = 0 \tag{8.83}$$

It follows from Eq. (8.82) that the elastic unloading criterion $\partial\varphi/\partial\boldsymbol{\epsilon} : d\boldsymbol{\epsilon} < 0$ in strain space corresponds exactly to the criterion $\partial f/\partial\boldsymbol{\sigma} : d\boldsymbol{\sigma} < 0$ in stress space and also that the neutral loading criterion $\partial\varphi/\partial\boldsymbol{\epsilon} : d\boldsymbol{\epsilon} = 0$ in strain space corresponds exactly to $\partial f/\partial\boldsymbol{\sigma} : d\boldsymbol{\sigma} = 0$ in stress space. In other words, the criteria for the elastic state, elastic unloading and neutral loading in stress space and in strain space are precisely equivalent.

The loading condition $\partial f/\partial\boldsymbol{\sigma} : d\boldsymbol{\sigma} > 0$ in stress space, however, does not necessarily imply $\partial\varphi/\partial\boldsymbol{\epsilon} : d\boldsymbol{\epsilon} > 0$, and vice versa. In this case $d\boldsymbol{\epsilon}^{\mathrm{p}}$ and hence $d\boldsymbol{\sigma}^{\mathrm{p}}$, $d\boldsymbol{\alpha}$, and $d\kappa$ are all nonzero so that

$$\begin{aligned} &\frac{\partial f}{\partial \boldsymbol{\sigma}} : d\boldsymbol{\sigma} + \frac{\partial f}{\partial \boldsymbol{\alpha}} : d\boldsymbol{\alpha} + \frac{\partial f}{\partial \kappa} d\kappa \\ &\quad = \frac{\partial \varphi}{\partial \boldsymbol{\epsilon}} : d\boldsymbol{\epsilon} + \frac{\partial \varphi}{\partial \boldsymbol{\sigma}^{\mathrm{p}}} : d\boldsymbol{\sigma}^{\mathrm{p}} + \frac{\partial \varphi}{\partial \boldsymbol{\alpha}} : d\boldsymbol{\alpha} + \frac{\partial \varphi}{\partial \kappa} d\kappa \end{aligned} \tag{8.84}$$

From this no general conclusion can be drawn regarding the relationship between $\partial f/\partial\boldsymbol{\sigma} : d\boldsymbol{\sigma} > 0$ and $\partial\varphi/\partial\boldsymbol{\epsilon} : d\boldsymbol{\epsilon}$. In this sense the stress space formulation and the strain space formulation cannot be considered equivalent. Naghdi and his coworkers (1975a, 1975b, 1975c, 1981, 1983a, 1983b, 1984) made a comprehensive study of this phenomenon and the interested reader can refer to their works for details.

A short remark regarding the normality flow rule in stress space is provided here before going to the next section. In Chapter 5 the normality flow rule

$$d\boldsymbol{\epsilon}^{\mathrm{p}} = \lambda \frac{\partial f}{\partial \boldsymbol{\sigma}} \tag{8.85}$$

is derived based on Drucker's postulate, which is valid for work-hardening materials. It is claimed there without proof that this normality flow rule holds true even for work-softening materials. Now it is easy to prove this statement by using the strain space theory just discussed. The proof below is valid for infinitesimal deformation and for materials without elastic-plastic coupling. In such cases

$$f(\boldsymbol{\sigma}, \boldsymbol{\alpha}, \kappa) = f(\mathbf{C}(\boldsymbol{\epsilon} - \boldsymbol{\epsilon}^{\mathrm{p}}), \boldsymbol{\alpha}, \kappa) = G(\boldsymbol{\epsilon}, \boldsymbol{\sigma}^{\mathrm{p}}, \boldsymbol{\alpha}, \kappa) \tag{8.86}$$

From this equation we can derive

$$\frac{\partial \varphi}{\partial \boldsymbol{\epsilon}} = \frac{\partial f}{\partial \boldsymbol{\sigma}} : \frac{\partial \boldsymbol{\sigma}}{\partial \boldsymbol{\epsilon}} = \mathbf{C} : \frac{\partial f}{\partial \boldsymbol{\sigma}} \tag{8.87}$$

where the symmetry $C_{ijkl} = C_{klij}$ has been used. In addition

$$d\boldsymbol{\sigma}^{\mathrm{p}} = \mathbf{C} : d\boldsymbol{\epsilon}^{\mathrm{p}} \tag{8.88}$$

Note that this equation is conceptually the same as Eq. (8.70) except that $d\boldsymbol{\sigma}^{\mathrm{p}}$ is assumed to be a deviator in Eq. (8.70) in conformity with the incompressibility of the plastic deformation. Substituting Eqs. (8.87) and (8.88) into Eq. (8.27) gives

$$\mathbf{C} : d\boldsymbol{\epsilon}^{\mathrm{p}} = \eta \mathbf{C} : \frac{\partial f}{\partial \boldsymbol{\sigma}} \tag{8.89}$$

which can be reduced to Eq. (8.85) after premultiplying both sides by $\mathbf{C}^{-1}$. Since the above derivation is based on the plasticity postulate which is valid for both work-hardening and work-softening materials, the resulting normality flow rule, Eq. (8.85), holds for both these materials.

8.6 FINITE PLASTICITY IN STRAIN SPACE

This section presents the purely mechanical plasticity theory for finite deformation developed by Naghdi and his coworkers. Selected concepts and derivations are discussed briefly. For simplicity and clarity, the following presentation considers only isotropic hardening behavior. For more details, readers are referred to the original papers.

8.6.1 Formulation of the Theory

Naghdi et al. (1975a, 1975b, 1975c, 1981, 1983a, 1983b, 1984) established this theory in response to the initial reference configuration. The symmetric Lagrangian strain defined by

$$\mathbf{E} = \frac{1}{2}(\mathbf{F}^T \cdot \mathbf{F} - \mathbf{I}) \tag{8.90}$$

is used as the strain measure in the theory, where $\mathbf{F}$ is the deformation gradient introduced in Section 2.2. Associated with $\mathbf{E}$ is the symmetric or second Piola-Kirchhoff stress which is chosen in formulating the theory:

$$\Sigma^{\mathrm{II}} = \det(\mathbf{F})\mathbf{F}^{-1} \cdot \boldsymbol{\sigma} \cdot (\mathbf{F}^{-1})^{\mathrm{T}} \tag{8.91}$$

where $\boldsymbol{\sigma}$ is the Cauchy stress tensor. In addition to the strain tensor $\mathbf{E}$ defined by Eq. (8.90) at each point of the continuum, it is assumed that there exist a second-order plastic strain tensor $\mathbf{E}^p$ and a work-hardening measure κ which is a scaler function. In the purely mechanical category it is assumed that the current state of a continuum subjected to finite elastic-plastic deformation can be uniquely described or specified by the current values of the variable set $\mathbf{E}$, $\mathbf{E}^p$, and κ. Therefore the stress response is described generally by a constitutive equation of the form

$$\mathbf{S} = \hat{\mathbf{S}}(\mathbf{E}, \mathbf{E}^p, \kappa) \tag{8.92}$$

For different materials $\hat{\mathbf{S}}$ may have different forms and should be determined from the experimental results. For example, in the case of infinitesimal deformation, as discussed in the previous section, $\hat{\mathbf{S}}$ can be assumed as

$$\mathbf{S} = \mathbf{C} : (\mathbf{E} - \mathbf{E}^p) \tag{8.93}$$

where $\mathbf{C}$ is the constant elastic stiffness tensor. In fact in this case there is no need to distinguish $\mathbf{S}$, $\mathbf{E}$, and $\mathbf{E}^p$ from $\boldsymbol{\sigma}$, $\boldsymbol{\epsilon}$, and $\boldsymbol{\epsilon}^p$, since the deformation is infinitesimal.

The derivative of $\hat{\mathbf{S}}$ with respect to $\mathbf{E}$ is denoted by $\mathscr{L}$:

$$\mathscr{L} = \frac{\partial \hat{\mathbf{S}}}{\partial \mathbf{E}} \tag{8.94}$$

which is a fourth-order tensor with the following symmetric properties:

$$\mathscr{L}_{klmn} = \mathscr{L}_{lkmn} = \mathscr{L}_{klnm} \tag{8.95}$$

The yield surface in strain space is described by the scaler-valued function

$$\varphi = \varphi(\mathbf{E}, \mathbf{E}^p, \kappa) = 0 \tag{8.96}$$

As usual, for fixed values of $\mathbf{E}^p$ and κ, the equation $\varphi = 0$ represents a closed orientable surface of dimension five enclosing the elastic region $\varphi < 0$ in strain space.

For the plastic strain $\mathbf{E}^p$, the following constitutive equation is proposed:

$$d\mathbf{E}^p = 0 \qquad \text{when } \varphi < 0 \tag{8.97}$$

$$d\mathbf{E}^p = 0 \qquad \text{when } \varphi = 0, \ \frac{\partial \varphi}{\partial \mathbf{E}} : d\mathbf{E} < 0 \tag{8.98}$$

$$d\mathbf{E}^p = 0 \qquad \text{when } \varphi = 0, \ \frac{\partial \varphi}{\partial \mathbf{E}} : d\mathbf{E} = 0 \tag{8.99}$$

$$d\mathbf{E}^p = \lambda \left(\frac{\partial \varphi}{\partial \mathbf{E}} : d\mathbf{E} \right) \rho \qquad \text{when } \varphi = 0, \ \frac{\partial \varphi}{\partial \mathbf{E}} : d\mathbf{E} > 0 \tag{8.100}$$

where $\boldsymbol{\rho}$ is a second-order, material tensorial function of $\mathbf{E}$, $\mathbf{E}^{\mathrm{p}}$, and κ. λ is a scaler factor to be determined from the consistency condition. It is assumed, without loss of generality, that $\lambda > 0$. The reason why $\partial\varphi/\partial\mathbf{E} : d\mathbf{E}$ appears in the constitutive equation for $\mathbf{E}^{\mathrm{p}}$ is that $d\mathbf{E}^{\mathrm{p}} > 0$ whenever $\partial\varphi/\partial\mathbf{E} : d\mathbf{E} > 0$, according to the loading criterion discussed previously. Recall that the four expressions in Eqs. (8.97)–(8.100) represent elastic state, unloading, neutral loading, and plastic loading, respectively. The evolution of the hardening parameter κ is expressed by

$$d\kappa = \mathbf{N} : d\mathbf{E}^{\mathrm{p}} \tag{8.101}$$

where $\mathbf{N}$ is another second-order material tensor. Equations (8.92), (8.96), and (8.101) are the four constitutive equations of the strain space plasticity theory developed by Naghdi and coworkers. It is seen that three tensor-valued constitutive functions, $\hat{\mathbf{S}}$, $\boldsymbol{\rho}$, and $\mathbf{N}$ and one scaler-valued constitutive function φ are involved in the general theory. These functions should be specified for particular materials before applying the theory to any practical problems. This determination, in principle, should be achieved through experimental studies. The value of the theoretical work is to provide some relations and restrictions between these functions and other known variables and further reduce the independent unknown variables or functions. This will be discussed later in connection with the physically plausible work assumption proposed by Naghdi and Trapp (1975a).

To determine the factor λ, the consistency condition is used

$$\frac{\partial\varphi}{\partial\mathbf{E}} : d\mathbf{E} + \frac{\partial\varphi}{\partial\mathbf{E}^{\mathrm{p}}} : d\mathbf{E}^{\mathrm{p}} + \frac{\partial\varphi}{\partial\kappa} d\kappa = 0 \tag{8.102}$$

which, after substituting Eqs. (8.99) and (8.97)–(8.100) into it, leads to

$$\frac{\partial\varphi}{\partial\mathbf{E}} : d\mathbf{E} + \lambda\frac{\partial\varphi}{\partial\mathbf{E}^{\mathrm{p}}} : \boldsymbol{\rho}\left(\frac{\partial\varphi}{\partial\mathbf{E}} : d\mathbf{E}\right) + \lambda\frac{\partial\varphi}{\partial\kappa}\mathbf{N} : \rho\left(\frac{\partial\varphi}{\partial\mathbf{E}} : d\mathbf{E}\right) = 0 \tag{8.103}$$

The common factor $\partial\varphi/\partial\mathbf{E} : d\mathbf{E} > 0$ in each term of the equation above can be eliminated to give

$$1 + \lambda\boldsymbol{\rho} : \left\{\frac{\partial\varphi}{\partial\mathbf{E}^{\mathrm{p}}} + \frac{\partial\varphi}{\partial\kappa}\mathbf{N}\right\} = 0 \tag{8.104}$$

λ can now be easily obtained from it as

$$\lambda = \frac{1}{\boldsymbol{\rho} : \left\{(\partial\varphi/\partial\mathbf{E}^{\mathrm{p}}) + \left(\dfrac{\partial\varphi}{\partial\kappa}\right)\mathbf{N}\right\}} \tag{8.105}$$

From Eqs. (8.92) and (8.97)–(8.100), it is seen that for elastic deformation, unloading, and neutral loading,

$$d\mathbf{S} = \mathscr{L} : d\mathbf{E} \tag{8.106}$$

During plastic loading

$$\begin{aligned} d\mathbf{S} &= \mathscr{L} : d\mathbf{E} + \frac{\partial \hat{S}}{\partial \mathbf{E}^p} : d\mathbf{E}^p + \frac{\partial \hat{\mathbf{S}}}{\partial \kappa} d\kappa \\ &= \left\{ \mathscr{L} + \lambda \boldsymbol{\beta} \frac{\partial \varphi}{\partial \mathbf{E}} \right\} : d\mathbf{E} \end{aligned} \tag{8.107}$$

where

$$\boldsymbol{\beta} = \left\{ \frac{\partial \hat{\mathbf{S}}}{\partial \mathbf{E}^p} + \frac{\partial \hat{\mathbf{S}}}{\partial \kappa} \mathbf{N} \right\} : \boldsymbol{\rho} \tag{8.108}$$

Equations (8.97)–(8.100) and (8.101) have been introduced to simplify the expression. Tensor $\boldsymbol{\beta}$ is a very important quantity that will be discussed later.

Note that $\mathbf{E}^p$ is defined as the plastic strain through the constitutive equations (8.97)–(8.100); it has no kinematic explanation of the type Eq. (8.90). It is an independent variable expressed as E. $\mathbf{E} - \mathbf{E}^p$, for finite deformation, is not the elastic strain in general, since the additive decomposition does not hold in this case, as pointed out in Chapter 7. $\mathbf{E} - \mathbf{E}^p = \mathbf{E}^e$ only when the deformation is infinitesimal.

8.6.2 Work Assumption by Naghdi and Trapp (1975a)

In developing the constitutive restrictions for the plasticity theory in strain space within the purely mechanical framework, Naghdi and Trapp proposed a physically plausible work assumption. With particular reference to elastic-plastic materials subjected to finite deformation, the work assumption may be stated as (Naghdi and Trapp 1975; Casey and Tseng 1984): The external work done on an elastic-plastic body by surface traction and body forces in any sufficiently smooth homogeneous cycle of deformation is nonnegative.

As defined by Naghdi and Trapp (1975a), a *homogeneous motion* is one whose deformation gradient **F** is independent of the material coordinates **X**. The definition of a closed cycle of deformation is given below (Casey and Tseng 1984).

Suppose that at some time t_0, a homogeneous elastic-plastic body has values of strain, plastic strain and work-hardening measure specified by

$$\mathbf{E} = {}^0\mathbf{E}, \quad \mathbf{E}^p = {}^0\mathbf{E}^p, \quad \kappa = {}^0\kappa \qquad (t = t_0) \tag{8.109}$$

The strain ${}^0\mathbf{E}$ may be in the elastic region $(\varphi({}^0\mathbf{E}, {}^0\mathbf{E}^p, {}^0\kappa) < 0)$ or on the yield surface $(\varphi({}^0\mathbf{E}, {}^0\mathbf{E}^p, {}^0\kappa) = 0)$.

The deformation process is said to be a closed cycle if at the end of the deformation, $t = t_f$, $\mathbf{E}$ reverts to its original value ${}^0\mathbf{E}$, while $\mathbf{E}^p$ and κ may change to ${}^*\mathbf{E}^p$ and ${}^*\kappa$, since plastic loading may possibly occur during the cycle. Therefore

$$\mathbf{E} = {}^0\mathbf{E}, \quad \mathbf{E}^p = {}^*E^p, \quad \kappa = {}^*\kappa \qquad (t = t_f) \tag{8.110}$$

With this definition, the work assumption of Naghdi and Trapp can be expressed mathematically as

$$\int_{t_0}^{t_f} \mathbf{S} : \dot{\mathbf{E}}\, dt \geq 0 \tag{8.111}$$

Since ${}^0\mathbf{E}$ remains constant, this can also be written

$$\int_{t_0}^{t_f} \mathbf{S} : \frac{d}{dt}\left[\mathbf{E} - {}^0\mathbf{E}\right] dt \geq 0 \tag{8.112}$$

Integrating this by parts and using Eqs. (8.109) and (8.110) yields

$$\int_{t_0}^{t_f} \dot{\mathbf{S}} : \left[\mathbf{E} - {}^0\mathbf{E}\right] dt \leq 0 \tag{8.113}$$

Comparing the work assumption Eq. (8.111) with Il'yushin's plasticity postulate Eq. (8.23) indicates that these two expressions are conceptually the same. However, the following differences should be observed:

1. Il'yushin's plasticity postulate is confined to small deformations, while Naghdi and Trapp's work assumption is valid for finite deformations.
2. Il'yushin considered only a specific class of elastic-plastic materials for which the linearized theory Eq. (8.19) applies. In contrast, Naghdi and Trapp's work assumption is for the most general elastic-plastic materials, the stress response of which is characterized by the general nonlinear constitutive equation (8.92).

In addition Il'yushin put more emphasis on the definition of the plastic deformation through the introduction of inequality Eq. (8.23), while Naghdi and coworkers focused their attention on deriving the supplementary restrictions on material response functions, $\hat{\mathbf{S}}$, $\boldsymbol{\rho}$, $\mathbf{N}$, and φ, from the work assumption Eq. (8.111). In fact more general results have been obtained by Naghdi and coworkers.

Like Drucker's stability and Il'yushin's plasticity postulates, Naghdi and Trapp's work assumption is the cornerstone of the plasticity theory for finite deformation in strain space. Several important consequences have been explored based on their assumptions, including the following three most significant ones:

1. *Existence of a potential function* (Casey and Naghdi 1984). The work assumption Eq. (8.109) implies the existence of a potential $\hat{\psi}(\mathbf{E}, \mathbf{E}^p, \kappa)$ from which the stress function $\hat{\mathbf{S}}$ can be derived in the form

$$\hat{S} = \frac{\partial \hat{\Psi}}{\partial \mathbf{E}} \tag{8.114}$$

The proof of this provided by Casey and Naghdi (1984) is as follows: Consider any smooth homogeneous cycle of deformation in which no plastic deformation occurs. According to Eqs. (8.97)–(8.100) and (8.101), $\mathbf{E}^p$ and κ will remain unchanged during this cycle, and the inequality (8.109) reduces to

$$\int_{t_0}^{t_f} \mathbf{S} : \dot{\mathbf{E}}\, dt = 0 \tag{8.115}$$

which hold for all homogeneous cycles of deformation within the current yield surface $\varphi = 0$ in strain space. Equation (8.114) is the direct result of Eq. (8.115) by the theorem on line integrals (Truesdell and Noll 1965). In view of Eq. (8.114) the following symmetric property can be added to the fourth-order tensor $\mathscr{L}$ defined by Eq. (8.94):

$$\mathscr{L}_{klmn} = \frac{\partial^2 \hat{\Psi}}{\partial E_{kl}\, \partial E_{mn}} = \frac{\partial^2 \hat{\psi}}{\partial E_{mn}\, \partial E_{kl}} = \mathscr{L}_{mnkl} \tag{8.116}$$

Now the problem of finding the tensor-valued stress function $\hat{\mathbf{S}}$ has been changed to the problem of determining a scaler-valued potential $\hat{\psi}$.

2. *Normality of β to φ* (Naghdi and Trapp 1975a). Naghdi and Trapp [1975a] showed, using an infinite sequence of special finite cycle of strain, that the tensor $\boldsymbol{\beta}$, defined by Eq. (8.108), is directed along the inward normal to the yield surface $\varphi = 0$ in strain space:

$$\boldsymbol{\beta} = -\gamma^* \frac{\partial \varphi}{\partial \mathbf{E}}, \qquad \gamma^* > 0 \tag{8.117}$$

where γ^* is an undetermined scaler-valued function of $\mathbf{E}$, $\mathbf{E}^p$, and κ. The proof of this equation is fairly long and complicated; hence it will not be given here. In view of Eq. (8.108) the significance of the normality condition (8.117) can be immediately seen: The tensor-valued plastic strain function $\boldsymbol{\rho}$

can be determined in terms of the stress function $\hat{\mathbf{S}}$, hardening function $\mathbf{N}$, as well as the yield function φ so that the independent unknown material functions are effectively reduced by one. Equation (8.117) plays the same important role as Eq. (8.42) but is more general. Equation (8.117) in fact is a generalized normality condition in the sense that the tensor $\boldsymbol{\beta}$ is more general than $d\boldsymbol{\sigma}^p$, as in infinitesimal deformation. The normality of $\mathscr{L}:d\mathbf{E}^p$ to the yield surface $\varphi = 0$, which corresponds to Eq. (8.42), will be discussed later in connection with a special class of materials. Note here that the introduction of the scaler function γ^* causes no extra unknown functions. In fact γ^* and the scaler factor λ, introduced in Eqs. (8.97)–(8.100), always appear in multiplication form and therefore can be represented by a single function $\lambda^* = \lambda\gamma^*$, which can be determined by the consistency condition. This also will be detailed later.

3. *Convexity of the yield surface* (Naghdi and Trapp 1975b; Casey 1984). The convexity of the yield surface, both in stress space and in strain space, holds true only for a class of special materials. It has been proved by Naghdi and Trapp (1975b) and Casey (1984) that for materials for which the potential ψ can be expressed as

$$\psi = \bar{\psi}(\mathbf{E} - \mathbf{E}^p) \tag{8.118}$$

the following inequality can be derived from the work assumption:

$$\left\{\mathbf{S} - {}^0\mathbf{S}\right\} : \frac{\partial f}{\partial \mathbf{S}} \geq 0 \tag{8.119}$$

where f is the yield surface in stress space, and $\mathbf{S}$ and ${}^0\mathbf{S}$ are stresses on and inside the yield surface, respectively. Equation (8.119) ensures the convexity of the yield surface in stress space. Note that this conclusion is consistent with the discussion in Section 5.7, since in infinitesimal deformation $\Psi(\mathbf{E} - \mathbf{E}^p) = \Psi(\mathbf{E}^e)$, implying no elastic-plastic coupling. Therefore the yield surface in stress space is convex whenever there is no elastic-plastic coupling, a conclusion already obtained from Drucker's postulate in Section 5.7.

If it is further assumed that the material has a linear stress response with constant coefficients, the potential Ψ is then expressed by

$$\Psi = \frac{1}{2}(\mathbf{E} - \mathbf{E}^p) : \mathscr{L} : (\mathbf{E} - \mathbf{E}^p) \tag{8.120}$$

where $\mathscr{L}$ is a constant coefficient tensor. From Eq. (8.107)

$$\mathbf{S} = \mathscr{L} : (\mathbf{E} - \mathbf{E}^p) \tag{8.121}$$

and ${}^0\mathbf{S}$ is given by

$${}^0\mathbf{S} = \mathscr{L} : \left({}^0\mathbf{E} - \mathbf{E}^p\right) \tag{8.122}$$

where ${}^0\mathbf{E}$ is the strain corresponding to ${}^0\mathbf{S}$, and the plastic strain $\mathbf{E}^p$ corresponding to $\mathbf{S}$ and ${}^0\mathbf{S}$ are the same. It follows from Eqs. (8.121) and (8.122) that

$$\left(\mathbf{S} - {}^0S\right) = \mathscr{L} : \left(\mathbf{E} - {}^0\mathbf{E}\right) \tag{8.116}$$

On the other hand,

$$f(\mathbf{S}, \mathbf{E}^p, \kappa) = f\left(\mathbf{S}(\mathbf{E}, \mathbf{E}^p, \kappa), \mathbf{E}^p, \kappa\right) = \varphi(\mathbf{E}, \mathbf{E}^p, \kappa) \tag{8.124}$$

Therefore

$$\frac{\partial \varphi}{\partial \mathbf{E}} = \frac{\partial f}{\partial \mathbf{S}} : \frac{\partial \mathbf{S}}{\partial \mathbf{E}} = \mathscr{L} : \frac{\partial f}{\partial \mathbf{S}} \tag{8.125}$$

where the symmetry of Eq. (8.116) has been used. Substituting Eqs. (8.123) and (8.125) into Eq. (8.119) gives

$$\left\{\mathbf{E} - {}^0\mathbf{E}\right\} : \frac{\partial \varphi}{\partial \mathbf{E}} \geq 0 \tag{8.126}$$

When Ψ is given by Eq. (8.120), the yield surface in strain space is also convex. This, again, is consistent with the conclusion on convexity discussed in Section 8.2.

Although the consequences above were obtained assuming homogeneous motion, it should be emphasized that they also hold true for nonhomogeneous motion (Naghdi and Trapp 1975a).

8.6.3 A Special Class of Elastic-Plastic Materials

In the preceding discussion the purely mechanical state of an elastic-plastic material is assumed to be determined uniquely by a set of variables $\mathbf{E}$, $\mathbf{E}^p$, and κ. The general stress response function $\hat{\mathbf{S}}$ and the potential ψ are expressed in terms of these variables as

$$\mathbf{S} = \hat{\mathbf{S}}(\mathbf{E}, \mathbf{E}^p, \kappa) \tag{8.127}$$

$$\psi = \hat{\psi}(\mathbf{E}, \mathbf{E}^p, \kappa) \tag{8.128}$$

However, sometimes it is more convenient to write these functions in terms of variables $\mathbf{E} - \mathbf{E}^p$, $\mathbf{E}^p$ and κ:

$$\mathbf{S} = \bar{\mathbf{S}}(\mathbf{E} - \mathbf{E}^p, \mathbf{E}^p, \kappa) \tag{8.129}$$

$$\psi = \bar{\psi}(\mathbf{E} - \mathbf{E}^p, \mathbf{E}^p, \kappa) \tag{8.130}$$

Obviously expressions of Eqs. (8.129) and (8.130) are equivalent to Eqs. (8.127) and (8.128), respectively. In addition the following equations can be determined from Eqs. (8.127)–(8.130) (Casey and Tseng 1984):

$$\frac{\partial \bar{\mathbf{S}}}{\partial(\mathbf{E}-\mathbf{E}^{\mathrm{p}})} = \frac{\partial \hat{\mathbf{S}}}{\partial \mathbf{E}} = \mathscr{L} \tag{8.131}$$

$$\frac{\partial \bar{\mathbf{S}}}{\partial \mathbf{E}^{\mathrm{p}}} = \mathscr{L} + \frac{\partial \hat{\mathbf{S}}}{\partial \mathbf{E}^{\mathrm{p}}} \tag{8.132}$$

$$\frac{\partial \bar{\mathbf{S}}}{\partial \kappa} = \frac{\partial \hat{\mathbf{S}}}{\partial \kappa} \tag{8.133}$$

$$\frac{\partial \bar{\psi}}{\partial(\mathbf{E}-\mathbf{E}^{\mathrm{p}})} = \frac{\partial \hat{\psi}}{\partial \mathbf{E}} = \mathbf{S} \tag{8.134}$$

$$\frac{\partial \bar{\psi}}{\partial \mathbf{E}^{\mathrm{p}}} = \mathbf{S} + \frac{\partial \hat{\psi}}{\partial \mathbf{E}^{\mathrm{p}}} \tag{8.135}$$

$$\frac{\partial \bar{\psi}}{\partial \kappa} = \frac{\partial \hat{\psi}}{\partial \kappa} \tag{8.136}$$

We have chosen the variable $\mathbf{E} - \mathbf{E}^{\mathrm{p}}$ because it reduces to elastic strain $\mathbf{E}^{\mathrm{e}}$ in infinitesimal deformation. Substituting Eqs. (8.132) and (8.136) into Eq. (8.108) yields

$$\bar{\boldsymbol{\beta}} = \left\{\frac{\partial \bar{\mathbf{S}}}{\partial \mathbf{E}} + \frac{\partial \bar{\mathbf{S}}}{\partial \kappa}\mathbf{N}\right\} : \boldsymbol{\rho} = \boldsymbol{\beta} + \mathscr{L} : \boldsymbol{\rho} \tag{8.137}$$

or

$$\boldsymbol{\beta} = \bar{\boldsymbol{\beta}} - \mathscr{L} : \boldsymbol{\rho} \tag{8.138}$$

The normality condition Eq. (8.117) can be rewritten as

$$\bar{\boldsymbol{\beta}} - \mathscr{L} : \boldsymbol{\rho} = -\gamma^* \frac{\partial \varphi}{\partial \mathbf{E}} \tag{8.139}$$

Now let us consider the special class of materials for which the constitutive function can be given by

$$\mathbf{S} = \bar{\mathbf{S}}(\mathbf{E} - \mathbf{E}^{\mathrm{p}}) \tag{8.140}$$

$$\psi = \bar{\psi}(\mathbf{E} - \mathbf{E}^{\mathrm{p}}) \tag{8.141}$$

For infinitesimal deformation these expressions become

$$\mathbf{S} = \bar{\mathbf{S}}(\boldsymbol{\epsilon}^e), \quad \psi = \bar{\psi}(\boldsymbol{\epsilon}^e) \tag{8.142}$$

where the difference between the Lagrange strain tensor and the infinitesimal strain tensor is ignored. These equations indicate no elastic-plastic coupling during the elastic-plastic deformation. In other words, the elastic properties will not be altered by the plastic deformation.

It is clear from Eqs. (8.138)–(8.141) that for this class of special materials

$$\boldsymbol{\beta} = -\mathscr{L} : \boldsymbol{\rho} \tag{8.143}$$

As a result the normality condition becomes

$$\mathscr{L} : \boldsymbol{\rho} = \gamma^* \frac{\partial \varphi}{\partial \mathbf{E}} \tag{8.144}$$

Multiplying both sides by $\lambda(\partial\varphi/\partial\mathbf{E} : d\mathbf{E})$, we have

$$\mathscr{L} : d\mathbf{E}^p = \lambda\gamma^* \left(\frac{\partial \varphi}{\partial \mathbf{E}} : d\mathbf{E} \right) \frac{\partial \varphi}{\partial \mathbf{E}} \tag{8.145}$$

Note that $\lambda\gamma^*$ in this equation can be replaced by a single scaler function $\lambda^* = \lambda\gamma^*$. Introducing the symbol

$$d\mathbf{S}^p = \mathscr{L} : d\mathbf{E}^p \tag{8.146}$$

we can rewrite Eq. (8.145) as

$$d\mathbf{S}^p = \lambda^* \left(\frac{\partial \varphi}{\partial \mathbf{E}} : d\mathbf{E} \right) \frac{\partial \varphi}{\partial \mathbf{E}} \tag{8.147}$$

Therefore $d\mathbf{S}^p$ is normal to the yield surface in strain space, and Eq. (8.147) is a counterpart conclusion of Eq. (8.42) in finite elastic-plastic deformation. If we apply Eq. (8.125) to Eq. (8.147), we have

$$\mathscr{L} : d\mathbf{E}^p = \lambda^* \left(\frac{\partial \varphi}{\partial \mathbf{E}} : d\mathbf{E} \right) \mathscr{L} : \frac{\partial f}{\partial \mathbf{S}} \tag{8.148}$$

which, after multiplying both sides by $\mathscr{L}^{-1}$, gives

$$d\mathbf{E}^p = \lambda^* \left(\frac{\partial \varphi}{\partial \mathbf{E}} : d\mathbf{E} \right) \frac{\partial f}{\partial \mathbf{S}} \tag{8.149}$$

Thus, for the class of materials being considered, the plastic strain rate is directed along the normal to the yield surface in stress space. This is a generalization of Drucker's normality flow rule to the case of finite elastic-plastic deformation in terms of the Lagrange description.

In summary, we have presented important constitutive results in this section:

1. General nonlinear material:

$$\psi = \hat{\psi}(\mathbf{E}, \mathbf{E}^{\mathrm{p}}, \kappa) = \bar{\psi}(\mathbf{E} - \mathbf{E}^{\mathrm{p}}, \mathbf{E}^{\mathrm{p}}, \kappa) \qquad \text{(Potential)} \tag{8.150}$$

$$\mathbf{S} = \frac{\partial \hat{\psi}}{\partial \mathbf{E}} = \frac{\partial \bar{\psi}}{\partial(\mathbf{E} - \mathbf{E}^{\mathrm{p}})} \qquad \text{(Stress function)} \tag{8.151}$$

$$\varphi = \varphi(\mathbf{E}, \mathbf{E}^{\mathrm{p}}, \kappa) \qquad \text{(Yield function)} \tag{8.152}$$

$$d\mathbf{E}^{\mathrm{p}} = \lambda\left(\frac{\partial \varphi}{\partial \mathbf{E}} : d\mathbf{E}\right)\rho, \quad \frac{\partial \varphi}{\partial \mathbf{E}} : d\mathbf{E} > 0 \qquad \text{(Flow rule)} \tag{8.153}$$

$$d\kappa = \mathbf{N} : d\mathbf{E}^{\mathrm{p}} \qquad \text{(Hardening rule)} \tag{8.154}$$

$$\boldsymbol{\beta} = -\gamma^* \frac{\partial \varphi}{\partial \mathbf{E}} \qquad \text{(Normality condition)} \tag{8.155}$$

where

$$\beta = \left\{\frac{\partial \hat{\mathbf{S}}}{\partial \mathbf{E}^{\mathrm{p}}} + \frac{\partial \hat{\mathbf{S}}}{\partial \kappa} : \mathbf{N}\right\} : \boldsymbol{\rho} \tag{8.156}$$

The three materials functions to be determined in this general theory are potential ψ, yield surface φ, and the hardening function **N**. If work-hardening is assumed

$$d\kappa = \mathbf{S} : d\mathbf{E}^{\mathrm{p}} \tag{8.157}$$

then $\mathbf{N} = \mathbf{S}$.

2. Special materials:

$$\psi = \bar{\psi}(\mathbf{E} - \mathbf{E}^{\mathrm{p}}) \qquad \text{(Potential)} \tag{8.158}$$

$$\mathbf{S} = \frac{\partial \bar{\psi}}{\partial(\mathbf{E} - \mathbf{E}^{\mathrm{p}})} = \bar{\mathbf{S}}(\mathbf{E} - \mathbf{E}^{\mathrm{p}}) \qquad \text{(Stress function)} \tag{8.159}$$

$$\varphi = \varphi(\mathbf{E}, \mathbf{E}^{\mathrm{p}}, \kappa) \qquad \text{(Yield function)} \tag{8.160}$$

$$d\mathbf{S}^{\mathrm{p}} = \lambda^*\left(\frac{\partial \varphi}{\partial \mathbf{E}} : d\mathbf{E}\right)\frac{\partial \varphi}{\partial \mathbf{E}} \qquad \text{(Normality flow rule)} \tag{8.161}$$

where

$$dS^p = \mathscr{L} : dE^p, \quad \mathscr{L} = \frac{\partial \bar{S}}{\partial E} \tag{8.162}$$

$$d\kappa = N : dE^p \qquad \text{(Hardening rule)} \tag{8.163}$$

The three materials functions are $\bar{\psi}$, φ, and **N**. The scaler factors λ in Eq. (8.153) and λ^* in Eq. (8.161) can be determined from the consistency condition.

REFERENCES

Casey, J. 1984. A simple proof of a result in finite plasticity. *Quart. J. Mech. Appl. Math.* **42**:61.

Casey, J., and P. M. Naghdi. 1984. Further constitutive result in finite plasticity. *Quart. J. Mech. Appl. Math.* **37**:231.

Casey, J., and P. M. Naghdi. 1983a. On the nonequivalence of the stress space and strain space formulation of plasticity theory. *J. Appl. Mech.* **50**:380.

Casey, J., and P. M. Naghdi. 1983b. A remark on the definition of hardening, softening and perfectly plastic behavior. *Acta Mech.* **48**:91.

Casey, J., and P. M. Naghdi. 1981. On the characterization of strain-hardening in plasticity. *J. Appl. Mech.* **48**:285.

Casey, J., and M. Tseng. 1984. A constitutive restriction related to convexity of yield surfaces in plasticity. *J. Appl. Math. Phys.* **35**:478.

Il'yushin, A. A. 1961. On the postulate of plasticity. *PMM* **25**:503.

Il'yushin, A. A. 1960. On the increments of plastic deformations and the yield surface. *PMM* **24**:663.

Naghdi, P. M., and J. A. Trapp. 1975a. The significance of formulating plasticity theory with reference to loading surfaces in strain space. *Int. J. Eng. Sc.* **13**:785.

Naghdi, P. M., and J. A. Trapp. 1975b. Restrictions on constitutive equations of finitely deformed elastic-plastic materials. *Quart. J. Mech. Appl. Math.* **28**:25.

Naghdi, P. M., and J. A. Trapp. 1975c. On the nature of normality of plastic strain rate and convexity of yield surfaces in plasticity. *J. Appl. Mech.* **42**:61.

Yoder, R. J. 1981. A strain-space plasticity theory and numerical implementation. Ph.D. dissertation. California Institute of Technology.

Yoder, R. J., and W. D. Iwan. 1981. On the formulation of strain-space plasticity with multiple loading surfaces. *J. Appl. Mech.* **48**:773.

9

INTRODUCTION TO DISLOCATION THEORY

In the preceding chapters we presented mathematical theories of plasticity. We showed that these theories are based on macroscopic or phenomenalogical observations of plastic flows of metals. In these theories metal is viewed as a continuum. However, as we pointed out in Chapter 1, the plastic deformation of metals is not uniform throughout a body. Metals have a crystalline structure, so the plastic flow of metals is confined to certain crystallographic planes in certain directions. These are details of plastic flow at the microscopic level that are not considered by the mathematical plasticity theories. On the other hand, a physical theory of plasticity starts with these microscopic details and attempts to explain why and how the plastic flow occurs. Mathematical simplicity may be lost but the physical background of the plastic deformation is fully explained.

The study of crystalline plasticity started in the 1890s. Ewing and Rosenhain (1899, 1900) published a series of papers that summarized their metallographic studies of plastic deformation of polycrystalline metals. Their conclusions had far-reaching significance and provided a remarkably accurate picture of crystalline plasticity. They observed the "slip steps" on the surface of the plastically deformed metals and attributed these steps to the emergence of "slip bands" inside the metals. Ewing and Rosenhain's findings can be summarized as follows:

1. Metals and alloys are crystalline and composed of aggregates of single crystals or grains.
2. Plastic deformation takes place by simple shearing caused by the sliding of only certain families of crystal planes over each other in certain crystallographic directions lying on the plane.

3. The particular crystalline structure of metal is preserved during the plastic deformation, which implies that the slip happens in whole multiples of lattice spacings.
4. Except for the sliding of the crystallographic planes, the plastic deformation may occur by twinning, another crystallographic phenomenon.

These findings form the basis of any recent physical plasticity theory. In the following chapters we will introduce physical plasticity theories. In this chapter we begin with dislocation theory. Then in Chapters 10 and 11 we will discuss the plasticity theories for single crystals and polycrystals, respectively.

In this chapter we discuss the crystalline structure of metals first in order to provide the necessary knowledge to understand the microscopic mechanisms of plastic deformation. Then we show how the theoretical strength (yield strength) is several orders of magnitude greater than the strengths of real crystals measured by experiments. This large discrepancy can be explained by the presence of dislocations, or the atomic defects, or elementary dislocation theory which we briefly present. More detailed treatment of the topic can be found elsewhere (e.g., see Honeycombe 1968).

9.1 CRYSTALLINE STRUCTURE OF METALS

Let us consider the atomic structure of metals or other solid bodies. It is well known today that metals are crystalline solids that consist of atoms arranged in a pattern that is repeated periodically. In an ideal representation of the crystalline structure, the atoms are represented by points in space. The patterns in which the atoms or points are arranged are called a *crystal lattice*. Therefore a crystal or space lattice is an infinite, three-dimensional array of points with periodic structure. This structure depends on the material.

9.1.1 Elementary Cells

The periodic nature of a crystal lattice allows us to fully describe its structure by the smallest crystal unit known as the *elementary cell*. The cell, and consequently the lattice structure, can take one of a number of different shapes—cubic, tetragonal, hexagonal, orthorhombic, monoclinic, and triclinic. The most common lattice types are the following three:

1. *Face-centered-cubic (FCC) lattice*. The unit cell of this lattice is shown in Fig. 9.1. The length of every edge of the cell is equal and is denoted by a. There is one atom at each of the eight corners of the cubic cell, and one atom in the middle position on each of the six faces of a cell. Many pure metals crystalize in this lattice structure, such as aluminum, copper, gold, silver, and nickel.

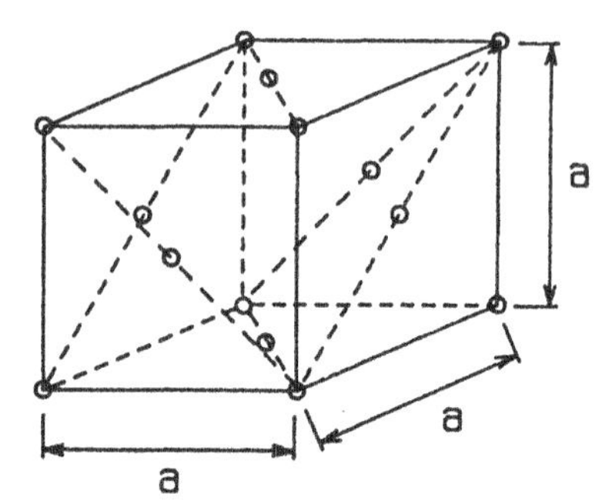

Fig. 9.1 Face-centered-cubic cell

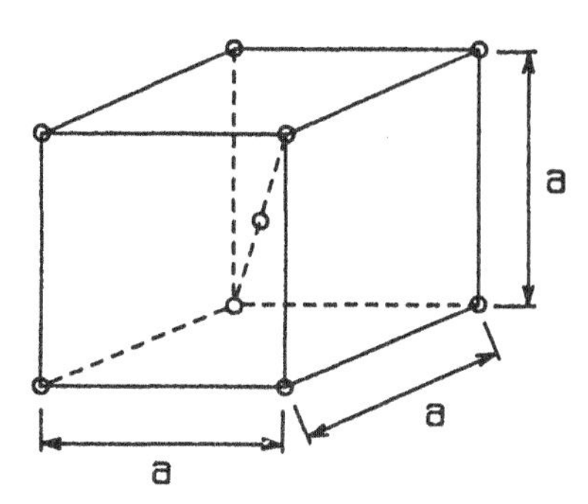

Fig. 9.2 Body-centered-cubic cell

2. *Body-centered-cubic (BCC) lattice.* The unit cell of this lattice structure is depicted in Fig. 9.2. Like FCC crystals the lengths of the edges of a cell are equal, and there is one atom at each of the eight corners of a cell. There is one extra atom at the centroid of the cell and no atom on the faces. Metals crystalizing in this structure include iron, niobium, tantalum, molybdenum, and tungsten.
3. *Hexagonal or closed-packed hexagonal (HCP) lattice.* The elementary unit of this lattice structure is shown in Fig. 9.3. As can be seen from the figure, the upper and lower basal planes are regular hexagon with side length a. The distance between the basal planes is given by c. The

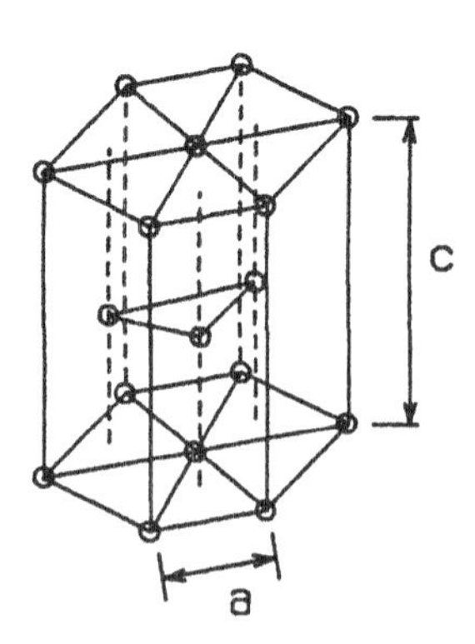

Fig. 9.3 Closed-packed hexagonal cell

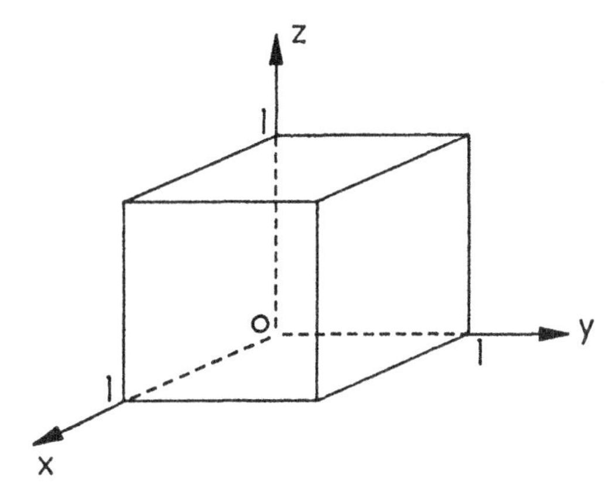

Fig. 9.4 Face-centered-cubic cell

ratio c/a is a very important parameter in determining the slip plane, which we will discuss later. The examples of HCP metals are zinc, cadmium, and magnesium. Lattice structures other than these three will not be considered here, since most of the pure metals crystalize in one of these three lattices and many alloys have these structures too.

9.1.2 Miller Indices

Miller indices are used to describe the directions and planes in a crystal. We discuss these briefly below.

1. *Indices of a Lattice Plane.* A plane in space can be described by the equation

$$\frac{x}{c_1} + \frac{y}{c_2} + \frac{z}{c_3} = 1 \tag{9.1}$$

where c_1, c_2, and c_3 are the intercepts of this plane on the x, y, and z axes, respectively. To describe the crystallographic planes, the axes are taken along three noncoplane edges of the unit cell and the intercepts are measured in terms of a unit length, which is usually the length of one edge of the atomic cell. These axes and the unit length for a cubic cell are shown in Fig. 9.4. Note that the edge length is assigned to be one unit although its actual length is a, as shown in Fig. 9.1.

To find the Miller indices for a crystallographic plane, we proceed in three steps:

a. Find the three intercepts c_1, c_2, and c_3.
b. Calculate the reciprocals of them ($1/c_1$, $1/c_2$, and $1/c_3$).
c. Reduce the reciprocals into smallest integers by dividing with the largest common factor.

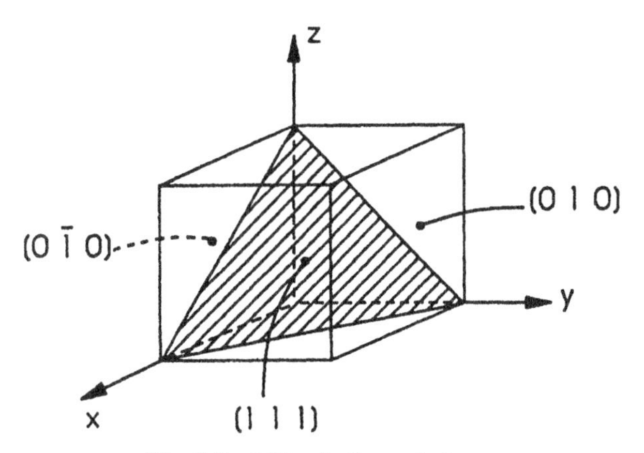

Fig. 9.5 Miller indices of planes

The indices so obtained for a crystallographic plane are called *Miller indices* and are enclosed in a parentheses as (hkl). Several planes are shown in Fig. 9.5. The indices of the opposite face of the cube are obtained by translating the origin of the coordinate system along y axis to another corner of the cell and then determining the reciprocals of the intercepts of the opposite plane to be $0, -1, 0$. To indicate -1, a bar is placed above 1. So the Miller indices for this plane are $(0\,\bar{1}\,0)$.

Because of the symmetry of the crystal structure, several planes can have the same atomic distributions, such as planes $(1\,0\,0)$, $(\bar{1}\,0\,0)$, $(0\,1\,0)$, $(0\,\bar{1}\,0)$, $(0\,0\,1)$, and $(0\,0\,\bar{1})$. These are called *crystallogaphically equivalent planes* and are represented by the indices of one of the planes, enclosed in braces $\{1\,0\,0\}$. Another example of crystallographically planes is $\{1\,1\,1\}$, representing a family of planes in FCC crystals, which represents eight planes $(1\,1\,1)$, $(1\,\bar{1}\,1)$, $(1\,1\,\bar{1})$, $(\bar{1}\,1\,1)$, $(1\,\bar{1}\,\bar{1})$, $(\bar{1}\,\bar{1}\,1)$, $(\bar{1}\,1\,\bar{1})$, and $(\bar{1}\,\bar{1}\,\bar{1})$.

2. *Indices of a lattice direction*. The indices of a direction are simply the vector components of the direction resolved along each of the coordinate axes and reduced to smallest integers. These indices are enclosed in a bracket as $[UVW]$. Several examples are shown in Fig. 9.6. Again, a bar an integer represents a negative number. Directions with the same atomic distributions are called *crystallographically equivalent* and are represented by carats $\langle UVW \rangle$. For example, $\langle 1\,1\,0 \rangle$ includes $[1\,1\,0]$, $[1\,0\,1]$, and $[0\,1\,1]$.

3. *Miller-Bravais indices for hexagonal crystals*. An alternate indexing system, called *Miller-Bravais indices*, has four numbers in each set of indices, and it is used for hexagonal crystals. Since four numbers are used, the relationship between the indices and the symmetry of the hexagonal lattice is made more obvious. Figure 9.7 shows a hexagonal cell and the four axes, one along the axis of the hexagonal prism and three in the base, 120° apart.

The Miller-Bravais indices of a plane are denoted by h, k, i, and l, enclosed in parentheses $(hkil)$. These indices are the reciprocals of intercepts

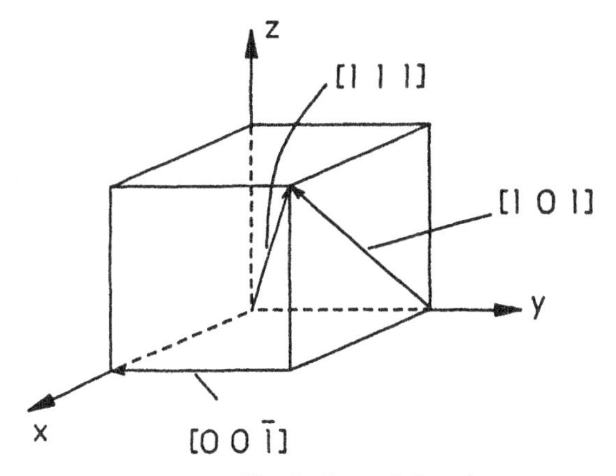

Fig. 9.6 Miller indices of directions

of the plane on a_1, a_2, a_3, and a_4 axes, respectively. As with Miller indices the reciprocals are usually divided by the largest common factor. Because only three noncoplanar axes are needed to specify a plane in space, the four indices above cannot be independent. It is obvious that the three indices on the base must satisfy the condition.

$$h = k = -i \tag{9.2}$$

Several planes are shown in Fig. 9.7.

The Miller-Bravais indices for a direction are the vector components of the direction resolved along each of the four axes and reduced to smallest

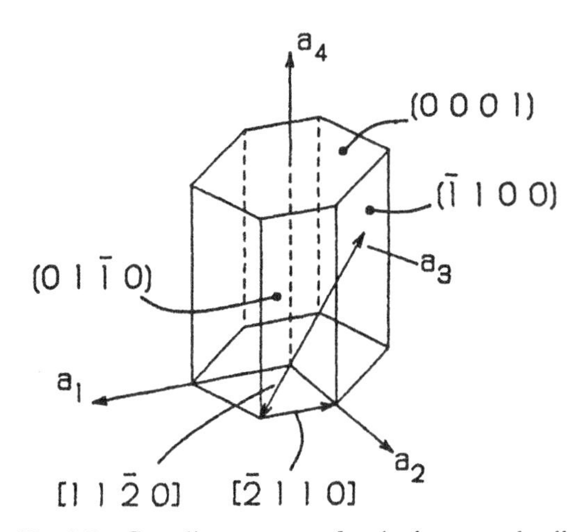

Fig. 9.7 Coordinate system for the hexagonal cell

integers. They are denoted by $[UVIW]$ and

$$U + V = -I \tag{9.3}$$

Directions $[11\bar{2}0]$ and $[\bar{2}110]$ are also shown in Fig. 9.7.

9.1.3 Stereographic Projection

The orientation of a metal crystal is an important variable. Consider a single crystal rod. Usually the principal axes of the crystal (axes along crystal cell edges) do not coincide with the axis of the rod. The usual way to describe the orientation of a crystal is to use the stereographic projection characteristic of the crystal structure.

Consider a unit cell of crystal sitting at the origin of a sphere. The normals of various planes of the crystal cell are drawn to intercept the sphere at various points. Each point represents a particular plane or a particular direction along the normal to the plane. Then all the points are projected onto a plane surface, producing a circular plot called *stereographic* projection. Every point representing a particular plane normal falls within this circular plot. The angles between various plane normals can then be measured with a circular stereographic net.

For cubic crystals the usual practice is to use the standard projection which is produced in such a way that the center of the projection is the normal of the plane [0 0 1]. A standard [0 0 1] stereographic projection of FCC crystal is shown in Fig. 9.8, where some other important planes or

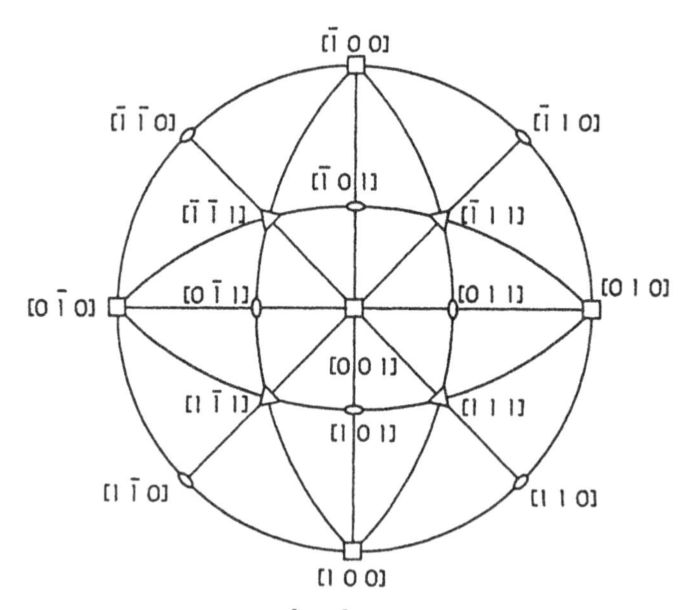

Fig. 9.8 Standard [0 0 1] stereographic projection

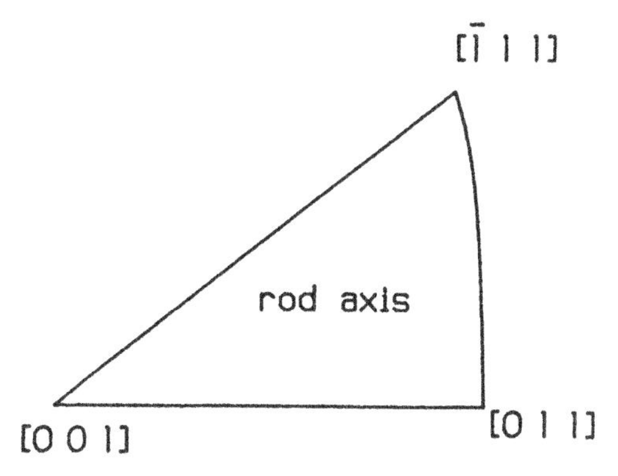

Fig. 9.9 Standard triangle

directions are labeled. Note that each point in the plot represents a plane or a direction along that plane normal.

It is seen that from Fig. 9.8 that the projection is divided by great circles into 24 unit stereographic triangles, which, due to the symmetry of the crystal structure, are crystallographically identical. Therefore in the specification of a crystal orientation only one triangle is normally used and it is the triangle bounded by [0 0 1], [0 1 1], and [$\bar{1}$ 1 1] in the center of the projection. All possible orientations of crystals of FCC structure can then be specified by plotting the position of the rod axis within this triangle or along its boundaries. In practice, to represent the relative orientations of a crystal rod, the angles between the rod axis and at least two of the [0 0 1], [0 1 1], and [$\bar{1}$ 1 1] directions are measured, and then the rod axis can be spotted in the standard triangle by using these measurements and a stereographic net, as shown in Fig. 9.9.

9.1.4 Slip System

One of the important conclusions obtained by Ewing and Rosenhain in the 1890s is that the plastic deformation occurs by slip on certain crystallographic planes in certain crystallographic directions. Such crystallographic planes are called *slip* or *glide planes*, while the directions along which slip occurs are referred to as *slip* or *glide directions*. The combination of any one of the slip planes with any one of the slip directions on that plane is then called a *slip* or *glide system*.

Experimental observations show that in most metals, the slip planes are usually those planes with the closest atomic packing, while the slip directions are always the closest packed ones on the slip planes. In FCC metals the slip occurs on {1 1 1} planes in ⟨1 1 0⟩ directions, while in BCC metal the slip directions are ⟨1 1 1⟩ but the slip plane may be one of the {1 1 0}, {1 1 2} and {1 2 3}. The situation for HCP metals is more complex. It has been found that

Table 9.1 Packing densities and slip plans of hexagonal metals

Metal	c/a	Packing Densities (0001)	$\{10\bar{1}0\}$	$\{10\bar{1}1\}$	Observed Slip Planes in Order of Ease of Operation		
Cadmium	1.886	1.000	0.918	0.816	(0001)	$(1\bar{1}00)$	$(10\bar{1}1)$
Zinc	1.856	1.000	0.933	0.846	(0001)	$(\bar{1}\bar{1}00)$	$(\bar{1}\bar{1}22)$
Magnesium	1.624	1.000	1.066	0.940	(0001)	$(10\bar{1}1)$	$(1\bar{1}00)$
Titanium	1.587	1.000	1.092	0.959	$(1\bar{1}00)$	(0001)	$(\bar{1}011)$

Source: Honeycombe (1968).

the ratio c/a, where c is the height and a the side length of the hexagonal cell as shown in Fig. 9.3, plays a very important role in determining the slip plane. At large c/a ratio, the slip occurs on the basal plane $\{0001\}$ in any of three closed-packed directions $\langle 11\bar{2}0\rangle$. However, as the ratio c/a decreases, the slip on other planes, such as the prism planes $\{10\bar{1}0\}$ and pyramidal planes $\{10\bar{1}1\}$, becomes possible. Table 9.1 (Honeycombe 1968) gives the relative packing densities of the basal, prismatic, and pyramidal planes, where the density of the basal plane is assumed to be unity, for several hexagonal metals of different c/a ratios. The slip direction in these cases is one of the $\langle 11\bar{2}0\rangle$ directions on the slip plane. From this table it can be seen that the predominance of the basal slip is only true with high c/a ratio.

9.2 THEORETICAL STRENGTH OF SINGLE CRYSTAL

The elementary theoretical estimation of the shear strength of a perfect crystal was first given by Frenkel (1926). Based on the concept of slip plane, Frenkel considered a simple model of shearing two rows of atoms as shown in Fig. 9.10. The spacing between the rows is a, while the interatomic distance in the slip direction is b. It was assumed that when slip occurred, the atomic rows would move over each other as two rigid blocks. Denote the slip or shear displacement of the upper row over the lower row by x. It is obvious that the shear stress required to maintain the configurations at $x = 0$, and b

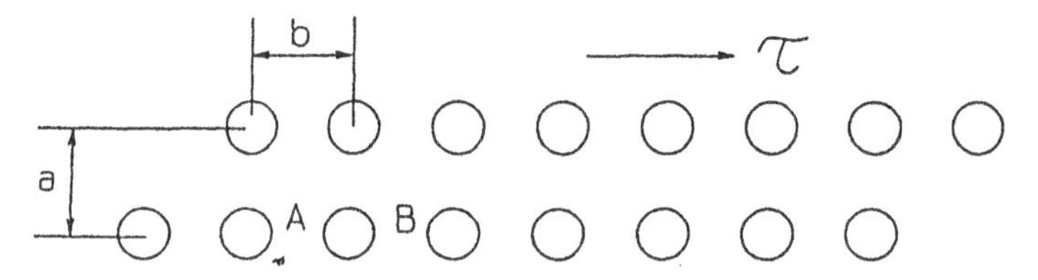

Fig. 9.10 Two rows of atoms

is zero since these are the normal lattice sites. At the halfway position $x = b/2$, this stress is also zero because of the symmetry of the configuration. During the slip over the lower row, the atoms of the upper row are subjected to a shear stress which varies from zero to a finite value. As a first approximation, Frenkel assumed that the periodicity of the energy was sinusoidal and therefore that the shear stress could be represented by

$$\tau = k \sin\left(\frac{2\pi x}{b}\right) \tag{9.4}$$

where k can be determined by specifying that initial slope in the region of small shear x/a corresponds to the shear modulus G of the crystal. When x/a is small, the shear stress τ can be written using Hooke's law:

$$\tau = G\frac{x}{a} \tag{9.5}$$

From Eq. (9.4) it follows that for small x,

$$\tau = k\frac{2\pi x}{b} \tag{9.6}$$

Therefore the coefficient k can be identified by comparing Eq. (9.5) with Eq. (9.6):

$$k = \frac{Gb}{2\pi a} \tag{9.7}$$

Hence Eq. (9.4) can be expressed as

$$\tau = \frac{Gb}{2\pi a} \sin\left(\frac{2\pi x}{b}\right) \tag{9.8}$$

The maximum theoretical shear strength obtained from this simple model is

$$\tau_c = \frac{Gb}{2\pi a} = k \tag{9.9}$$

It can be seen that τ_c is minimum when the atomic spacing a is maximum and that the interatomic distance b is minimum. This means that the weakest plane and direction under shear is that with highest atomic density or the closely packed plane in a closely packed direction. For most metal crystals, $b \approx a$ (e.g., in FCC crystals, $b/a = \sqrt{3}/\sqrt{2}$), the shear strength is thus approximately

$$\tau_c = \frac{G}{2\pi} \tag{9.10}$$

Calculations taking account of more realistic interatomic force laws in place of Eq. (9.4) have been done by Mackenzie (1962) and Tyson (1966). Their results only reduced the theoretical strength τ_c to about $G/30$. On the other hand, many experiments have been conducted to measure the yield strength of single crystals under shear. The observed shear strength of annealed crystals of various metals, including FCC crystals, is of the order of $10^{-6} \sim 10^{-4}G$. Therefore the estimated strength of crystals is several orders higher than the observed one.

The large disparity discussed above inevitably suggests that the simple model proposed by Frenkel does not represent the real behavior of crystals during plastic deformation. The real crystal must contain defects that reduce the strength of the crystal. Indeed the disparity we noted above has been satisfactorily explained by Taylor (1934), Polanyi (1934), and Orowan (1934) using the concept of dislocations.

9.3 EDGE AND SCREW DISLOCATIONS

Dislocations are one-dimensional or line-defects in crystals which are responsible for the low strength of real crystals. There are two basic types of dislocations: edge dislocation and screw dislocation. They are discussed separately below.

9.3.1 Edge Dislocation

Edge dislocation was introduced by Orowan (1934), Polanyi (1934), and Taylor (1934) to explain the lower strength of real crystals than the theoretical estimation. A schematic show of an edge dislocation is given in Fig. 9.11 by line AB. Small circles in the figure represent atoms. It can be seen that all the atoms are regularly arranged except those on the half plane $ABCD$. In fact in simple cubic lattice, as shown in Fig. 9.11, edge dislocation is indicated by inserting an extra half-plane into the perfect crystal. The name edge dislocation comes from the fact that the line AB, called *dislocation line*, is the edge of the extra half-plane $ABCD$.

Now the plastic deformation can be explained by the movement of dislocations. This is shown in Fig. 9.12. An edge dislocation starts moving from the left surface along AB toward right. As it moves, it causes the atoms above AB to slip over the atoms below AB by a distance of one atomic space. This slip happens gradually over the whole plane AB as the dislocation passes through the plane, as shown by Fig. 9.12(b)–(d). The stress required to move a dislocation line through one atomic spacing is much less than that required to move the whole block at the same time, which therefore explains the lower strength of real crystals.

To define the edge dislocation more precisely, a Burgers vector $\mathbf{b}$ has to be used. The Burgers vector $\mathbf{b}$ is introduced to describe the slip direction of a

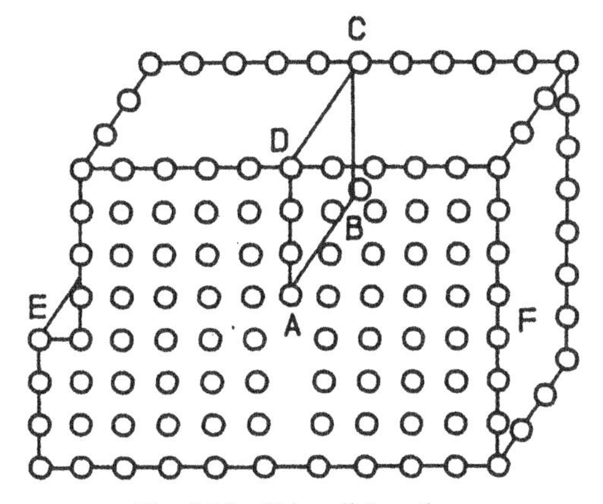

Fig. 9.11 Edge dislocation

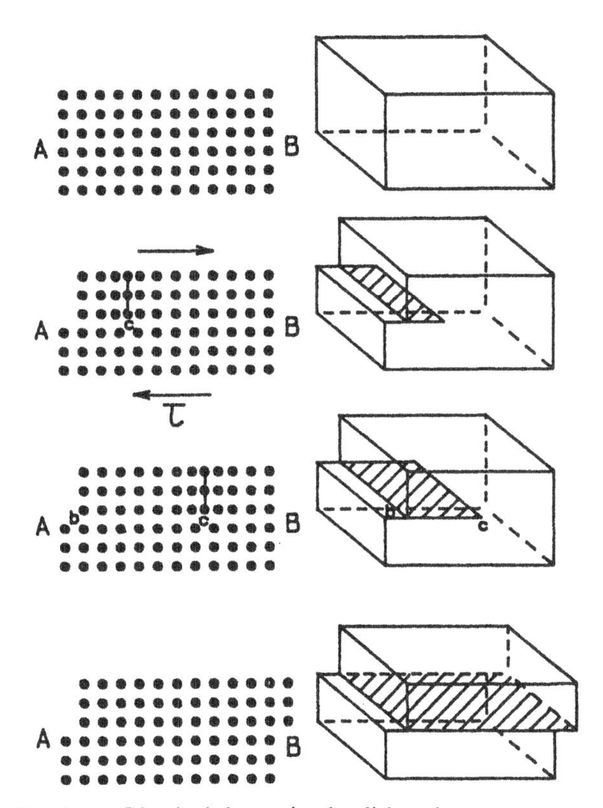

Fig. 9.12 Plastic deformation by dislocation movement

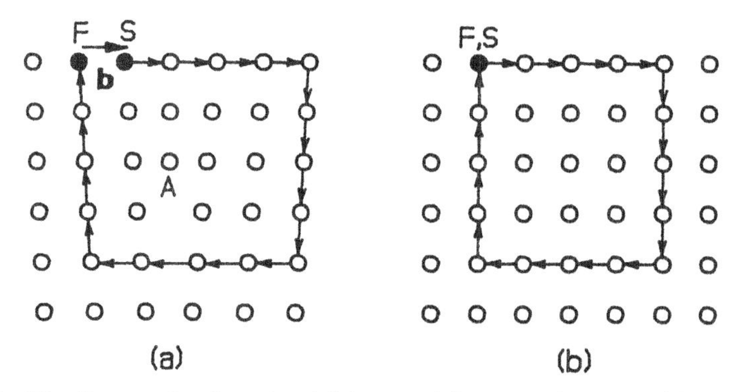

Fig. 9.13 The Burger circuit paths: (*a*) in a real (imperfect) crystal; (*b*) in a perfect crystal

dislocation, and it is determined by the Burger circuit, a closed path involving two lattice directions and surrounding the dislocation line. One such Burger circuit is shown in Fig. 9.13(*a*). Note that point *A* represents an edge dislocation perpendicular to the paper. Imagine an atomic path starting from point *S* and going four atoms to the right, then four atoms down, then four atoms to the left, and finally four atoms up to point *F*. Due to the presence of the edge dislocation, this path cannot be closed, and point *F* is one atom spacing apart from point *S*. The vector required to close the circuit from *F* to *S* defines the Burger vector **b**, as shown in Fig. 9.13(*a*). If the path above is formed in a perfect crystal, point *S* and point *F* will obviously meet, as shown in Fig. 9.13(*b*). The direction of the Burgers vector defines the slip direction of the dislocation and the magnitude of it is equal to one atomic spacing along a slip direction. With the Burgers vector the slip plane can be defined more precisely. A slip plane of a dislocation is the one that contains the Burgers vector and the dislocation line, such as plane *EF* in Fig. 9.11 and *AB* in Fig. 9.12.

Usually the symbol $\perp$ is used to represent the edge dislocation, as shown in Fig. 9.12, where the horizontal line marks the slip plane while the vertical line denotes the extra half-plane.

9.3.2 Screw Dislocation

A screw dislocation is shown in Fig. 9.14. It is formed by cutting a perfect crystal along a lattice plane *ABCD* down along AD and CB, then pulling point *D* up and pushing point *C* down by one atomic spacing, and finally rejoining the crystal. Line *AB* is the screw dislocation line.

The Burgers vector for a screw dislocation can also be determined by forming a Burgers circuit. This is shown in Fig. 9.15. At the screw dislocation in the figure, the starting point *S* and the final point *F* are marked and the

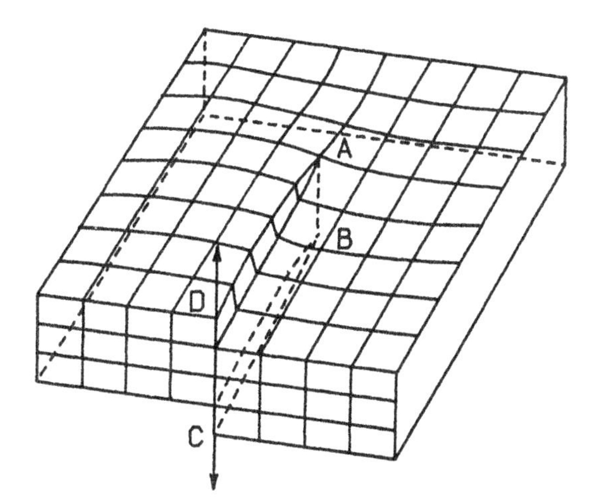

Fig. 9.14 Screw dislocation

Burgers vector is indicated. It should be pointed out that for screw dislocation, the Burgers vector is parallel to its dislocation line, while for edge dislocation the Burger vector is perpendicular to the dislocation line. Usually this observation is used as definitions for edge and screw dislocations.

9.3.3 Motion of Dislocations

Dislocations will move inside the crystal under the action of stresses. The most important movement of dislocations is the slip or glide. Slip or glide of dislocations is confined on the slip planes of the crystal and are considered as the most important mechanism for plastic deformation.

For edge dislocations the slip planes are well defined and unique, since the Burgers vector is perpendicular to the dislocation line. In other words, there is only one slip plane for a particular edge dislocation on which the slip will occur. A screw dislocation, on the other hand, is able to glide on different planes that contain the dislocation line, since its Burgers vector is

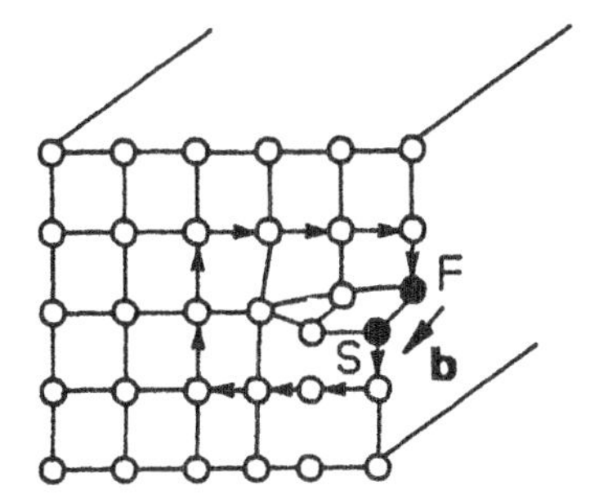

Fig. 9.15 The Burger circuit for screw dislocation

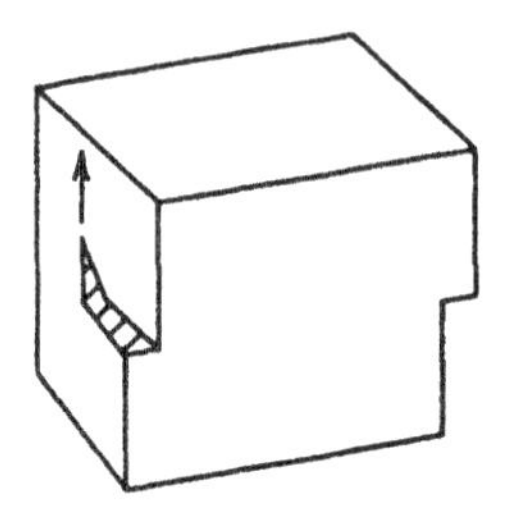

Fig. 9.16 Cross-slip of a screw dislocation

parallel to the dislocation line and therefore the slip plane is not uniquely defined. As a result a screw dislocation can cross-slip on any plane out of the original slip plane, as shown in Fig. 9.16.

The motion of an edge dislocation normal to its slip plane is possible at elevated temperatures by self-diffusion during which the atoms along the dislocation line transfer to other lattice vacancies. This process is called *dislocation climb*, shown in Fig. 9.17.

9.3.4 Dislocation Loops

In real imperfect crystals dislocations exist not only in straight configuration, such as edge or screw dislocations, but also in the form of a ring or a loop, as shown by Fig. 9.18. There are two regions on the slip plane: the one enclosed by the loop *ABCD*, over which the atoms above the slip plane have displaced one atomic spacing relative to the atoms below this plane, and the one outside the loop, over which no displacement or slip has occurred between the atoms. The loop separating these two regions is the dislocation line called *dislocation loop* because of its particular configuration in this case. In fact a dislocation line is always a boundary between two areas in the crystal lattice where slip has occurred in one of the two areas.

It should be noted that a dislocation loop is a combination of edge dislocations and screw dislocations. Consider the dislocation loop in Fig. 9.18. At points A and C the Burgers vector is perpendicular to the dislocation line

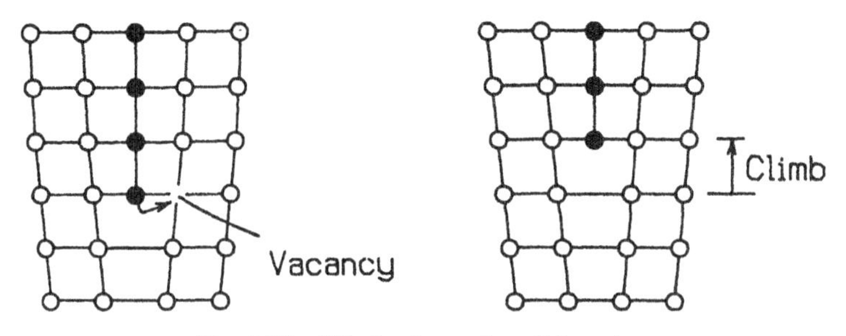

Fig. 9.17 Climb of an edge dislocation

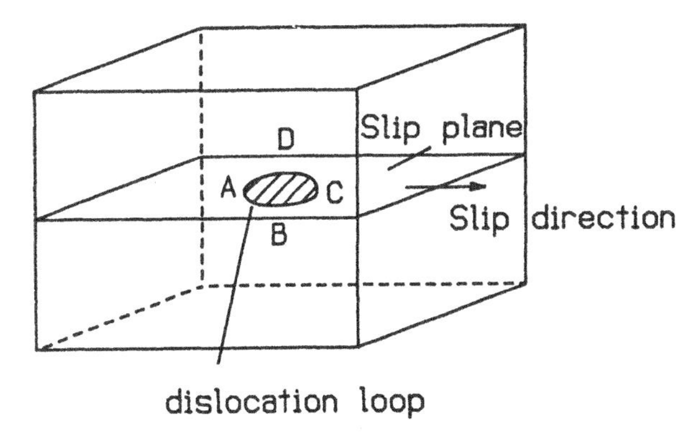

Fig. 9.18 Dislocation loop

(more precisely, perpendicular to the tangential line to the dislocation at these points); therefore the loop forms edge dislocation at these points. On the other hand, the loop forms screw dislocation at point *B* and *D*, since the Burger vector is parallel to the dislocation direction at these points. At points other than *A*, *B*, *C*, and *D*, the Burgers vector is neither perpendicular nor parallel to the dislocation line; therefore the loop forms mixed type of dislocation at these points.

A dislocation of mixed type can be decomposed into two components: the edge and screw components. This is accomplished by resolving its Burger vector **b** into two mutually orthogonal components, as shown in Fig. 9.19. The angle between the dislocation line and its Burgers vector is denoted by θ. Then the Burgers vector **b** can be considered as the sum of two Burgers vectors:

$$\mathbf{b} = \mathbf{b}_1 + \mathbf{b}_2 \tag{9.11}$$

where

$$|\mathbf{b}_1| = |\mathbf{b}|\sin\theta \tag{9.12}$$

$$|\mathbf{b}_2| = |\mathbf{b}|\cos\theta \tag{9.12}$$

and $\mathbf{b}_1$ and $\mathbf{b}_2$ are perpendicular and parallel to the dislocation line, respectively.

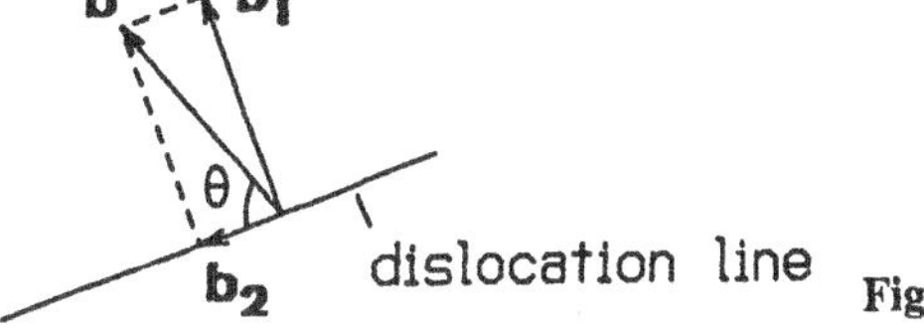

Fig. 9.19 Mixed type of dislocation

9.4 PROPERTIES OF DISLOCATIONS

Basic properties of dislocations will be briefly introduced in this section. More detailed treatment of this topic can be found in the works by Hull (1975) or Friedel (1964).

9.4.1 Stress Field around a Dislocation

A dislocation is a defect in crystal, a disturbance of the regular atomic order. This disturbance of the atomic order causes the elastic distortion around a dislocation line that can be described by linear elasticity theory. Since different distortions are caused by edge and screw dislocations, the displacements and stresses are different in nature.

9.4.2 Screw Dislocation

Fig. 9.20(*a*) shows a screw dislocation in a crystal. To calculate the stress field around it, a simplified model consisting of a distorted cylindrical ring of isotropic material, shown in Fig. 9.20(*b*), is used. The ring is cut radially, and then the cut surfaces are pulled relative to each other along the z axis to produce one atomic spacing distortion. Because of this distortion a uniform shear strain $\epsilon_{\theta z}$ is produced throughout the ring. This shear strain is calculated by dividing the distortion magnitude b, which is equal one atomic

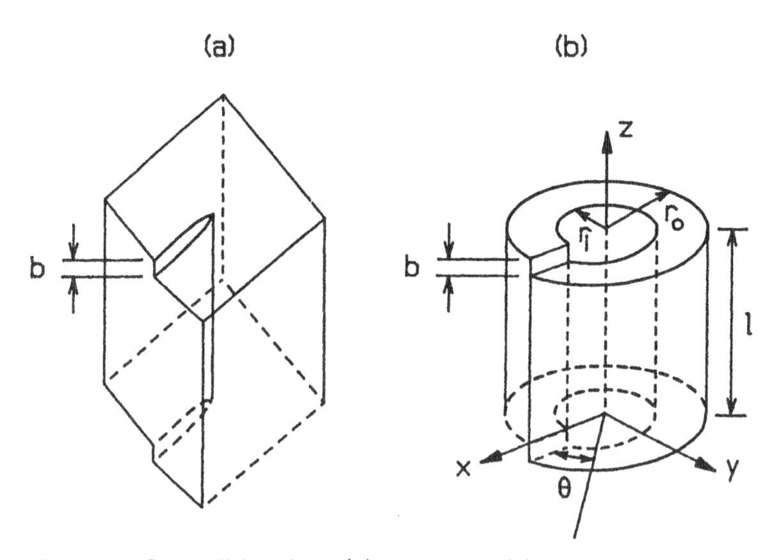

Fig. 9.20 Screw dislocations: (*a*) in a crystal; (*b*) in an elasticity model

spacing, by the perimeter $2\pi r$ of a cylindrical element with radius r:

$$\gamma_{\theta z} = \gamma_{z\theta} = \frac{b}{2\pi r} \tag{9.14}$$

The stress corresponding to it can be obtained as

$$\tau_{\theta z} = \tau_{z\theta} = \frac{Gb}{2\pi r} \tag{9.15}$$

where G is the shear modulus. The displacements due to the distortion can be shown to be

$$u = v = 0 \tag{9.16}$$

$$w = \frac{b}{2\pi}\theta \tag{9.17}$$

where the angle θ is shown in Fig. 9.20(b). Therefore the only nonzero stress component is $\tau_{\theta z}$ given above and

$$\sigma_r = \sigma_\theta = \sigma_z = \tau_{r\theta} = \tau_{\theta r} = \tau_{rz} = \tau_{zr} = 0 \tag{9.18}$$

These stress and displacement fields are valid at locations where $r > r_i \approx 2b$. On the dislocation line, or $r_i < 2b$, the situation is more complex due to the stress concentration, so the linear elasticity theory does not apply. The stress field given is axisymmetric, which is consistent with the fact that screw dislocation can cross-slip on any plane parallel to the dislocation line and no preferred slip plane exists.

In cartesian coordinate system these stress and displacement fields can be expressed as

$$\sigma_{yz} = \sigma_{zy} = \frac{Gb}{2\pi}\frac{x}{x^2 + y^2} \tag{9.19}$$

$$\sigma_{xz} = \sigma_{zx} = -\frac{Gb}{2\pi}\frac{y}{x^2 + y^2} \tag{9.20}$$

$$\sigma_x = \sigma_y = \sigma_z = \sigma_{xy} = \sigma_{yx} = 0 \tag{9.21}$$

$$u = v = 0, \quad w = \frac{b}{2\pi}\tan^{-1}\frac{y}{x} \tag{9.22}$$

The stress field around the dislocation, while interacting with other stress fields caused by other dislocations, will produce the resistance to the movement of the dislocation line.

9.4.3 Edge Dislocation

The stress field around an edge dislocation is more complicated than that around a screw dislocation, but it still can be calculated by linear elasticity theory using a cylindrical model as shown in Fig. 9.21. The details are omitted here and only the results are given.

$$\sigma_x = -Dy\frac{(3x^2 + y^2)}{(x^2 + y^2)^2} \tag{9.23}$$

$$\sigma_y = Dy\frac{(x^2 - y^2)}{(x^2 + y^2)^2} \tag{9.24}$$

$$\sigma_z = \nu(\sigma_x + \sigma_y) \tag{9.25}$$

$$\tau_{xy} = \tau_{yx} = Dx\frac{(x^2 - y^2)}{(x^2 + y^2)^2} \tag{9.26}$$

$$\tau_{yz} = \tau_{zy} = \tau_{xz} = \tau_{zx} = 0 \tag{9.27}$$

$$D = \frac{Gb}{2\pi(1 - \nu)} \tag{9.28}$$

where ν is the Poisson's ratio.

$$u = -\frac{b}{8\pi}\left[\tan^{-1}\frac{y}{x} + \frac{1}{2(1 - \nu)}\cdot\frac{xy}{x^2 + y^2}\right] \tag{9.29}$$

$$v = \frac{b}{2\pi(1 - \nu)}\left[(1 - 2\nu)\ln(x^2 + y^2) + \frac{x^2 - y^2}{x^2 + y^2}\right] \tag{9.30}$$

$$w = 0$$

Note that the solutions above apply to the case where the Burgers vector is in x direction, the dislocation line is along the z axis, and the extra half-plane is on positive y axis, as shown in Figs. 9.21 and 9.22. Obviously, due to the insertion of the extra plane, the atoms above the slip plane are in compression, while those below the slip plane are in tension. The maximum compressive stress σ_x occurs immediately above the slip plane $y = 0$, and the maximum tensile stress σ_x acts immediately below the slip plane. As we noted for screw dislocation, the equations given here apply only in the range $r > r_i \approx 2b$.

The stress field around a mix type of dislocation can be obtained as follows: First, the Burgers vector **b** is resolved into its edge component $\mathbf{b}_1$ and screw component $\mathbf{b}_2$. The stress fields corresponding to these two component

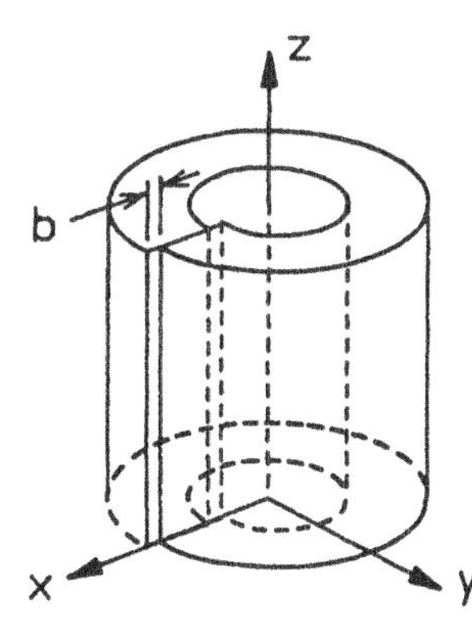

Fig. 9.21 Elasticity model of an edge dislocation

dislocations are then calculated by substituting $\mathbf{b}_1$ into Eqs. (9.23)–(9.30) and $\mathbf{b}_2$ into Eqs. (9.14)–(9.22), and the stress field for $\mathbf{b}$ can be obtained by superimposing these two solutions.

9.4.4 Stored Energy of a Dislocation

Associated with stress field around a dislocation, there will be elastic energy stored in that region. The elastic energy can easily be calculated from the known stress field. In screw dislocation the only nonzero strain and stress components are

$$\gamma_{\theta z} = \frac{b}{2\pi r} \quad \text{and} \quad \tau_{\theta z} = \frac{Gb}{2\pi r}$$

Therefore the elastic energy per unit volume is given by

$$U_0 = \frac{1}{2}\tau_{\theta z}\gamma_{\theta z} = \frac{1}{2}G\left(\frac{b}{2\pi r}\right)^2 \tag{9.31}$$

The total stored energy is then obtained by integrating U_0 over the volume of

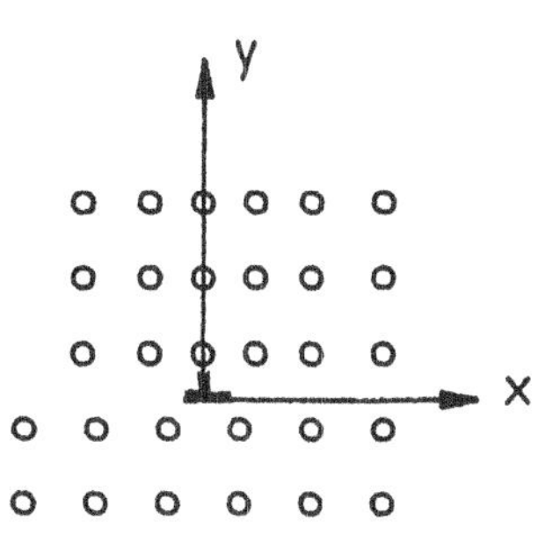

Fig. 9.22 Coordinate system for an edge dislocation

the cylindrical ring shown in Fig. 9.20(*b*), where $r_i \approx 2b$ and r_o is chosen so that the disturbance caused by the dislocation can be ignored when $r > r_o$. Therefore the total stored energy U is calculated as

$$\begin{aligned} U &= \int_v u_o dv = \int_{r_i}^{r_o} \frac{1}{2} G \left(\frac{b}{2\pi r} \right)^2 2\pi r l dr \\ &= \frac{Gb^2 l}{4\pi} \int_{r_i}^{r_o} \frac{dr}{r} \\ &= \frac{Gb^2 l}{4\pi} \ln\left(\frac{r_o}{r_i} \right) \end{aligned} \tag{9.32}$$

where l is the length of the cylindrical ring as shown in Fig. 9.20. The energy per unit length of the dislocation, called *line energy* U_2 for a screw dislocation, is then given by

$$U_L = \frac{Gb^2}{4\pi} \ln\left(\frac{r_o}{r_i} \right) \tag{9.33}$$

Similarly the line energy can be calculated for an edge dislocation

$$U_L = \frac{Gb^2}{4\pi(1 - \nu)} \ln\left(\frac{r_o}{r_i} \right) \tag{9.34}$$

By comparing the result above with Eq. (9.33), we can see that the stored energy of an edge dislocation is larger than that of a screw dislocation. Typically $\nu = 0.3$, and it follows from Eq. (9.33) and Eq. (9.34) that

$$U_{L(edge)} \approx 1.43 U_{L(screw)} \tag{9.35}$$

For both edge and screw dislocations, the line energy is proportional to $\mathbf{b}^2$.

9.4.5 Forces Acting on Dislocations

The basic concept in dislocation theory is that when a sufficiently high stress is applied to a crystal, the dislocations inside the crystal will move and cause the plastic deformation of the crystal. Therefore the result of applying a stress to a crystal is equivalent to applying to force on a dislocation line. This problem was first considered by Mott and Nabarro (1948), which is introduced below.

Consider a rectangular slip plane which is l_1 wide and l_2 long. A shear stress τ is acting on the plane. The total force on this plane is therefore

$$T = \tau l_1 l_2 \tag{9.36}$$

Now imagine that a dislocation line which is l_1 long with Burgers vector **b** moves from one end of the slip plane across this plane. The total movement of the dislocation is l_2, and displacement of the atoms above the plane relative to those below the plane is b, one atomic spacing. During this process the work done by the total applied force T is

$$W_1 = T_b = \tau l_1 l_2 b \tag{9.37}$$

On the other hand, the dislocation moves as if there were a force acting on it. Denote the force per unit length by F; then the work done by this force during the same process is

$$W_2 = F l_1 l_2 \tag{9.38}$$

These two expressions—Eq. (9.37) for the slip plane as a whole and Eq. (9.38) for the movement of only the dislocation—should be the same, since they correspond to the same physical process. Therefore

$$\tau l_1 l_2 b = F l_1 l_2 \tag{9.39}$$

which gives the force per unit length acting on the dislocation line

$$F = \tau b \tag{9.40}$$

In words, the force on a dislocation line equals to the product of the shear stress and the magnitude of the Burgers vector. This force causes the dislocation to glide. Therefore it always acts in the slip plane and is normal to the dislocation line. Another definition for F is the work done when unit length of dislocation moves unit distance

$$F = \frac{\partial^2 W_1}{\partial l_1 \, \partial l_2} \tag{9.41}$$

which gives the same result.

9.4.6 Line Tension of a Dislocation

Dislocations seen in crystals are rarely straight. For example, when a segment of dislocation line is pinned at point A and B by impurities in the crystal and this segment is driven by a force $F = \tau b$, it will tend to bend, as shown in Fig. 9.23. The line tension T always tries to straighten the dislocation and therefore to reduce its length to attain a minimum energy, since as shown by Eq. (9.32) the stored energy of a dislocation is proportional to its length. The driving force τb and the line tension T will reach an equilibrium at a certain configuration where the segment ds will have a radius of ρ. The shear stress

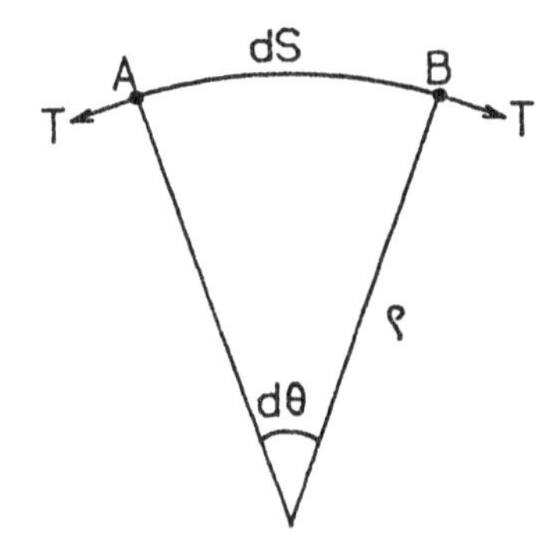

Fig. 9.23 Line tension of a dislocation

τ_ρ needed to maintain this configuration can be calculated as follows: The force equilibrium in the normal direction of ds leads to

$$\tau_\rho b\, ds = 2T \sin\left(\frac{d\theta}{2}\right) \approx T d\theta \tag{9.42}$$

since $d\theta$ is very small. Recall that $ds = \rho\, d\theta$; therefore

$$\tau_\rho = \frac{T}{b\rho} \tag{9.43}$$

The determination of T has proved to be a difficult task. It has been found that T depends on the shape of the dislocation line in a fairly complex way. Usually the following approximation has been used:

$$T = \alpha G b^2 \tag{9.44}$$

where α is a proportional factor. Substituting this into Eq. (9.43) leads to

$$\tau_\rho = \frac{\alpha G b}{\rho} \quad \text{or} \quad \rho = \frac{\alpha G b}{\tau_\rho} \tag{9.45}$$

The radius ρ increases with Gb and decreases with τ_ρ, which represents the rigidity of a dislocation line. The equation above is useful in understanding the Frank-Read dislocation multiplication source, which we will discuss later.

9.4.7 Forces between Dislocations

Due to the presence of the stress field around a dislocation line, the dislocations in a crystal will interact with each other, resulting in a configuration of reduced total stored strain energy. This can be shown in the following

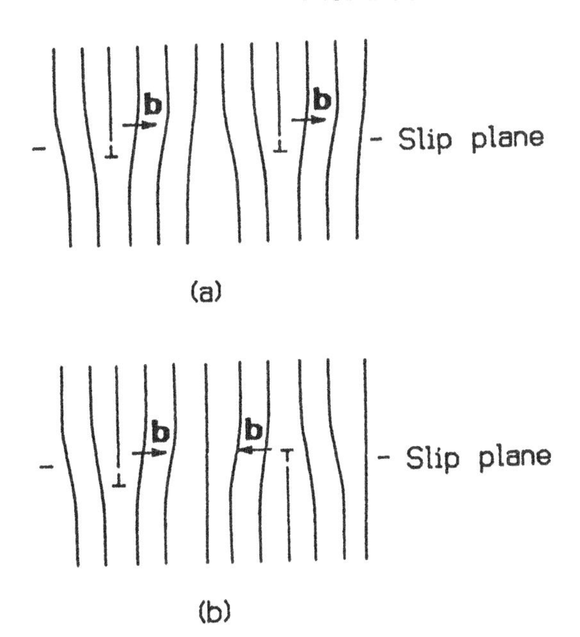

Fig. 9.24 Interaction of edge dislocations with common slip planes

way (Hull 1975): For an edge dislocation the stored energy per unit length is given by Eq. (9.34). Consider the situation shown in Fig. 9.24(a) where two edge dislocations are on the same slip plane with the same Burgers vector **b**. The total stored energy, if the two dislocations are considered separate, is given by

$$U = \frac{2Gb^2}{4\pi(1-\nu)} \ln\left(\frac{r_o}{r_i}\right) \tag{9.46}$$

While the total stored energy, if these two dislocations are very close to each other approximately forming a dislocation with Burgers vector 2**b**, is

$$U = \frac{G(2b)^2}{4\pi(1-\nu)} \ln\left(\frac{r_o}{r_i}\right) \tag{9.47}$$

which is twice as much as that given by Eq. (9.46). Therefore these two dislocations will "repel" each other until the distance between them is large enough so that they can be considered separately. On the other hand, in case of Fig. 9.24(b), where two edge dislocations are on the same plane but with opposite Burgers vectors, the total stored energy is given by Eq. (9.46) if they are separated and will be zero if they are close to each other, since their Burgers vectors are in opposite direction (or opposite sign). Therefore these

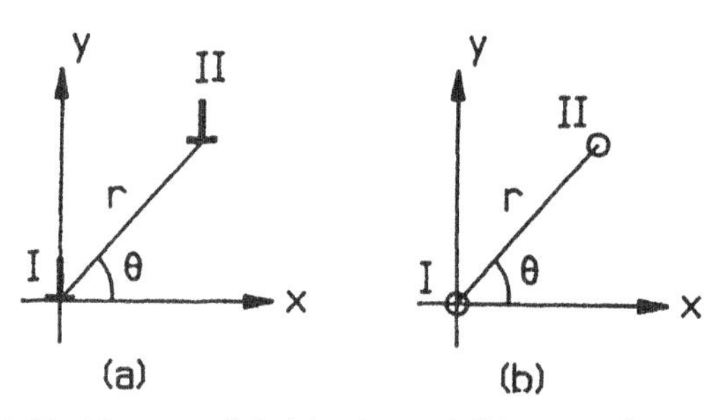

Fig. 9.25 Two parallel: (*a*) edge and (*b*) screw dislocations

two "opposite" dislocations tend to attract each other and then ultimately annihilate each other. The explanations preceding give examples of the interaction between dislocations.

The interaction between dislocation can be described by the forces between them. In the following, the forces between two parallel edge dislocations or two parallel screw dislocations are introduced. One of the dislocations is at the origin and its slip plane is the *x*-*z* plane. Another dislocation is at location (x, y), parallel to the first one. This is shown in Fig. 9.25.

9.4.8 Forces between Two Edge Dislocations

The forces acting on the second dislocation at (x, y) due to the presence of the first dislocation at origin has been found to be

$$F_x = \frac{Gb^2}{2\pi(1-\nu)} \frac{x(x^2-y^2)}{(x^2+y^2)^2} \tag{9.48}$$

$$F_y = -\frac{Gb^2}{2\pi(1-\nu)} \frac{y(3x^2-y^2)}{(x^2+y^2)^2} \tag{9.49}$$

where F_x and F_y are components in x and y directions, respectively.

9.4.9 Forces between Two Screw Dislocations

In the case of two parallel screw dislocations, the force acting on the second dislocation at (x, y) due to the stress field of the first dislocation is simply

$$F_r = \tau_{\theta z} b \tag{9.50}$$

where $\tau_{\theta z}$ is given by Eq. (9.15). Substituting Eq. (9.15) into the equation above gives

$$F_r = \frac{Gb^2}{2\pi r} \tag{9.51}$$

In cartesian coordinates F_r can be resolved into two components F_x and F_y, in x and y directions, respectively:

$$F_x = \frac{Gb^2 x}{2\pi(x^2 + y^2)} \tag{9.52}$$

$$F_y = \frac{Gb^2 y}{2\pi(x^2 + y^2)} \tag{9.53}$$

These results are obtained assuming that the Burgers vectors of the two screw dislocations are in the same direction.

9.4.10 Stress to Move a Dislocation

It has been discussed that the estimated theoretical shear strength of perfect crystals is several orders higher than the observed one. The estimation was based on the assumption that the shear plane would move as a whole and all the bonds between atoms were overcome simultaneously. The introduction of dislocation theory allows the slip of the whole slip plane in two steps: the local displacement when the dislocation line moves through only a portion of the slip plane, and the overall slip after the dislocation line has moved across the entire plane. The atomic bond is broken in a series of local movements. The force required to push a dislocation line through the atomic bond over the slip plane step by step is obviously much smaller than that required to move the entire slip plane in one step.

Peierls (1940) first calculated the magnitude of the shear stress needed to move a dislocation across a slip plane. He used the energy method, and assuming that the resistance to the shear deformation changed sinusoidally with the slip movement, he found the shear stress to be

$$\tau_c = \frac{2G}{(1-\nu)} e^{-2\pi a/b(1-\nu)} \tag{9.54}$$

where G, ν, and a are the shear modulus, the Poisson's ratio, and the lattice spacing between slip planes, respectively, and b is the lattice spacing in the slip direction. Substituting a typical value of $\nu = 0.3$ and assuming that $a = b$ provided an estimation

$$\tau_c \approx 2 \times 10^{-4} G \tag{9.55}$$

This value is a fairly close approximation to the shear strength of crystals, although it is still somewhat higher than the actual values.

Note that τ_c decreases with the increase of a and with the decrease of b. This means that the shear strength is smaller on the closely packed planes and in the directions with dense packing or shorter Burgers vectors. This is consistent with what has been discussed before: The slip occurs on the closely packed planes in the closely packed directions.

9.4.11 Dislocation Multiplication: Frank-Read Source

As we mentioned earlier, the deformation band or step observed on the surface of a specimen is caused by the dislocation movement on the slip plane. However, experimental observations have revealed that the dislocation population or density in an annealed metal is not enough to generate the observed slip steps on the deformed metals. Therefore there must be some sources that will generate new dislocations and thus increase the dislocation density within the crystal during the course of deformation.

Various sources exist in real crystals. We will discuss one of the common sources, called the *Frank-Read source*. The arrangement of the Frank-Read source was proposed by Frank and Read (1950) and observed experimentally by Dash (1956).

The Frank-Read source is a short piece of movable dislocation with a length L, as shown in Fig. 9.26(a). This piece of dislocation is locked at both ends A and B by immovable dislocation knots. When a shear stress τ is applied on the slip plane which contains AB, a force $F = \tau b$ will exert on the dislocation line AB, as we noted earlier. Because the dislocation is locked at A and B, the force τb pushes it to bend into curves. When τ reaches the value

$$\tau = \frac{2T}{bL} = \frac{2\alpha Gb}{L} \tag{9.56}$$

the dislocation will form a half-circle, as shown in Fig. 9.26(b), since the radius at this moment is

$$\rho = \frac{\alpha Gb}{\tau} = \frac{L}{2} \tag{9.57}$$

This is an unstable equilibrium position. Any small increase of the shear stress τ will cause the dislocation to move further and to rotate around the locked points A and B, as shown in Fig. 9.26(c). Then the curved dislocation line will meet and react on the reverse side. As a result a new dislocation loop is formed, Figs. 9.26(d)–(f), and curved line continues to spread outward creating a new piece of dislocation line AB which, pushed by the shear stress τ, can repeat the procedure to generate more dislocation loops.

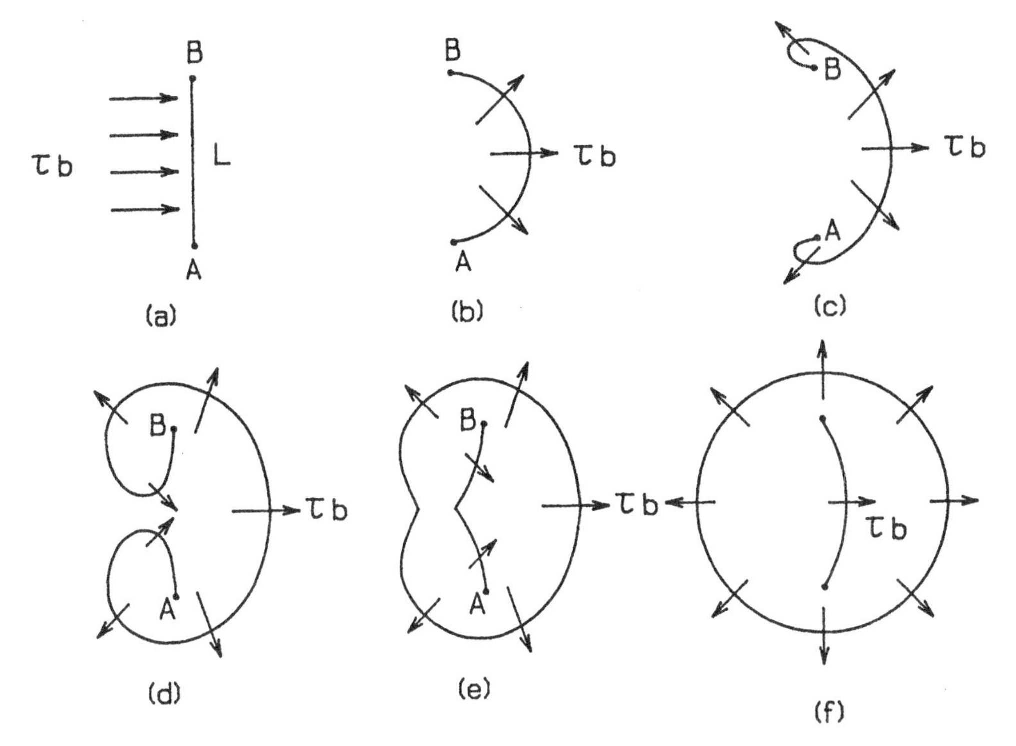

Fig. 9.26 Formation of the Frank-Read loop

The dislocation loops generated this way are called the *Frank-Read loops*.

This mechanism of dislocation multiplication proposed by Frank and Read can generate a large number of dislocation loops under very small shear stress τ. In fact more than 100 dislocation loops have been found to be produced by a single Frank-Read source.

9.4.12 Dislocation Networks

Dislocations can react with each other, as we mentioned earlier. If two dislocations attract each other, they will meet to form a node if they are not parallel to each other. This is the way that dislocation networks are formed, as shown in Fig. 9.27. Frank (1950) first pointed out that the equilibrium distribution of dislocations in a crystal should be a three-dimensional network.

The criterion governing the Burgers vectors of dislocation is that where they meet, at a node or knot of the network, the resultant Burgers vector should be zero. In other words, the vectorial sum of the Burgers vectors for all dislocations that meet at a knot must be zero. Figure 9.28 gives an example of this criterion. Two dislocations, D_1 with $\mathbf{b}_1 = [1\bar{1}1]/2$, and D_2,

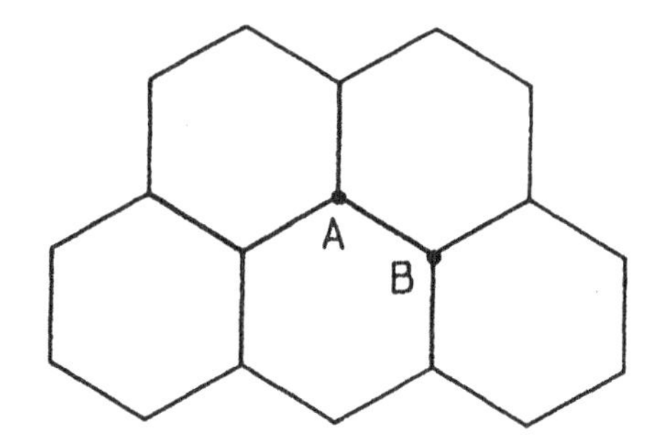

Fig. 9.27 Dislocation networks

with $\mathbf{b}_2 = [1\,1\,1]/2$ are initially separated, as shown in Fig. 9.28(*a*). The stress field around D_1 and D_2 will set up an attraction between these two dislocations. The two dislocations will meet and form a node *A* and a third dislocation D_3. According to the criterion mentioned above, the following relation holds between $\mathbf{b}_1$, $\mathbf{b}_2$, and $\mathbf{b}_3$:

$$\mathbf{b}_1 + \mathbf{b}_2 - \mathbf{b}_3 = 0 \qquad (9.58)$$

The minus sign before $\mathbf{b}_3$ is because D_3 is pointed out of the node while D_1 and D_2 are pointed toward the node. The Burgers vector $\mathbf{b}_3$ is thus obtained

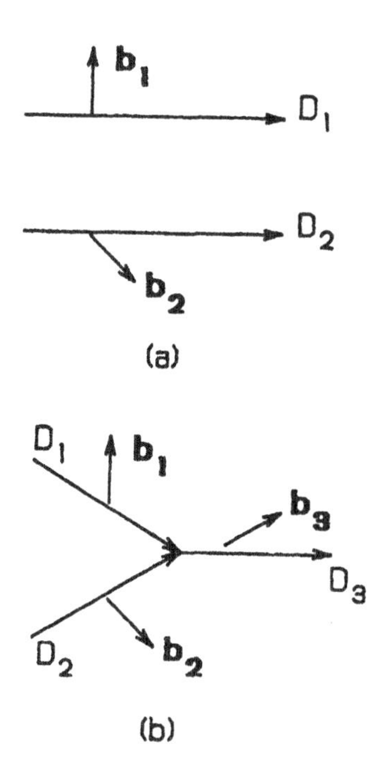

Fig. 9.28 Reaction of two dislocations

as

$$\mathbf{b}_3 = [1\,0\,1] \tag{9.59}$$

It is noted here that the Frank-Read source discussed previously can be thought as one segment of the dislocation network such as AB in Fig. 9.27, while the locked points A and B can be the knots in the network.

9.4.13 Dislocation Pileups

There are always some obstacles inside the crystal against the movement of dislocations. One such obstacle is the grain boundary. If dislocations generated from a source such as a Frank-Read source meet an obstacle, they will pile up against this obstacle, as shown in Fig. 9.29. Therefore the dislocations near the obstacles will be more closely spaced than those far away from the obstacle. The dislocations stopped by the obstacle will exert a force on the obstacle, while the obstacle is exerting a force on these dislocations. This force (per unit length), denoted by $\tau_l b$ can be obtained by the following method: Recall that if the first dislocation moves forward by δx, all the following dislocations will move the same amount. For the work done during this movement δx, we write

$$W = n \cdot F\delta x = n\tau b\delta x \tag{9.60}$$

where n is the total number of dislocations piled up against the obstacle. On the other hand, since the reaction force $\tau_l b$ is applied to the first dislocation

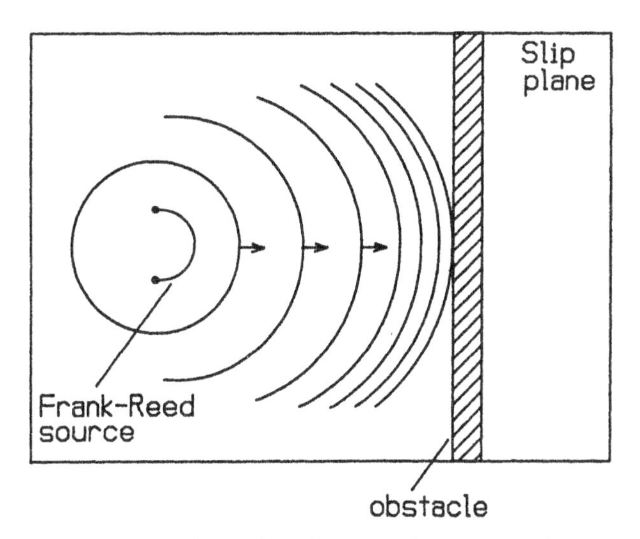

Fig. 9.29 Dislocation pileups against an obstacle

which moves forward against τ_l by δx, this work done by the first dislocation in overcoming τ_l is

$$W = \tau_l b \delta x \tag{9.61}$$

Comparing Eq. (9.60) with Eq. (9.61) results in

$$\tau_l = n\tau \tag{9.62}$$

This means that in a process under equilibrium, the internal stress, characterized by τ_l, at the obstacle caused by dislocation pileup is equal to n times the applied stress. The internal stress causes the crystal to harden by exerting a force called *back stress* on the dislocation source. Therefore the dislocation pileups are responsible for the strain hardening and the Bauschinger effect observed in experiments. In addition, in case of grain boundary obstacles the stress concentration expressed by Eq. (9.62) can lead the plastic deformation in the neighboring grain by activating dislocation movement within it, or it can cause the failure of the crystal boundary, sometimes called *grain boundary cracking*. The dislocation distribution in front of an obstacle was studied by Eshelby, Frank, and Nabarro (1951). By their calculation the number of dislocations that can pile up within a length l_0 on the slip plane under the action of the shear stress τ was found to be given by

$$n = \frac{\pi l_0 \tau}{Gb} \tag{9.63}$$

for edge dislocations and

$$n = \frac{\pi l_0 \tau (1 - \nu)}{Gb} \tag{9.64}$$

for screw dislocations. It can be seen that the number of dislocations that pile up against an obstacle is proportional to the applied stress τ and inversely proportional to the shear modulus G and Burgers vector **b**.

9.4.14 Dislocation Jogs

If two dislocations are not on the same slip plane when they slip across each other, a new dislocation line, called a *dislocation jog* with a length of a Burgers vector, can be formed. This is shown in Fig. 9.30 and is explained below. Figure 9.30(a) shows two edge dislocations on different slip planes. When dislocation D_1 slips across the plane, it will intersect dislocation D_2. Then the atoms on two sides of the D_1 slip plane will be displaced by a distance $|\mathbf{b}_1|$, which is an atomic spacing in the $\mathbf{b}_1$ direction on the slip plane. After passing through D_2, the plane containing D_2 will be split by a step

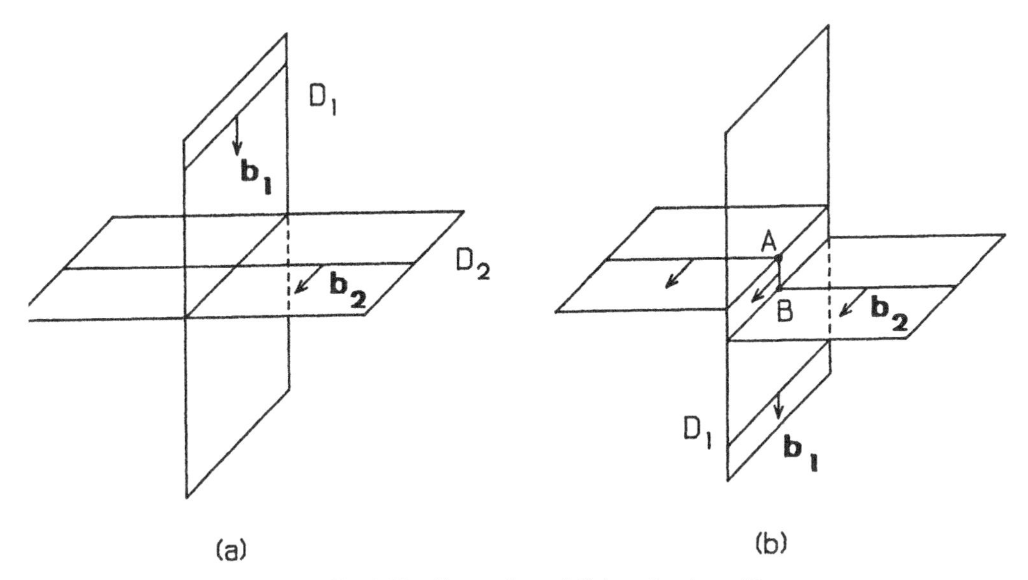

Fig. 9.30 Formation of dislocation jog *AB*

$|\mathbf{b}_1|$, as shown in Fig. 9.30(*b*), resulting in a dislocation jog *AB*. Usually the jog *AB* is not able to move with the dislocation D_2 because the plane containing it is not its slip plane. In these circumstances movement in the given direction is only possible by climb. This climb is assisted by the creation of vacancies, a process that usually happens at high temperature. Generally, this way the formation of dislocation jog will lock a movable dislocation and form one of the knots on a dislocation network.

REFERENCES

Dash, W. C. 1956. *J. Appl. Phys.* **27**:1193.

Eshelby, J. D., F. C. Frank, and F. R. N. Nabarro. 1951. *Phil. Mag.* **42**:351.

Ewing, J. A., and W. Rosenhain. 1900. The crystalline structure of metals, *Philos. Trans. R. Soc., London* **193**:353–375.

Ewing, J. A., and W. Rosenhain. 1899. Experiments in micro-metallography: Effects of strain, preliminary notice. *Proc. R. Soc., London* **65**:85–90.

Frank, F. C. 1950. Pittsburgh conference on plastic deformation of crystals. Carnegie Institute of Technology, p. 100.

Frank, F. C., and W. T. Read, Jr. 1950. Multiplication processes for slow moving dislocations. *Phys. Rev.* **79**:722–724.

Frenkel, J. 1926. Zur Theorie der Tlastizitätsgrenze und dert gestigheit kristallinischer Körper. *Z. Phys.* **37**:572–609.

Friedel, J. 1964. Dislocations. Oxford: Pergamon.

Honeycombe, R. W. K. 1968. The plastic deformation of metals. London: Butter and Tanner Ltd.

Hull, D. 1975. *Introduction to Dislocations*. Oxford: Pergamon.

Mackenzie, J. K. 1962. A theory of sintering and theoretical yield strength of solids. Ph.D. dissertation. University of Bristol.

Mott, N. F., and F. R. N. Nabarro. 1948. Report on strength of solids. Physical Society, London.

Orowan, W. 1934. Zur Kristallplastizität III. Über den Mechanismus des gleituorganges. *Z. Phys.* **89**:634–659.

Peierls, R. 1940. The size of dislocation. *Proc. Phys. Soc.* **50**:34.

Polanyi, von M. 1934. Über eine Art gitterstorung die einen Kristall plastisch machen konnte. *Z. Phys.* **89**:660–664.

Taylor, G. I. 1934. The mechanism of plastic deformation of crystals. Part I: Theoretical. *Proc. R. Soc., London. Sec. A* **145**:362–387.

Tyson, W. R. 1966. *Phil. Mag.***14**:925–936.

10

PLASTIC DEFORMATION OF SINGLE CRYSTALS

The physics of single-crystal plasticity was established during early in this century, in 1900 to 1938, with the contributions of Ewing and Rosenhain (1900), Bragg (1933), Taylor and coworkers (1923, 1925, 1934, 1938), Polanyi (1922), Schmid (1924), and others. Their experimental measurements established that at room temperature the major source for plastic deformation is the dislocation movements through the crystal lattice. These dislocation motions occur on certain crystal planes in certain crystallographic directions, and the crystal structure of metals is not altered by the plastic flow. The mathematical representation of these physical phenomena of plastic deformation in single crystals was pioneered by Taylor (1938) when he investigated the plastic deformation of polycrystalline metals in terms of single-crystal deformation. More rigorous and rational formulations of single-crystal plasticity have been provided by Hill (1966), Hill and Rice (1972), Asaro and Rice (1977), and more recently by Hill and Havner (1982). A comprehensive review of this subject can be found in Asaro (1983).

In this chapter we present the fundamentals of plasticity theory of single crystals. We will start by introducing several elementary concepts and experimental observations, useful for mathematical modeling. Then we will discuss the general kinematics of crystalline deformation and elastic-plastic constitutive equations. We will progress to a full discussion of several hardening rules for crystalline slip later in the chapter and finish off with a detailed presentation of rigid plasticity theory, its theoretical representation, and uniaxial tension experiments.

10.1 ELEMENTARY CONSIDERATIONS

This section considers some basic concepts and experimental observations that will be implemented in formulating single-crystal plasticity theory.

10.1.1 Schmid's Law

The most widely used experimental technique for single-crystal deformation is uniaxial tension or uniaxial compression. Early studies found that the axial load initiating plastic flow (initial yield stress) varied from sample to sample, and this variation depended on, among other things, the relative orientation of the crystal lattice to the loading axis. To determine the criterion that governs the initial plastic yield, Schmid (1924) carried out an analysis of the experimental data on a zinc single crystal under uniaxial tension. He found that it was the shear stress component that resolved along the slip direction on the slip plane (called *resolved shear stress*) rather than the applied axial tensile stress, which initiated the plastic deformation. As a result Schmid postulated that yield would begin on a slip system (a combination of the slip plane and the slip direction) when the resolved shear stress on this slip plane and in the slip direction reached a critical value, independent of the tensile axial stress and the other normal stress on the lattice plane. This statement is now commonly referred to as *Schmid's law*.

At about the same time as Schmid, Taylor and Elam (1925) had independently established this law in their studies on aluminum single crystals. In summarizing their experimental observations on single crystal aluminum, they stated: "Of the twelve crystallographically similar possible models of shearing, the one for which the component of shear-stress in the direction of shear was greatest was the one which actually occurred." Note that this is a somewhat different statement of Schmid's law.

In quantitative terms the Schmid law can be represented as follows: Consider a single crystal bar subjected to tension stress σ, as shown in Fig. 10.1. The shaded plane in the figure is the slip plane. Also shown are the unit normal vector $\mathbf{n}$ to the slip plane and the unit slip direction vector $\mathbf{s}$. It is obvious that the shear stress component resolved on this slip plane in the slip direction is given by

$$\tau = \mathbf{n} \cdot \boldsymbol{\sigma} \cdot \mathbf{s} = \sigma \cos \phi \cos \lambda \tag{10.1}$$

where ϕ and λ are angles that the slip plane normal $\mathbf{n}$ and slip direction $\mathbf{s}$ make with the loading axis, respectively, as shown in Fig. 10.1. The factor $\cos \phi \cos \lambda$ in the equation is usually called the *Schmid factor*.

For a given test sample of single-crystal metal, if the angles that the normal to the slip plane and the slip direction of each slip system make with the loading axis are known, Eq. (10.1) can be used to calculate the resolved shear stress for a particular slip system. The equation holds for all slip

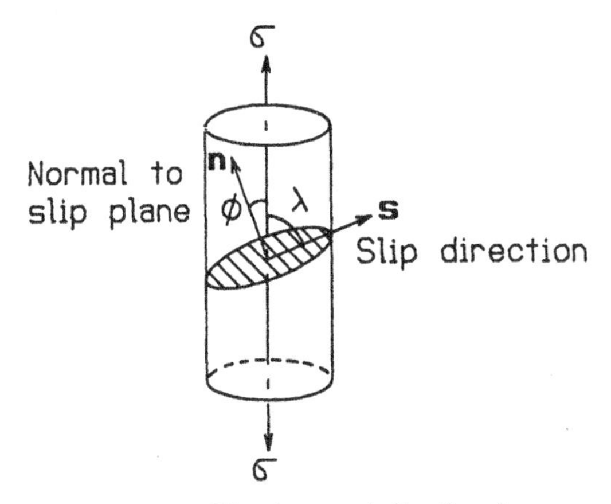

Fig. 10.1 Slip plane and slip direction

systems. Although the axial stress σ is constant, the resolved shear stresses are different for different slip systems due to the different orientation of each slip system relative to the loading axis. According to Schmid's law, the slip system on which the plastic slip actually occurs is the one for which τ is greatest and has reached a critical value τ_c, which is a material property and has to be determined experimentally. From Eq. (10.1), the critical axial stress σ_c is given by

$$\sigma_c = \frac{\tau_c}{\cos\phi \cos\lambda} \tag{10.2}$$

Therefore, although the critical resolved stress τ_c is a constant, the axial stress that results in the plastic yield varies depending on the orientation of the slip system.

A plot of σ_c versus $\cos\phi \cos\lambda$, the Schmid factor, is shown in Fig. 10.2, where the experimental data for magnesium crystals, obtained by Burke and Hibbard (1952), are also included (circles). It can be seen that Schmid's law provides an excellent explanation for the variation of σ_c with the orientation of slip system.

It has been pointed out by Asaro (1983) that although experiments conducted in uniaxial tension or compression often provide an approximate confirmation of Schmid's law, the micromechanics of plastic slip (dislocation movement) suggests that deviation from Schmid's law should exist. For example, in modeling for cross-slip of a screw dislocation and climb of an edge dislocation, the stress components other than the resolved shear stress should be considered. The interested readers can refer to Asaro (1983) for more details.

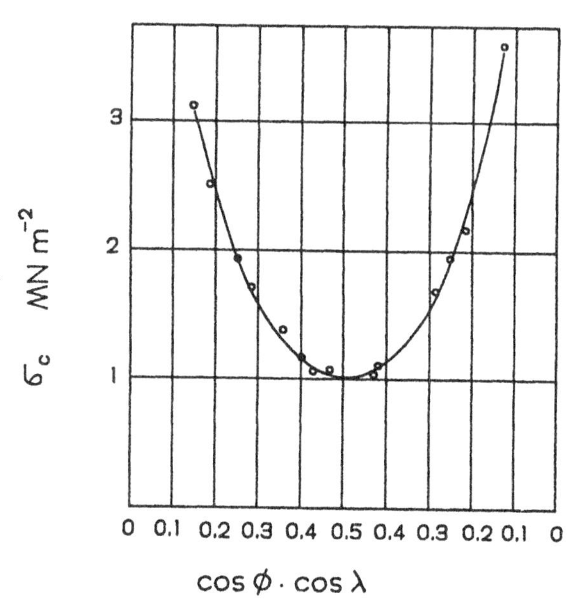

Fig. 10.2 The variation of σ_c with orientation

10.1.2 Shear Stress–Shear Strain Curves

In the preceding discussion we assumed that the critical shear stress τ_c is constant. However, τ_c is actually a function of plastic strain and usually will increase due to the material hardening caused by plastic deformation. This hardening behavior is described by τ_c-γ curves for different materials, where γ is the resolved shear strain of the slip system. Based on the experimental studies on aluminum crystals, Taylor and Elam (1936) found that the resolved shear stress-shear strain curves of FCC crystals can be approximately represented by the parabolic equation

$$\tau_c = \beta\gamma^{1/2} \tag{10.3}$$

where β is a material constant. In addition they showed that the data taken from uniaxial tension and uniaxial compression experiments fell on the same curve given by Eq. (10.3), while the nominal stress σ versus nominal strain ϵ curves for these two experiments were different.

Now it has been found that the τ_c-γ curve is quite different from what Taylor and Elam (1936) observed and cannot be described completely by Eq. (10.3). In general, there are three well-defined regions of the τ_c-γ curve in the deformation process of FCC metal crystals, as shown in Fig. 10.3. The first region, stage I, is usually called the *easy glide region*, for there the hardening rate is low. This region is followed by stage II, representing a much higher, linear hardening process. Stage II is terminated by stage III

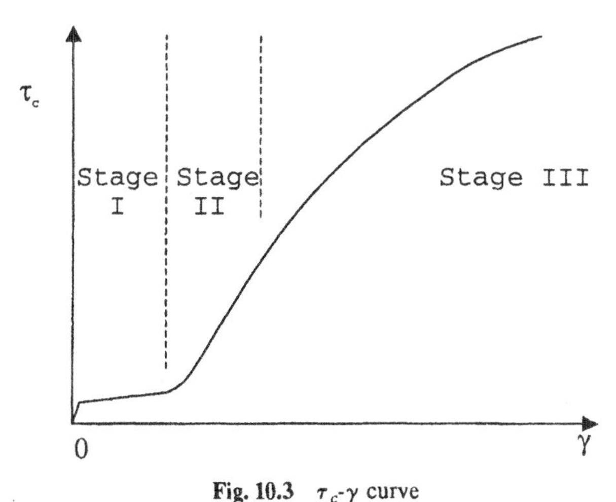

Fig. 10.3 τ_c-γ curve

characterizing a dccreased hardening rate. The hardening behavior in stage III can be approximated by Taylor and Elam's parabolic Eq. (10.3).

It should be pointed out that it is not necessary that all three stages always be present. There are several conditions that decide whether a particular stage will occur and its relative importance to the other stages. These include the orientation of the crystals, the purity, the temperature during deformation, the material, the grain size, the surface condition, and the strain rate. For example, stage I hardening observed in aluminum crystals at room temperature ends at 4–5% shear strain, whereas in copper, under the same testing condition, the shear strain in stage I can reach as high as 20%; in gold and silver it can be as much as 50%. If fewer pure aluminum crystals are used as specimen, as in Taylor and Elam's studies, stage I and stage II deformation may disappear and stage III will dominate the whole deformation process. Thus Taylor and Elam (1936) only observed stage III deformation.

10.1.3 Micromechanics Consideration of Taylor and Elam's Equation

The first model describing the hardening behavior of single crystals based on dislocation theory was proposed by Taylor (1934). Using a dislocation model, Taylor assumed that the dislocations in crystals were regularly arranged arrays, as shown in Fig. 10.4. Dislocations would move along the slip plane and the maximum movement would be L, which was either the dimension of the crystal or the distance to a boundary. The average distance dislocations moved was $L/2$. The average spacing between active slip planes was $d/2$, while the spacing between dislocations was a (see Fig. 10.4). Therefore the average dislocation density ρ was $2/ad$. The shear strain in this model was

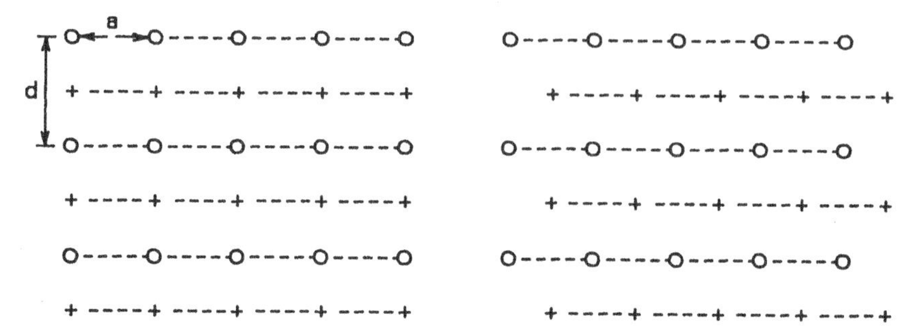

Fig. 10.4 Taylor's dislocation model for hardening of single crystals

thus

$$\gamma = \frac{L \cdot (L/2) \cdot b \cdot \rho}{L} = \frac{Lb}{ad} \tag{10.4}$$

where b is the magnitude of Burgers vector. On the other hand, from the elastic analysis of forces between two dislocations (see Chapter 11), the stress caused by a dislocation was proportional to $1/a$, and therefore proportional to $\rho^{1/2}$ if a and d are comparable. Thus, using a proportionality factor K, Taylor obtained

$$\tau_c = KGb\rho^{1/2} = KG\left(\frac{b\gamma}{L}\right)^{1/2} \tag{10.5}$$

which gave a parabolic law for the resolved shear stress–resolved shear strain curve and provided a micromechanics basis for Taylor and Elam's empirical equation (10.3). The analysis above was based on the elastic interaction between dislocations and the dimensional arguments. Mott (1952) generalized Taylor's model by considering the dislocation pileup. The equation was modified as

$$\tau_c = \frac{G}{2\pi}\left(\frac{\gamma n b}{L}\right)^{1/2} \tag{10.6}$$

10.1.4 Rotation of the Crystal Lattice

A very important concept of crystal deformation is that the crystal lattice will rotate relative to a fixed coordinate frame (e.g., the loading machine or loading axis which is fixed to the ground) during the process of large plastic deformation. This is illustrated in Fig. 10.5 for the case of tension experiment and in Fig. 10.6 for the case of compression experiment.

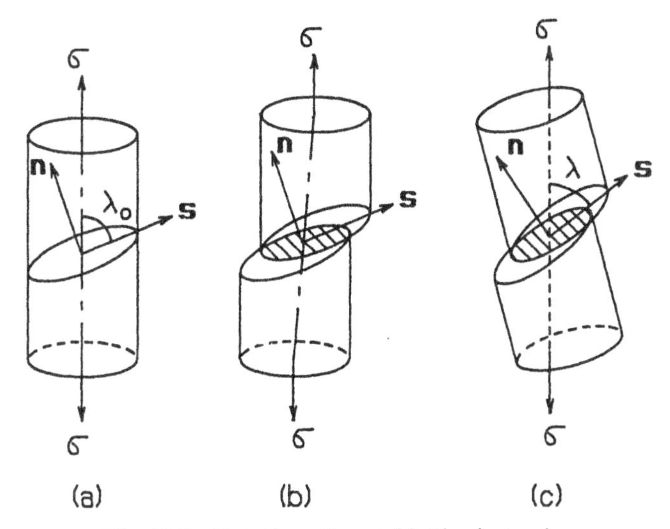

Fig. 10.5 Rotation of crystal lattice in tension

Fig. 10.5(a) shows a single-crystal specimen in tension. It is oriented so that the resolves shear stress is the greatest on the slip system shown in the figure. Therefore the slip will occur only on this slip system. This called *single slip*, since only one slip system is activated. During the slip the orientation of the tensile axis will change with respect to the crystal axes **s** and **n**; the tensile axis will pass through the centroid of the contacted area (shaded area) of the upper and the lower part of the specimen on the slip plane. As a result the tensile axis would rotate toward the slip direction **s**.

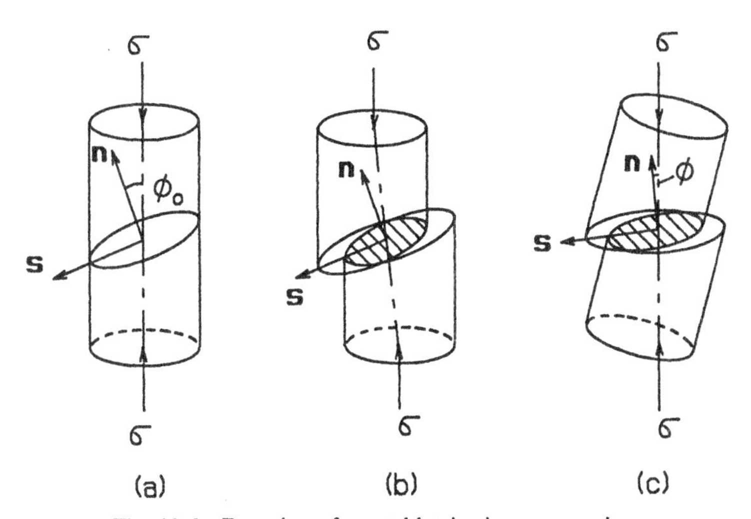

Fig. 10.6 Rotation of crystal lattice in compression

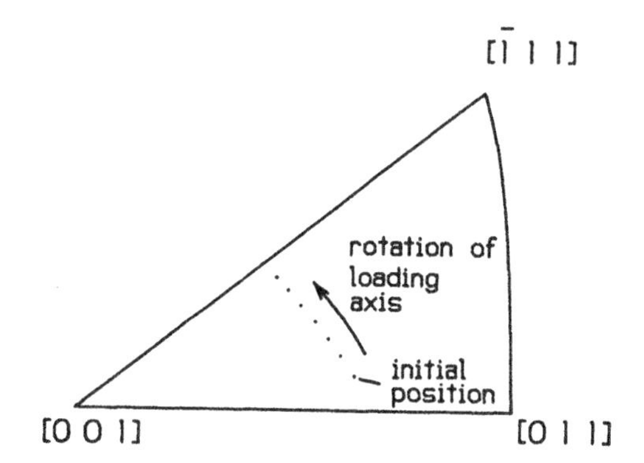

Fig. 10.7 Representation of rotation of loading axis in standard stereographic triangle

However, since the tension axis is constrained by the loading machine so that the tensile axis is fixed with respect to the laboratory frame, it is the crystal lattice that undergoes the actual rotation, as can be seen from Fig. 10.5(*c*). Hence, during a large tensile plastic deformation the slip direction will rotate toward the tensile axis so that $\lambda < \lambda_0$, where λ_0 and λ are the initial and current (deformed) angles between the tensile axis and slip direction, respectively.

Similarly the crystal lattice will rotate in an uniaxial compression experiment. In this case, however, the normal to the slip plane will rotate toward the compression axis, resulting in $\phi < \phi_0$, where ϕ_0 and ϕ are, respectively, the angles between the initial and current normal to the slip plane and the compression axis. This rotation can be best shown on a stereographic projection, where, as discussed in Chapter 9, the loading axis of a crystal rod is represented by a point in the standard triangle (see Fig. 9.9). At any instant of the deformation, using the current relative orientation of the loading (tensile or compressive) axis to the crystal lattice, one point can be found in the standard triangle that represents the loading axis. Therefore the loading axis during the deformation can be traced by a series points in this triangle, as shown in Fig. 10.7.

It should be emphasized that in a stereographic projection, the rotation of the loading axis is represented. In practice, however, it is the crystal lattice and not the loading axis that rotates. The presentation of stereographic projection shows a relative measure of this rotation. The modeling and calculation of this lattice rotation is very important in crystal plasticity and will be discussed in detail later in this chapter.

10.1.5. Single-Slip and Multislip Modes

In our previous discussion we assumed that only one slip system is active (or activated), namely that the resolved shear stress on all other slip systems is

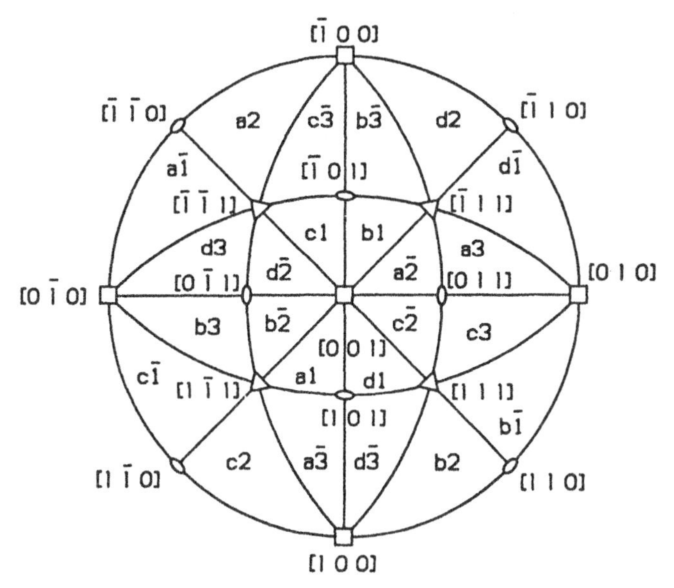

Fig. 10.8 Standard [0 0 1] stereographic projection of FCC crystals

smaller than the initial critical yield stress or strength, and therefore no slip occurs on these systems. This situation is called *single slip*, and the other slip systems on which no slip occurs are referred to as *inactive*.

According to Schmid's law, whether or not a slip system is active depends on the Schmid factor $\cos\phi \cos\lambda$, which in turn depends on the orientation of the crystal relative to the loading axis. In FCC crystals there are 12 slip systems, four {1 1 1} planes with three ⟨0 1 1⟩ directions each; reverse slips are not considered as the independent slip systems. These 12 slip systems are best shown in the stereographic projection of Fig. 10.8, where the whole projection is divided into 24 triangles each representing a region in which a particular slip system operates (Havner et al. 1979). The slip system designations are given in Table 10.1. It should be noted that in Fig. 10.8 a slip system with a bar over its number represents a system that is on the same slip plane but in the opposite slip direction as an unbarred system. Such a system is called a *reversed slip system*. For example, system $a_{\bar{1}}$ is on (1 1 1) plane but in $[0\,1\,\bar{1}]$ direction.

Suppose that all slip systems have the same initial critical stress τ_c. Then, if the loading axis is within any one of the triangles, only one slip system will operate. For example, in a triangle with [0 0 1], [0 1 1], and $[\bar{1}\,1\,1]$ as three vertices, system $a_{\bar{2}}$, $(1\,1\,1)[\bar{1}\,0\,1]$ will be active, causing single slip. However, if the loading axis lies on the boundaries of the adjacent triangles, such as the line connecting [0 0 1] and $[\bar{1}\,1\,1]$ directions, two slip systems, b_1 and $a_{\bar{2}}$ will be active at the same time according to Schmid's law, since the Schmid factors for these two systems are the same in this case. This is called a *double*

Table 10.1 Designation of slip systems in FCC crystals

Plane		(111)			$(\bar{1}\bar{1}1)$	
Direction	$[0\bar{1}1]$	$[10\bar{1}]$	$[\bar{1}10]$	$[011]$	$[\bar{1}01]$	$[1\bar{1}0]$
System	a_1	a_2	a_3	b_1	b_2	b_3
Plane		$(\bar{1}11)$			$(1\bar{1}1)$	
Direction	$[0\bar{1}1]$	$[\bar{1}0\bar{1}]$	$[110]$	$[011]$	$[10\bar{1}]$	$[\bar{1}\bar{1}0]$
System	c_1	c_2	c_3	d_1	d_2	d_3

slip, since two slip systems are operative. It is possible that more than two slip systems can be operative at the same time. For example, if the loading axis is in $[010]$ direction, four slip systems, $d_{\bar{1}}$, a_3, c_3, and $b_{\bar{1}}$ will be active at the same time according to Schmid's law. Similarly six slip systems, d_2, $b_{\bar{3}}$, b_1, $a_{\bar{2}}$, a_3, and $d_{\bar{1}}$, will be operative if the loading axis is in the direction $[\bar{1}11]$, and slips will occur simultaneously on eight systems, c_1, $d_{\bar{2}}$, $b_{\bar{2}}$, a_1, d_1, $c_{\bar{2}}$, $a_{\bar{2}}$, and b_1, if the loading axis coincides with the direction $[001]$. The slip mode in which more than one system is active is called *multislip*.

10.1.6 Overshooting and Latent Hardening Effect

Earlier discussion on multislip is based on Schmid's law and the assumption that the initial yield stresses of all the slip systems are the same. Under such conditions merely geometrical consideration will completely determine whether a slip system will be operative. However, experimental results do not support this pure geometrical or orientation-dependent consideration. Taylor and Elam traced the tensile axis using stereographic projection, as discussed above, in a series of experiments on aluminum crystals (Taylor and Elam 1925). Their results are schematically shown in Fig. 10.9. At the beginning of each experiment, only one slip system, $a_{\bar{2}}$, is active, and this slip system is called the *primary system*. As discussed previously, during the test the slip

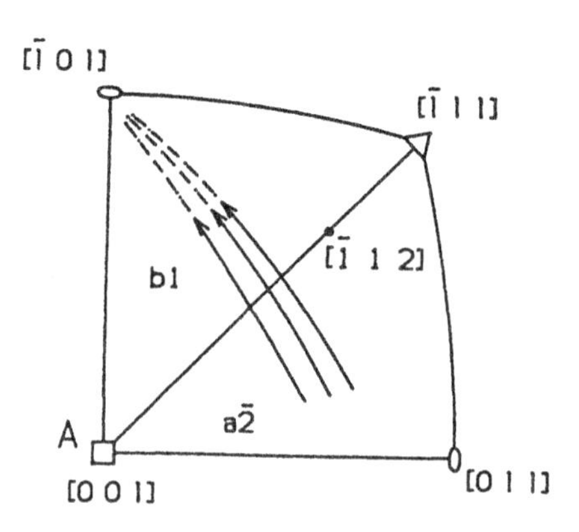

Fig. 10.9 Overshooting of the tensile axis

direction will rotate toward the tensile axis. In stereographic projection the axis tends to rotate toward the slip direction $[\bar{1}01]$, as shown in Fig. 10.9. This rotation brings the tensile axis to the boundary AB, which is symmetric to both system $a_{\bar{2}}$ and b_1. According to Schmid's law, double slip will now occur because the Schmid factors for these two systems is the same if the tensile axis lies along the symmetric boundary AB. System b_1 is called *secondary slip system* or *conjugate slip system to* $a_{\bar{2}}$. Theoretically the tensile axis should not pass through the symmetric line AB because the rotation effects due to slip on system $a_{\bar{2}}$ and slip on system b_1 will cancel each other, so the tensile axis will be stable along the symmetric line AB. However, the X-ray measurement obtained by Taylor and Elam show that the tensile axis overshoots the symmetric boundary, which leads to a larger resolved shear stress on the conjugate (or latent) system. In addition the experimental results also indicated that the rotation of the tensile axis was slowed down once it passed through the symmetric boundary. This means that the conjugate slip did occur. But higher resolved shear stress than that on the primary slip system was necessary to activate the conjugate slip, since the Schmid's factor on the conjugate slip system was higher than that on the primary slip system when the tensile axis passed through the symmetric boundary. Based on these observations, Taylor and Elam concluded the following:

1. Slip systems are hardened by slip on other systems (whether they themselves are active or not), and
2. The latent hardening rate is at least comparable in magnitude to the self-hardening rate (hardening of the active system by slip on itself).

In fact the phenomenon of overshooting suggests that the latent hardening is stronger than the self-hardening. Otherwise, the tensile axis would not rotate over the symmetric boundary and that the double slip would cause the tensile axis to move along AB to direction $[\bar{1}\,1\,2]$ and remain there, since the $[\bar{1}\,1\,2]$ direction is the exact midway orientation between slip directions $[0\,1\,1]$ and $[\bar{1}\,0\,1]$. Because of the stronger latent hardening effect, the tensile axis has to overshoot the symmetric boundary and produce a larger resolved shear stress on the conjugate slip system to activate it.

Overshooting as described above is only an indirect indication of latent hardening. Direct measurements of latent hardening have been obtained for aluminum, iron, copper, silver, and other crystals (Jackson and Basinski 1967). These will be discussed in more detail later.

10.2 KINEMATICS OF SINGLE-CRYSTAL DEFORMATION

In this section we describe the kinematics for the single crystal deformation. To make this presentation more general, we consider finite deformation for which the rotation of the crystal lattice can be formulated.

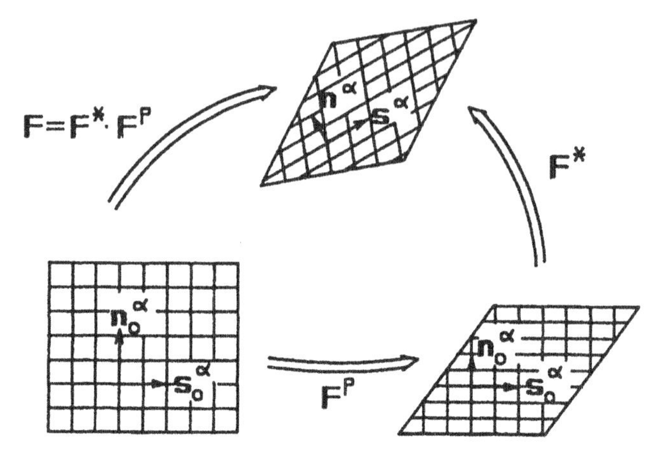

Fig. 10.10 Kinematics of single-crystal deformation

One of the most important observations of plastic deformation of metal crystals during the 1930s was that the plastic deformation is due to the movements of dislocations through the crystal lattice; the crystal lattice retains its shape and orientation during this plastic flow. In addition to the plastic deformation, there is an elastic deformation under which no dislocation movement is involved and which is due to the deformation of the crystal lattice. This is schematically shown in Fig. 10.10.

Consider an infinitesimal element of a single crystal as shown in the lower left of Fig. 10.10. The mutually perpendicular grids represent the crystal lattice; $\mathbf{s}_0^\alpha$ and $\mathbf{n}_0^\alpha$ are the initial unit vector in the slip direction and the initial unit normal vector to the slip plane of the α slip system, respectively ($\alpha = 1, 2, 3, \ldots, n$, where n is the total number of slip systems). After the elastic-plastic deformation this element takes the shape in the upper part of Fig. 10.10. The infinitesimal line elements in these two configurations are related to each other by

$$d\mathbf{x} = \mathbf{F} \cdot d\mathbf{X} \tag{10.7}$$

where $\mathbf{x}$ and $\mathbf{X}$ are the current and initial coordinates of a material particle, respectively, and $\mathbf{F}$ is the deformation gradient. By our earlier observations of crystal deformation, we can assume this total deformation gradient to take place in the following two steps (Asaro and Rice 1979):

1. The crystal undergoes a plastic (shear) deformation due to the dislocation movements through the crystal lattice. During this plastic deformation the orientation and crystal lattice spacing remain unchanged, as shown by the lower right part of Fig. 10.10. Using $\mathbf{F}^\mathrm{p}$ to denote the deformation gradient of

this plastic deformation, we write

$$d\mathbf{p} = \mathbf{F}^{\mathrm{p}} \cdot d\mathbf{X} \tag{10.8}$$

where $d\mathbf{p}$ is the corresponding infinitesimal line element after deformation. Since the plastic deformation has taken place by shear or slip γ^α (defined as the relative displacement of two slip planes with a unit distance apart) on slip systems $\mathbf{n}_0^\alpha$ and $\mathbf{s}_0^\alpha$, the displacement $\mathbf{u}^{\mathrm{p}}$ (caused by plastic flow) of a material particle, denoted by coordinates $\mathbf{X}$ in the reference or undeformed configuration, can be expressed in terms of γ^α. In the general case of multislip, $\mathbf{u}^{\mathrm{p}\alpha}$ can be expressed by (Asaro 1983)

$$\mathbf{u}^{\mathrm{p}\alpha} = \gamma^\alpha(\mathbf{n}_0^\alpha \cdot \mathbf{X})\mathbf{s}_0^\alpha = \gamma^\alpha \mathbf{s}_0^\alpha \mathbf{n}_0^\alpha \cdot \mathbf{X} \tag{10.9}$$

where the summation is taken over all of the active systems. (This is generally true and will not be restated again.) $\mathbf{F}^{\mathrm{p}}$ is then calculated from Eq. (10.9) to be

$$\mathbf{F}^{\mathrm{p}\alpha} = \mathbf{I} + \mathbf{s}_0^\alpha \mathbf{n}_0^\alpha \gamma^\alpha \tag{10.10}$$

2. Following the plastic deformation, the lattice and material deform elastically together with no relative movement between the material particles and the lattice. During this process the lattice will be stretched or shortened, and even distorted. This elastic deformation can be described by

$$d\mathbf{x} = \mathbf{F}^{\mathrm{e}} \cdot d\mathbf{p} \tag{10.11}$$

where $\mathbf{F}^{\mathrm{e}}$ is the elastic deformation gradient. It should be noted that an arbitrary rigid body rotation has been implicitly included in F^{e}. From Eqs. (10.7), (10.8), and (10.11), the total deformation gradient F can then be expressed by

$$\mathbf{F} = \mathbf{F}^{\mathrm{e}} \cdot \mathbf{F}^{\mathrm{p}} \tag{10.12}$$

which can be considered as the physical basis of the multiplicative decomposition of the deformation gradient proposed by Lee (1966). After elastic unloading the lattice returns to its original position and orientation, and the deformation remaining is then characterized by $(\mathbf{F}^{\mathrm{e}})^{-1} \cdot \mathbf{F} = \mathbf{F}^{\mathrm{p}}$.

The vectors attached to the crystal lattice, such as $\mathbf{s}^\alpha$, $\mathbf{n}^\alpha$ ($\mathbf{s}_0^\alpha$ and $\mathbf{n}_0^\alpha$ initially), will deform and rotate due to the deformation. According to the preceding discussion, it is obvious that this deformation and rotation are caused only by the elastic deformation gradient $\mathbf{F}^{\mathrm{e}}$ including a rigid body rotation. Therefore it is a natural practice to define the vector $\mathbf{s}^\alpha$ in the slip direction of the αth slip system at the current state in the following way:

$$\mathbf{s}^\alpha = \mathbf{F}^e \cdot \mathbf{s}_0^\alpha \tag{10.13}$$

However, to keep vector $\mathbf{n}^\alpha$ always normal to the slip plane, the following relation has to be introduced:

$$\mathbf{n}^\alpha = \mathbf{n}_0^\alpha \cdot (\mathbf{F}^e)^{-1} \tag{10.14}$$

Since $\mathbf{n}_0^\alpha$ and $\mathbf{s}_0^\alpha$ are unit vectors and perpendicular to each other, it is easy to show that

$$\mathbf{n}^\alpha \cdot \mathbf{s}^\alpha = \mathbf{s}^\alpha \cdot \mathbf{n}^\alpha = 0 \tag{10.15}$$

that is, $\mathbf{s}^\alpha$ is perpendicular to $\mathbf{n}^\alpha$. It should be noted that $\mathbf{n}^\alpha$ and $\mathbf{s}^\alpha$ are no longer unit vectors, as can be seen from Eqs. (10.13) and (10.14).

The velocity gradient $\mathbf{L}$ in the current configuration can be calculated by the following equation

$$\mathbf{L} = \dot{\mathbf{F}} \cdot \mathbf{F}^{-1} = \dot{\mathbf{F}}^e \cdot \mathbf{F}^{e-1} + \mathbf{F}^e \cdot \dot{\mathbf{F}}^p \cdot (\mathbf{F}^p)^{-1} \cdot (\mathbf{F}^e)^{-1} \tag{10.16}$$

$\mathbf{L}$ can be decomposed into the deformation rate and the spin tensor $\mathbf{D}$ and $\mathbf{W}$ as

$$\mathbf{L} = \mathbf{D} + \mathbf{W} \tag{10.17}$$

Based on this discussion about elastic-plastic deformation of crystals, $\mathbf{D}$ and $\mathbf{W}$ can be written as

$$\mathbf{D} = \mathbf{D}^e + \mathbf{D}^p, \quad \mathbf{W} = \mathbf{W}^e + \mathbf{W}^p \tag{10.18}$$

where $\mathbf{D}^p$ and $\mathbf{W}^p$ are plastic deformation rate and plastic spin, respectively, due to dislocation slip, and $\mathbf{D}^e$ and $\mathbf{W}^e$ are the elastic deformation rate and elastic spin due to the crystal lattice deformation and rotation. From Eq. (10.16), it is seen that

$$\mathbf{D}^e + \mathbf{W}^e = \dot{\mathbf{F}}^e \cdot (\mathbf{F}^e)^{-1} = \mathbf{L}^e \tag{10.19}$$

$$\mathbf{D}^p + \mathbf{W}^p = \mathbf{F}^e \cdot \dot{\mathbf{F}}^p \cdot (\mathbf{F}^p)^{-1} \cdot (\mathbf{F}^e)^{-1} = \mathbf{F}^e \cdot \mathbf{L}_0^p \cdot (\mathbf{F}^e)^{-1} = \mathbf{L}^p \tag{10.20}$$

where

$$\mathbf{L}_0^p = \dot{\mathbf{F}}^p \cdot (\mathbf{F}^p)^{-1} = \mathbf{D}_0^p + \mathbf{W}_0^p \tag{10.21}$$

$$\mathbf{D}_0^p = \tfrac{1}{2}\left(\mathbf{L}_0^p + \mathbf{L}_0^{p^T}\right), \quad \mathbf{W}_0^p = \tfrac{1}{2}\left(\mathbf{L}_0^p - \mathbf{L}_0^{p^T}\right) \tag{10.22}$$

$\mathbf{D}^{\mathrm{p}}$ and $\mathbf{W}^{\mathrm{p}}$ should not be confused with $\mathbf{D}_0^{\mathrm{p}}$ and $\mathbf{W}_0^{\mathrm{p}}$. $\mathbf{D}^{\mathrm{p}}$ and $\mathbf{W}^{\mathrm{p}}$ are the quantities resulting from plastic deformation that are measured in the current configuration, while $\mathbf{D}_0^{\mathrm{p}}$ and $\mathbf{W}_0^{\mathrm{p}}$ are the corresponding quantities that are measured in the intermediate configuration in which no lattice deformation ever occurred. In fact from Eqs. (10.20) and (10.22) it is easy to show that

$$\mathbf{D}^{\mathrm{p}} = \mathbf{F}^{\mathrm{e}} \cdot \mathbf{D}_0^{\mathrm{p}} \cdot (\mathbf{F}^{\mathrm{e}})^{-1}, \quad \mathbf{W}^{\mathrm{p}} = \mathbf{F}^{\mathrm{e}} \cdot \mathbf{W}_0^{\mathrm{p}} \cdot (\mathbf{F}^{\mathrm{e}})^{-1} \tag{10.23}$$

Since the plastic deformation is considered to be caused by dislocation slip, it has been shown that in the reference configuration (e.g., see Asaro 1983)

$$\mathbf{L}_0^{\mathrm{p}} = \sum_{\alpha=1}^{n} \dot{\gamma}^{\alpha} \mathbf{s}_0^{\alpha} \mathbf{n}_0^{\alpha} \tag{10.24}$$

where $\dot{\gamma}^{\alpha}$ is the shear rate on the αth slip system, measured relative to the lattice, and n is the total number of active slip systems. In FCC crystals, for example, $n \leq 12$. Equation (10.24) can be easily verified for a single-slip case. From Eq. (10.10) it follows that

$$\mathbf{F}_0^{\mathrm{p}\alpha^{-1}} = \mathbf{I} - \gamma^{\alpha} \mathbf{s}_0^{\alpha} \mathbf{n}_0^{\alpha} \tag{10.25}$$

and

$$\dot{\mathbf{F}}_0^{\mathrm{p}\alpha} = \dot{\gamma}^{\alpha} \mathbf{s}_0^{\alpha} \mathbf{n}_0^{\alpha} \tag{10.26}$$

since $\dot{\mathbf{n}}_0 = \dot{\mathbf{s}}_0 = \mathbf{0}$. Using definition (10.21) and $\mathbf{n}_0 \cdot \mathbf{s}_0 = 0$, it is easy to show that

$$\mathbf{L}_0^{\mathrm{p}\alpha} = \dot{\gamma} \mathbf{s}_0^{\alpha} \mathbf{n}_0^{\alpha} \tag{10.27}$$

Equation (10.24) is a generalization of this result for single slip to the case of multislip. The plastic deformation rate and spin in the reference configuration are

$$\mathbf{D}_0^{\mathrm{p}} = \sum_{\alpha=1}^{n} \mathbf{P}_0^{\alpha} \dot{\gamma}^{\alpha}, \quad \mathbf{W}_0^{\mathrm{p}} = \sum_{\alpha=1}^{n} \boldsymbol{\Omega}_0^{\alpha} \dot{\gamma}^{\alpha} \tag{10.28}$$

where

$$\mathbf{P}_0^{\alpha} = \tfrac{1}{2}(\mathbf{s}_0^{\alpha} \mathbf{n}_0^{\alpha} + \mathbf{n}_0^{\alpha} \mathbf{s}_0^{\alpha}) \tag{10.29}$$

$$\boldsymbol{\Omega}_0^{\alpha} = \tfrac{1}{2}(\mathbf{s}_0^{\alpha} \mathbf{n}_0^{\alpha} - \mathbf{n}_0^{\alpha} \mathbf{s}_0^{\alpha}) \tag{10.30}$$

Equations corresponding to Eqs. (10.24), (10.28), (10.29), and (10.30) in the current configuration are obtained by using Eqs. (10.20) and (10.23). They are

$$\mathbf{L}^{\mathrm{p}} = \sum_{\alpha=1}^{n} \dot{\gamma} \mathbf{s}^{\alpha} \mathbf{n}^{\alpha} \tag{10.31}$$

$$\mathbf{D}^{\mathrm{p}} = \sum_{\alpha=1}^{n} \mathbf{P}^{\alpha} \dot{\gamma}, \quad \mathbf{W}^{\mathrm{p}} = \sum_{\alpha=1}^{n} \boldsymbol{\Omega}^{\alpha} \dot{\gamma} \tag{10.32}$$

$$\mathbf{P}^{\alpha} = \tfrac{1}{2}(\mathbf{s}^{\alpha}\mathbf{n}^{\alpha} + \mathbf{n}^{\alpha}\mathbf{s}^{\alpha}) \tag{10.33}$$

$$\boldsymbol{\Omega}^{\alpha} = \tfrac{1}{2}(\mathbf{s}^{\alpha}\mathbf{n}^{\alpha} - \mathbf{n}^{\alpha}\mathbf{s}^{\alpha}) \tag{10.34}$$

where Eqs. (10.13) and (10.14) have been used to obtain these results. It is easy to prove that $\mathrm{tr}(\mathbf{D}^{\mathrm{p}}) = \mathrm{tr}(\mathbf{D}_0^{\mathrm{p}}) = 0$, which is the expression for the incompressibility condition.

As noted previously that the orientation of the crystal lattice will change, and thus the resolved shear stress on each slip system will be altered during the process of the large deformation. Therefore it is important to determine the changing rate of both vectors $\mathbf{s}^{\alpha}$ and $\mathbf{n}^{\alpha}$ due to the elastic deformation of the crystalline lattice. This can be accomplished by using Eqs. (10.13), and (10.14) as follows:

$$\dot{\mathbf{s}}^{\alpha} = \dot{\mathbf{F}}^{\mathrm{e}} \cdot \mathbf{s}_0^{\alpha} = \dot{\mathbf{F}}^{\mathrm{e}} \cdot (\mathbf{F}^{\mathrm{e}})^{-1} \cdot \mathbf{s}^{\alpha} = \mathbf{L}^{\mathrm{e}} \cdot \mathbf{s}^{\alpha} = (\mathbf{D}^{\mathrm{e}} + \mathbf{W}^{\mathrm{e}}) \cdot \mathbf{s}^{\alpha} \tag{10.35}$$

$$\dot{\mathbf{n}}^{\alpha} = \mathbf{n}_0^{\alpha} \cdot \left(\dot{\mathbf{F}}^{\mathrm{e}}\right)^{-1} = \mathbf{n}_0^{\alpha} \cdot \left(-(\mathbf{F}^{\mathrm{e}})^{-1} \cdot \mathbf{L}^{\mathrm{e}}\right) = -\mathbf{n}^{\alpha} \cdot \mathbf{L}^{\mathrm{e}} = -\mathbf{n}^{\alpha} \cdot (\mathbf{D}^{\mathrm{e}} + \mathbf{W}^{\mathrm{e}}) \tag{10.36}$$

These are the two differential equations governing the rotation of $\mathbf{s}^{\alpha}$ and $\mathbf{n}^{\alpha}$. Starting from the initial $\mathbf{s}_0^{\alpha}$ and $\mathbf{n}_0^{\alpha}$, the current $\mathbf{s}^{\alpha}$ and $\mathbf{n}^{\alpha}$ can be determined by integrating Eqs. (10.35) and (10.36), provided that the elastic deformation history is known. In real applications of course these integrations have to be performed numerically.

Eqs. (10.13), (10.14), (10.35), and (10.36) can be simplified in cases where the total deformation is large and the elastic deformation is small compared with the plastic deformation and thus can be ignored (e.g., in the case of metal forming). This is shown below.

If elastic deformation is small and can be ignored, the elastic deformation gradient $\mathbf{F}^{\mathrm{e}}$ can be approximated by

$$\mathbf{F}^{\mathrm{e}} = \mathbf{R}^{\mathrm{e}} \cdot \mathbf{U}^{\mathrm{e}} \simeq \mathbf{R}^{\mathrm{e}}, \quad \mathbf{U}^{\mathrm{e}} \simeq \mathbf{I} \tag{10.37}$$

where $\mathbf{R}^{\mathrm{e}}$ is an orthogonal tensor, and $\mathbf{R}^{\mathrm{e}} \cdot \mathbf{R}^{\mathrm{eT}} = \mathbf{I}$ represents the rigid body rotation of the crystal. Then

$$\mathbf{L}^{\mathrm{e}} = \dot{\mathbf{F}}^{\mathrm{e}} \cdot (\mathbf{F}^{\mathrm{e}})^{-1} \simeq \dot{\mathbf{R}}^{\mathrm{e}} \cdot \mathbf{R}^{\mathrm{eT}} \tag{10.38}$$

Since $\dot{\mathbf{R}}^e \cdot \mathbf{R}^{eT}$ is an antisymmetric tensor, as pointed out in Chapter 2, it follows that

$$\mathbf{L}^e = \mathbf{W}^e = \dot{\mathbf{R}}^e \cdot \mathbf{R}^{eT}, \quad \mathbf{D}^e = 0 \tag{10.39}$$

Therefore Eqs. (10.13), (10.14), (10.35), and (10.36) can be simplified as

$$\mathbf{s}^\alpha = \mathbf{R} \cdot \mathbf{s}_0^\alpha, \quad \mathbf{n}^\alpha = \mathbf{n}_0^\alpha \cdot \mathbf{R}^T \tag{10.40}$$

$$\dot{\mathbf{s}}^\alpha = \mathbf{W}^e \cdot \mathbf{s}^\alpha, \quad \dot{\mathbf{n}}^\alpha = \mathbf{W}^e \cdot \mathbf{n}^\alpha \tag{10.41}$$

In most applications Eqs. (10.40) and (10.41), rather than Eqs. (10.13), (10.14), (10.35), and (10.36), are used because of their simplicity. It can be shown that $\mathbf{s}^\alpha$ and $\mathbf{n}^\alpha$, described by Eqs. (10.40) and (10.41), are unit vectors and perpendicular to each other at any stage of the deformation if the elastic deformation is ignored. It has to be pointed out here that Eqs. (10.35) and (10.36), or (10.41) represent only one way of defining $\dot{\mathbf{s}}^\alpha$ and $\dot{\mathbf{n}}^\alpha$. Different ways of defining $\dot{\mathbf{s}}^\alpha$ and $\dot{\mathbf{n}}^\alpha$ have been discussed by Asaro and Rice (1970), and interested readers can refer to their work for more details.

In closing this section, we summarize the main results:

$$\mathbf{L} = \mathbf{L}^e + \sum_{\alpha=1}^{n} \dot{\gamma}^\alpha \mathbf{s}^\alpha \mathbf{n}^\alpha \tag{10.42}$$

$$\mathbf{D} = \mathbf{D}^e + \sum_{\alpha=1}^{n} \mathbf{P}^\alpha \dot{\gamma}^\alpha \tag{10.43}$$

$$\mathbf{W} = \mathbf{W}^e + \sum_{\alpha=1}^{n} \mathbf{\Omega}^\alpha \dot{\gamma}^\alpha \tag{10.44}$$

$$\mathbf{s}^\alpha = \mathbf{F}^e \cdot \mathbf{s}_0^\alpha, \quad \mathbf{n}^\alpha = \mathbf{n}_0^\alpha \cdot (\mathbf{F}^e)^{-1} \tag{10.45}$$

$$\dot{\mathbf{s}}^\alpha = \mathbf{L}^e \cdot \mathbf{s}^\alpha, \quad \dot{\mathbf{n}}^\alpha = -\mathbf{n}^\alpha \cdot \mathbf{L}^e \tag{10.46}$$

where $\mathbf{L}^e$, $\mathbf{D}^e$, and $\mathbf{W}^e$ are caused by the elastic deformation of a crystal lattice, including rigid body rotation. If the elastic deformation is small and can be ignored, then

$$\mathbf{L} = \mathbf{W}^e + \sum_{\alpha=1}^{n} \dot{\gamma}^\alpha \mathbf{s}^\alpha \mathbf{n}^\alpha \tag{10.47}$$

$$\mathbf{D} = \sum_{\alpha=1}^{n} \mathbf{P}^\alpha \dot{\gamma}^\alpha \tag{10.48}$$

$$\mathbf{W} = \mathbf{W}^e + \sum_{\alpha=1}^{n} \mathbf{\Omega}^\alpha \dot{\gamma}^\alpha = \dot{\mathbf{R}} \cdot \mathbf{R}^T + \sum_{\alpha=1}^{n} \mathbf{\Omega}^\alpha \dot{\gamma}^\alpha \tag{10.49}$$

$$\mathbf{s}^\alpha = \mathbf{R}^e \cdot \mathbf{s}_0^\alpha, \quad \mathbf{n}^\alpha = \mathbf{n}_0^\alpha \cdot \mathbf{R}^{eT} \tag{10.50}$$

$$\dot{\mathbf{s}}^\alpha = \mathbf{W}^e \cdot \mathbf{s}^\alpha, \quad \dot{\mathbf{n}}^\alpha = \mathbf{W}^e \cdot \mathbf{n}^\alpha \tag{10.51}$$

$$\mathbf{F} = \mathbf{R}^e \cdot \mathbf{F}^p \tag{10.52}$$

10.3 ELASTIC-PLASTIC CONSTITUTIVE EQUATIONS FOR SINGLE CRYSTALS

In this section we will study the general mathematical formulations of elastic-plastic constitutive equations for single crystals using the knowledge of kinematics of the crystalline deformation we described previously. Since the crystal lattice is not altered by the dislocation slip, it is assumed that the elastic behavior of the crystal is not changed during the elastic-plastic deformation. Under this assumption, Hill and Rice (1972) suggested that the elastic law of the crystals under large deformation can be written

$$\overset{\circ}{\boldsymbol{\sigma}}{}^{\mathrm{e}} + \boldsymbol{\sigma}\,\mathrm{tr}(\mathbf{D}^{\mathrm{e}}) = \mathbf{C} : \mathbf{D}^{\mathrm{e}} \tag{10.53}$$

where $\mathbf{C}$ is a fourth-order tensor of elastic moduli, whose components are formed relative to the crystal lattice. C_{ijkl} is symmetric in indices i, j and k, l and generally has the following symmetry:

$$C_{ijkl} = C_{klij} \tag{10.54}$$

assuming the existence of an elastic strain energy function. The number of independent components of C depends on the symmetry property of a particular crystal. If the crystal is isotropic, then there are only two independent elastic constants.

In Eq. (10.53) $\overset{\circ}{\boldsymbol{\sigma}}{}^{\mathrm{e}}$ is the Jaumann rate of Cauchy stress based on the axes that spin together with the lattice:

$$\overset{\circ}{\boldsymbol{\sigma}}{}^{\mathrm{e}} = \dot{\boldsymbol{\sigma}} - \mathbf{W}^{\mathrm{e}} \cdot \boldsymbol{\sigma} + \boldsymbol{\sigma} \cdot \mathbf{W}^{\mathrm{e}} \tag{10.55}$$

where $\dot{\boldsymbol{\sigma}}$ is the material time rate of $\boldsymbol{\sigma}$. It is clear from this equation that $\overset{\circ}{\boldsymbol{\sigma}}{}^{\mathrm{e}}$ is the corotational rate of $\boldsymbol{\sigma}$ based on axes that rotate rigidly at the lattice spin $\mathbf{W}^{\mathrm{e}}$; namely the axes move together with lattice axes. The left-hand side of Eq. (10.53) represents the so-called Jaumann rate of Kirchhoff stress:

$$\overset{\circ}{\boldsymbol{\tau}}{}^{\mathrm{e}} = \overset{\circ}{\boldsymbol{\sigma}}{}^{\mathrm{e}} + \boldsymbol{\sigma}\,\mathrm{tr}(\mathbf{D}^{\mathrm{e}}) \tag{10.56}$$

where the reference configuration is chosen to coincide instantaneously with the current configuration. $\boldsymbol{\tau}$ is the Kirchhoff stress tensor defined by

$$\boldsymbol{\tau} = \frac{\rho_0}{\rho}\boldsymbol{\sigma} = \det(\mathbf{F})\boldsymbol{\sigma} \tag{10.57}$$

where ρ_0 and ρ are the mass densities in the reference and current configurations, respectively.

The constitutive equation (10.53) is expressed in terms of elastic quantities $\mathbf{D}^{\mathrm{e}}$ and $\mathbf{W}^{\mathrm{e}}$. However, in application it is more convenient to express it in terms of the total quantities **D** and **W** and slips $\dot{\gamma}^\alpha$. Using Eqs. (10.43) and (10.44), Eq. (10.53) can be written

$$\mathring{\boldsymbol{\sigma}} + \boldsymbol{\sigma}\,\mathrm{tr}(\mathbf{D}) = \mathbf{C}:\mathbf{D} - \sum_{\alpha=1}^{n} [\mathbf{C}:\mathbf{P}^\alpha + \boldsymbol{\beta}^\alpha]\dot{\gamma}^\alpha \tag{10.58}$$

where

$$\boldsymbol{\beta}^\alpha = \boldsymbol{\Omega}^\alpha \cdot \boldsymbol{\sigma} - \boldsymbol{\sigma} \cdot \boldsymbol{\Omega}^\alpha \tag{10.59}$$

and $\mathring{\boldsymbol{\sigma}}$ is the Jaumann rate of Cauchy stress $\boldsymbol{\sigma}$, based on the total spin of the continuum, as in Eq. (3.105). In deriving Eq. (10.58), the following relation has been used:

$$\mathrm{tr}(\mathbf{D}^{\mathrm{p}}) = 0, \quad \mathrm{tr}(\mathbf{D}) = \mathrm{tr}(\mathbf{D}^{\mathrm{e}}) \tag{10.60}$$

If $\dot{\gamma}^\alpha$s are determined in terms of $\mathring{\boldsymbol{\sigma}}$ or **D**, they can be substituted into Eq. (10.58), which completes the elastic-plastic constitutive equations for single crystals. This depends on the constitutive equation for slips on each slip system:

$$\dot{\tau}_c^\alpha = \sum_{\beta=1}^{n} h_{\alpha\beta}\dot{\gamma}^\beta \tag{10.61}$$

where τ_c^α is the current yield strength of the αth slip system, $h_{\alpha\beta}$ is the hardening modulus matrix characterizing the material behavior of the crystalline solids. Modulus $h_{\alpha\alpha}$ is self-hardening, while modulus $h_{\alpha\beta}$ $(\alpha \neq \beta)$ represents the latent hardening effect. Equation (10.61), where summation on β is over all active slip systems, is very important in formulating time-independent, elastic-plastic theory for single crystals. Modulus $h_{\alpha\beta}$ represents material properties and is different for single crystals of different materials. Therefore the determination of $h_{\alpha\beta}$ has long attracted attention from material scientists and mechanics researchers. Several experimental studies have been done in order to measure $h_{\alpha\beta}$ for different crystals. Also various models have been proposed (e.g., Havner and Shalaby 1977; Peirce et al. 1982; Weng 1987). These models will be discussed in detail next. For the purpose of representing the general form of constitutive equations, it is sufficient here to use the general form as given in Eq. (10.61).

To proceed further, we need to understand the following loading/unloading criteria based on Schmid's law (see Section 10.1):

$$\begin{aligned}
&\dot{\gamma}^\alpha = 0 && \text{if } 0 \le \tau_\alpha < \tau_c^\alpha \\
&\dot{\gamma}^\alpha = 0 && \text{if } 0 < \tau_\alpha = \tau_c^\alpha;\ \dot{\tau}_\alpha < \dot{\tau}_c^\alpha = \sum_{\beta=1}^{n} h_{\alpha\beta}\dot{\gamma}^\beta \\
&\dot{\gamma}^\alpha \ge 0 && \text{if } 0 < \tau_\alpha = \tau_c^\alpha;\ \dot{\tau}_\alpha = \dot{\tau}_c^\alpha = \sum_{\beta=1}^{n} h_{\alpha\beta}\dot{\gamma}^\beta
\end{aligned} \tag{10.62}$$

where

$$\tau_\alpha = \mathbf{s}^\alpha \cdot \boldsymbol{\sigma} \cdot \mathbf{n}^\alpha = \boldsymbol{\sigma} : \mathbf{P}^\alpha \tag{10.63}$$

is the resolved shear stress (Schmid stress) on the αth slip system. These criteria assume that stress components other than the resolved shear stress have no effect on the plastic slip. τ_α and $\dot{\tau}_\alpha$ should not be confused with τ_c^α and $\dot{\tau}_c^\alpha$. τ_α and $\dot{\tau}_\alpha$ are the resolved shear stress with its rate on the αth slip system produced by the applied stress $\boldsymbol{\sigma}$, as shown by Eq. (10.63), while τ_c^α and $\dot{\tau}_c^\alpha$ are the current shear yield strength with its rate caused by slips of the αth slip system, which are the material properties. $\dot{\tau}_\alpha$ has to be calculated from Eq. (10.63). Taking the derivative with respect to time of both sides of Eq. (10.63), we obtain

$$\dot{\tau}_\alpha = \dot{\mathbf{s}}^\alpha \cdot \boldsymbol{\sigma} \cdot \mathbf{n}^\alpha + \mathbf{s}^\alpha \cdot \dot{\boldsymbol{\sigma}} \cdot \mathbf{n}^\alpha + \mathbf{s}^\alpha \cdot \boldsymbol{\sigma} \cdot \dot{\mathbf{n}}^\alpha \tag{10.64}$$

where $\dot{\mathbf{s}}^\alpha$ and $\dot{\mathbf{n}}^\alpha$ have to be obtained by either Eq. (10.46) or Eq. (10.51). For infinitesmal elastic deformation Eq. (10.51) is used in the following derivation. Using Eq. (10.51), we get the following form for Eq. (10.64):

$$\dot{\tau}_\alpha = \mathbf{s}^\alpha \cdot \left(\dot{\boldsymbol{\sigma}} - \mathbf{W}^{\mathrm{e}} \cdot \boldsymbol{\sigma} + \boldsymbol{\sigma} \cdot \mathbf{W}^{\mathrm{e}}\right) \cdot \mathbf{n}^\alpha = \mathring{\boldsymbol{\sigma}}^{\mathrm{e}} : \mathbf{P}^{\mathrm{e}} \tag{10.65}$$

Now the criteria given in Eq. (10.62) can be rewritten as

$$\begin{aligned}
&\dot{\gamma}^\alpha = 0 \qquad \text{if } 0 \le \tau_\alpha \le \tau_c^\alpha \\
&\dot{\gamma}^\alpha = 0 \qquad \text{if } 0 < \boldsymbol{\sigma} : \mathbf{P}^\alpha = \tau_c^\alpha;\ \mathring{\boldsymbol{\sigma}}^{\mathrm{e}} : \mathbf{P}^\alpha < \sum_{\beta=1}^{n} h_{\alpha\beta}\dot{\gamma}^\beta \\
&\dot{\gamma}^\alpha \ge 0 \qquad \text{if } 0 < \boldsymbol{\sigma} : \mathbf{P}^\alpha = \tau_c^\alpha;\ \mathring{\boldsymbol{\sigma}}^{\mathrm{e}} : \mathbf{P}^\alpha = \sum_{\beta=1}^{n} h_{\alpha\beta}\dot{\gamma}^\beta
\end{aligned} \tag{10.66}$$

Note that

$$\mathring{\boldsymbol{\sigma}}^{\mathrm{e}} : \mathbf{P}^\alpha = \mathring{\boldsymbol{\sigma}} : \mathbf{P}^\alpha + \sum_{\beta=1}^{n} \boldsymbol{\beta}^\beta : \mathbf{P}^\alpha \dot{\gamma}^\beta \tag{10.67}$$

where $\boldsymbol{\beta}^\alpha$ and $\mathbf{P}^\alpha$ are defined by Eqs. (10.59) and (10.33), respectively. Equation (10.66) can be further revised as

$$\begin{aligned}
&\dot{\gamma}^\alpha = 0 \qquad \text{if } \mathring{\boldsymbol{\sigma}} : \mathbf{P}^\alpha < \sum_{\beta=1}^{n} k_{\alpha\beta}\dot{\gamma}^\beta \\
&\dot{\gamma}^\alpha > 0 \qquad \text{if } \mathring{\boldsymbol{\sigma}} : \mathbf{P}^\alpha = \sum_{\beta=1}^{n} k_{\alpha\beta}\dot{\gamma}^\beta
\end{aligned} \tag{10.68}$$

where

$$k_{\alpha\beta} = h_{\alpha\beta} - \mathbf{P}^{\alpha} : \boldsymbol{\beta}^{\beta} \tag{10.69}$$

Expressions similar to those in Eq. (10.68), but in terms of the total material deformation rate $\mathbf{D}$, can be derived in the following way: From the elastic constitutive equation (10.53) and kinematic relation (10.43), we derive the relation

$$\begin{aligned} \overset{\circ}{\boldsymbol{\sigma}}{}^{e} &= \mathbf{C} : \mathbf{D} - \mathbf{C} : \sum_{\beta=1}^{n} \mathbf{P}^{\beta} \dot{\gamma}^{\beta} - \boldsymbol{\sigma}\, \mathrm{tr}(\mathbf{D}) \\ &= (\mathbf{C} - \boldsymbol{\sigma}\mathbf{I}) : \mathbf{D} - \mathbf{C} : \sum_{\beta=1}^{n} \mathbf{P}^{\beta} \dot{\gamma}^{\beta} \end{aligned} \tag{10.70}$$

Substituting this equation into Eq. (10.66), we obtain criteria similar to that in Eq. (10.68):

$$\begin{aligned} \dot{\gamma}^{\alpha} &= 0 \qquad \text{if } \mathbf{P}^{\alpha} : (\mathbf{C} - \boldsymbol{\sigma}\mathbf{I}) : \mathbf{D} < \sum_{\beta=1}^{n} g_{\alpha\beta} \dot{\gamma}^{\beta} \\ \dot{\gamma}^{\alpha} &> 0 \qquad \text{if } \mathbf{P}^{\alpha} : (\mathbf{C} - \boldsymbol{\sigma}\mathbf{I}) : \mathbf{D} = \sum_{\beta=1}^{n} g_{\alpha\beta} \dot{\gamma}^{\beta} \end{aligned} \tag{10.71}$$

where

$$g_{\alpha\beta} = h_{\alpha\beta} + \mathbf{P}^{\alpha} : \mathbf{C} : \mathbf{P}^{\beta} \tag{10.72}$$

We should emphasize that Eqs. (10.68), (10.69), (10.71), and (10.72) are derived by neglecting the elastic deformation of the crystal lattice.

The matrices $g_{\alpha\beta}$ and $k_{\alpha\beta}$ were introduced by Hill and Rice (1972) in discussing the sufficient condition to uniquely determine $\dot{\gamma}^{\alpha}$'s. They pointed out that

1. $\dot{\gamma}^{\alpha}$'s are uniquely determined for a prescribed $\mathbf{D}$ if the matrix $g_{\alpha\beta}$ is positive definite.
2. $\dot{\gamma}^{\alpha}$'s are uniquely determined for a prescribed $\overset{\circ}{\sigma}$ if the matrix $k_{\alpha\beta}$ is positive definite.

Unfortunately, the preceding requirement of positive definiteness in $g_{\alpha\beta}$ and $k_{\alpha\beta}$ is too severe to be satisfied by real crystalline materials, and therefore the uniqueness is generally not assured. This will be discussed later. At this stage it is simply assumed that $g_{\alpha\beta}$ and $k_{\alpha\beta}$ are positive definite and thus can be inverted. Under this condition $\dot{\gamma}^{\alpha}$'s can be determined with a prescribed deformation rate **D** to be

$$\dot{\gamma}^{\alpha} = \sum_{\beta=1}^{n} g_{\alpha\beta}^{-1} \mathbf{P}^{\beta} : (\mathbf{C} - \sigma \mathbf{I}) : \mathbf{D} \tag{10.73}$$

or with a prescribed $\overset{\circ}{\sigma}$ to be

$$\dot{\gamma}^{\alpha} = \sum_{\beta=1}^{n} k_{\alpha\beta}^{-1} \mathbf{P}^{\beta} : \overset{\circ}{\boldsymbol{\sigma}} \tag{10.74}$$

Substitution of $\dot{\gamma}^{\alpha}$, as determined by Eq. (10.73), into Eq. (10.58) leads to

$$\overset{\circ}{\boldsymbol{\sigma}} = \mathbf{C}^{\mathrm{ep}} : \mathbf{D} \tag{10.75}$$

where

$$\mathbf{C}^{\mathrm{ep}} = \mathbf{C} - \sigma \mathbf{I} - \sum_{\alpha=1}^{n} \sum_{\beta=1}^{n} [\mathbf{C} : \mathbf{P}^{\alpha} + \boldsymbol{\beta}^{\alpha}] g_{\alpha\beta}^{-1} \mathbf{P}^{\beta} : [\mathbf{C} - \sigma \mathbf{I}] \tag{10.76}$$

On the other hand, if $\dot{\gamma}^{\alpha}$, as determined by Eq. (10.74), is substituted into Eq. (10.58), the following constitutive equation can be obtained

$$\mathbf{D} = \mathbf{S}^{\mathrm{ep}} : \overset{\circ}{\boldsymbol{\sigma}} \tag{10.77}$$

where

$$\mathbf{S}^{\mathrm{ep}} = [\mathbf{C} - \sigma \mathbf{I}]^{-1} : \left\{ \mathbf{I} + \sum_{\alpha=1}^{n} \sum_{\beta=1}^{n} [\mathbf{C} : \mathbf{P}^{\alpha} + \boldsymbol{\beta}^{\alpha}] k_{\alpha\beta}^{-1} \mathbf{P}^{\beta} \right\} \tag{10.78}$$

Equation (10.75) or Eq. (10.77) is the elastic-plastic constitutive equation of single crystals based on the consideration of dislocation slip and Schmid's law. In literature this equation is usually written in terms of Kirchhoff stress $\boldsymbol{\tau}$ as

$$\overset{\circ}{\boldsymbol{\tau}} = \mathbf{C}^{\mathrm{ep}} : \mathbf{D} \tag{10.79}$$

where

$$\mathbf{C}^{\mathrm{ep}} = \mathbf{C} - \sum_{\alpha=1}^{n} \sum_{\beta=1}^{n} [\mathbf{C} : \mathbf{P}^{\alpha} + \boldsymbol{\beta}^{\alpha}] g_{\alpha\beta}^{-1} \mathbf{P}^{\beta} : [\mathbf{C} - \sigma \mathbf{I}] \tag{10.80}$$

The normality condition derived from the Schmid's law is later. Note that in stress space

$$f^{\alpha}(\boldsymbol{\sigma}) = \boldsymbol{\sigma} : \mathbf{P}^{\alpha} = \tau_{c}^{\alpha} \tag{10.81}$$

is a set of planes which define the yield locus. Obviously

$$\mathbf{P}^{\alpha} = \frac{\partial f^{\alpha}}{\partial \boldsymbol{\sigma}} \tag{10.82}$$

is the outward normal to the yield plane f^α. However, as can be seen from Eq. (10.43), direction of the plastic deformation rate $\mathbf{D}^p$ is given by $\mathbf{P}^\alpha$ when the current stress $\boldsymbol{\sigma}$ is on the yield plane f^α. Therefore Eq. (10.82) is the normality condition of plastic deformation rate $\mathbf{D}^p$ for the yield surface derived from Schmid's law.

10.4 HARDENING RULES FOR SINGLE CRYSTALS

As previously noted, the hardening rule expressed by Eq. (10.61) is important in describing the elastic-plastic behaviors of single crystals. In this section several hardening models will be described. The general form of the hardening rule Eq. (10.61) can be represented as

$$d\tau_\alpha = h_{\alpha\beta} d\gamma^\beta \tag{10.83}$$

where the summation on index β is implied and the rates $\dot{\tau}_\alpha$ and $\dot{\gamma}^\beta$ are replaced by the increments $d\tau_\alpha$ and $d\gamma^\beta$ because of their equivalence in the time-independent theory. Before presenting various hardening rules (different forms of $h_{\alpha\beta}$), it is helpful to briefly review the experimental observations on hardening behaviors of crystals due to slip.

It has been discussed previously that a slip system will be hardened not only by the slip on this slip system (self-hardening) but also by slip on other slip systems (latent hardening). An indirect indication of latent hardening is the phenomenon called *overshooting*, observed by many researchers in early 1930s. The direct measurement of the latent hardening has been done by Kocks and Brown (1966), Jackson and Basinski (1967), and many others. The general conclusions are as follows:

1. The latent-hardening rate is much higher than the self-hardening rate. According to Kocks (1970) the latent hardening rate can be 20% to 40% higher than the self-hardening rate.
2. The slip systems on the same plane as the primary slip system (called *coplanar systems*) show almost the same hardening rate as the primary system.
3. The slip systems whose slip direction is perpendicular to that of the primary slip system have a higher hardening rate than the primary system but show few differences among themselves.
4. The other intersecting slip systems harden at the highest hardening rate of all the slip systems. Again, no difference of hardening rate is observed among them.

The preceding experimental results are the basis of any mathematical model.

10.4.1 Taylor's Model

Taylor (1935) proposed an isotropic hardening model for single crystals

$$d\tau_\alpha = h \sum_{\beta=1}^{n} d\gamma^\beta \tag{10.84}$$

where β runs over all the active slip systems, h is a constant. Comparing Eq. (10.82) with the general hardening rule (10.83), we get

$$h_{\alpha\beta} = h \tag{10.85}$$

for Taylor's model. Since h is same for all slip systems (active and inactive), Taylor's model implies that all the slip systems harden equally, irrespective of whether they are active or not. Therefore the crystal hardens isotropically.

Obviously Taylor's isotropic hardening model cannot describe the higher latent hardening effect observed experimentally, nor can it explain the "overshooting" phenomenon. Actually there is little experimental evidence that supports Taylor's model; nevertheless, because of its simplicity, it is still a commonly used model. It should be noted that in Eq. (10.84) the constant h has to be multiplied by the sum of the slips or shears occurring on all the active slip systems to obtain the increment of the yield strength of any active or inactive slip system.

10.4.2 Nakada and Keh's Two-Parameter Model

To describe the higher latent hardening rate, Nakada and Keh (1966) proposed an empirical two-parameter modification of Taylor's model:

$$d\tau_\alpha = H_1 \sum_{l} d\gamma^l + H_2 \sum_{m} d\gamma^m \tag{10.86}$$

$$\mathbf{n}_l = \mathbf{n}_\alpha, \quad \mathbf{n}_m \neq \mathbf{n}_\alpha, \quad H_2 > H_1 > 0$$

where H_1 and H_2 are two material constants. H_1 represents the hardening rate of the primary system and the coplanar systems, while H_2 denotes the hardening rate of other slip systems. The summation on l is over coplanar slip systems and on m is over other intersecting slip systems; $l + m = n$, the total number of active slip systems.

In Nakada and Keh's two-parametric model, since $H_2 > H_1$, the higher latent hardening rate can be described. Furthermore the coplanar systems are assumed to harden at the same rate as the primary systems, which is in accord with the experimental observations mentioned before. However, some aspects of crystal deformation, such as the Bauschinger effect, are not described by this model.

10.4.3 Havner and Shalaby's Model

Havner and Shalaby (1977) took a comprehensive theoretical analysis of the finite elastic-plastic deformation of single crystals and proposed a "mathematical theory of finite distortional latent hardening" in single crystals, according to which

$$d\tau_{\alpha} = \left(k_{\alpha\beta} + \mathbf{P}^{\alpha} : \boldsymbol{\beta}^{\beta}\right) d\gamma^{\beta} \tag{10.87}$$

where $k_{\alpha\beta}$ is a symmetric matrix and $\mathbf{P}^{\alpha}$ and $\boldsymbol{\beta}^{\beta}$ are defined by Eqs. (10.33) and (10.59), respectively. In the equation above, the summation of index β is implied. By comparing Eq. (10.87) with the general form of hardening rule (10.83), we can define the matrix of hardening moduli $h_{\alpha\beta}$ of Havner and Shalaby's model as

$$h_{\alpha\beta} = k_{\alpha\beta} + \mathbf{P}^{\alpha} : \boldsymbol{\beta}^{\beta} \tag{10.88}$$

In fact these results can be derived from Eq. (10.69).

Havner and Shalaby further simplified their mathematical model (10.87) by specifying that

$$\dot{k}_{\alpha\beta} = h \tag{10.89}$$

where h is a constant. Therefore Eqs. (10.87) and (10.88) are reduced to

$$d\tau_{\alpha} = \left(h + \mathbf{P}^{\alpha} : \boldsymbol{\beta}^{\beta}\right) d\gamma^{\beta} \tag{10.90}$$

$$h_{\alpha\beta} = h + \mathbf{P}^{\alpha} : \boldsymbol{\beta}^{\beta} \tag{10.91}$$

which are called the *simple theory of latent hardening of single crystals* by Havner and Shalaby.

According to model (10.88) and (10.91), the higher latent hardening rate, or generally, the anisotropic hardening of single crystalline deformation is caused by the relative rotation of the crystalline material with respect to the underlying lattice. Without this relative rotation, the hardening would be isotropic. That is, all slip systems will harden equally. This can be made clear as we will show next (see Havner et al. 1979).

From Eq. (10.44) the total rotation of crystalline material in the process of finite elastic-plastic deformation is the sum of $\mathbf{W}^{e}$ and $\Sigma\boldsymbol{\Omega}^{\alpha}\dot{\gamma}^{\alpha}$. Since $\mathbf{W}^{e}$ is the rotation of crystal lattice, $\Sigma\boldsymbol{\Omega}^{\alpha}\dot{\gamma}^{\alpha}$ is therefore the relative rotation of material with respect to the lattice. We denote this relative spin by $\boldsymbol{\omega}$. It is easy to show that

$$\begin{aligned}\sum_{\beta=1}^{n} \mathbf{P}^{\alpha} : \boldsymbol{\beta}^{\beta} d\gamma^{\beta} &= \mathbf{P}^{\alpha} \cdot \left[\sum_{\beta=1}^{n} \boldsymbol{\Omega}^{\beta} d\gamma^{\beta} \cdot \boldsymbol{\sigma} - \boldsymbol{\sigma} : \sum_{\beta=1}^{n} \boldsymbol{\Omega}^{\beta} d\gamma^{\beta}\right] \\ &= \mathbf{P}^{\alpha} \cdot (\boldsymbol{\omega} \cdot \boldsymbol{\sigma} - \boldsymbol{\sigma} \cdot \boldsymbol{\omega}) = 2\mathbf{P}^{\alpha} \cdot \boldsymbol{\omega} \cdot \boldsymbol{\sigma}\end{aligned} \tag{10.92}$$

and thus that

$$d\tau_\alpha = h \sum_{\beta=1}^{n} d\gamma^\beta + 2\mathbf{P}^\alpha \cdot \boldsymbol{\omega} \cdot \boldsymbol{\sigma} \tag{10.93}$$

A comparison of the equation above with Taylor's isotropic hardening model of Eq. (10.84) shows that Havner and Shalaby's simple theory, as described by Eq. (10.91), reduces to Taylor's isotropic hardening model for the following two cases:

1. The deformation is infinitesimal and the relative rotation $\boldsymbol{\omega}$ can be ignored.
2. There is no relative rotation at all ($\boldsymbol{\omega} = \mathbf{0}$).

Havner and Shalaby's simple theory can therefore predict almost the same hardening rate for all the slip systems in the initial portion of deformation (corresponding to infinitesimal deformation), which is the above-mentioned first case. In the second case yield strength τ_0 in all systems is subjected to axial loading in the 4-, 6-, or 8-fold symmetry orientation. Because of equal multislip there is no rotation of the loading axis away from the initial crystallographic position.

When $\boldsymbol{\omega} = \mathbf{0}$ or is negligibly small, the second term on the right-hand side of Eq. (10.92) vanishes, and Eq. (10.92) becomes Taylor's isotropic hardening model. For this reason Havner and Shalaby's model is also called the *rotation-dependent theory*.

Havner and coworkers have made a series of analytical studies of the single slip, double slip, and multislip of single crystal deformation using the simple mathematical theory discussed above. Various features of this theory have been comprehensively explored. Interested readers can refer to Havner (1984) and the references cited there for more details.

10.4.4 Peirce, Asaro, and Needleman's Model

It has been pointed out by Peirce, Asaro, and Needleman (1982) that Havner and Shalaby's simple theory of latent hardening of single crystals predicts very strong latent hardening, which has been used to describe "overshooting." To obtain a less severe latent hardening effect that is more in accord with experimental results, they suggested the following modified form of the latent hardening model:

$$d\tau_\alpha = \left[k_{\alpha\beta} + \tfrac{1}{2}\left(\boldsymbol{\beta}^\beta : \mathbf{P}^\alpha - \boldsymbol{\beta}^\alpha : \mathbf{P}^\beta\right)\right] d\gamma^\beta \tag{10.94}$$

$$h_{\alpha\beta} = k_{\alpha\beta} + \tfrac{1}{2}\left(\boldsymbol{\beta}^\beta : \mathbf{P}^\alpha - \boldsymbol{\beta}^\alpha : \mathbf{P}^\beta\right) \tag{10.95}$$

Although this hardening rule introduces a less severe latent hardening effect than Havner and Shalaby's simple theory, it still predicts a latent hardening effect much stronger than experimental observations indicate. This is because the off-diagonal terms in Eq. (10.88) and Eq. (10.95) include the effects of the current Cauchy stress $\boldsymbol{\sigma}$ [see definition for β^{α} in (10.59)] and thus will become much larger than the diagonal or self-hardening terms when the stress σ becomes larger than the hardening moduli $h_{\alpha\beta}$. As noted by Mitchell (1964) and Chang and Asaro (1981), when the ratio of stress to the slip hardening moduli $k_{\alpha\beta}$ becomes significantly larger than unity (of order 10–100 in FCC crystals after finite strains), these stress terms in Eq. (10.88) and Eq. (10.95) can dominate so that the physical description of material behavior, which enters via the $k_{\alpha\beta}$'s, is lost. A detailed discussion on this matter can be found in Asaro (1983).

10.4.5 Weng's Model

To describe the isotropic hardening, kinematic hardening, combined isotropic and kinematic hardening, Bauschinger effect, and latent hardening, and so on, Weng (1987), using his kinematic hardening model, proposed the following anisotropic hardening law for single crystals:

$$d\tau_{\alpha} = \sum_{\beta=1}^{n} \lambda_{\alpha\beta} d\tau_{(\beta)} \tag{10.96}$$

$$\lambda_{\alpha\beta} = \alpha_1 + (1 - \alpha_1)\cos\theta_{\alpha\beta}\cos\phi_{\alpha\beta} + \left(\alpha_2 \sin\theta_{\alpha\beta} + \alpha_3 \sin\phi_{\alpha\beta}\right) \tag{10.97}$$

where

$$\cos\theta_{\alpha\beta} = s_{\alpha} \cdot s_{\beta}, \quad \sin\theta_{\alpha\beta} = \left(1 - \cos^2\theta_{\alpha\beta}\right)^{1/2} \tag{10.98}$$

$$\cos\phi_{\alpha\beta} = n_{\alpha} \cdot n_{\beta}, \quad \sin\phi_{\alpha\beta} = \left(1 - \cos^2\phi_{\alpha\beta}\right)^{1/2} \tag{10.99}$$

In Eq. (10.97), no sum is over α and β; α_1, α_2, and α_3 are three constants. As will be shown later, α_1 is the degree of isotropy in work hardening, α_2 represents the anisotropic departure for the latent hardening of coplanar systems ($\phi_{\alpha\beta} = 0$), and α_3 the additional departure for the intersecting systems ($\phi_{\alpha\beta} \neq 0$). The parenthesis around the subscript β in Eq. (10.96) indicates that $d\tau_{(\beta)}$ is related to its own slip by

$$d\tau^{\beta} = h_{\beta\beta} d\gamma^{\beta} \tag{10.100}$$

where no sum is over β and where $d\gamma^{\beta}$ is the slip on the βth slip system and $h_{\beta\beta}$ is the self-hardening modulus of this system. Substituting Eq. (10.100) into Eq. (10.96) results in

$$d\tau_\alpha = \sum_{\beta=1}^{n} \lambda_{\alpha\beta} h_{\beta\beta} d\gamma^\beta \tag{10.101}$$

the matrix $h_{\alpha\beta}$ in the general form of hardening rule (10.83) is given by

$$h_{\alpha\beta} = \lambda_{\alpha\beta} h_{\beta\beta} \tag{10.102}$$

where no sum is implied over β.

From Eq. (10.97), $\lambda_{\beta\beta} = 1$, and therefore $h_{\alpha\beta} = h_{\beta\beta}$ when $\alpha = \beta$. It is obvious that the latent hardening is governed by coefficients $\lambda_{\alpha\beta}$ and that self-hardening is governed by $h_{\beta\beta}$. Therefore $\lambda_{\alpha\beta}$ and $h_{\beta\beta}$ are called the *latent hardening coefficients* and self-hardening modulus, respectively. Since the latent hardening coefficients are established by Eq. (10.97), the problem has been reduced to determine the self-hardening modulus $h_{\beta\beta}$. Before discussing the particular form of $h_{\beta\beta}$, the various features of the latent hardening coefficients $\lambda_{\alpha\beta}$ are presented.

As discussed by Weng (1987), the general form of $\lambda_{\alpha\beta}$ reduces to the following special cases:

1. Fully isotropic,

$$\lambda_{\alpha\beta} = 1 \qquad \text{when } \alpha_1 = 1,\ \alpha_2 = \alpha_3 = 0 \tag{10.103}$$

2. Fully kinematic,

$$\lambda_{\alpha\beta} = \cos\theta_{\alpha\beta} \cos\phi_{\alpha\beta} \qquad \text{when } \alpha_1 = \alpha_2 = \alpha_3 = 0 \tag{10.104}$$

3. Coplanar isotropic, and stronger latent hardening for the intersecting systems,

$$\lambda_{\alpha\beta} = 1 + \alpha_3 \sin\phi_{\alpha\beta}, \qquad \text{when } \alpha_1 = 1,\ \alpha_2 = 0,\ \alpha_3 > 0 \tag{10.105}$$

Since $\alpha_3 > 0$ and $\sin\phi_{\alpha\beta} = 0$ for coplane systems, and $0 < \sin\phi_{\alpha\beta} < 1$ for noncoplane systems, $\lambda_{\alpha\beta}$ for noncoplane systems are larger than those for coplane systems, according to Eq. (10.105), by an amount depending on the relative orientations of the slip planes and the constant α_3.

4. Coplanar kinematic and stronger hardening for the intersecting systems,

$$\lambda_{\alpha\beta} = \cos\theta_{\alpha\beta} \cos\phi_{\alpha\beta} + \alpha_3 \sin\phi_{\alpha\beta} \qquad \text{when } \alpha_1 = \alpha_2 = 0,\ \alpha_3 > 0 \tag{10.106}$$

Again, the stronger hardening for the intersecting systems is governed by $\phi_{\alpha\beta} = \mathbf{n}^\alpha \cdot \mathbf{n}^\beta$ (the relative orientation between slip planes) and α_3. The kinematic hardening effect as given by $\cos\theta_{\alpha\beta}\cos\phi_{\alpha\beta}$ and will be explained shortly.

5. Stronger Bauschinger effect results in a decrease of the elastic range for the active slip system. This can be obtained by assigning $\alpha_1 < 0$.

More detailed discussions on the above-mentioned features of Weng's latent hardening coefficients can be found in Weng (1987), where the determination of the material constants α_1, α_2, and α_3 from experimental data is also presented. Here we only detail the kinematic hardening effect produced by Eq. (10.104). It should be mentioned that Weng had derived this kinematic hardening model from the dislocation interaction Weng (1979). To show the kinematic hardening effect of Eq. (10.104), it is sufficient to show that it can predict the Bauschinger effect; that is, the increase of the yield stress in the slip direction causes a decrease of the yield stress in the reverse direction by the same amount. Consider the case of single slip. Denote the slip system by 1 and the system in the reversed direction by $\bar{1}$. Then the Bauschinger effect is expressed by

$$d\tau_{\bar{1}} = -d\tau_1 \tag{10.107}$$

Now consider the prediction, using Eq. (10.104) for this case. Note that

$$\mathbf{s}^{\bar{1}} = -\mathbf{s}^1, \quad \mathbf{n}^{\bar{1}} = \mathbf{n}^1 \tag{10.108}$$

$$\begin{aligned} d\tau_{\bar{1}} &= \mathbf{P}^{\bar{1}} : \mathbf{P}^1 d\tau \\ &= \left(\mathbf{s}^{\bar{1}} \cdot \mathbf{s}^1\right)\left(\mathbf{n}^{\bar{1}} \cdot \mathbf{n}^1\right) d\tau_1 \\ &= -d\tau_1 \end{aligned} \tag{10.109}$$

which is the Bauschinger effect given by Eq. (10.107). Therefore the Bauschinger effect, and hence kinematic hardening, is modeled by Eq. (10.104).

After exploring the features of the latent hardening coefficients $\lambda_{\alpha\beta}$, the attention is now turned to the determination of the self-hardening modulus $h_{\beta\beta}$ in Eq. (10.102). Weng (1987) suggested that the self-hardening modulus be associated with the latent hardening coefficients $\lambda_{\alpha\beta}$. He then proposed the following two forms for $h_{\beta\beta}$:

$$h_{\beta\beta} = \left[1 + \sum_{k=1}^{n} \left(1 - \lambda_{\beta k}\right)\right] h\left(\gamma^\beta\right) \tag{10.110}$$

$$h_{\beta\beta} = h\left(\gamma^\beta\right) + \sum_{k=1}^{n} h\left(\frac{\gamma^\beta}{-\lambda_{\beta k}}\right) \tag{10.111}$$

where no sum over β is implied, and the function h is calculated from the experimentally measured $\tau = \tau(\gamma)$ curve under single slip by

$$h(\gamma) = \frac{d\tau(\gamma)}{d\gamma} \tag{10.112}$$

In the case of single slip, Eqs. (10.101) and (10.110) reduce to

$$d\tau = h(\gamma)d\gamma \tag{10.113}$$

on the active slip systems ($\lambda_{\alpha\beta} = 1$), and Eqs. (10.101) and (10.111) will also reduce to Eq. (10.113) if $h(\infty)$ approaches zero.

In this section several latent hardening models have been discussed. All of these models (except Taylor's model) provide a higher latent hardening rate that is in accord with the experimental observations. Unfortunately, one problem caused by this higher latent hardening rate (i.e., off-diagonal terms are larger than the diagonal terms in the matrix $h_{\alpha\beta}$) is that the matrix $g_{\alpha\beta}$ or $k_{\alpha\beta}$ is not positive definite and therefore cannot be inverted to uniquely determine the incremental slip $d\gamma^{\beta}$'s. Actually this is an inherent difficulty of time-independent theory of crystalline plasticity (see Asaro and Needleman 1984). Generally an energy postulate (e.g., Taylor's minimum-work postulate) is necessary to select the combination of slip systems that conforms to a prescribed strain or stress increment. Even though the selection is still not unique (Havner 1981, 1985), consistent results can be obtained. It is for this reason of nonuniqueness that Asaro and Needleman (1984) suggested that a time-dependent theory should be used to determine the crystalline deformation. More detailed discussion about this matter is given in next chapter on the plasticity theory of polycrystals.

10.5 RIGID PLASTICITY; UNIAXIAL TENSION

The general elastic-plastic constitutive equations for multiple slip of single crystals have been given in Section 10.3. In this section we consider a rigid plasticity model that is a simplified version of the general theory. After deriving the equations for rigid plasticity, we will discuss the uniaxial tension of a single crystal specimen, using the derived equations with the condition of single slip. Then we will look at several interesting results that can be used to analyze the experimental data of the uniaxial tension.

Constitutive equations for rigid plasticity of single crystal can be obtained from the general theory established in Section 10.3 as follows: Substitution of Eq. (10.74) into Eq. (10.58) leads to

$$\mathring{\boldsymbol{\sigma}} + \boldsymbol{\sigma}\,\mathrm{tr}(\mathbf{D}) = \mathbf{C}:\mathbf{D} - \sum_{\alpha=1}^{n}\sum_{\beta=1}^{n}[\mathbf{C}:\mathbf{P}^{\alpha} + \boldsymbol{\beta}^{\alpha}]k_{\alpha\beta}^{-1}\mathbf{P}^{\beta}:\mathring{\boldsymbol{\sigma}} \tag{10.114}$$

multiplying both sides of this equation by $\mathbf{C}^{-1}$ gives

$$\mathbf{C}^{-1}:[\mathring{\boldsymbol{\sigma}} + \boldsymbol{\sigma}\,\mathrm{tr}(\mathbf{D})] = \mathbf{D} - \sum_{\alpha=1}^{n}\sum_{\beta=1}^{n}[\mathbf{P}^{\alpha} + \mathbf{C}^{-1}:\boldsymbol{\beta}^{\alpha}]k_{\alpha\beta}^{-1}\mathbf{P}^{\beta}:\mathring{\boldsymbol{\sigma}} \quad (10.115)$$

In the case of rigid plasticity, the elastic deformation is ignored. Therefore the equation reduces to

$$\mathbf{D} = \sum_{\alpha=1}^{n}\sum_{\beta=1}^{n}\mathbf{P}^{\alpha}k_{\alpha\beta}^{-1}\mathbf{P}^{\beta}:\mathring{\boldsymbol{\sigma}} \quad (10.116)$$

which is the constitutive equation for rigid plasticity. Note that in this case

$$\mathbf{D}^{\mathrm{e}} = 0, \quad \mathbf{D} = \mathbf{D}^{\mathrm{p}} \quad (10.117)$$

Equation (10.116) can be directly derived by substituting Eq. (10.74) into Eq. (10.32). In addition, the plastic spin tensor is given by

$$\mathbf{W}^{\mathrm{p}} = \sum_{\alpha=1}^{n}\sum_{\beta=1}^{n}\boldsymbol{\Omega}^{\alpha}k_{\alpha\beta}^{-1}\mathbf{P}^{\beta}:\mathring{\boldsymbol{\sigma}} \quad (10.118)$$

Note that generally

$$\mathbf{W} \neq \mathbf{W}^{\mathrm{p}} \quad (10.119)$$

This is because the crystal lattice may be subjected to rigid rotation even though the elastic deformation is infinitesimal and can be ignored.

For single slip, Eqs. (10.116) and (10.118) reduce to

$$\mathbf{D} = k_{11}^{-1}(\mathbf{P}^{1}:\mathring{\boldsymbol{\sigma}})\mathbf{P}^{1}; \qquad \mathbf{W}^{\mathrm{p}} = k_{11}^{-1}(\mathbf{P}^{1}:\mathring{\boldsymbol{\sigma}})\boldsymbol{\Omega}^{1} \quad (10.120)$$

where

$$\mathbf{P}^{1} = \tfrac{1}{2}(\mathbf{s}^{1}\mathbf{n}^{1} + \mathbf{n}^{1}\mathbf{s}^{1}), \quad \boldsymbol{\Omega}^{1} = \tfrac{1}{2}(\mathbf{s}^{1}\mathbf{n}^{1} - \mathbf{n}^{1}\mathbf{s}^{1}) \quad (10.121)$$

$$k_{11} = h - \mathbf{P}^{1}:\boldsymbol{\beta}^{1} = h - \mathbf{P}^{1}:(\boldsymbol{\Omega}^{1}\cdot\boldsymbol{\sigma} - \boldsymbol{\sigma}\cdot\boldsymbol{\Omega}^{1}) \quad (10.122)$$

Here $\mathbf{s}^1$ and $\mathbf{n}^1$ are the unit vector in the slip direction and the unit normal vector to the slip plane, respectively; h is the experimentally obtained hardening modulus in single slip, as defined by Eq. (10.113).

Now consider the deformation of a single crystal specimen subjected to uniaxial tension. Figure 10.11 shows a single crystal specimen under uniaxial tension. Due to the slip on the slip plane, the lattice will rotate, as previously explained. However, if the lattice is imagined to be held fixed, the loading axis will rotate relative to the slip plane. It is shown in the figure that the

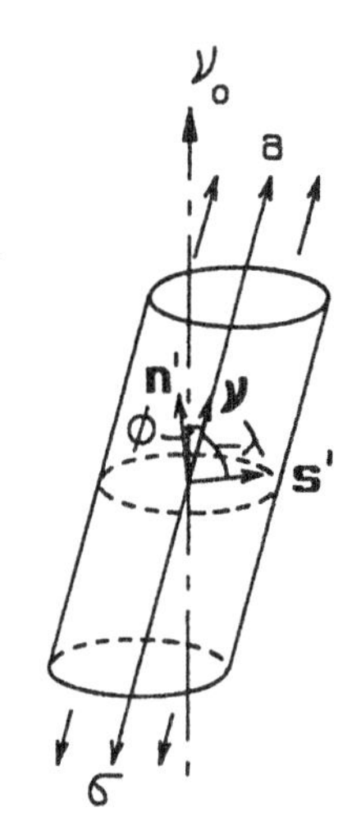

Fig. 10.11 Uniaxial tension

loading axis, denoted by $\boldsymbol{\nu}$, rotates relative to the unit slip direction vector $\mathbf{s}^1$ and unit normal to the slip plane $\mathbf{n}^1$.

In this case the Cauchy stress tensor can be expressed as

$$\boldsymbol{\sigma} = \sigma \boldsymbol{\nu}\boldsymbol{\nu}, \qquad \sigma > 0 \tag{10.123}$$

where $\boldsymbol{\nu}$ is the unit vector in the loading direction, as shown in Fig. 10.11. Since in the uniaxial tension experiment the specimen is restrained by the grips of the loading machine, the total rotation of the continuum is zero. Therefore

$$\mathbf{W} = \mathbf{0}, \qquad \overset{\circ}{\boldsymbol{\sigma}} = \dot{\boldsymbol{\sigma}} \tag{10.124}$$

Using Eqs. (10.121)–(10.124), we calculate the following results:

$$\boldsymbol{\beta}^1 = \boldsymbol{\Omega}^1 \cdot \boldsymbol{\sigma} - \boldsymbol{\sigma} \cdot \boldsymbol{\Omega}^1 = 2\boldsymbol{\Omega}^1 \cdot \boldsymbol{\sigma} = \sigma\left[\mathbf{s}^1\boldsymbol{\nu}(\mathbf{n}^1 \cdot \boldsymbol{\nu}) - \mathbf{n}^1\boldsymbol{\nu}(\mathbf{s}^1 \cdot \boldsymbol{\nu})\right] \tag{10.125}$$

$$\mathbf{P}^1 : \boldsymbol{\beta}^1 = \frac{\sigma}{2}\left[(\mathbf{n}^1 \cdot \boldsymbol{\nu})^2 - (\mathbf{s}^1 \cdot \boldsymbol{\nu})^2\right] \tag{10.126}$$

$$k_{11} = h - \frac{\sigma}{2}\left[(\mathbf{n}^1 \cdot \boldsymbol{\nu})^2 - (\mathbf{s}^1 \cdot \boldsymbol{\nu})^2\right] \tag{10.127}$$

$$\mathbf{P}^1 : \overset{\circ}{\boldsymbol{\sigma}} = \sigma(\mathbf{n}^1 \cdot \boldsymbol{\nu})(\mathbf{s}^1 \cdot \boldsymbol{\nu}) \tag{10.128}$$

By using these relations, we can easily obtain the relation between the increment of the axial stress σ and the axial strain $\dot{\epsilon}$:

$$\dot{\sigma} = \left[\frac{h}{\cos^2\lambda \cos^2\phi} - \frac{\sigma(\cos^2\lambda - \cos^2\phi)}{2\cos^2\lambda\cos^2\phi}\right]\dot{\boldsymbol{\epsilon}} \tag{10.129}$$

where

$$\dot{\epsilon} = \boldsymbol{\nu} \cdot \mathbf{D} \cdot \boldsymbol{\nu} \tag{10.130}$$

is the axial component of the deformation rate $\mathbf{D}$, and λ and ϕ are the angles between the loading direction and the slip direction, and the normal to the slip plane, respectively, as shown in Fig. 10.11.

Equation (10.129) reveals a very interesting phenomenon in crystalline deformation. It is seen that the effective modulus in uniaxial tension of a single crystal specimen for single slip cases consists of two terms. The first term includes the hardening modulus h and is always positive for hardening materials ($h > 0$). However, the second term is determined by the relative orientation of the slip system with respect to the loading axis. Since the orientation of the slip system will change due to the plastic deformation, this term may be positive, negative, or zero depending on the relative orientation. Furthermore, if the lattice rotates to such an angle that the condition

$$\cos^2 \lambda - \cos^2 \phi > \frac{2h}{\sigma} \tag{10.131}$$

holds, the effective hardening modulus will become negative. This indicates the axial stress decreases with the axial extension (i.e., the material softens although $h > 0$). Since this softening is caused by the geometrical effect (the rotation of the lattice), not by the material property, it is called *geometrical softening*.

The relationship between axial stress and shear stress on the slip system have already been derived before in general cases using Schmid's law [see Eqs. (10.63) and (10.65)]. In uniaxial tension

$$\tau^1 = \sigma \cos \lambda \cos \phi, \quad \dot{\tau}^1 = \dot{\sigma} \cos \lambda \cos \phi \tag{10.132}$$

The relation between the axial strain and the slip on the slip system are given as

$$\begin{aligned} \dot{\epsilon} &= \boldsymbol{\nu} \cdot \mathbf{D} \cdot \boldsymbol{\nu} = \boldsymbol{\nu} \cdot \mathbf{P}^1 \cdot \boldsymbol{\nu} \dot{\gamma}^1 \\ &= \dot{\gamma}^1 \cos \lambda \cos \phi \end{aligned} \tag{10.133}$$

From Eqs. (10.132) and (10.133) the work rate per unit volume by the axial stress σ is given by

$$\dot{w} = \sigma \dot{\epsilon} = \tau^1 \dot{\gamma}^1 \tag{10.34}$$

This equation can be generalized to the case of multislip

$$\dot{w} = \boldsymbol{\sigma} : \dot{\boldsymbol{\epsilon}} = \sum_{\alpha=1}^{n} \tau^\alpha \dot{\gamma}^\alpha \tag{10.135}$$

which implies that the work rate done in terms of axial stress σ and the axial strain rate $\dot{\epsilon}$ equals to the sum of the work rates done based on τ^α and γ^α on all the slip systems.

To obtain a relation between the axial extension and the total slip γ^1, Eq. (10.133) has to be integrated. Note that $\dot{\epsilon} = \dot{l}/l$; here l is the current length of the specimen. Equation (10.133) is then rewritten as

$$\frac{dl}{l} = \cos\lambda \cos\phi d\gamma^1 \tag{10.136}$$

The integration cannot be performed at this stage, since $\cos\lambda$ and $\cos\phi$ will change during the deformation, and the variation of them with deformation has not yet been established. This variation is obtained next.

For the specimen shown in Fig. 10.11, the displacement of any material point, denoted by $\mathbf{X}$, is only caused by the slip on the slip plane, since the elastic deformation is ignored and the slip plane is assumed to be fixed by the loading machine. Therefore

$$\mathbf{u} = \gamma^1 \mathbf{s}^1 \mathbf{n}^1 \cdot \mathbf{X} \tag{10.137}$$

Then, denoting the initial length l_0 of the specimen by a vector,

$$\mathbf{l}_0 = l_0 \boldsymbol{\nu}_0 \tag{10.138}$$

the current length (after deformation) can be expressed by

$$\mathbf{l} = l\boldsymbol{\nu} = \mathbf{l}_0 + \gamma^1 \mathbf{s}^1 \mathbf{n}^1 \cdot \mathbf{l}_0 \tag{10.139}$$

Using Eqs. (10.138) and (10.139), we obtain

$$\mathbf{s}^1 \cdot \mathbf{l} = l\cos\lambda = l_0 \cos\lambda_0 + l_0 \gamma^1 \cos\phi_0 \tag{10.140}$$

$$\mathbf{n}^1 \cdot \mathbf{l} = l\cos\phi = l_0 \cos\phi_0 \tag{10.141}$$

Therefore

$$\cos\lambda = \frac{l_0}{l}(\cos\lambda_0 + \gamma^1 \cos\phi_0) \tag{10.142}$$

$$\cos\phi = \frac{l_0}{l}\cos\phi_0 \tag{10.143}$$

Substituting Eqs. (10.142) and (10.143) into Eq. (10.136), we get

$$l dl = l_0^2(\cos\phi_0 \cos\lambda_0 + \cos^2\phi_0 \gamma^1) d\gamma^1 \tag{10.144}$$

Since $\cos \phi_0$ and $\cos \lambda_0$ are constant during deformation, the equation above can now be integrated to give

$$l^2 - l_0^2 = 2l_0^2 \cos \phi_0 \cos \lambda_0 \gamma^1 + l_0^2 \cos^2 \phi_0 (\gamma^1)^2 \qquad (10.145)$$

or

$$\left(\frac{l}{l_0}\right)^2 = 1 + 2\cos \phi_0 \cos \lambda_0 \gamma^1 + \cos^2 \phi_0 (\gamma^1)^2 \qquad (10.146)$$

This is an expression for the extension l/l_0 in terms of the slip on the slip plane. For a given slip system, $\cos \phi_0$ and $\cos \lambda_0$ are known, and the current length of the specimen can be obtained if the slip is known. Eq. (10.146) can be inverted to give

$$\gamma^1 = \frac{1}{\cos \phi_0} \left\{ \sqrt{\left(\frac{l}{l_0}\right)^2 - \sin^2 \lambda_0} - \cos \lambda_0 \right\} \qquad (10.147)$$

which can be used to calculate the slip on the slip system provided that the axial extension and the initial orientation of the slip system are known.

REFERENCES

Asaro, R. J. 1983. Micromechanics of crystals and polycrystals. *Advances in Applied Mechanics*, vol. 23. San Diego: Academic Press.

Asaro, R. J., and J. R. Rice. 1977. Strain localization in ductile single crystals. *J. Mech. Phys. Solid* **25**:309.

Bragg, W. H., and W. L. Bragg. 1933. *The Crystalline State*. London: Bell.

~~Burke, E. C. and W. R. Hibbard. 1952. *Trans. AIME*. **194**:295.~~

Chang, Y. W., and R. J. Asaro. 1981. An experimental study of shear localization in aluminum-copper single crystals. *Acta Metall.* **29**:241.

Ewing, J. A., and W. Rosenhain. 1900. The crystalline structure of metals. *Phil. Trans. R. Soc., London* **193**:353.

Fuh, S., and K. S. Havner. 1986. On uniqueness of multiple-slip solutions in constrained and unconstrained F.C.C. crystal deformation problems. *Int. J. Plast.* **2**:329.

Havner, K. S. 1984. First- and second-order analysis of axially loaded crystals in n-fold symmetry. *Phil. Trans. R. Soc., London* **A311**:46.

Havner, K. S. 1981. A theoretical analysis of finitely deforming F.C.C. crystals in the sixfold symmetry position. *Proc. R. Soc., London* **A378**:329.

Havner, K. S., G. S. Baker, and R. F. Vause. 1979. Theoretical latent hardening in crystals. Part I: General equations for tension and compression with application of F.C.C. crystals in tension. *J. Mech. Phys. Solid* **27**:33.

Havner, K. S., and A. H. Shalaby. 1977. A simple mathematical theory of finite distortional latent hardening in single crystal. *Proc. R. Soc., London* **A358**:47.

Hill, R. 1966. Generalized constitutive relations for incremental deformation of metals by multislip. *J. Mech. Phys. Solid* **14**:99.

Hill, R., and K. S. Havner. 1982. Perspectives in the mechanics of elastoplastic crystals. *J. Mech. Phys. Solid* **30**:5.

Hill, R., and J. R. Rice. 1972. Constitutive analysis of elasto-plastic crystals at arbitrary strain. *J. Mech. Phys. Solid* **20**:401.

Jackson, P. J., and Z. S. Basinski. 1967. Latent hardening and the flow stress in copper single crystals. *Can. J. Phys.* **45**:707.

Kocks, U. F. 1970. The relation between polycrystal deformation and single crystal deformation. *Metall. Trans.* **1**:1121.

Kocks, U. F., and T. J. Brown. 1966. Latent hardening in aluminum. *Acta Metall.* **9**:155.

Lee, E. H. 1966. Elastic-plastic waves of one-dimensional strain. *Proc. 5th U.S. Nat. Congr. Appl. Mech.*, ASME, p. 405.

Mitchell, T. E. 1964. Dislocations and plasticity in single crystals of face-centered cubic metals and alloys. *Prog. Appl. Mat. Res.* **6**:119.

Mott, N. F. 1952. A theory of work-hardening of metal crystals. *Phil. Mag.* **43**:1151.

Nakada, Y., and A. S. Keh. 1966. Latent hardening in iron single crystals. *Acta Metall.* **14**:961.

Peirce, D., R. J. Asaro, and A. Needleman. 1983. An analysis of non-uniform and localized deformation in ductile single crystals. *Acta Metall.* **31**:1951.

Polanyi, von M. 1922. Rontgenographische Bestimmung von Ksistallanordnunge. *Naturwissenschaften* **10**:411.

Schmid, E. 1924. *Proc. Int. Congr. Appl. Mech.* (Delft), p. 342.

Taylor, G. I. 1938. Plastic strain in metals. *J. Inst. Metals* **62**:307.

Taylor, G. I. 1934. The mechanism of plastic deformation of crystals. *Proc. R. Soc., London* **A165**:362.

Taylor, G. I., and C. F. Elam. 1925. The plastic extension and fracture of aluminum crystals. *Proc. Roy. Soc.* **A108**:28.

Taylor, G. I., and C. F. Elam. 1935. The distortion of an aluminum crystal during a tensile test. *Proc. Roy. Soc.* **A102**:647.

Weng, G. J. 1987. Anisotropic hardening in single crystals and the plasticity of polycrystals. *Int. J. Plast.* **3**:315.

Weng, G. J. 1979. Kinematic hardening rule in single crystals. *Int. J. Solids Struct.* **15**:861.

11

POLYCRYSTAL PLASTICITY THEORY

A polycrystal is an aggregate of a great number of single-crystal grains with different orientations. Polycrystal plasticity relates the macromechanical behavior of polycrystal metals to the fundamental mechanisms of single-crystal deformation. The responses of a polycrystal metal to an external load are derived by using an appropriate average procedure based on the knowledge of the behavior of single crystals under loading. The polycrystal plasticity theory is sometimes called the *physical theory of plasticity* because it is based on the consideration of the physics of the single-crystal plastic deformation.

Several objectives can be achieved by the study of polycrystal plasticity. First of all, as pointed out by Bishop and Hill (1951), the main micromechanisms that are responsible for the observed macroscopic phenomenon of plastic deformation can be identified; second, based on this understanding of the causes for plastic deformation, more reasonable modifications than indicated by limited experimental observations can be added to the mathematical theory of plasticity. This in turn can provide better predictions for the macroscopic plastic deformation of metals. In addition several phenomena can be described only by the polycrystal plasticity theory. Included among these are such items as internal stress, the texture development due to large plastic deformation, and the effect of grain size and shape.

The primary task of a polycrystal plasticity theory is to formulate the relations between the macroscopic and microscopic quantities, such as stress and strain. Early studies were originated by Sachs (1928), Cox and Sopmith (1937), and Taylor (1938). The basic assumption used by Sachs (1928) is that all single-crystal grains with aggregate or polycrystal experience the same state of stress. In this model the equilibrium condition across the grain boundaries is satisfied, while the compatibility condition between grains is violated. On the other hand, the basic assumption used by Taylor (1938) is that all grains within the aggregate experience the same state of deformation. The compatibility condition is ensured, but the equilibrium across the grain boundaries is violated in Taylor's model. To ensure compatibility as well as equilibrium between the grains, a self-consistent model was proposed by Kroner (1958, 1961) and extended by Budiansky and Wu (1962). In this self-consistent model the interaction between a single-crystal grain and other parts of the aggregate is approximately determined by regarding it as an elastic-plastic spherical inclusion embedded in a homogeneous, isotropic elastic infinite body, and solving the problem using Eshelby's solution (Eshelby 1957). A simple relation between the overall stress and strain, defined for the aggregate material, and the local stress and strain, defined within each single crystal grain, is established which conveys the information from microscale to macroscale. This self-consistent model was extended later on by Hill (1965a, 1965b) and Lin (1966, 1974, 1984). Hill introduced the concept of a constraint tensor for the aggregate to the single crystal grain and considered the elastic-plastic properties of the aggregate, which are simply assumed to be elastic in Kroner, Budiansky, and Wu's model. Lin, on the other hand, introduced the concept of equivalent body force and formulated the relation between the overall stress and strain and the local stress and strain in an integrated equation using the green function. Lin's formulation is more accurate but the calculation is more complicated. A modified version of Kroner, Budiansky, and Wu's model was proposed by Berveiller and Zaoui (1979). In deriving the constraint tensor as in Hill's self-consistent model, they used a reasonable, though not rigorous, approximation that characterizes the elastic-plastic behavior of the aggregate as isotropic. This model greatly reduced the complexity of Hill's model, since it assumed plastic isotropy and generalized the Kroner, Budiansky, and Wu model by introducing a scalar plastic accommodation function.

A systematic study of polycrystal plasticity at large finite deformation was more recently done by Iwakuma and Nemat-Nasser (1984) and Lipinski and Berveiller (1989). Iwakuma and Nemat-Nasser (1984) generalized Hill's self-consistent model, while Lipinski and Berveiller (1989) presented a systematic approach to the problem of determination of elastic-plastic properties of metallic polycrystals at finite deformation using the Green function.

In this chapter, the polycrystal models proposed by Taylor (1938), Kroner (1958, 1961), Budiansky and Wu (1962), Hill (1965a, 1965b), and Berveiller and Zaoui (1979) are discussed. Attention will focus on the formulation of the relation between the overall or macroscopic stress and strain and the local or microscopic stress and strain. Several calculated results will be presented to show the relative merits of each model. Since the determination of slip systems of a single crystal subjected to the prescribed loading or deformation is essential to every polycrystal model, slip systems will be studied prior to examining the polycrystal plasticity models.

11.1 DETERMINATION OF SLIP SYSTEMS

In Chapter 10 we mentioned the determination of slip systems but did not extensively discuss them. In Eqs. (10.73) and (10.74) the slips were formally expressed in terms of the deformation rate or the objective stress rate for which the matrices $k_{\alpha\beta}$ and $g_{\alpha\beta}$ were assumed to be positive definite. However, as we noted in the last chapter, the slip systems usually are not uniquely defined by a given deformation rate or an objective stress rate, and some other criterion, such as the minimum work postulate proposed by Taylor (1938), is generally required to select the appropriate combination of slip systems that corresponds to the prescribed deformation or stress rate. The selection procedure is presented in detail in this section. To make the discussion specific, we consider only FCC crystals. Therefore there are 12 (if the reverse and forward slip systems are considered as the same and thus are not considered as independent from each other) or 24 (if forward and reverse slip systems are considered independent) active slip systems when the crystal is subjected to a given deformation rate or stress rate. In this section we explain how to select the set of slip systems from the total 12 (or 24) slip systems due to the prescribed deformation rate or stress rate. The rigid plasticity model and the elastic-plastic models will be covered separately.

11.1.1 Selection of Slip Systems for Rigid Plasticity Models

For a large finite deformation, the elastic deformation rate is usually small compared to the plastic deformation rate and thus can be ignored. The reasoning behind this is as follows: Since the elastic deformation is confined by the ratio of stress to elastic modulus, for perfectly plastic materials the equivalent stress is equal to the initial yield stress. However, for most metals, the initial yield stress is two or three orders less than the elastic (Young's) modulus. As a result the elastic deformation in such cases is two or three orders less than 1, which is the order of the large finite plastic deformation. Even for work-hardening or strain-hardening materials the subsequent yield stress is in the same order as the initial yield stress, while the elastic strain remains two or three orders less than the large finite plastic strain. The material behavior can be reasonably modeled by the rigid plasticity model as characterized by

$$\mathbf{D} = \mathbf{D}^{p} = \sum_{\alpha=1}^{n} \mathbf{P}^{\alpha}\dot{\gamma}^{\alpha} \tag{11.1}$$

where the summation is taken over all the active slip systems and n is less or equal to 12 (or 24). Note that $\mathbf{D}$ can be expressed by six components, since it is a symmetric tensor. In addition only five of the six components are independent because of the incompressibility condition of the plastic deformation caused by the dislocation slip

$$\mathrm{tr}(\mathbf{D}) = D_{ii} = 0 \tag{11.2}$$

Therefore the number of possible active slip systems is larger than the independent components of the plastic deformation rate. Mathematically any combination of five independent slip systems that satisfy Eq. (11.2) can cause the prescribed deformation rate. Obviously more than one possibility exists that can satisfy Eq. (11.2). The selection of slip systems for a prescribed deformation rate is described next.

Minimum Work Principle To determine the combination of shears or slips that will occur when any prescribed strain is induced on a crystal by an external agency, Taylor (1938) proposed the following minimum work, or least possible energy dissipation, principle: of all possible combinations of the 12 shears that can produce the assigned strain, only that combination for which the energy dissipation is least is operative.

Let $\dot{\gamma}^\alpha$ be the actually activated slips or shears that produce the forced deformation rate $\mathbf{D}$, and let $\dot{\gamma}_*^\alpha$ be any set of slips that are geometrically equivalent to the prescribed $\mathbf{D}$ [$\dot{\gamma}_*^\alpha$ satisfy Eq. (11.2)] but are not actually operated by the corresponding stress that satisfies the yield condition (10.62). Taylor's minimum work principle can then be expressed mathematically as

$$\sum_{\alpha=1}^{n} \tau_c^\alpha \dot{\gamma}^\alpha \le \sum_{\alpha=1}^{n} \tau_{c*}^\alpha \dot{\gamma}_*^\alpha \tag{11.3}$$

where τ_c^α is the current shear strength of the αth actually activated slip system that is compatible with the externally applied stress and τ_{c*}^α is the current shear strength of the αth geometrically possible slip system that may not be compatible with the externally applied stress. In Taylor's work the slip systems are assumed to harden at the same rate. Therefore τ_c^α and τ_{c*}^α are the same for all the slip systems. Equation (11.3) is hence equivalent to

$$\sum_{\alpha=1}^{n} |\dot{\gamma}^\alpha| \le \sum_{\alpha=1}^{n} |\dot{\gamma}_*^\alpha| \tag{11.4}$$

where the absolute values of slips are used since the reverse shear is considered as the same slip system as the forward one and the total number of slip systems is 12. Since Eqs. (11.3) and (11.4) are equivalent in Taylor's model under the assumption of isotropic hardening, Taylor's minimum work principle can be restated as follows: of all possible combinations of the 12 shears that can produce an assigned strain, only that combination is operative for which the sum of the absolute values of shears is least. Because of this statement and Eq. (11.4), Taylor's minimum work principle is generally called the *minimum shear principle* or simply *minimum principle*.

It should be noted that the minimum shear principle proposed by Taylor is only a postulate and has no priori justification. Taylor proposed this postulate based on the analysis of the dynamics of a nonconservative mechanical system consisting of several boards. The minimum principle was rigorously proved by Bishop and Hill (1951) as follows:

Since both $\dot{\gamma}^\alpha$ and $\dot{\gamma}_*^\alpha$ are geometrically equivalent to **D**,

$$\mathbf{D} = \sum_{\alpha=1}^{n} \mathbf{P}^\alpha \dot{\gamma}^\alpha = \sum_{\alpha=1}^{n} \mathbf{P}_*^\alpha \dot{\gamma}_*^\alpha \tag{11.5}$$

Multiplying the equation by the Cauchy stress tensor yields

$$\sum_{\alpha=1}^{n} \tau_\alpha \dot{\gamma}^\alpha = \sum_{\alpha=1}^{n} \tau_{\alpha *} \dot{\gamma}_*^\alpha \tag{11.6}$$

where

$$\tau_\alpha = \mathbf{P}^\alpha : \boldsymbol{\sigma}, \quad \tau_{\alpha *} = \mathbf{P}_*^\alpha : \boldsymbol{\sigma} \tag{11.7}$$

and the resolved shear stress or Schmid stress on the corresponding slip systems. It should be noted that Eq. (11.7) is different from Eq. (11.3). In Eq. (11.3) τ_c^α and τ_{c*}^α are the current yield stresses on slip systems and are material properties, while τ_α and $\tau_{\alpha *}$ in Eq. (11.7) are Schmid stresses and are dependent on the applied stress $\boldsymbol{\sigma}$.

Since $\dot{\gamma}^\alpha$'s are actually operated slips, but $\dot{\gamma}_*^\alpha$ are not; τ_α and $\tau_{\alpha *}$ satisfy the following relation:

$$\tau_\alpha = \tau_c^\alpha = \tau_c, \qquad \tau_{\alpha *} \le \tau_c \tag{11.8}$$

where the same hardening rate is assumed for all the slip systems. Therefore

$$\sum_{\alpha=1}^{n} \tau_\alpha \dot{\gamma}^\alpha = \tau_c \sum_{\alpha=1}^{n} |\dot{\gamma}^\alpha| \tag{11.9}$$

$$\sum_{\alpha=1}^{n} \tau_{\alpha *} \dot{\gamma}_*^\alpha \le \tau_c \sum_{\alpha=1}^{n} |\dot{\gamma}_*^\alpha| \tag{11.10}$$

Substitution of Eqs. (11.9) and (11.10) into Eq. (11.6) immediately leads to Taylor's minimum principle (11.4). Taylor's minimum principle will now be used to determine the combination of slip systems that will occur when a prescribed deformation rate is forced on a crystal by an external agency.

Single Slip The simplest case of uniaxial tension or compression is considered first. In this simplest case the only independent deformation rate is the axial strain rate given by

$$\dot{\epsilon} = \mathbf{l} \cdot \mathbf{D} \cdot \mathbf{l} \tag{11.11}$$

where $\mathbf{l}$ is the unit vector in the loading direction. Obviously only one of the 12 slip systems could be operative to satisfy the geometrical restriction of Eq. (11.1). This shear or slip can be easily calculated by using Eqs. (11.11) and (11.5):

$$\dot{\gamma} = \frac{\dot{\epsilon}}{\cos\lambda \cos\phi} \tag{11.12}$$

where λ is the angle between the slip direction and loading axis and ϕ is the angle between the slip plane normal and loading axis, respectively, as defined in the previous chapter. According to Taylor's minimum principle, the actual operative slip $\dot{\gamma}$ of the 12 slip systems is the one for which $\dot{\gamma}$ is the least when $\dot{\epsilon}$ is prescribed. From Eq. (11.12) we can see that the least $\dot{\gamma}$ can be achieved if the Schmid factor, $\cos\lambda \cos\phi$, for this system is the greatest. Since $\tau = \sigma \cos\lambda \cos\phi$, the condition that $\dot{\gamma}$ should be the least is equivalent to the condition that the resolved shear stress or Schmid stress on the system should be the greatest. This is actually Schmid's law. It is shown here that it can be derived from Taylor's minimum principle. The active slip system in single slip (corresponding to the simplest uniaxial loading case) is thus determined according to Schmid's law. Since

$$\frac{\sigma}{\tau} = \frac{\dot{\gamma}}{\dot{\epsilon}} = \frac{1}{\cos\lambda \cos\phi} = m \tag{11.13}$$

it follows that

$$\sigma = m\tau(\gamma) = m\tau(m\epsilon) \tag{11.14}$$

where $\tau(\gamma)$ is the resolved stress-resolved shear or slip curve. It should be noted that this equation is derived based on the assumption that m is independent of the amount of strain. This is only an approximation, since m is actually a function of strain, as discussed at the end of Chapter 10.

Multislip Now consider the general case where the deformation rate is arbitrarily given. Due to incompressibility the number of independent components of the deformation rate is five. The determination of the combination of shears or slips that are geometrically equivalent to the prescribed deformation rate can be achieved as follows:

Consider an FCC crystal, and take the axes of the coordinates to coincide with the crystal axes, as shown in Fig. 11.1. In this coordinate system the normals to the 4 slip planes and the 12 slip directions in the positive sense are given in Table 10.1. Note that vectors in Table 10.1 are not unit vectors. They have to be normalized to provide unit vectors.

Let $\dot{\gamma}_{a1}, \dot{\gamma}_{a2}, \ldots, \dot{\gamma}_{d3}$ represent the slip rates on the slip systems $a_1, a_2, \ldots, d_3$, respectively. The following relations between the components of the deformation rate $\mathbf{D}$ and these slip rates can be easily obtained by using Eq. (11.1):

$$2\sqrt{6}\, D_{xy} = 2\sqrt{6}\, \mathbf{e}_x \cdot \mathbf{D} \cdot \mathbf{e}_y$$

$$= -\dot{\gamma}_{a1} + \dot{\gamma}_{a2} - \dot{\gamma}_{b1} + \dot{\gamma}_{b2} + \dot{\gamma}_{c1} - \dot{\gamma}_{c2} + \dot{\gamma}_{d1} - \dot{\gamma}_{d2} \qquad (11.15a)$$

$$2\sqrt{6}\, D_{yz} = 2\sqrt{6}\, \mathbf{e}_y \cdot \mathbf{D} \cdot \mathbf{e}_z$$

$$= -\dot{\gamma}_{a2} + \dot{\gamma}_{a3} + \dot{\gamma}_{b2} - \dot{\gamma}_{b3} - \dot{\gamma}_{c2} + \dot{\gamma}_{c3} + \dot{\gamma}_{d2} - \dot{\gamma}_{d3} \qquad (11.15b)$$

$$2\sqrt{6}\, D_{zx} = 2\sqrt{6}\, \mathbf{e}_z \cdot \mathbf{D} \cdot \mathbf{e}_y$$

$$= -\dot{\gamma}_{a3} + \dot{\gamma}_{a1} + \dot{\gamma}_{b3} - \dot{\gamma}_{b1} + \dot{\gamma}_{c3} - \dot{\gamma}_{c1} - \dot{\gamma}_{d3} + \dot{\gamma}_{d1} \qquad (11.15c)$$

$$\sqrt{6}\, D_{xx} = \sqrt{6}\, \mathbf{e}_x \cdot \mathbf{D} \cdot \mathbf{e}_x = \dot{\gamma}_{a2} - \dot{\gamma}_{a3} + \dot{\gamma}_{b2} - \dot{\gamma}_{b3} + \dot{\gamma}_{c2} - \dot{\gamma}_{c3} + \dot{\gamma}_{d2} - \dot{\gamma}_{d3} \qquad (11.15d)$$

$$\sqrt{6}\, D_{yy} = \sqrt{6}\, \mathbf{e}_y \cdot \mathbf{D} \cdot \mathbf{e}_y = \dot{\gamma}_{a3} - \dot{\gamma}_{a1} + \dot{\gamma}_{b3} - \dot{\gamma}_{b1} + \dot{\gamma}_{c3} - \dot{\gamma}_{c1} + \dot{\gamma}_{d3} - \dot{\gamma}_{d1} \qquad (11.15e)$$

$$\sqrt{6}\, D_{zz} = \sqrt{6}\, \mathbf{e}_z \cdot \mathbf{D} \cdot \mathbf{e}_z = \dot{\gamma}_{a1} - \dot{\gamma}_{a2} + \dot{\gamma}_{b1} - \dot{\gamma}_{b2} + \dot{\gamma}_{c1} - \dot{\gamma}_{c2} + \dot{\gamma}_{d1} - \dot{\gamma}_{d2} \qquad (11.15f)$$

where $\mathbf{e}_x$, $\mathbf{e}_y$, and $\mathbf{e}_z$ are unit vectors in the x, y, and z axes, respectively.

The Equations (11.15a–f) can be easily verified by considering the $\dot{\gamma}_{c3}$ terms and calculating as follows: For system c_3 the unit vector in the slip direction and the unit normal vector to the slip plane are (see Table 10.1)

$$\mathbf{n} = \frac{1}{\sqrt{3}}(-1,1,1), \quad \mathbf{s} = \frac{1}{\sqrt{2}}(1,1,0)$$

The contribution of shear on system c_3 to the deformation rate $\mathbf{D}$ is

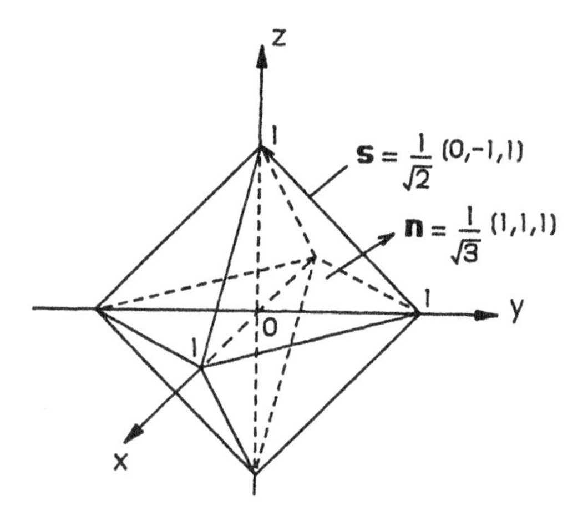

Fig. 11.1 Crystal axes and slip systems

calculated as

$$\frac{1}{2}(\mathbf{sn} + \mathbf{ns})\dot{\gamma}_{c3} = \frac{\dot{\gamma}_{c3}}{2\sqrt{6}}\begin{bmatrix} -2 & 0 & 1 \\ 0 & 2 & 1 \\ 1 & 1 & 0 \end{bmatrix} \tag{11.16}$$

From this result, and noting that $\mathbf{e}_x = (1, 0, 0)$, $\mathbf{e}_y = (0, 1, 0)$, and $\mathbf{e}_z = (0, 0, 1)$, the contribution of $\dot{\gamma}_{c3}$ to Eq. (11.15) can be obtained. Similarly the contribution of shear rate on other slip systems can be calculated.

Equations (11.15a–f) provide 6 relations between the components of the deformation rate **D** and the 12 shear or slip rates on 12 possible slip systems. Since the plastic deformation caused by shear is incompressible, Eq. (11.2) holds. Therefore only 5 independent relations can be obtained from Eqs. (11.15a–f). Even though there are 12 shear rates in Eqs. (11.15a–f), only 5 relations are sufficient to uniquely decide the actual combination of 5 shears among the 12 possible slip systems.

Mathematically, 5 independent shear rates are necessary in order to satisfy the 5 relations provided by Eqs. (11.15a–f). It was pointed out by Taylor (1937) that the minimum principle would correspond to a combination of slip systems in which 7 of the 12 shear rates were assumed to be 0. It is necessary therefore to solve Eqs. (11.15a–f) for a given deformation rate **D** using all possible combinations of the 5 remaining shear rates. The valid combination of shear rates is the one for which the sum of their absolute values is minimum. This combination of the shear rates should be chosen as the solution to the prescribed deformation rate **D**.

The procedure to determine the slip combination for a given **D**, as discussed above, was first proposed by Taylor (1937, 1938). It was a formidable task because there are 792 ways in which 5 of 12 shears can be chosen.

Taylor (1937, 1938) found that these 792 ways were not independent of each other. For example, of three slip systems on the same slip plane, only two are independent, and the third one can be derived from these two because only two independent vector bases lie on a plane. It is easy to show that

$$\mathbf{a}_1 = -\tfrac{1}{2}(\mathbf{a}_2 + \mathbf{a}_3) \tag{11.17}$$

where $a_1 = (0, -1, 1)$, $a_2 = (1, 0, -1)$, and $a_3 = (-1, 1, 0)$, as given in Table 10.1. According to Taylor (1937, 1938) the number of independent shear combinations that have to be considered to obtain a solution to a given deformation rate can be eventually reduced to 96. The details of this reduction from 792 to 96 are omitted here, but the interested reader can refer to Taylor's work (1937, 1938).

In summary, to determine the actual combination of shear rates that produces a prescribed deformation rate **D**, 5 shear or slip rates out of 12 possibilities have to be selected that satisfy Eqs. (11.5a–f). This calculation must be repeated many times to cover the 96 independent combinations to obtain all possible solutions. Finally, the actual solution(s) must satisfy Taylor's minimum shear principle; that is, they must be the least sum of the absolute values of the shear rates.

It should be pointed out that although Taylor's minimum shear principle provides a criterion with which we can select the actual combination of shears from a great number of possibilities, it does not ensure that the selection is unique. Actually it is not unique; it is possible to find 2 or more combinations of 5 shear rates that satisfy Eqs. (11.15a–f) and have the same minimum sum of the absolute values of the 5 shear rates.

11.1.2 Selection of Slip Systems for Elastic Plastic Materials

For elastic-plastic materials the total deformation rate **D** no longer equals the plastic deformation rate, since the elastic deformation rate is significant to the plastic deformation rate and cannot be ignored. Therefore Eq. (11.1) and Eqs. (11.15a–f) cannot be used to determine the combination of slip systems for such materials.

Consider the case where the deformation rate **D** is prescribed to the elastic-plastic crystal by an external agent. As discussed in Chapter 10, the constitutive equations that have to be satisfied by both active and inactive slip systems are the following [see Eqs. (10.71) and (10.72)]:

For $\dot{\gamma}^\alpha = 0$,

$$\mathbf{P}^\alpha : (\mathbf{C} - \sigma\mathbf{I}) : \mathbf{D} < \sum_{\beta=1}^{n} g_{\alpha\beta}\dot{\gamma}^\beta \quad (\alpha = 1, 2, \ldots, n) \tag{11.18}$$

For $\dot{\gamma}^\alpha > 0$,

$$\mathbf{P}^\alpha : (\mathbf{C} - \sigma\mathbf{I}) : \mathbf{D} = \sum_{\beta=1}^{n} g_{\alpha\beta}\dot{\gamma}^\beta \quad (\alpha = 1, 2, \ldots, n) \tag{11.19}$$

Note that in these expressions the slip systems in the reverse direction and the forward direction are considered as two separate slip systems so that $\dot{\gamma}^\alpha > 0$ whenever the system is operative. Otherwise, if the reverse slip and forward slip are considered as the same system, $\dot{\gamma}^\alpha > 0$ should be replaced by $|\dot{\gamma}^\alpha| > 0$ whenever a slip system is operative. Here, for computational convenience, the forward and reverse slips are considered to be different and to total 24 slip systems for FCC crystals.

Because of the incompressibility constraint (11.2), only 5 independent slip rates are necessary to determine arbitrary plastic deformation rates. Equation (11.19) provides 5 relations between the components of total deformation rate **D** and the 5 independent slip rates. It is obvious that choosing 5 slip rates out of 24 possible slip rates is not unique and that there are certainly other possible combinations of the 5 slip rates that satisfy Eq. (11.19). Other criteria must be invoked to determine which combination (among all possibilities) corresponds to **D**. While there are several different methods to accomplish this, only two are described below:

1. The actual slip systems can be determined by using Taylor's minimum shear principle. Among all the combinations which satisfy Eq. (11.19), this principle provides the actual combination for which the sum of the five shear rates is the least. The procedure requires considerable computational effort to find all the solutions that satisfy Eq. (11.19), but the number of independent solutions can be greatly reduced by remembering that not all the possibilities are independent of each other, as was noted by Taylor (1937, 1938) for the case of rigid plasticity. Actually with the help of a modern, powerful computer, this procedure can now be carried out very easily.

2. The determination of the slip systems can also be formulated into a nonlinear programming problem, as suggested by Havner (1971). We multiply Eqs. (11.18) and (11.19) by $\dot{\gamma}^\alpha$ and sum over α to obtain

$$\sum_{\alpha=1}^{n} \mathbf{P}^\alpha : (\mathbf{C} - \sigma\mathbf{I}) : \mathbf{D}\dot{\gamma}^\alpha \leq \sum_{\alpha=1}^{n}\sum_{\beta=1}^{n} g_{\alpha\beta}\dot{\gamma}^\alpha\dot{\gamma}^\beta \qquad (\dot{\gamma}^\alpha \geq 0) \tag{11.20}$$

or

$$I(\dot{\gamma}^\alpha) = \sum_{\alpha=1}^{n}\sum_{\beta=1}^{n} g_{\alpha\beta}\dot{\gamma}^\alpha\dot{\gamma}^\beta - \sum_{\alpha=1}^{n} \mathbf{P}^\alpha : (\mathbf{C} - \sigma\mathbf{I}) : \mathbf{D}\dot{\gamma}^\alpha \geq 0 \qquad (\dot{\gamma}^\alpha \geq 0) \tag{11.21}$$

In Eq. (11.21), the function $I(\dot{\gamma}^\alpha)$ is zero if Eq. (11.19) is satisfied. Therefore our determination of slip rates that satisfy Eq. (11.19) is equivalent to determining a set of $\dot{\gamma}^\alpha$ that minimizes the objective function $I(\dot{\gamma}^\alpha)$ subjected to the linear constraint $\dot{\gamma}^\alpha \geq 0$. This is a standard nonlinear programming problem and thus can be solved using standard methods.

The reason for taking the forward and reverse slip systems as two different systems is clear now from Eq. (11.21). The constraint can be simply written $\dot{\gamma}^\alpha \geq 0$. Otherwise, the physical constraint $|\dot{\gamma}^\alpha| \geq 0$ is not a mathematically convenient constraint.

We should point out that the solution to this nonlinear programming problem is not unique. Still, to select the appropriate solution from all the possible solutions, the use of Taylor's minimum principle is necessary. If the hardening rates for different slip systems are not the same (e.g., higher latent hardening is assumed), then Taylor's original minimum work principle Eq. (11.3), instead of the minimum shear principle, Eq. (11.4), has to be used. Readers interested in other methods for determining the active slip systems corresponding to a given deformation rate **D** can refer to Guldenpfenning and Clifton's (1976) linear programming procedure.

Now consider a single crystal subjected to a prescribed Jaumann rate of the Cauchy stress $\overset{\circ}{\boldsymbol{\sigma}}$. The constitutive equations which have to be satisfied by slip systems are given by Eq. (10.68) and are reproduced below for easy reference:

$$\overset{\circ}{\boldsymbol{\sigma}} : \mathbf{P}^\alpha < \sum_{\beta=1}^{n} k_{\alpha\beta}\dot{\gamma}^\beta \qquad (\dot{\gamma}^\alpha = 0) \tag{11.22}$$

$$\overset{\circ}{\boldsymbol{\sigma}} : \mathbf{P}^\alpha = \sum_{\beta=1}^{n} k_{\alpha\beta}\dot{\gamma}^\beta \qquad (\dot{\gamma}^\alpha > 0) \tag{11.23}$$

Again, the forward and reverse slip systems are considered as different slip systems.

Consider next how to determine 5 independent slip or shear rates (chosen from 24 slip systems) that correspond to the deformation pattern if the crystal is subjected to an arbitrarily prescribed stress rate $\overset{\circ}{\boldsymbol{\sigma}}$. Mathematically this problem is similar to that for which the deformation rate **D** is prescribed. Since the procedures for the solution of $\dot{\gamma}^\alpha$'s in that case apply here, they will not be repeated.

After determination of the slip systems for single crystals subjected to the prescribed deformation rate or stress rate, the plastic deformation rate and plastic spin can be obtained by using Eq. (10.32). The lattice rotation relative to a fixed coordinate system can be found by the using Eq. (10.32) and the following spin

$$\mathbf{W}^{\mathrm{e}} = \mathbf{W} - \mathbf{W}^{\mathrm{p}} \tag{11.24}$$

where $\mathbf{W}$ and $\mathbf{W}^p$ are the total continuum spin tensor and the plastic spin tensor, respectively. This equation can be used to calculate the lattice rotation of a single crystal due to deformation and also the texture development caused by this lattice rotation. This calculation of texture development is very important in analyzing the elastic-plastic behavior of polycrystals under large finite deformation.

11.2 TAYLOR'S RIGID PLASTIC MODEL FOR POLYCRYSTALS

The earliest model that successfully described the stress-strain relation and the texture development of polycrystal metals under uniaxial loading in terms of single crystal's constitutive behavior was proposed by Taylor (1937, 1938). Taylor's model is based on the following assumptions:

1. The elastic deformation is small compared to the plastic strain and can be ignored.
2. Each grain of the single crystal is subjected to the same homogeneous deformation (displacement or velocity field) as the polycrystal aggregate. In terms of infinitesimal strain tensor $\boldsymbol{\epsilon}$, this assumption of strict compatibility among the grains can be written as

$$\boldsymbol{\epsilon} = \boldsymbol{\epsilon}^*, \quad \dot{\boldsymbol{\epsilon}} = \dot{\boldsymbol{\epsilon}}^* \tag{11.25}$$

   where ϵ and $\dot{\epsilon}$ are the strain and strain rate within a single crystal grain (local quantities), while ϵ^* and $\dot{\epsilon}^*$ are the strain and strain rate of the polycrystal aggregate (overall quantities). In the case of large deformation, Eq. (11.25) is replaced by

$$\mathbf{v} = \mathbf{v}^*, \quad \mathbf{D} = \mathbf{D}^* \tag{11.26}$$

   where $\mathbf{v}$ and $\mathbf{D}$ are the velocity field and deformation rate within a single-crystal grain, while $\mathbf{v}^*$ and $\mathbf{D}^*$ are the corresponding quantities of the polycrystal aggregate.
3. The material behavior is rate independent, and the hardening rates for all the slip systems are assumed to be the same (isotropic hardening):

$$h_{\alpha\beta} = h \tag{11.27}$$

   and

$$d\tau_c^1 = d\tau_c^2 = \cdots = h \sum_{\beta=1}^{n} d\gamma^\beta \qquad (d\gamma^\beta > 0) \tag{11.28}$$

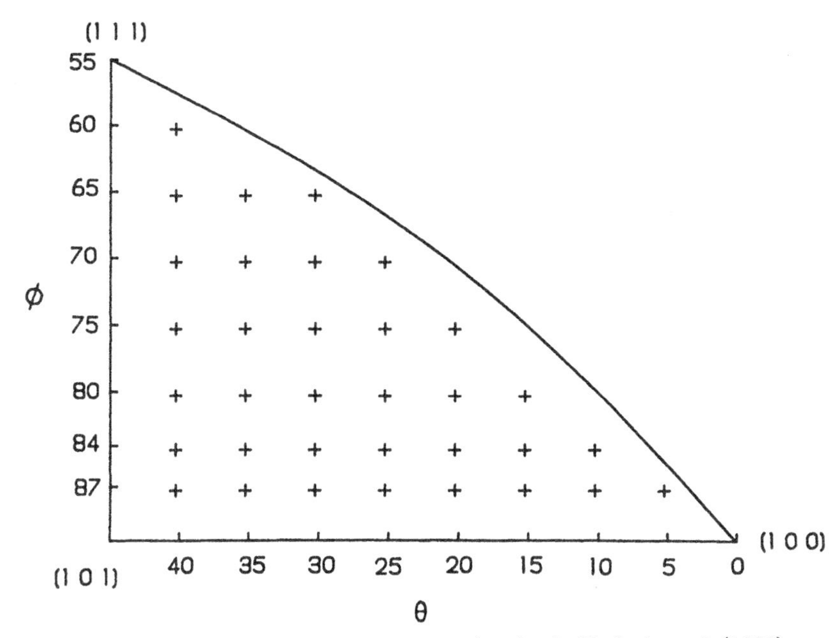

Fig. 11.2 Orientations of single-crystal grains in Taylor's work (1938)

Taylor solved the problem of a polycrystalline FCC aluminum subjected to uniform uniaxial tension, using the rigid plastic model based on the foregoing assumptions. The deformation is considered to be axial symmetric. The polycrystal aluminum specimen was regarded as an aggregate of a large number of single-crystal grains. The orientations of grains were equally distributed in a standard triangle of stereographic projection, as shown in Fig. 11.2, where ϕ and θ are angles as defined in Fig. 11.4. The crosses within the stereographic projection triangle represent the orientations of the tensile axis relative to the single-crystal lattice axes of the grains. The particular steps of Taylor's analysis are described below.

1. For a prescribed overall strain increment $d\epsilon^*$, the combination of shear increments $d\gamma^\alpha$ within any single crystal grain are calculated from (note $d\epsilon = d\epsilon^*$ for each grain)

$$d\boldsymbol{\epsilon}^* = \sum_{\alpha=1}^{n} \mathbf{P}^\alpha \, d\gamma^\alpha \tag{11.29}$$

and

$$\sum_{\alpha=1}^{n} |d\gamma^\alpha| = minimum \tag{11.30}$$

It is worth noting here that although the single crystal grain is assumed to be subjected to the same uniaxial tension as the aggregate, the single slip mode is not allowed because the compatibility condition will be violated if every grain is in single slip mode. Multislip has to be assumed to satisfy the compatibility condition.

2. The incremental work associated with the incremental slips $d\gamma^\alpha$ in the ith grain is

$$dw_i = \sum_{\alpha=1}^{n} \tau_\alpha \, d\gamma_i^\alpha = \tau \sum_{\alpha=1}^{n} d\gamma_i^\alpha \qquad (d\gamma_i^\alpha > 0) \tag{11.31}$$

where τ is the current yield stress on the active slip system. The last equation holds in this expression because it is assumed that all the slip systems harden at the same rate. For the whole aggregate the total incremental work is the sum of dw_i over all N grains:

$$dw = \sum_{i=1}^{N} dw_i = \tau \sum_{i=1}^{N} \sum_{\alpha=1}^{n} d\gamma_i^\alpha \tag{11.32}$$

3. The total incremental work done on the aggregate by the overall tensile stress σ^* through the overall tensile strain increment $d\epsilon^*$ is

$$dw = N\sigma^* \, d\epsilon^* \tag{11.33}$$

where the multiplier N is introduced because it is assumed that the volume of each grain is equal to unity. Equations (11.32) and (11.33) lead to

$$\sigma^* \, d\epsilon^* = \frac{\tau}{N} \sum_{i=1}^{N} \sum_{\alpha=1}^{n} d\gamma_i^\alpha \tag{11.34}$$

Since $d\epsilon^*$ is given ($d\epsilon^*$ is a component of the prescribed incremental strain tensor $\mathbf{d\epsilon}^*$), the corresponding stress is then calculated to be

$$\sigma^* = \frac{\tau}{N} \sum_{i=1}^{N} \sum_{\alpha=1}^{n} \frac{d\gamma_i^\alpha}{d\epsilon^*} \tag{11.35}$$

In this way the overall stress-strain curve $\sigma^*(\epsilon^*)$ can be calculated in terms of the resolved shear stress-resolved shear strain curve $\tau(\gamma)$ of a single crystal.

4. The texture development due to the rotation of the lattice of the crystal grain is calculated using Taylor's model, with all grains subjected to the same displacement or velocity field as the aggregate. Therefore the total spin (or rotation) of each grain is the same as that of the aggregate. In the case of

uniaxial tension

$$\mathbf{W} = \mathbf{W}^* = 0 \tag{11.36}$$

that is, the total rotation of the continuum is zero since the specimen is constrained by the loading machine. Therefore from Eq. (11.24),

$$d\mathbf{W}^e = -d\mathbf{W}^p = \sum_{\alpha=1}^{n} \mathbf{\Omega}^\alpha \, d\gamma^\alpha \tag{11.37}$$

As discussed earlier, $\mathbf{W}^e$ represents the spin of the lattice. Therefore the reorientation of all the grains of the aggregate relative to the tensile axis due to this spin can be obtained after $d\mathbf{W}^e$ has been determined.

The stress-strain curve $\sigma^*(\epsilon^*)$ calculated by following preceding procedures for polycrystal aluminum is shown in Fig. 11.3 by circles. The experimental data are indicated in the figure by crosses. Clearly the agreement between the prediction and the experimental data is excellent. Also shown in the figure is the resolved shear stress-resolved shear strain curve of the single crystal aluminum obtained by Elam.

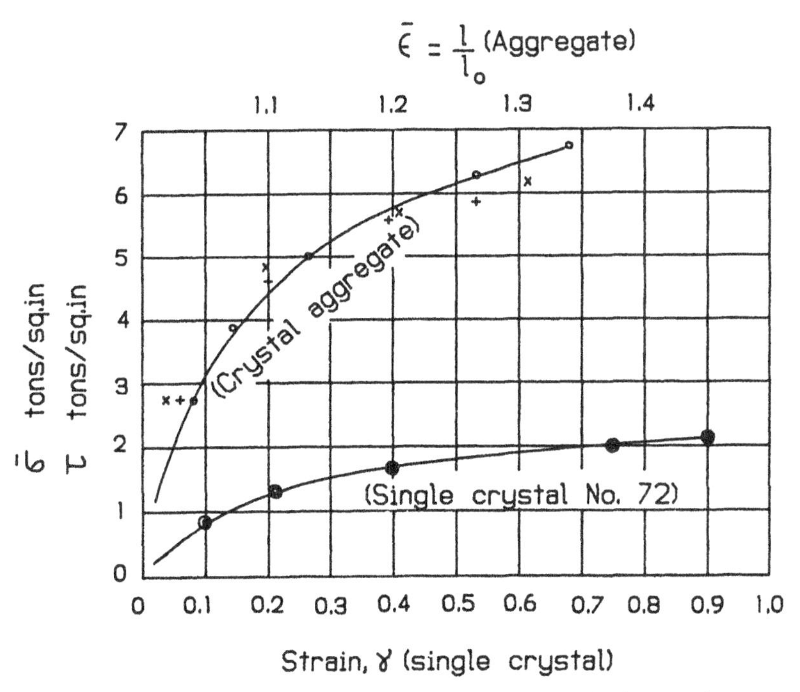

Fig. 11.3 Prediction of polycrystal stress-strain curve from a single-crystal stress-strain curve (after Taylor 1937)

Thus far, from our calculation we have found that the overall stress-strain curve of the polycrystal can be given by

$$\sigma^* = m^* \tau(m^* \epsilon^*) \tag{11.38}$$

where $\tau(\gamma)$ is the resolved shear stress-resolve shear strain curve for a single crystal. Since the single-crystal grains are in multislip, the parameter m^* is an average value of the Schmid factor ($\cos\lambda \cos\phi$) of all the grains. By Taylor's calculation the value of m^* for FCC polycrystal metals is 3.1.

The rotation of the specimen axis relative to the lattice axes at an extension of 2.37% is shown in Fig. 11.4 for each crystal grain of different orientations. It is seen that the whole triangle is divided into several different

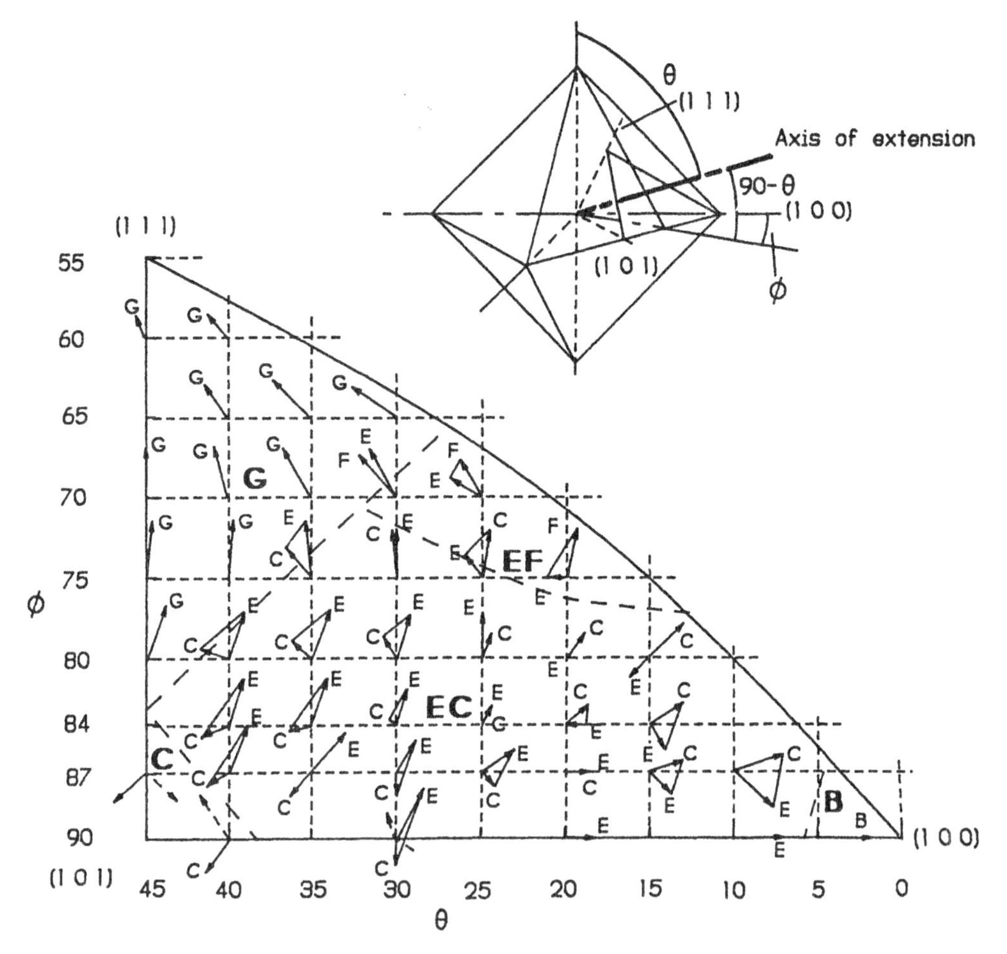

Fig. 11.4 Rotation of single crystal grains at an extension of 2.37% (after Taylor 1937)

areas. Within each area one or two combinations of five shears or slips are activated, and each combination is denoted by a letter in the figure. For example, in area G the combination of shears for all grains with different orientations is the same and is denoted by letter G, while in area EC two combinations are equally possible because these two different combinations of five shears correspond to exactly the same sum of the absolute values of shears (this result verifies the nonuniqueness of the combination of shears corresponding to a prescribed strain, as discussed earlier). The rotation of the tensile axis relative to the crystal lattice for each grain (corresponding to every point in Fig. 11.4) is shown by an arrow starting from each point in Fig. 11.4. The starting points of the arrows represent the original orientations of the tensile axis and the ending points of the arrows represent the orientations of the tensile axis after an extension of 2.37%. In areas where two combinations of slip systems are possible, two arrows appear at each point, each arrow representing one combination of the five shears. The angle between the arrows indicates the range of possible movement of the tensile axis relative to the crystal lattice axes. Thus one direct result of nonuniqueness of the shear combination is that the lattice rotation is nonunique.

According to Taylor's calculation, as shown in Fig. 11.4, the result of the reorientation is that "the aggregate will tend to attain a state in which the crystal axes of grains have either a (1 1 1) or a (1 0 0) axis, but no grains with a (1 1 0) axis in the direction of extension." Taylor (1937, 1938) pointed out that the calculation of the texture is in accord with the experimental results. However, Bishop (1954) found that the texture predicted by Taylor's model did not agree with the experimental results for α-brass, and he attributed this to the effect of strong latent hardening.

Taylor's model is the simplest approximation of the elastic-plastic behavior of polycrystal metals and can describe the basic features of polycrystalline deformation, such as texture development and the stress-strain curve. However, the main criticism of Taylor's model is that the equilibrium condition across the grain boundaries is violated by forcing all the single grains to deform the same as the aggregate. To satisfy both equilibrium and compatibility conditions between grain boundaries, several self-consistent models have been proposed, as mentioned previously. A basic and important tool used in most of these self-consistent theories is Eshelby's solution of an ellipsoidal inclusion embedded in an infinitely extended homogeneous, elastically isotropic media. This solution is briefly presented in the next section.

11.3 ESHELBY'S SOLUTION FOR ELLIPSOIDAL INCLUSION

Consider an ellipsoidal domain Ω in a homogeneous, infinitely extended material D, as shown in Fig. 11.5. The material, including the ellipsoidal domain Ω, is assumed to be isotropic and linear elastic. Since in the elastic range, the deformation is very small, the infinitesimal strain tensor ϵ is used,

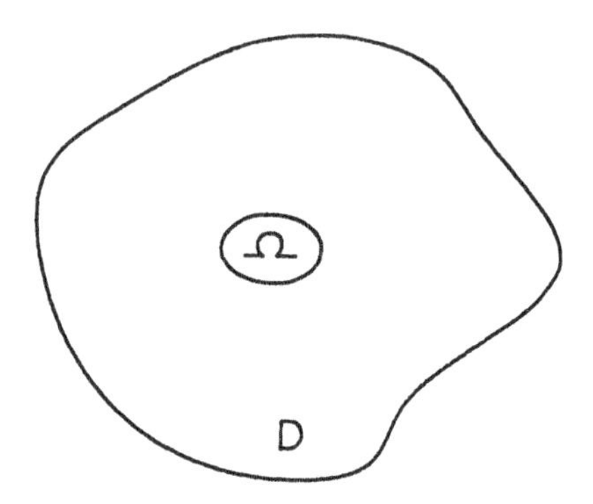

Fig. 11.5 Ellipsoidal inclusion in D

and no distinction between the deformed and the undeformed configurations is necessary. Suppose that the ellipsoidal domain Ω is removed from the material D and undergoes an arbitrary homogeneous strain ϵ' without any constraint on its boundary. The ellipsoidal domain Ω can not be fitted freely into D now because of ϵ'. If Ω is forced back to its original size and shape and put into D and the applied force (which was necessary to get Ω back to its original size and shape) is released, a stress field will be produced inside Ω as well as in the region outside D–Ω because of the misfit. In response to this stress field, there will be a strain field in Ω as well as in D–Ω. Obviously the actual strain attained in Ω will be different from ϵ', since ϵ' would be the strain of Ω if it had deformed freely. The actual strain within Ω will be the sum of ϵ' and the elastic strain ϵ^e caused by the stress field due to the misfit.

The strain ϵ' described here was called the *stress-free transformation strain* by Eshelby (1957). He called the ellipsoidal domain Ω, with this stress-free transformation strain ϵ' embedded in D, an *ellipsoidal inclusion*. The source of ϵ' could be thermal expansion, phase transformation, plastic strain, and so on. Eshelby obtained the solutions for the elastic stress and strain fields in the infinite extended isotropic elastic material D (including Ω) caused by an inclusion Ω (with stress-free transformation strain ϵ') embedded in D. There is no other external force acting on the boundary of D or inside of D. Eshelby's work is briefly presented below. A detailed derivation can be found in Eshelby (1957) or in Mura (1987).

The solution $\boldsymbol{\epsilon}$ and $\boldsymbol{\sigma}$ should satisfy all the equations of linear elasticity. At first, stress $\boldsymbol{\sigma}$ must be self-equilibrated:

$$\sigma_{ij,j} = 0 \qquad (\mathbf{X} \in D) \tag{11.39}$$

The strain field should be compatible:

$$\epsilon_{ij} = \tfrac{1}{2}(u_{i,j} + u_{j,i}) \qquad (\mathbf{X} \in D) \tag{11.40}$$

where $\mathbf{u}$ is the displacement field in D. The Hooke's law should be written separately for Ω and for D–Ω as

$$\sigma_{ij} = \lambda\delta_{ij}(\epsilon_{mm} - \epsilon'_{mm}) + 2G(\epsilon_{ij} - \epsilon'_{ij}) \qquad (\mathbf{X} \in \Omega) \tag{11.41}$$

$$\sigma_{ij} = \lambda\delta_{ij}\epsilon_{mm} + 2G\epsilon_{ij} \qquad (\mathbf{X} \in D - \Omega) \tag{11.42}$$

The boundary conditions are given by

$$\sigma_{ij} = 0 \qquad (\mathbf{X} \to \infty) \tag{11.43}$$

Except for the equations above, the solutions σ and ϵ must be so determined that across the boundary of the inclusion Ω, the equilibrium and compatibility conditions are satisfied. Eshelby (1957) found that when ϵ' is uniform inside Ω, the strain ϵ caused inside Ω of the preceding problem is also uniform and is related to the stress-free transformation strain ϵ' in the form

$$\boldsymbol{\epsilon} = \mathbf{S} : \boldsymbol{\epsilon}' \tag{11.44}$$

where $\mathbf{S}$ is a fourth-order tensor, called *Eshelby's tensor*, and has the following symmetry property:

$$S_{ijkl} = S_{jikl} = S_{ijlk} \tag{11.45}$$

The expressions of S_{ijkl} for a sphere inclusion are

$$S_{ijkl} = \left(1 - \frac{4(4-5\nu)}{15(1-\nu)}\right)\delta_{ij}\delta_{kl} + \frac{4-5\nu}{15(1-\nu)}(\delta_{ik}\delta_{jl} + \delta_{il}\delta_{jk}) \tag{11.46}$$

where ν is the poisson's ratio. In particular,

$$S_{1111} = S_{2222} = S_{3333} = \frac{7-5\nu}{15(1-\nu)} \tag{11.47}$$

$$S_{1122} = S_{2233} = S_{3311} = S_{1133} = S_{2211} = S_{3322} = \frac{5\nu - 1}{15(1-\nu)} \tag{11.48}$$

$$S_{1212} = S_{2323} = S_{3131} = \frac{4-5\nu}{15(1-\nu)} \tag{11.49}$$

and all other components [except those obtained by using symmetry Eq. (11.45)] are zero. The stress components σ_{ij} can be obtained by using Eqs. (11.41) and (11.46). For example,

$$\sigma_{11} = -G\frac{16}{15(1-\nu)}\epsilon_{11} - 2G\frac{5\nu+1}{15(1-\nu)}\epsilon'_{22} - 2G\frac{5\nu+1}{15(1-\nu)}\epsilon'_{33} \quad (11.50)$$

It should be emphasized that these solutions to the inclusion problem are obtained under the following assumptions:

1. The general expressions for S_{ijkl} for linear elastic, isotropic materials with an other shape of inclusion, such as ellipse and penney shape, can be found in Eshelby's work (1957) or in Mura (1987).
2. The stress σ and ϵ obtained here satisfy all the equations of elasticity theory, including the equilibrium and compatibility conditions on the inclusion boundary.

11.4 KRONER, BUDIANSKY, AND WU'S MODEL

In Taylor's early work, as discussed before, the strain ϵ within single-crystal grains (micro-strain or local strain) is simply assumed to be the same as the overall strain ϵ^* of the polycrystal metals (macro-strain) to ensure the compatibility condition across the grain boundaries. The cost of this forced compatibility is that the equilibrium between grains is violated.

To ensure both compatibility and equilibrium conditions on grain boundaries, Budiansky and Wu (1962) proposed a self-consistent model that relates strains and stresses within the single crystal grains to the overall strain and stress. They used a simple model for the polycrystals. In their model they approximated the interaction of a single-crystal grain with the surrounding materials of the polycrystal by a sphere embedded in a homogeneous polycrystal matrix or aggregate, as shown in Fig. 11.6. It is further assumed that the single-crystal grain and aggregate are elastically isotropic, have the same linear elastic property, and can be described by an elastic stiffness tensor $\mathbf{C}$ with the inverse $\mathbf{C}^{-1}$.

Since the size of a single crystal is very small, the homogeneous aggregate or matrix containing it can be considered to be infinitely extended. The aggregate is subjected to the overall stress $\boldsymbol{\sigma}^*$ producing the overall strain $\boldsymbol{\epsilon}^*$ and the overall plastic strain $\boldsymbol{\epsilon}^{*p}$. Note that $\boldsymbol{\sigma}^*$, $\boldsymbol{\epsilon}^*$, and $\boldsymbol{\epsilon}^{*p}$ can be considered as the average values of the local quantities σ, ϵ, and ϵ^p over all the randomly distributed single-crystal grains. Due to the application of $\boldsymbol{\sigma}^*$, the single-crystal grain contained in the aggregate will deform. It is obvious that if σ^* is small so that no plastic deformation is produced in either the grain or the aggregate. The strain (elastic strain) will be uniform since the

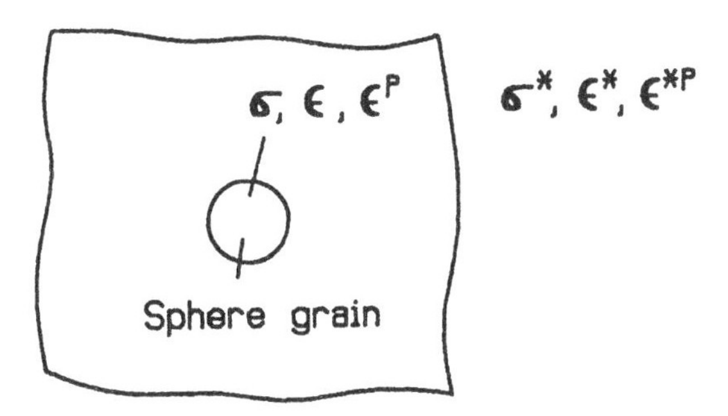

Fig. 11.6 Simple model of polycrystals

single-crystal grain and aggregate have the same elastic deformation parameters. However, as σ^* increases, the plastic deformation will occur either in the aggregate matrix (due to the plastic slips of other single crystal grains) or in the sphere grain under consideration, and the deformation will become nonuniform because the plastic deformation parameters of the single-crystal grain and that of the polycrystal aggregate are different. The plastic deformation parameters of the aggregate are the unknown property and must be derived from the constitutive behavior of single crystals under plastic deformation. The plastic deformation constitutive equations of the single crystal grain are the known input and are used for the analysis. Therefore σ, ϵ, and ϵ^p in the sphere grain are different from σ^*, ϵ^*, and ϵ^{*p} because of the inhomogeneity of the plastic deformation. However, σ, ϵ, and ϵ^p have to satisfy the following three restrictions: (1) equilibrium on the grain boundary, (2) compatibility on the grain boundary, (3) averages of σ, ϵ, and ϵ^p over all the grains must yield σ^*, ϵ^*, and ϵ^{*p}.

To determine the relation between σ, ϵ, and ϵ^p and σ^*, ϵ^*, and ϵ^{*p} in a polycrystal model Kroner, Budiansky, and Wu concluded that it is necessary to find the stress σ, strain ϵ, and plastic strain ϵ^p in a sphere embedded in a matrix (with the same elastic property as the sphere), and they subjected these to overall stress σ^*, strain ϵ^*, and plastic strain ϵ^{*p}. The solutions σ, ϵ, and ϵ^p should satisfy the above-mentioned restrictions as well as other conventional conditions such as equilibrium and compatibility. This problem can be solved using Eshelby's solution as follows: Separate the problem shown in Fig. 11.6 into the two problems shown in Fig. 11.7:

1. The aggregate and grain are subjected to the overall quantities σ^*, ϵ^*, and ϵ^{*p}. In this case, the strain in the sphere grain is assumed to be the same in the aggregate and is given by

$$\epsilon = \mathbf{C}^{-1} : \sigma^* + \epsilon^{*p} \tag{11.51}$$

where $\mathbf{C}$ is the elastic stiffness tensor which is the same for both the sphere grain and the aggregate.

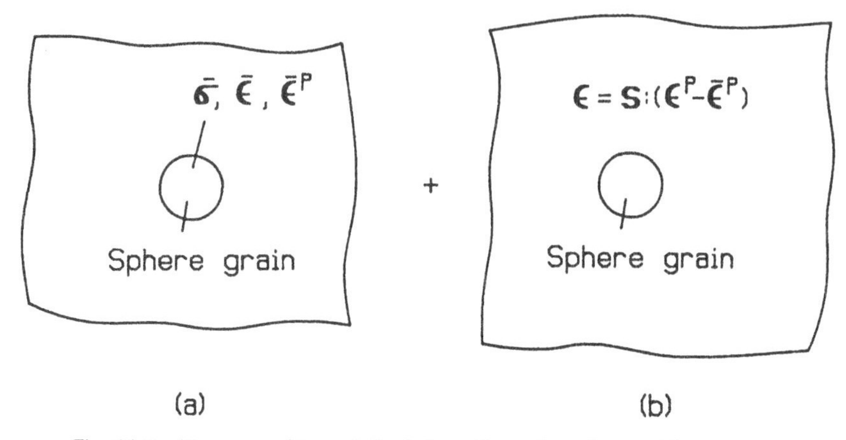

Fig. 11.7 Decomposition of the interaction of a sphere with aggregate

2. The sphere grain has a stress-free transformation strain $\boldsymbol{\epsilon}' = \boldsymbol{\epsilon}^{\mathrm{p}} - \boldsymbol{\epsilon}^{*\mathrm{p}}$, and it is embedded in a homogeneous aggregate with the same elastic property. There is no other load applied on the aggregate. Since the spherical single-crystal grain is very small compared with the aggregate, the aggregate is considered infinitely extended. According to Eshelby (1957) the strain inside the sphere grain due to the elastic interaction between the grain and the aggregate caused by the stress-free transformation strain $\boldsymbol{\epsilon}'$ is given by

$$\boldsymbol{\epsilon} = \mathbf{S} : \boldsymbol{\epsilon}' = \mathbf{S} : (\boldsymbol{\epsilon}^{\mathrm{p}} - \boldsymbol{\epsilon}^{*\mathrm{p}}) \tag{11.52}$$

where $\mathbf{S}$ is the Eshelby's tensor given by Eqs. (11.46)–(11.48) for a sphere inclusion in an isotropic elastic aggregate.

The equivalence of these two problems shown in Fig. 11.7 with the original problem shown in Fig. 11.6 can be justified easily. As discussed earlier, under the application of the overall stress $\boldsymbol{\sigma}^*$ on the aggregate, the induced $\boldsymbol{\sigma}$, $\boldsymbol{\epsilon}$, and $\boldsymbol{\epsilon}^{\mathrm{p}}$ inside the grain are different from $\boldsymbol{\sigma}^*$, $\boldsymbol{\epsilon}^*$, and $\boldsymbol{\epsilon}^{*\mathrm{p}}$ for the aggregate. This difference is considered to be caused by the different plastic deformation within the sphere grain from the plastic deformation of the aggregate due to the difference of the plastic behavior between them. This justifies the use of $\boldsymbol{\epsilon}' = \boldsymbol{\epsilon}^{\mathrm{p}} - \boldsymbol{\epsilon}^{*\mathrm{p}}$ as the stress-free transformation strain because it is the source of difference. The actual strain inside the sphere grain is then equal to the sum of the strain corresponding to the uniform deformation given by Eqs. (11.51) and the strain caused by the difference of the plastic strain, as given by Eq. (11.52):

$$\boldsymbol{\epsilon} = \mathbf{C}^{-1} : \boldsymbol{\sigma}^* + \boldsymbol{\epsilon}^{*\mathrm{p}} + \mathbf{S} : (\boldsymbol{\epsilon}^{\mathrm{p}} - \boldsymbol{\epsilon}^{*\mathrm{p}}) \tag{11.53}$$

Using, $\mathbf{S}$, given by Eq. (11.46), and noting that $\boldsymbol{\epsilon}^{\mathrm{p}}$ and $\boldsymbol{\epsilon}^{*\mathrm{p}}$ are deviatoric tensors, it is easy to calculate that

$$\mathbf{S}:\left(\boldsymbol{\epsilon}^{\mathrm{p}} - \boldsymbol{\epsilon}^{*\mathrm{p}}\right) = \beta\left(\boldsymbol{\epsilon}^{\mathrm{p}} - \boldsymbol{\epsilon}^{*\mathrm{p}}\right) \tag{11.54}$$

where the coefficient β is given by

$$\beta = \frac{2(4 - 5\nu)}{15(1 - \nu)} \tag{11.55}$$

and ν is the elastic Poisson's ratio of the aggregate (as well as the grain). Equation (11.53) can now be rewritten as

$$\boldsymbol{\epsilon} = \mathbf{C}:\boldsymbol{\sigma}^{*} + \boldsymbol{\epsilon}^{*\mathrm{p}} + \beta\left(\boldsymbol{\epsilon}^{\mathrm{p}} - \boldsymbol{\epsilon}^{*\mathrm{p}}\right) \tag{11.56}$$

From this equation it follows that the stress inside the sphere is

$$\boldsymbol{\sigma} = \mathbf{C}:\boldsymbol{\epsilon}^{\mathrm{e}} = \mathbf{C}:\left(\boldsymbol{\epsilon} - \boldsymbol{\epsilon}^{\mathrm{p}}\right) = \boldsymbol{\sigma}^{*} - 2G(1 - \beta)\left(\boldsymbol{\epsilon}^{\mathrm{p}} - \boldsymbol{\epsilon}^{*\mathrm{p}}\right) \tag{11.57}$$

where the isotropic form of $\mathbf{C}$ has been used to arrive at the final result.

In plasticity theory the constitutive equations are usually given in the rate or incremental form. Therefore Eq. (11.57) is usually given by

$$\dot{\boldsymbol{\sigma}} = \dot{\boldsymbol{\sigma}}^{*} - 2G(1 - \beta)\left(\dot{\boldsymbol{\epsilon}}^{\mathrm{p}} - \dot{\boldsymbol{\epsilon}}^{*\mathrm{p}}\right) \tag{11.58}$$

Equation (11.57) was first derived by Kroner (1961) from a different procedure than discussed here. Therefore Eq. (11.57) and Eq. (11.58) are now called the *Kroner, Budiansky, and Wu model*. Both relations provide a connection between the microscale quantities (σ, ϵ, and ϵ^{p}) and the macroscale quantities (σ^{*}, ϵ^{*}, and $\epsilon^{*\mathrm{p}}$). They replace Eq. (11.25) in Taylor's model and render a more reasonable consideration of the interaction between the single-crystal grains and the polycrystal aggregate. Basically Eqs. (11.56) and (11.57) are the solutions to a sphere inclusion embedded in a aggregate subjected to $\boldsymbol{\sigma}^{*}$ at infinity. Clearly these solutions consist of two parts, Eqs. (11.51) and (11.52); each satisfies both equilibrium and compatibility conditions on the grain boundary. These conditions are also satisfied by Eqs. (11.56), (11.57), and (11.58). This is why this model is called a self-consistent model. It is easy to verify that Eqs. (11.56), (11.57), and (11.58) are consistent with the following relations:

$$\boldsymbol{\sigma}^{*} = (\sigma)_{\mathrm{ave}}, \quad \dot{\boldsymbol{\sigma}}^{*} = (\dot{\sigma})_{\mathrm{ave}} \tag{11.59}$$
$$\boldsymbol{\epsilon}^{*\mathrm{p}} = \left(\boldsymbol{\epsilon}^{\mathrm{p}}\right)_{\mathrm{ave}}, \quad \dot{\boldsymbol{\epsilon}}^{*\mathrm{p}} = \left(\dot{\boldsymbol{\epsilon}}^{\mathrm{p}}\right)_{\mathrm{ave}}$$

where $(\ )_{\mathrm{ave}}$ represents the average over all grains.

Equation (11.58) provides a way to transform the constitutive behavior of single crystals into that of polycrystals, and vice versa. The application of this equation to the determination of mechanical behavior of polycrystal metals based on the knowledge of single-crystal plasticity will be discussed later after first introducing some other models.

It is interesting to note that the influence of the grain shape can be estimated by the self-consistent mode discussed above. For example, if the shape of the grain in Fig. 11.6 is an ellipsoidal sphere rather than a regular sphere, Eq. (11.58) can be modified as

$$\dot{\sigma} = \dot{\sigma}^* - 2G(\mathbf{I} - \mathbf{S}) : (\epsilon^{\mathrm{p}} - \epsilon^{*\mathrm{p}}) \tag{11.60}$$

where $\mathbf{I}$ is a fourth-order unit tensor and $\mathbf{S}$ is the Eshelby's tensor for ellipsoidal inclusion embedded in a finitely extended matrix. The exact form can be found in Mura (1987). It is well known that one of the reasons for anisotropy of plastic deformation is the irregular shape of grains. This is called morphologic anisotropy. It is seen that the influence of morphologic anisotropy can be estimated by introducing appropriate Eshelby's tensor in Kroner, Budiansky, and Wu's model.

It should be emphasized here that the Kroner, Budiansky and Wu model was derived by assuming an elastic interaction between the grain and the polycrystal aggregate. This is clear from Eq. (11.52) where Eshelby's solution is used to obtain a linear elastic inclusion embedded in an infinitely extended matrix with the same elastic property. As a result the constraint provided by the elastic aggregate surrounding the grain is much stronger than it would be in actual cases where the surrounding aggregate is plastically deformed to some extent. Therefore the prediction of the overall stress σ^* by this model is an upper bound to the real stress and will be discussed later. Since the polycrystal aggregate is also plastically deformed to a certain extent, the actual restraint to the single-crystal grain should be "softer" than what is provided by the elastic surroundings, used in Kroner, Budiansky, and Wu's model. A modification of Kroner, Budiansky, and Wu's model motivated by this consideration was proposed by Berveiller and Zaoui (1970). They considered the elastic-plastic interaction between the single-crystal grains and the polycrystal aggregate. Their model will be discussed next.

11.5 MODIFICATION OF KRONER, BUDIANSKY, AND WU'S MODEL

Berveiller and Zaoui (1979) modified Kroner, Budiansky, and Wu's model by introducing a scalar function α_{T} as follows (Beradi, Berveiller, and Lipinski 1987):

$$\dot{\sigma} = \dot{\sigma}^* - 2\alpha_{\mathrm{T}}G(1 - \beta)(\dot{\epsilon}^{\mathrm{p}} - \dot{\epsilon}^{*\mathrm{p}}) \tag{11.61}$$

This scalar function α_T, called the *plastic accommodation factor*, is given by

$$\alpha_T = \frac{9K_T + 8G_T}{9K + 8G} \frac{5G_T(3K + 4G)}{6G(K_T + 2G_T) + G_T(9K_T + 8G_T)} \tag{11.62}$$

where K and G are the elastic bulk and shear moduli, respectively, while K_T and G_T are the respective tangent bulk and shear moduli in the elastic-plastic region defined by

$$\dot{\mathbf{S}}^* = 2G_T\left(\dot{\boldsymbol{\epsilon}}^* - \frac{\boldsymbol{\epsilon}_m^*}{3}\mathbf{I}\right) \tag{11.63}$$

$$\dot{\sigma}_m^* = 3K_T\boldsymbol{\epsilon}_m^* \tag{11.64}$$

$\mathbf{S}^*$ is the overall deviatoric stress tensor and σ_m^* and ϵ_m^* are the overall mean stress and mean strain, respectively.

From Eq. (11.61), it is seen that the only quantities introduced to describe the elastic-plastic interaction between the single crystal grains and the polycrystal aggregate are G_T and K_T. For cases of elastic deformation or elastic unloading of the aggregate, $G_T = G$ and $K_T = K$, and it can be easily verified from Eq. (11.62) that $\alpha_T = 1$. Therefore Eq. (11.61) reduces to Kroner, Budiansky, and Wu's model.

Beradi, Berveiller, and Lipinski (1987) pointed out that for most metals

$$\beta \approx \frac{1}{2}, \quad \alpha_T \approx \frac{1}{100} \tag{11.65}$$

Equation (11.61) can therefore be simplified to

$$\dot{\boldsymbol{\sigma}} = \dot{\boldsymbol{\sigma}}^* - \frac{G}{100}(\dot{\boldsymbol{\epsilon}}^p - \dot{\boldsymbol{\epsilon}}^{*p}) \tag{11.66}$$

In the case of plastic flow, the value of α_T is estimated (Beradi, Berveiller, and Lipinski 1987) to be in the range

$$0.01 < \alpha_T < 0.1 \tag{11.67}$$

for most metals. It should be pointed out here that Eq. (11.61) was derived by Berveiller and Zaoui (1979) by their assuming the isotropy of the plastic response. In other words, they assumed that

$$\dot{\boldsymbol{\sigma}}^* = \mathbf{C}^{*ep} : \dot{\boldsymbol{\epsilon}}^{*p} \tag{11.68}$$

where $\mathbf{C}^{*ep}$ is the elastic-plastic tangential stiffness tensor, and they assumed that it is isotropic:

$$C_{ijkl}^{*\text{ep}} = G_{\text{T}}\left[\frac{2\nu_{\text{T}}}{1-2\nu_{\text{T}}}\delta_{ij}\delta_{kl} + \delta_{ik}\delta_{jl} + \delta_{il}\delta_{jk}\right] \tag{11.69}$$

where ν_{T} is the tangential elastic-plastic poisson's ratio and is related to G_{T} and K_{T} by

$$\nu_{\text{T}} = \frac{3K_{\text{T}} - 2G_{\text{T}}}{2(3K_{\text{T}} + G_{\text{T}})} \tag{11.70}$$

An equivalent assumption for Eqs. (11.68) and (11.69) is

$$\dot{\boldsymbol{\epsilon}}^* = f\dot{\mathbf{S}}^* \tag{11.71}$$

where f is a proportionality factor. Obviously these assumptions oversimplify the plastic behavior of the polycrystal aggregate, but it was demonstrated by Weng (1982) that under proportional loading such a modification provides a rather close approximation. The resulting behavior is softer than those obtained by using the Kroner, Budiansky, and Wu model since the "weakening" effect of the surrounding constraint to the single-crystal grain due to the plastic deformation of the aggregate has been approximated through the plastic accommodation factor α_{T}. The calculated results will be presented later.

Sometimes the interaction relation (11.61) is given in terms of total quantities rather than in terms of rates or increments, for example,

$$\boldsymbol{\sigma} = \boldsymbol{\sigma}^* - 2\alpha_{\text{s}}G(1-\beta)(\boldsymbol{\epsilon}^{\text{p}} - \boldsymbol{\epsilon}^{*\text{p}}) \tag{11.72}$$

when the plastic accommodation is given by the secant moduli K_{s} and G_{s}

$$\alpha_{\text{s}} = \frac{9K_{\text{s}} + 8G_{\text{s}}}{9K + 8G}\,\frac{5G_{\text{s}}(3K + 4G)}{6G(K_{\text{s}} + 2G_{\text{s}}) + G_{\text{s}}(9K_{\text{s}} + 8G_{\text{s}})} \tag{11.73}$$

where K and G are elastic bulk and shear moduli respectively; K_{s} and G_{s} are the respective secant bulk and shear moduli defined in the same way as Eqs. (11.63) and (11.64), except that all the rate quantities must be replaced by the total quantities. Equation (11.72) reduces to the following assumption:

$$\boldsymbol{\epsilon}^{*\text{p}} = f\mathbf{S}^* \tag{11.74}$$

This is a more realistic statement than Eq. (11.71).

11.6 HILL'S SELF-CONSISTENT MODEL

A self-consistent model using a different approach was proposed by Hill (1965a, 1965b). As in previously discussed models, an ellipsoidal single-crystal grain is imagined to be embedded in the polycrystal aggregate. It is assumed that $\boldsymbol{\epsilon}$ and $\boldsymbol{\sigma}$ in the single-crystal grain and $\boldsymbol{\epsilon}^*$ and $\boldsymbol{\sigma}^*$ of the aggregate can be related by

$$\dot{\boldsymbol{\sigma}} = \mathbf{C}^{ep} : \dot{\boldsymbol{\epsilon}} \tag{11.75}$$

$$\dot{\boldsymbol{\sigma}}^* = \mathbf{C}^{*ep} : \dot{\boldsymbol{\epsilon}}^* \tag{11.76}$$

where $\mathbf{C}^{ep}$ is the tangential elastic plastic stiffness tensor of the single crystal, discussed in detail in the last chapter for the case of finite deformation, and $\mathbf{C}^{*ep}$ is the tangential elastic-plastic stiffness tensor of the polycrystal aggregate. It should be noted that $\mathbf{C}^{*ep}$ is initially unknown and that the purpose of the polycrystal plasticity is to derive or calculate $\mathbf{C}^{*ep}$ in terms of $\mathbf{C}^{ep}$, the behavior of the single crystals.

Based on Eshelby's solution of an ellipsoidal inclusion embedded in an infinite homogeneous media, Hill (1965a, 1965b) introduced an overall constraint tensor $\mathbf{L}^*$ as

$$\dot{\boldsymbol{\sigma}} - \dot{\boldsymbol{\sigma}}^* = -\mathbf{L}^* : \left(\dot{\boldsymbol{\epsilon}} - \dot{\boldsymbol{\epsilon}}^*\right) \tag{11.77}$$

where the constraint tensor $\mathbf{L}^*$ is a fourth-order tensor dependent on $\mathbf{C}^{*ep}$ and the geometrical aspects of the ellipsoidal inclusion.

Using Eqs. (11.75) and (11.76), Eq. (11.77) becomes

$$\mathbf{L}^* : \dot{\boldsymbol{\epsilon}} + \mathbf{C}^{ep} : \dot{\boldsymbol{\epsilon}} = \left(\mathbf{L}^* + \mathbf{C}^{*ep}\right) : \dot{\boldsymbol{\epsilon}} \tag{11.78}$$

from which the strain rate in the single crystal $\dot{\boldsymbol{\epsilon}}$ can be expressed in terms of the strain rate aggregate $\dot{\boldsymbol{\epsilon}}^*$:

$$\dot{\boldsymbol{\epsilon}} = \left(\mathbf{L}^* + \mathbf{C}^{ep}\right)^{-1} : \left(\mathbf{L}^* + \mathbf{C}^{*ep}\right) : \dot{\boldsymbol{\epsilon}}^* \tag{11.79}$$

If $\mathbf{L}^*$, $\mathbf{C}^{ep}$, and $\mathbf{C}^{*ep}$ have inverses of $\mathbf{L}^{*-1}$, $\mathbf{C}^{ep-1}$, and $\mathbf{C}^{*ep-1}$, a similar expression for $\dot{\boldsymbol{\sigma}}$ can also be derived:

$$\dot{\boldsymbol{\sigma}} = \left(\mathbf{L}^{*-1} + \mathbf{C}^{ep^{-1}}\right)^{-1} : \left(\mathbf{L}^{*-1} + \mathbf{C}^{*ep^{-1}}\right) : \dot{\boldsymbol{\sigma}}^* \tag{11.80}$$

Equations (11.79) and (11.80) furnish the strain and stress rates in the single-crystal grain in terms of the overall strain and stress rate. If the constraint tensor $\mathbf{L}^*$ and $\mathbf{C}^{*ep}$ are known, the strain and stress rates in the single-crystal grain can be obtained easily in terms of $\dot{\boldsymbol{\epsilon}}^*$ and $\dot{\boldsymbol{\sigma}}^*$.

Unfortunately, the calculation of $\mathbf{L}^*$ is very difficult in the general cases of elastic-plastic deformation. This will be seen from the following discussions.

The overall constraint tensor $\mathbf{L}^*$ can be calculated in terms of Eshelby's tensor (in a generalized sense). It has been shown by Hill (1965a, 1965b) that $\mathbf{L}^*$ is related to the Eshelby's tensor $\mathbf{S}$ in the following way:

$$\mathbf{L}^* : \mathbf{S} = \mathbf{C}^{*ep} : (\mathbf{I} - \mathbf{S}), \quad (\mathbf{I} - \mathbf{S}) : \mathbf{L}^{*-1} = \mathbf{S} : \mathbf{C}^{*ep^{-1}} \tag{11.81}$$

where $\mathbf{I}$ is the fourth-order unit tensor defined by

$$I_{ijkl} = \tfrac{1}{2}(\delta_{ki}\delta_{lj} + \delta_{li}\delta_{kj}) \tag{11.82}$$

and $\mathbf{S}$ is the Eshelby's tensor. It should be pointed out that Hill considered the plastically deformed polycrystal aggregate to be an anisotropic material, and therefore the Eshelby's tensor used here is obtained by considering an ellipsoidal inclusion with stress-free transformation strain $\boldsymbol{\epsilon}'$ embedded in an anisotropic material characterized by the tangential stiffness tensor $\mathbf{C}^{*ep}$ (instead of embedded in an isotropic linear elastic material, as discussed in Section 11.3). The calculation of $\mathbf{S}$ for general anisotropic materials for three-dimensional cases was sketched by Eshelby (1951) and was discussed in great detail by Mura (1987).

From Eq. (11.81), the constraint tensor $\mathbf{L}^*$ can be expressed as

$$\mathbf{L}^* = \mathbf{C}^{*ep} : (\mathbf{S}^{-1} - \mathbf{I}) \tag{11.83}$$

from which, the difficulty of calculating $\mathbf{L}^*$ can be easily seen. To calculate the overall constraint tensor $\mathbf{L}^*$, the overall tangential stiffness tensor $\mathbf{C}^{*ep}$ has to be known in advance. In addition Eshelby's tensor $\mathbf{S}$, which is required in order to calculate $\mathbf{L}^*$, according to Eq. (11.83), is also dependent on a knowledge of $\mathbf{C}^{*ep}$, as pointed out previously. It is $\mathbf{C}^{*ep}$ that has to be found through the analysis of the polycrystal plasticity, and it is not known before the analysis is complete. Therefore accurate determination of $\mathbf{L}^*$ requires an iterative process, and this procedure is very complicated. However, an approximate estimation of $\mathbf{L}^*$ can be obtained by introducing additional assumptions.

If $\mathbf{C}^{*ep}$ is assumed to be isotropic, as given by Eq. (11.69), and the inclusion (grain) is a sphere, then Hill's self-consistent model reduces to the modified Kroner, Budiansky, and Wu model (Berveiller and Zaoui 1979) discussed in the last section. This is shown below.

Since $\mathbf{C}^{*ep}$ is given by Eq. (11.69), Eshelby's tensor $\mathbf{S}$ can be written in the simple form

$$S_{ijkl} = (1 - 2\beta_T)\delta_{ij}\delta_{kl} + \tfrac{1}{2}\beta_T(\delta_{ik}\delta_{jl} + \delta_{il}\delta_{jk}) \tag{11.84}$$

where

$$\beta_{\mathrm{T}} = \frac{2(4 - 5\nu_{\mathrm{T}})}{15(1 - \nu_{\mathrm{T}})} \tag{11.85}$$

ν^{T} is the tangential poisson's ratio and is related to K^{T} and G^{T} through Eq. (11.70). Equation (11.84) can be directed obtained from Eq. (11.46) by simply replacing ν with ν^{T}. Substituting Eqs. (11.69) and (11.84) into Eq. (11.83), we get

$$L^*_{ijkl} = \frac{G_T(3 - 5\nu_{\mathrm{T}})}{4 - 5\nu_{\mathrm{T}}}\delta_{ij}\delta_{kl} + \frac{G_T(7 - 5\nu_{\mathrm{T}})}{2(4 - 5\nu)}(\delta_{ik}\delta_{jl} + \delta_{il} + \delta_{jk}) \tag{11.86}$$

Using this expression, Eq. (11.77) becomes

$$\dot{\boldsymbol{\sigma}} - \dot{\boldsymbol{\sigma}}^* = \frac{2GG_{\mathrm{T}}(7 - 5\nu_{\mathrm{T}})}{G_{\mathrm{T}}(7 - 5\nu_{\mathrm{T}}) + 2G(4 - 5\nu)}(\dot{\boldsymbol{\epsilon}}^{\mathrm{p}} - \dot{\boldsymbol{\epsilon}}^{*\mathrm{p}}) \tag{11.87}$$

which is exactly Eq. (11.61) if relation (11.70) is introduced. Actually, this is how Berveiller and Zaoui (1979) derived their modified Kroner, Budiansky, and Wu's model presented in the last section. In deriving Eq. (11.87), the following relations are used:

$$\dot{\boldsymbol{\epsilon}} = \dot{\boldsymbol{\epsilon}}^{\mathrm{e}} + \dot{\boldsymbol{\epsilon}}^{\mathrm{p}}, \quad \dot{\boldsymbol{\epsilon}}^* = \dot{\boldsymbol{\epsilon}}^{*\mathrm{e}} + \dot{\boldsymbol{\epsilon}}^{*\mathrm{p}} \tag{11.88}$$

$$\dot{\epsilon}^{\mathrm{e}}_{kk} = \dot{\epsilon}^{*\mathrm{e}}_{kk} \tag{11.89}$$

$$\dot{\boldsymbol{\sigma}} - \dot{\boldsymbol{\sigma}}^* = 2G(\dot{\boldsymbol{\epsilon}}^{\mathrm{e}} - \dot{\boldsymbol{\epsilon}}^{*\mathrm{e}}) \tag{11.90}$$

Equations (11.89) and (11.90) imply that the elastic properties of the single-crystal grain (inclusion) and the aggregate are the same and isotropic.

Obviously, if the isotropic, elastic stiffness tensor $\mathbf{C}$ is used in place of $\mathbf{C}^{*\mathrm{ep}}$, then Hill's model (11.77) reduces to the original Kroner, Budiansky, and Wu model (11.58). Therefore the self-consistent models (11.58) and (11.61) can be viewed as simplified versions of Hill's general model (11.77).

It is useful to derive an expression for the overall tangential stiffness tensor $\mathbf{C}^{*\mathrm{ep}}$. This can be done by using Eq. (11.79) and noting that Eq. (11.79) can be rewritten as

$$\dot{\boldsymbol{\sigma}} = \mathbf{C}^{\mathrm{ep}} : (\mathbf{C}^{\mathrm{ep}} + \mathbf{L}^*)^{-1} : (\mathbf{L}^* + \mathbf{C}^{*\mathrm{ep}}) : \dot{\boldsymbol{\epsilon}}^* \tag{11.91}$$

Since $\dot{\boldsymbol{\sigma}}^* = (\dot{\boldsymbol{\sigma}})_{\mathrm{ave}}$, the averages of this equation over all grains gives

$$\dot{\boldsymbol{\sigma}}^* = \left(\mathbf{C}^{\mathrm{ep}} : (\mathbf{C}^{\mathrm{ep}} + \mathbf{L}^*)^{-1} : (\mathbf{L}^* + \mathbf{C}^{*\mathrm{ep}})\right)_{\mathrm{ave}} : \dot{\boldsymbol{\epsilon}}^* \tag{11.92}$$

Comparison of this equation with Eq. (11.76) yields

$$\mathbf{C}^{*ep} = \left(\mathbf{C}^{ep} : (\mathbf{C}^{ep} + \mathbf{L}^*)^{-1} : (\mathbf{L}^* + \mathbf{C}^{*ep})\right)_{ave} \tag{11.93}$$

The derivation of $\mathbf{C}^{*ep}$ from Eq. (11.92) is not rigorous because, as pointed out by Hutchinson (1970), the tangential stiffness tensor $\mathbf{C}^{ep}$ of a single crystal depends on the value of $\dot{\epsilon}$ or $\dot{\sigma}$, which is related to the overall strain rate $\dot{\epsilon}^*$. This means that $\mathbf{C}^{ep} : (\mathbf{C}^{ep} + \mathbf{L}^*)^{-1} : (\mathbf{L}^* + \mathbf{C}^{*ep})$ is not constant for arbitrary strain rate $\dot{\epsilon}$. However, Eq. (11.93) is still widely used, since it provides a reasonable estimation of $\mathbf{C}^{*ep}$. From Eq. (11.93) it is seen that the determination of $\mathbf{C}^{*ep}$ is very complicated, since it is also included in the right-hand side of the equation. The calculation of $\mathbf{C}^{*ep}$ is thus iterative in that at a given state of deformation an initial guess has to be made for $\mathbf{C}^{*ep}$; then Eq. (11.93) can be used to obtain an improved value of $\mathbf{C}^{*ep}$. This procedure must be repeated many times until a convergent $\mathbf{C}^{*ep}$ is obtained.

According to our previous discussion, the overall tangential stiffness tensor obtained by using Kroner, Budiansky, and Wu's model is

$$\mathbf{C}^{*ep} = \left(\mathbf{C}^{ep} : (\mathbf{C}^{ep} + \mathbf{L}^*)^{-1} : (\mathbf{L}^* + \mathbf{C})\right)_{ave} \tag{11.94}$$

where $\mathbf{L}^*$ is given by Eq. (11.86) with ν^{T} and G^{T} replaced by ν and G. For the modified Kroner, Budiansky, and Wu model, $\mathbf{C}^{*ep}$ is given by Eq. (11.93) with $\mathbf{C}^{*ep}$ on the right-hand side of the equation given by Eq. (11.69) and $\mathbf{L}^*$ given by Eq. (11.86).

Finally, we note that for Taylor's model, since $\dot{\epsilon} = \dot{\epsilon}^*$, it follows that

$$\mathbf{C}^{*ep} = (\mathbf{C}^{ep})_{ave} \tag{11.95}$$

which can be obtained by averaging of Eq. (11.75) over all grains. The application of Hill's self-consistent model will be discussed next.

11.7 CALCULATED RESULTS OF SELF-CONSISTENT MODELS

In this chapter three self-consistent models have been studied: Kroner, Budiansky, and Wu's model, Berveiller and Zaoui's model (plastic accommodation model), and Hill's model. When any one of these models is applied to solve engineering problems, it has to be complemented by the yield criterion for every slip system of the single crystal. For finite deformation cases involving finite rotation of the crystal lattice, the required criterion was discussed and given in Eq. (10.62). Next we will demonstrate with Kroner, Budiansky, and Wu's model how the constitutive equations of single crystals can be used to derive the elastic-plastic response of a polycrystal aggregate subjected to uniaxial tension. This is a classical problem and has been studied by many researchers using different models (see Budiansky and Wu 1962; Berveiller and Zaoui 1979; Hutchinson 1964, 1970).

As we noted in Section 10.2, the polycrystal is viewed as an aggregate of numerous single-crystal grains with different orientations with respect to the loading axis. The orientations of these grains can be illustrated using a stereographic projection plot, as shown in Fig. 11.2, where the crosses represent the orientations of the loading axis with respect to the lattice axis of many single crystal grains. Usually the polycrystal is assumed to be initially isotropic, the orientations of the grains are assumed to be random, and the number of grains is very large. Obviously more accurate results can be obtained when the number of grains is increased.

For Kroner, Budiansky, and Wu's model the stress and plastic strain rates in a crystal grain are related to the overall stress and plastic strain rates by Eq. (11.58), rewritten below for convenience,

$$\dot{\boldsymbol{\sigma}} = \dot{\boldsymbol{\sigma}}^* - 2G(1 - \boldsymbol{\beta})(\dot{\boldsymbol{\epsilon}}^p - \dot{\boldsymbol{\epsilon}}^{*p}) \tag{11.96}$$

The yield criterion is expressed in terms of resolved stress τ_α and yield stress $\tau^\alpha{}_c$ on the slip system

$$\tau_\alpha = \mathbf{P}^\alpha : \boldsymbol{\sigma} = \tau_c^\alpha \qquad (\alpha = 1, 2, \ldots, n) \tag{11.97}$$

where n is the number of activated or active slip systems in a single-crystal grain. The slip system hardens according to

$$\dot{\tau}_c^\alpha = \sum_{\alpha=1}^{n} h_{\alpha\beta}\dot{\gamma}^\beta \tag{11.98}$$

where $h_{\alpha\beta}$ is the hardening moduli. In the case of infinitesimal deformation, the loading/unloading criterion can be determined from Eqs. (11.97) and (11.98) as

$$\begin{aligned} &\dot{\gamma}^\alpha = 0 \qquad \text{if } 0 \leq \tau_\alpha < \tau_c^\alpha \\ &\dot{\gamma}^\alpha = 0 \qquad \text{if } 0 < \tau_\alpha = \tau_c^\alpha; \qquad \mathbf{P}^\alpha : \dot{\boldsymbol{\sigma}} < \sum_{\beta=1}^{n} h_{\alpha\beta}\dot{\gamma}^\beta \\ &\dot{\gamma}^\alpha \geq 0 \qquad \text{if } 0 < \tau_\alpha = \tau_c^\alpha; \qquad \mathbf{P}^\alpha : \dot{\boldsymbol{\sigma}} = \sum_{\beta=1}^{n} h_{\alpha\beta}\dot{\gamma}^\beta \end{aligned} \tag{11.99}$$

These equations can also be derived from Eq. (10.62). It should be noted that for infinitesimal deformation, the rotation of the lattice can be ignored so that

$$\dot{\mathbf{P}}^\alpha = 0 \tag{11.100}$$

$$\dot{\tau}_c^\alpha = \mathbf{P}^\alpha : \dot{\boldsymbol{\sigma}} \tag{11.101}$$

Note that the stress rate $\dot{\boldsymbol{\sigma}}$ in the single crystal grain has been given by Eq. (11.96). Substitution of Eq. (11.96) into Eq. (11.99) gives

$$\sum_{\beta=1}^{n} h_{\alpha\beta}\dot{\gamma}^{\beta} + 2G(1-\beta)\mathbf{P}^{\alpha} : \dot{\boldsymbol{\epsilon}}^{\mathrm{p}} = \mathbf{P}^{\alpha} : \left[\dot{\boldsymbol{\sigma}} + 2G(1-\beta)\dot{\boldsymbol{\epsilon}}^{*\mathrm{p}}\right]$$

$$\alpha = 1, 2, \ldots, n \tag{11.102}$$

Since the plastic strain rate $\dot{\boldsymbol{\epsilon}}^{\mathrm{p}}$ in a single crystal is assumed to be caused only by slips on the slip plane, $\dot{\boldsymbol{\epsilon}}$ is expressed as [see Eq. (10.32)]

$$\dot{\boldsymbol{\epsilon}} = \sum_{\alpha=1}^{n} \mathbf{P}^{\alpha}\dot{\gamma}^{\alpha} \tag{11.103}$$

Substituting this equation into Eq. (11.102), we have

$$\sum_{\beta=1}^{n} \left[h_{\alpha\beta} + 2G(1-\beta)\mathbf{P}^{\alpha} : \mathbf{P}^{\beta}\right]\dot{\gamma}^{\beta} = \mathbf{P}^{\alpha} : \left[\dot{\boldsymbol{\sigma}}^{*} + 2G(1-\beta)\dot{\boldsymbol{\epsilon}}^{*\mathrm{p}}\right]$$

$$\alpha = 1, 2, \ldots, n \tag{11.104}$$

These are n equations to be used to determine the n slips $\dot{\gamma}^{\alpha}$ for a single-crystal grain corresponding to a given loading history $\boldsymbol{\sigma}^{*}(t)$. Since the plastic strain rate is included on the right-hand side of these equations, the calculation has to be an iterative process: for a prescribed rate of overall stress rate $\dot{\boldsymbol{\sigma}}^{*}$, a trial overall plastic strain rate $\dot{\boldsymbol{\epsilon}}^{*\mathrm{p}}$ is assumed, and these $\dot{\boldsymbol{\sigma}}^{*}$ and $\dot{\boldsymbol{\epsilon}}^{*\mathrm{p}}$ are substituted into the right-hand side of Eq. (11.104) to obtain the slips of a single-crystal grain. The methods discussed previously to determine the combination of active sip systems can be used here. The plastic strain rate $\dot{\boldsymbol{\epsilon}}^{\mathrm{p}}$ of a single-crystal grain is thus obtained by Eq. (11.103). After repeating this procedure for all grains with different orientations, $\dot{\boldsymbol{\epsilon}}^{\mathrm{p}}$ of each single-crystal grain can be calculated and the overall plastic strain rate goes from $\dot{\boldsymbol{\epsilon}}^{\mathrm{p}}$ to

$$\dot{\boldsymbol{\epsilon}}^{*\mathrm{p}} = \left(\dot{\boldsymbol{\epsilon}}^{\mathrm{p}}\right)_{\mathrm{ave}} \tag{11.105}$$

The most commonly used averaging formula for the equation above is

$$\dot{\boldsymbol{\epsilon}}^{*\mathrm{p}} = \frac{\int_{0}^{2\pi} d\phi \int_{-\pi/2}^{\pi/2} d\theta \int_{0}^{2\pi} \sin\theta\, \dot{\boldsymbol{\epsilon}}^{\mathrm{p}}\, d\eta}{8\pi^{2}} \tag{11.106}$$

where the angles ϕ, θ, and η are defined in Fig. 11.8. Note that ϕ and θ are the same as defined in Fig. 11.4. ϕ, θ, and η are called *Euler angles*. In Fig. 11.8, x, y, and z are the single-crystal lattice axes, while 1, 2, and 3 are the

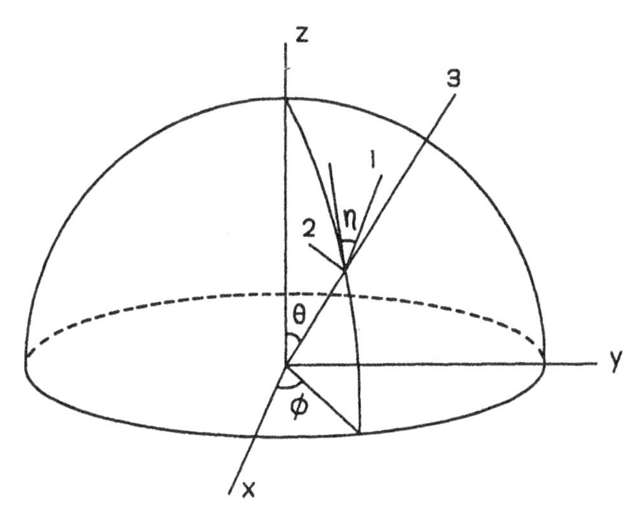

Fig. 11.8 Definition of Euler angles

global coordinate system of specimen axes. The calculated overall plastic strain rate $\dot{\boldsymbol{\epsilon}}^{*p}$ has to be compared with the estimated trial value. If these two are the same (or within an acceptable tolerance), then the correct solution has been found, and the calculation can move on to the next stress increment (or rate). If, on the other hand, these two tensors are quite different, the calculated $\dot{\boldsymbol{\epsilon}}^{*p}$ needs to be used as the new trial value, and the procedure must be repeated to obtain another new value of $\dot{\boldsymbol{\epsilon}}^{*p}$. This process continues until a satisfactory solution is obtained for the given $\dot{\boldsymbol{\sigma}}^{*}$.

Although the foregoing discussion is based on the Kroner, Budiansky, and Wu model, the procedures using Berveiller and Zaoui's model and Hill's model are very similar and do not need to be repeated here. Interested readers can obtain more detail from Berveiller and Zaoui (1979) and Hutchinson (1964, 1970).

Budiansky and Wu (1962) calculated the overall stress-strain curves of a polycrystal specimen subjected to uniaxial tension or simple shear. They assumed that $h^{\alpha\beta} = 0$ and $\tau^{\alpha}{}_{c} = \tau_Y$ in their calculations. The results obtained from uniaxial tension and simple shear are shown in Figs. 11.9 and 11.10. The dashed lines in these figures represent results obtained by Taylor's rigid plasticity model assuming $h_{\alpha\beta} = 0$ and

$$\sigma_Y = 2\tau_Y \tag{11.107}$$

$$\epsilon_Y = \frac{\sigma_Y}{E}, \quad \gamma_Y = \frac{\tau_Y}{G} \tag{11.108}$$

where E and G are the elastic Young's and shear modulus, respectively. It is seen that the elastic-plastic solution of these problems approaches Taylor's

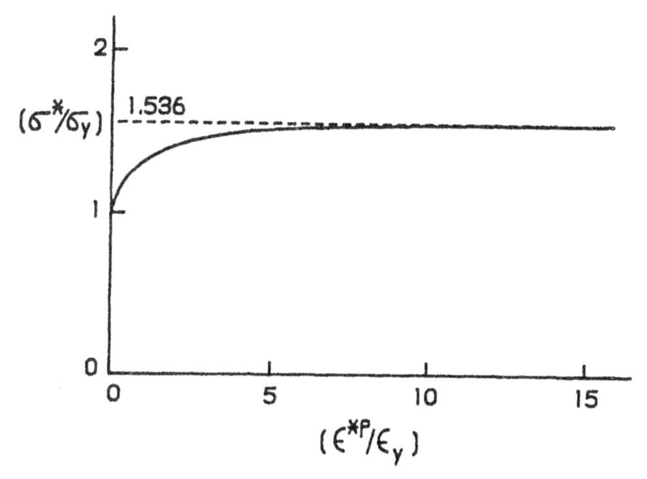

Fig. 11.9 Stress-plastic strain curve in uniaxial tension ($h_{\alpha\beta} = 0$, $\nu = 0.33$)

solutions when the strain becomes very large. It is interesting to note that

$$\frac{\sigma^*}{\sigma_Y} = 1.536, \quad \frac{\tau^*}{\tau_Y} = 1.658 \tag{11.109}$$

when the strain becomes large. Here σ^* and τ^* are the overall tensile and shear stresses at large strain, respectively, and σ_Y and τ_Y are the initial yield stresses of the single crystal. Therefore, although the single crystal is perfectly plastic, the polycrystal will show hardening behavior after initial yielding, and the ultimate strength of the polycrystal will be higher than that of the single crystals.

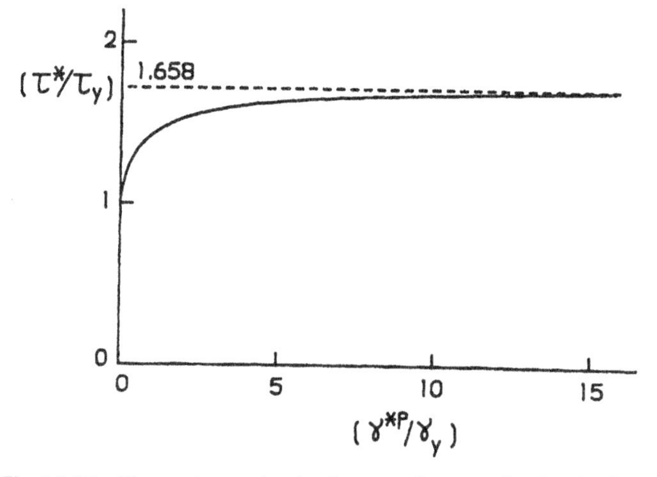

Fig. 11.10 Shear stress-plastic shear strain curve in simple shear

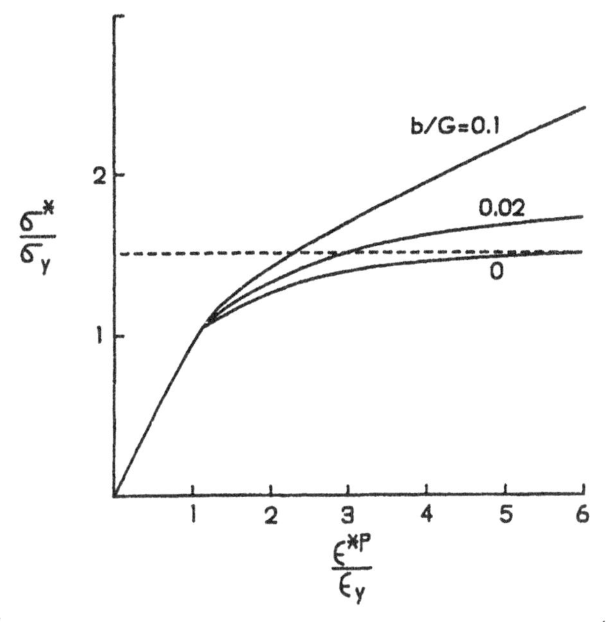

Fig. 11.11 Stress-plastic strain curves for different values of $b/G(\nu = 0.33)$

The calculation of the stress-strain curve of a polycrystal specimen from a hardening single-crystal response has been done by Hutchinson (1964). He assumed isotropic hardening for slip systems on single crystals, whereby

$$h_{\alpha\beta} = b \tag{11.110}$$

The calculated stress-strain curve of a polycrystal specimen subjected to tension is shown in Fig. 11.11. Hutchinson also did the calculation using Hill's model (Hutchinson 1970). The comparison of the results obtained by Kroner, Budiansky, and Wu's model and by Hill's model is shown in Fig. 11.12. It is seen that the stress-strain curve obtained by the Kroner, Budiansky, and Wu model lies above that obtained by the Hill model. This is expected since, as we mentioned earlier, much stronger restraint is provided for in Kroner, Budiansky, and Wu's model. Hill's model accommodates a "softer" restraint, since it uses an elastic-plastic stiffness tensor of the polycrystal aggregate to calculate the Eshelby tensor. In actual cases of elastic-plastic deformation of a polycrystal, the aggregate cannot be plastically deformed everywhere, and there should be some regions where the material remains elastic. Therefore the real restraint provided by the surrounding aggregate to the single-crystal grain should be softer than that provided by the elastic restraint, (as in Kroner, Budiansky, and Wu's model)

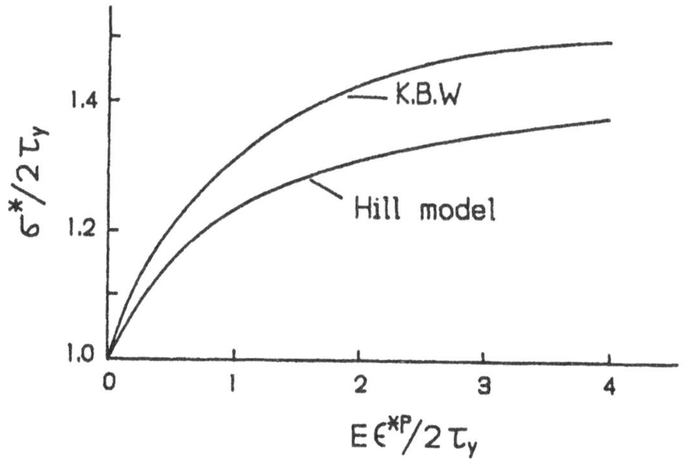

Fig. 11.12 Comparison of K.B.W. model and Hill's model ($h_{\alpha\beta} = 0$, $\nu = 0.33$)

but stronger than that provided by plastically deformed aggregate (as in Hill's model). As pointed out by Hutchinson (1970), the real stress-strain curve should be somewhere between the curves predicted by the Kroner, Budiansky, and Wu model and the Hill's model.

Berveiller and Zaoui (1979) obtained solutions of uniaxial tension of polycrystal specimen using their plastic accommodation model. The specific

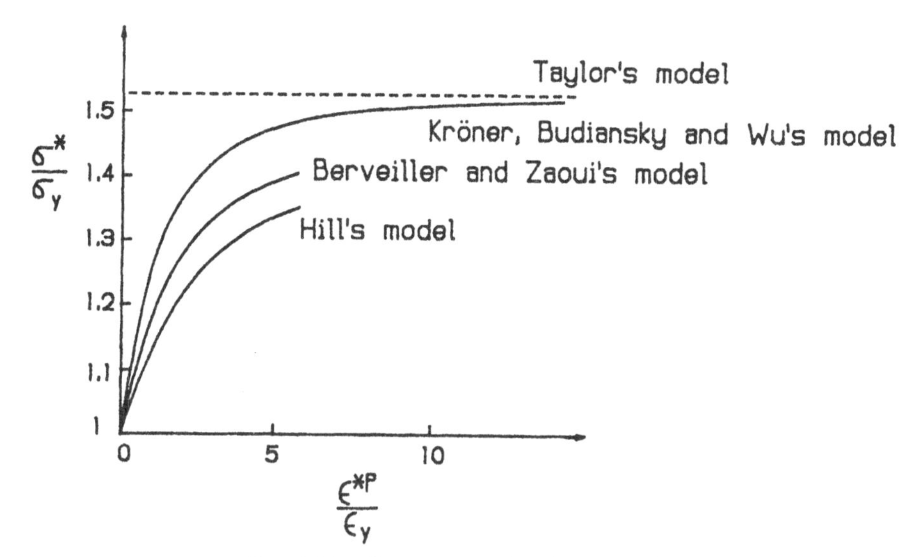

Fig. 11.13 Comparison of various models

expression for the plastic accommodation factor α_T used in their study is

$$\alpha_T \simeq \frac{1}{1 + (3G/2)G(\epsilon^{*p}/\sigma^*)} \tag{11.111}$$

where $\boldsymbol{\sigma}^*$ and $\boldsymbol{\epsilon}^{*p}$ are the overall stress and overall plastic strain, respectively. A typical result of their study is shown in Fig. 11.13. The figure shows that with the help of the plastic accommodation factor α_T, an improved result can be achieved. The curve will lie between the curves obtained by the Kroner, Budiansky, and Wu model and the Hill model.

REFERENCES

Beradi, Ch., M. Berveiller, and P. Lipinski. 1987. Plasticity of metallic polycrystals under complex loading paths. *Int. J. Plast.* **3**:143.

Berveiller, M., and A. Zaoui. 1979. An extension of the self-consistent scheme to plastically-flowing polycrystals. *J. Mech. Phys. Solids* **26**:325.

Bishop, J. F. W. 1954. A theory of the tensile and compressive textures of face-centered cubic metals. *J. Mech. Phys. Solids* **3**:130.

Bishop, J. F. W., and R. Hill. 1951. A theory of plastic distortion of a polycrystalline aggregate under combined stress. *Phil. Mag.* **42**:414.

Budiansky, B., and T. Y. Wu. 1962. Theoretical prediction of plastic strains of polycrystals. *Proc. 4th U.S. Nat. Congr. Appl. Mech.* 1175.

Cox, H. L., and D. E. Sopmith. 1937. Effect of orientation on stresses in single crystals and of random orientation on strength of polycrystalline aggregates. *Proc. Phys. Soc., London* **49**:134.

Eshelby, J. D. 1957. The determination of the elastic field of an ellipsoidal inclusion and related problems. *Proc. R. Soc., London* **A241**:398.

Eshelby, J. D. 1951. The force on an elastic singularity. *Phil. Trans. R. Soc.* **A244**:87.

Guldenpfennig, J., and R. J. Clifton. 1977. On the computation of plastic stress-strain relations for polycrystalline metals. *Comp. Meth. Appl. Mech. Eng.* **10**:141.

Havner, K. S. 1971. A discrete model for the prediction of subsequent yield surface in polycrystalline plasticity. *Int. J. Solids. Struct.* **7**:719.

Hill, R. 1965a. Continuum micro-mechanics of elastoplastic polycrystals. *J. Mech. Phys. Solids* **13**:89.

Hill, R. 1965b. A self-consistent mechanics of composite materials. *J. Mech. Phys. Solids* **13**:213.

Hutchinson, J. W. 1970. Elastic-plastic behavior of polycrystalline metals and composite. *Proc. R. Soc., London* **A319**:247.

Hutchinson, J. W. 1964. Plastic stress-strain relations of F.C.C. polycrystalline metals hardening according to Taylor's rule. *J. Mech. Phys. Solids* **12**:11.

Iwakuma, T., and S. Nemat-Nasser. 1984. Finite elastic plastic deformation of polycrystalline metals and composites. *Proc. Roy. Soc., London* **A394**:87.

Kroner, E. 1961. Zur plastischen verformung des vielkristalls. *Acta Met.* **9**:153.

Kroner, E. 1958. Kontinuumstheorie der Versetzungen and Eigenspannungen. Berlin: Springer-Verlag.

Lin. T. H. 1984. A physical theory of plasticity and creep. *J. Eng. Mater. Tech.* **106**:290.

Lin, T. H., Y. M. Ito, and C. L. Yu. 1974. A new slip theory of plasticity. *J. Appl. Mech.* **41**:587.

Lin, T. H., and Y. M. Ito. 1966. Theoretical plastic stress-strain relationships of a polycrystal and comparison with von Mises' and Tresca's plasticity theories. *Int. J. Eng. Sci.* **4**:543.

Lipinski, P., and M. Berveiller. 1989. Elastoplasticity of micro-inhomogeneous metals at large strains. *Int. J. Plast.* **5**:149.

Mura, T. 1987. Micromechanics of defects in solids. Deventer, The Netherlands: Martinus-Nijhoff.

Sache, G. 1928. Zur ableilung einer fleissbedingung. *VDI Z.* **72**:734.

Taylor, G. I. 1938. Plastic strain in metals. *I. Inst. Metals* **62**:307.

Weng. G. I. 1982. A unified, self-consistent theory for the plastic-creep deformation of metals. *J. Appl. Mech.* **49**:728.

INDEX

CPSIA information can be obtained at www.ICGtesting.com
Printed in the USA
BVOW06*1104261215

431007BV00008B/62/P

9 780471 310433